Compendium of Thermophysical Property Measurement Methods

Volume 2

Recommended Measurement
Techniques and Practices

Compendium of Thermophysical Property Measurement Methods

Volume 2

Recommended Measurement Techniques and Practices

Edited by

K. D. Maglić

Boris Kidrič Institute of Nuclear Sciences
Belgrade, Yugoslavia

A. Cezairliyan

National Institute of Standards and Technology
Gaithersburg, Maryland, USA

and

V. E. Peletsky

Institute for High Temperatures
Moscow, Russian Federation

Springer Science+Business Media, LLC

Library of Congress Cataloging in Publication Data

(Revised for vol. 2)
Compendium of thermophysical property measurement methods: recommended measurement techniques and practices

Includes bibliographical references and indexes.
Contents: v. 1. Survey of measurement techniques — v. 2. Recommended measurement techniques and practices.
1. Solids — Thermal properties — Measurement — Handbooks, manuals, etc. I. Maglić, K. D. (Kosta D.) II. Cezairliyan, A. III. Peletsky, V. E.

QC176.8.T4C66 1984	530.4′12	84-3268

ISBN 978-1-4613-6445-0   ISBN 978-1-4615-3286-6 (eBook)
DOI 10.1007/978-1-4615-3286-6

ISBN 978-1-4613-6445-0

© 1992 Springer Science+Business Media New York
Originally published by Plenum Press, New York in 1992
Softcover reprint of the hardcover 1st edition 1992

Contributors

PIERRE ABÉLARD, Ecole Nationale Supérieure de Céramiques Industrielles, Limoges Cédex, France

BERNARD CALÈS, Céramiques Techniques Desmarquest, Evreux Cédex, France

ARED CEZAIRLIYAN, Thermophysics Division, National Institute of Standards and Technology, Gaithersburg, Maryland, USA

VITALIY YA. CHEKHOVSKOI, Institute for High Temperatures, USSR Academy of Sciences, Moscow, USSR

J.W. COOKE, Energy Programs Division, Oak Ridge Field Office, Department of Energy, Oak Ridge, Tennessee

JOHN M. CORSAN, Division of Quantum Metrology, National Physical Laboratory, Teddington, Middlesex, England

W.R. DAVIS, 28 Joseph Crescent, Alsager, Stoke-on-Trent, England

RUDOLF DE CONINCK, Materials Science Department, SCK/CEN, Mol, Belgium

FRANCESCO DE PONTE, Istituto di Fisica Tecnica, Università di Padova, Padova, Italy

DAVID A. DITMARS, National Institute of Standards and Technology, Gaithersburg, Maryland, USA

DANIEL R. FLYNN, National Institute of Standards and Technology, Gaithersburg, Maryland, USA

R.K. KIRBY, Retired from National Institute of Standards and Technology, Gaithersburg, Maryland, USA

SORIN KLARSFELD, Saint-Gobain Recherche, Aubervilliers, France

YA.A. KRAFTMAKHER, Institute of Inorganic Chemistry, USSR Academy of Sciences, Novosibirsk, USSR

S.N. KRAVCHUN, Physics Department, Moscow State University, Moscow, USSR

CATHERINE LANGLAIS, Isover Saint-Gobain, Centre de Recherches Industrielles, Rantigny, France

K.D. MAGLIĆ, Boris Kidrič Institute of Nuclear Sciences, Institute of Thermal Engineering and Energy Research, Belgrade, Yugoslavia

A.G. MOZGOVOI, Institute for High Temperatures, USSR Academy of Sciences, Moscow, USSR

V.E. PELETSKY, Institute for High Temperatures, USSR Academy of Sciences, Moscow, USSR

L.P. PHYLIPPOV, Physics Department, Moscow State University, Moscow, USSR

E.S. PLATUNOV, Physics Department, Leningrad Technological Institute of the Refrigerating Industry, Leningrad, USSR

MICHAEL J. RICHARDSON, Division of Materials Metrology, National Physical Laboratory, Teddington, Middlesex, England

G. RUFFINO, Mechanical Engineering Department, Second University of Rome Tor Vergata, Rome, Italy

E.E. SHPIL'RAIN, Institute for High Temperatures, USSR Academy of Sciences, Moscow, USSR

RAYMOND E. TAYLOR, Thermophysical Properties Research Laboratory, Purdue University, West Lafayette, Indiana, USA

A.S. TLEUBAEV, Physics Department, Moscow State University, Moscow, USSR

RONALD P. TYE, Consultant, Sinku Riko Inc., Cohasset, Massachusetts, USA

ALFRED E. WECHSLER, Arthur D. Little, Inc., Cambridge, Massachusetts, USA

K.A. YAKIMOVICH, Institute for High Temperatures, USSR Academy of Sciences, Moscow, USSR

Preface

The two-volume reference work *Compendium of Thermophysical Property Measurement Methods* was intended to systematize and record the knowledge accumulated in the past, especially during the last three decades of intensive studies of thermophysical property measurement methods, and to serve as a guide in selecting the best technique to be employed in measuring the required property of a material. The first volume, *Survey of Measurement Techniques*, published in 1984, provided an exhaustive compilation of the currently used methods for the measurement of thermal and electrical conductivity, thermal diffusivity, specific heat, thermal expansion, and thermal radiative properties of solid materials, from room temperature to very high temperatures. The first volume also served as the basis for the second phase of the work, namely, the description of recommended practices for the determination of transport and thermodynamic properties of selected groups of materials.

This second volume, *Recommended Measurement Techniques and Practices*, presents the results of work in the second phase, with detailed coverage of selected recommended techniques and relevant information. Contributing authors were selected to represent various measurement methods developed and used in different parts of the world, keeping in mind that only techniques which have reached a mature state of development should be included in the first place. Contributions in this volume include information at a sufficient level of detail to permit potential users of the method to be guided in the construction of the equipment and its subsequent use. Particular attention was paid to the fundamental aspect of the measurements, i.e., the mismatch between the physical model and the actual measurement conditions, which is usually the principal source of error in thermophysical experiments.

The intention in composing this second volume was to record all significant aspects of thermophysical property measurements resulting from many years of dedicated studies and a high degree of specialization. There is little doubt that the experimental methods will continue to be developed particularly in the areas related to data acquisition and their computerized processing. The area related to the physics of the measurements and the sources of error arising from the deviations of real measurement conditions from those considered for the models are very actual, gaining importance with automation of the measurement processes. The increasing availability and use of automated apparatus enhance the importance of such a source book in assessing the nature and magnitude of errors resulting from the measurements and computational techniques.

The main portions of the chapters in this volume refer to a description of the individual apparatus and measurement procedures, written by authors who have designed and developed such instruments. While most of the apparatus described is for measuring solid materials, a few methods for molten metals or fluids have also been included to bridge the gap between measurement techniques used for the two aggregate states of matter. Two chapters have been devoted to an assessment of the state-of-the-art in electrical resistivity and thermal expansion measurements. The last chapter summarizes thermophysical property reference materials, being an update of the list provided in the first volume of the Compendium.

The present volume is composed of six major sections, with twenty two chapters in all. The first section presents eight chapters on methods for measuring thermal conductivity: the methods for metals, from room to very high temperatures, for loose fill, insulating, building and refractory materials, and a method appropriate for fluids. The second section is devoted to a review of methods for measuring electrical resistivity, applicable to medium and good conductors. The third section covers three methods for measuring thermal diffusivity of solid materials: the laser pulse technique, the modulated electron beam and thermal radiation method, and a group of methods for measuring both thermal diffusivity and thermal conductivity in the monotonic heating regime. The last chapter within this section deals with a method for measuring several properties of liquids and compressed gases. The fourth section contains five chapters describing equipment and measurement procedures for measuring specific heat; they include the method of mixtures, and its high-temperature variant, levitation calorimetry, modulation calorimetry, the subsecond pulse heating calorimetry, as well as instructions for the use of differential scanning calorimetry in specific heat measurements. The fifth section contains three chapters dealing with methods for the measurement of thermal expansion. The first is an overview of approaches for measuring thermal expansion. The other two describe an apparatus for interferometric measurement of thermal expansion of solids and an apparatus for the continuous measurement of thermal expansion and density of materials in the condensed phase at high temperatures and pressures. The sixth section contains an updated list of certified thermophysical property standard reference materials.

K.D. Maglić
A. Cezairliyan
V.E. Peletsky

Acknowledgments

The editors are indebted to many people who helped in the preparation of this book. Special thanks, however, must go to the individuals who contributed most, the authors of chapters.

In addition, thanks must be also extended to the referees of individual chapters, whose reviews contributed to the final form and quality of the manuscripts in this book. These are, in the order of appearance of particular chapters in the main section of the book:

For the section on thermal conductivity:

R.P. Tye, *Holometrix, Inc. Cambridge, Massachusetts, USA* (two chapters)

J.P. Moore, *Martin Marietta Energy Systems, Oak Ridge, Tennessee, USA*

W. Neumann, *Austrian Research Centre, Seibersdorf, Austria*

M.L. Laubitz, *National Research Council, Ottawa, Canada*

D.R. Smith, *National Institute of Standards and Technology, Boulder, Colorado, USA*

A.E. Wechsler, *Arthur D. Little, Inc., Cambridge, Massachusetts, USA*

D.L. McElroy, *Oak Ridge National Laboratory, Oak Ridge, Tennessee, USA*

For the section on electrical resistivity:

F. Cabannes, *Centre de Recherches sur la Physique des Hautes Températures, Orléans, France*

For the section on thermal diffusivity:

R.U. Acton, *Sandia National Laboratories, Albuquerque, New Mexico, USA*

V.E. Zinov'ev, *Mining Institute, Ural's Polytechnical Institute, Sverdlovsk, USSR*

K.D. Maglić, *Boris Kidrič Institute, Vinča, Belgrade, Yugoslavia*

S.C. Saxena, *University of Illinois at Chicago Circle, Chicago, Illinois, USA*

For the section on specific heat:

> E. Hanitzsch, *Physikalisch-Technische Bundesanstalt, Braunschweig, Germany*
> V.Ya. Chekhovskoi, *Institute for High Temperatures, Moscow, USSR*
> J.L. Margrave, *Rice University, Houston, Texas, USA*
> F. Righini, *Istituto di Metrologia "G. Colonnetti," Turin, Italy*
> R.E. Taylor, *Thermophysical Properties Research Laboratory, West Lafayette, Indiana, USA*

For the section on thermal expansion:

> G. Ruffino, *Second University of Rome, Rome, Italy*
> T.A. Hahn, *Naval Research Laboratory, Washington, D.C., USA*
> R. Ohse, *European Institute for Transuranium Elements, Karlsruhe, Germany*

For the section on standard reference materials:

> L.L. Sparks, *National Institute of Standards and Technology, Boulder, Colorado, USA*

The editors wish to express grateful appreciation to referees: T.A. Hahn and J.L. Margrave for voluntary lectorial editing of the refereed manuscripts of non-English origin.

Thanks are extended to all organizations represented by the authors and referees throughout the world. The contributions of the U.S.-Yugoslav Joint Board for Scientific and Technological Cooperation and the Soviet Organizing Committee for International Thermophysical Properties Conferences to the coordination of work on this volume are also gratefully acknowledged.

Contents

CHAPTER 4. REFERENCE GUARDED HOT PLATE APPARATUS
FOR THE DETERMINATION OF STEADY-STATE
THERMAL TRANSMISSION PROPERTIES
Francesco De Ponte, Catherine Langlais, and Sorin Klarsfeld

CHAPTER 5. APPARATUS FOR TESTING HIGH-TEMPERATURE
THERMAL-CONDUCTIVITY STANDARD
REFERENCE MATERIALS WITH CONDUCTIVITIES
ABOVE $1 \, W \, m^{-1} \, K^{-1}$ IN THE TEMPERATURE RANGE
400 TO 2500 K
V.E. Peletsky

CHAPTER 6. THE PROBE METHOD FOR MEASUREMENT OF
THERMAL CONDUCTIVITY
Alfred E. Wechsler

CHAPTER 7. B.S. 1902 PANEL TEST METHOD FOR THE MEASUREMENT OF THE THERMAL CONDUCTIVITY OF REFRACTORY MATERIALS
W.R. Davis

CHAPTER 8. THE VARIABLE-GAP TECHNIQUE FOR MEASURING THERMAL CONDUCTIVITY OF FLUID SPECIMENS
J.W. Cooke

II. ELECTRICAL RESISTIVITY

CHAPTER 9. METHODS FOR ELECTRICAL RESISTIVITY MEASUREMENT APPLICABLE TO MEDIUM AND GOOD ELECTRICAL CONDUCTORS
B. Calès and P. Abélard

III. THERMAL DIFFUSIVITY

CHAPTER 10. THE APPARATUS FOR THERMAL DIFFUSIVITY MEASUREMENT BY THE LASER PULSE METHOD
K.D. Maglić and R.E. Taylor

CHAPTER 11. MODULATED ELECTRON BEAM THERMAL DIFFUSIVITY EQUIPMENT
R. De Coninck

CHAPTER 12. INSTRUMENTS FOR MEASURING THERMAL CONDUCTIVITY, THERMAL DIFFUSIVITY, AND SPECIFIC HEAT UNDER MONOTONIC HEATING
E.S. Platunov

IV. SPECIFIC HEAT

CHAPTER 15. PHASE-CHANGE CALORIMETER FOR MEASURING
RELATIVE ENTHALPY IN THE TEMPERATURE
RANGE 273.15 TO 1200 K
David A. Ditmars

CHAPTER 16. APPARATUS FOR INVESTIGATION OF
THERMODYNAMIC PROPERTIES OF METALS BY
LEVITATION CALORIMETRY
Vitaliy Ya. Chekhovskoi

CHAPTER 17. A MILLISECOND-RESOLUTION PULSE HEATING
SYSTEM FOR SPECIFIC-HEAT MEASUREMENTS AT
HIGH TEMPERATURES
Ared Cezairliyan

CHAPTER 18. THE APPLICATION OF DIFFERENTIAL SCANNING CALORIMETRY TO THE MEASUREMENT OF SPECIFIC HEAT
M.J. Richardson

V. THERMAL EXPANSION

CHAPTER 19. METHODS OF MEASURING THERMAL EXPANSION
R.K. Kirby

CHAPTER 20. RECENT THERMAL EXPANSION INTERFEROMETRIC MEASURING INSTRUMENTS
G. Ruffino

VI. REVIEW OF CERTIFIED THERMOPHYSICAL PROPERTY
STANDARD REFERENCE MATERIALS

I

THERMAL CONDUCTIVITY

1

Axial Heat Flow Methods of Thermal Conductivity Measurement for Good Conducting Materials

J.M. CORSAN

1. INTRODUCTION

The advantages of a longitudinal or axial heat flow method of thermal conductivity measurement over a radial method depend to a large extent on the temperature range to be covered and the types of specimen to be measured. In general, the former technique is more flexible in that a wide range of conductivity values can be covered (e.g., from about $5 \mathrm{~W} \mathrm{~m}^{-1} \mathrm{~K}^{-1}$ to $400 \mathrm{~W} \mathrm{~m}^{-1} \mathrm{~K}^{-1}$ in measurements above ambient temperature) by choosing the diameter of the specimen to suit the conductivity expected, sample preparation is relatively simple, material in rod form is normally readily available, the location of thermocouples in the specimen is easily determined, and suitable reference materials are available to check performance if required.[1-3] On the negative side, heat losses due to radiation are more of a problem at high temperatures although they can usually be kept at a reasonable level if, in the areas where radiation is a problem, the temperatures of any shields are closely matched to the specimen temperature and a suitable insulation (density about $90 \mathrm{~kg} \mathrm{~m}^{-3}$) is used between the respective parts.

Much of the general background to thermal conductivity measuring techniques is available elsewhere,[4,5] together with specific analyses and discussion of the merits of the two principal methods for determining the conductivity of solid materials: the reader is advised to consult work of Laubitz[6,7] for a theoretical treatment of the axial method, and of McElroy and Moore[8] and Moore[9] for full particulars of the radial method.

J.M. CORSAN • Division of Quantum Metrology, National Physical Laboratory, Teddington, Middlesex, TW11 0LW, England.

The types of apparatus described here are all based on the axial method and have been developed at the National Physical Laboratory over the last ten years to meet the increasing demands of research and industry for more accurate thermal conductivity values over a wide range of temperatures and for the provision of calibrated reference materials. The equipment described is capable of giving values with an uncertainty of 4% or less in the temperature range for which it was designed and, subject to small modifications, could be further improved. For example, if only one type of material in a particular thermal conductivity range were being studied, each apparatus could be adapted to meet the specific characteristics of that material.

In the first part of this chapter a general-purpose apparatus operating in air between 320 K and 820 K is described.[10] The design incorporates such features as a reference specimen to check the validity of each measurement, a principle used extensively at NPL by Powell et al.,[11,12] a novel type of self-contained heater assembly, and a system of shields around the sample with closely matching temperatures. The apparatus has been used for measurements on titanium alloys,[13] stainless steels, aluminum alloys, and copper alloys and has proved both reliable and accurate. Two other forms of the axial method using longer specimens in a vacuum environment have also been developed. The first covers the temperature range 77 K to 373 K and has been used mainly for measurements on stainless steels, nickel steels, and a variety of aluminum alloys using specimens measuring 200 mm long by up to 10 mm diameter. The other, more recent apparatus, designed as a standard for characterizing reference materials over the temperature range 323 K to 1023 K, uses larger specimens 320 mm long by up to 25 mm diameter. Both types of apparatus have a number of similarities which include the use of a matched shield around the specimen, a subsidiary guard to aid temperature matching, and a high-vacuum system to limit heat transfer between specimen and shields.

The final section of the chapter deals with the measurement of electrical conductivity as a function of temperature and the usefulness of this parameter in predicting thermal conductivity values for metallic materials.

2. GENERAL-PURPOSE APPARATUS

2.1. Introduction

In this apparatus the specimen is firmly clamped between a guarded heater unit and a water-cooled reference specimen by means of a screw and spring arrangement, which allows for the length changes of the central column as the temperature is raised or lowered (see Fig. 1). Heat energy is supplied at a known rate at one end of the specimen using a guarded heater unit and constrained to flow axially through the specimen with minimum loss or gain

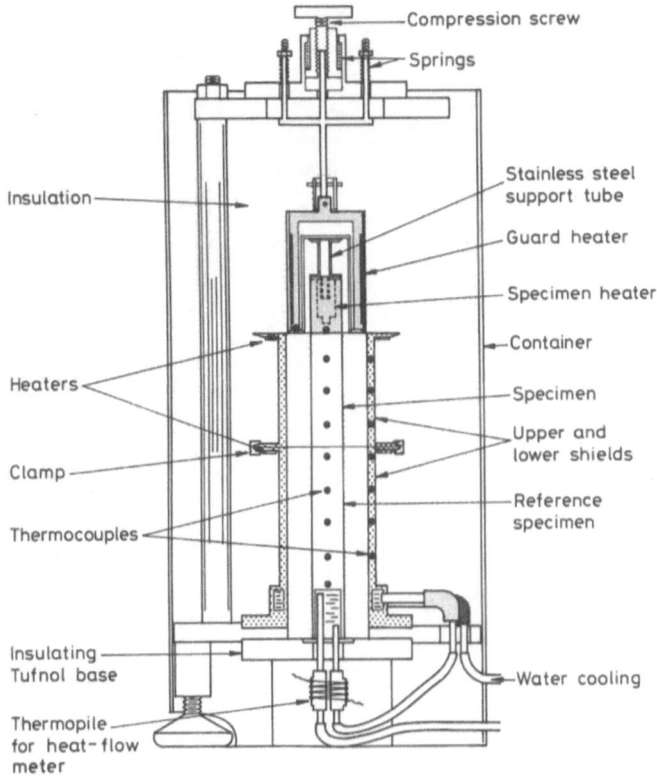

FIGURE 1. General-purpose thermal conductivity apparatus (310 K–773 K) showing specimen mounted in position.

by the use of heat shields at closely matching temperatures. The reference specimen acts as a heat sink for the heat conducted through the specimen and also serves as a heat-flow meter.

2.2. Heater Assemblies

Two heater assemblies of fairly similar design have been used in this apparatus: one made from copper (Fig. 2a) is suitable for measurements up to about 650 K while the other made from nickel (Fig. 2b) is preferred for measurements at higher temperatures because it avoids the corrosion problems encountered with copper.

The inner and outer parts of the copper assemblies have thick sections to give good temperature uniformity and are separated by a stainless steel tube 25 mm diameter by 30 mm long by 0.5 mm wall thickness with flanges at both ends, one of which is attached to three 3-mm-diameter stainless-steel

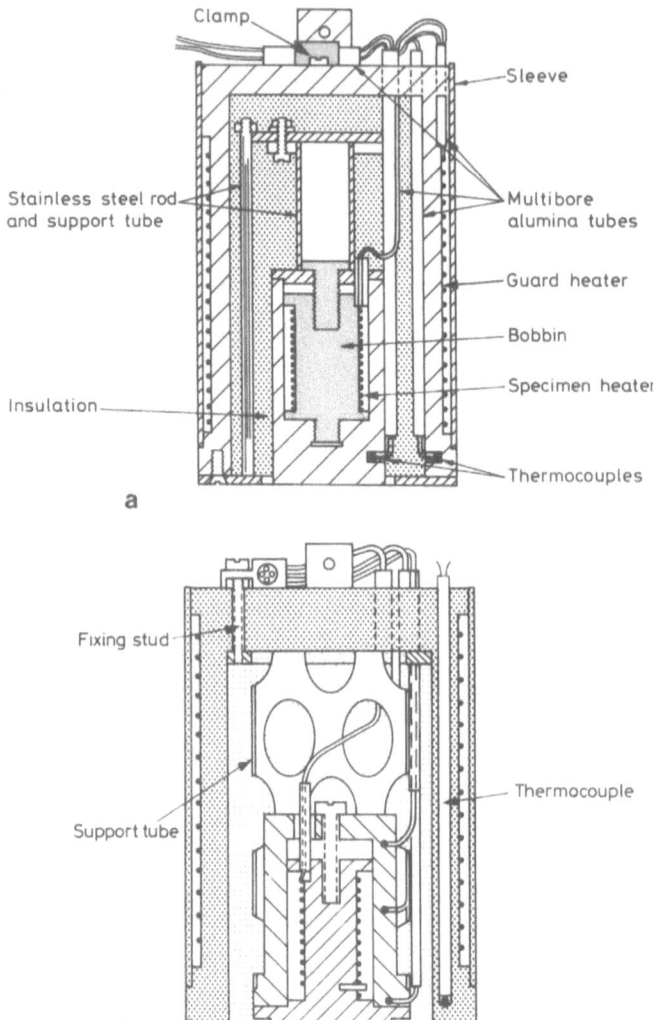

FIGURE 2. (a) Specimen heater assembly made from copper and stainless steel for measurements up to 673 K. (b) Heater assembly made from nickel and stainless steel for measurements above 673 K.

support rods. In order to limit heat transfer between the two parts when temperatures are not matched, the effective heat path length has been increased by boring twelve 10-mm-diameter holes through the support tube around its circumference. The conducted heat flow rate is estimated to be less than 50 mW K^{-1}.

The specimen heater bobbins and guard regions are electrically insulated using Refrasil tape stuck to the metal surfaces with a high-temperature adhesive

(Autostic). The heaters themselves are wound noninductively with 0.32-mm-diameter Nichrome wire using a doubled loop which locates around a ceramic post inserted in a radial hole. The windings are further insulated by the same technique so that the heaters fit snugly into metal sleeves and give good thermal contact. The resistance of the heaters is chosen to match the power supplies available and, in the case of the specimen, high enough to reduce measurement errors due to lead resistance. Potential leads are spot-welded to the specimen heater close to the bobbin, the current leads being thickened in the region between the specimen heater and guard to reduce the effects of stray heating.

In order to facilitate temperature matching, a differential thermocouple (or thermopile) is located between the inner and outer parts of the heater assembly at the same vertical height. This is constructed by inserting bare thermocouple wires through multibore alumina tubes and spot-welding the junctions alternately so that the thermocouple voltages add in series. All leads and thermocouple wires are taken through multibore alumina tubes and thermally anchored to the guard using small clamps, which have holes bored in them to match the size of the tubing. This procedure ensures that the temperatures at the ends of the wires are well defined in the guard region and enables the heat flow along the leads due to small temperature differences to be estimated more precisely.

The main differences in the heater unit made from nickel are the thicker walls of the guard to compensate partly for the lower thermal conductivity of nickel (about $65 \text{ W m}^{-1} \text{ K}^{-1}$ as opposed to $380 \text{ W m}^{-1} \text{ K}^{-1}$ for copper), and the inclusion of additional thermocouples in the specimen and guard heater sections. These enable the heat transferred between the two parts through the insulation to be estimated more accurately whenever heat is flowing into the specimen, since this causes a significant temperature drop along the length of the specimen heater, resulting in a small radial component of heat flow. A more compact design with a flatter heater and shorter overall length would perhaps be preferable (see elsewhere[6] for a discussion of this point) but it would be necessary to ensure sufficient path length in the supports between the inner and outer sections to minimize conduction between them, so some form of compromise is called for.

Materials having higher conductivity than nickel, such as silver or gold, could also be used but would prove expensive. In any case, a good estimate of the magnitude of the heat transferred can be made from a knowledge of the geometry involved, the temperature distribution attained, and the thermal conductivity of the materials used. Typical estimates of the rates of heat loss/gain for the heater unit made from copper, obtained by summing the heat flow along supports, etc., are 0.035 W per degree imbalance at 323 K, and 0.1 W per degree imbalance at 773 K, corresponding respectively to no more than 1.5% and 0.1% of the power supplied to a specimen of relatively low conductivity, such as stainless steel (about $15 \text{ W m}^{-1} \text{ K}^{-1}$). If the temperatures of the guard and heater sections are matched to better than 1 K, it is estimated that the uncertainty in the heat-flow rate into the specimen due to temperature mismatch is less than 0.2%. A further uncertainty of about 0.2% is associated with effects due to stray heating in the heater leads between the guard and heater. Uncertainties due to instabilities of the power supplies themselves are an order of magnitude smaller.

2.3. Specimens

Specimens used in the apparatus normally take the form of cylindrical rods 70 mm long by 20 mm diameter, with machined end-faces. Three equidistant 0.5-mm-diameter by 3-mm-deep holes are spark-machined (or drilled) along their length for inserting thermocouples, their separation being measured optically using a traveling microscope. This type of specimen is suitable for the conductivity range 5 W m^{-1} K^{-1} to about 65 W m^{-1} K^{-1}, but for materials of higher conductivity such as tungsten and alloys of aluminum and copper either a solid dumbbell-shaped specimen or a composite specimen with end caps is preferred (see Fig. 3). This dumbbell specimen arrangement has the advantage that the longitudinal temperature gradient needed for the measurement can be produced with a lower, more convenient heat-flux density. In addition, since the ratio of heat conducted through the sample compared to the radial heat loss is still relatively high, the accuracy of the measurement remains good.

When specimens are placed in the apparatus, they are held in position between the heater assembly and the reference specimen using a compression screw arrangement, care being taken to ensure that the ground surfaces at the

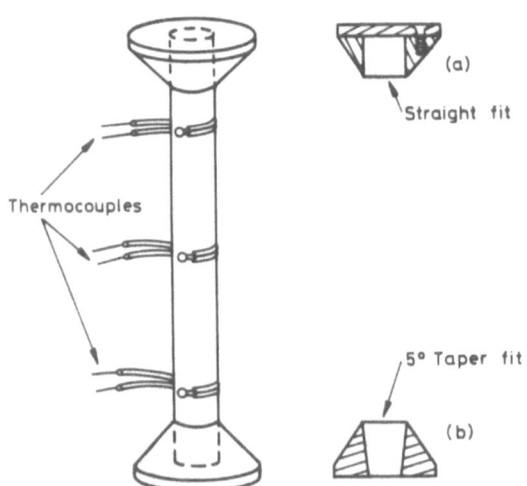

FIGURE 3. Methods for mounting specimens of small diameter and high conductivity.

interfaces are well aligned to avoid thermal contact problems. Some degree of lateral movement is provided at the point of attachment of the specimen heater to the push rod to allow for proper alignment.

The reference specimen on which the specimen stands acts as a heat sink for heat conducted through the specimen and also serves as a heat-flow meter. It is made of stainless steel whose conductivity is determined as a function of temperature by placing the heater unit directly in contact with its upper surface and carrying out measurements. During an experiment the upper two thermocouples in the reference specimen enable the axial heat-flow rate in that region to be determined using $\dot{q} = -\lambda(T)S(dT/dx)$, where S is the cross-sectional area of the reference specimen and dT/dx is the temperature gradient between the upper two thermocouples. The reference specimen and the shield surrounding it are both made from stainless steel, so the temperature gradients in them can be matched very closely, thereby ensuring negligible error in $\dot{q}$ due to radial heat transfer. Thus $\dot{q}$ provides an accurate measure of the heat-flow rate used in the determination of the conductivity of the specimen. The base of the reference specimen is water-cooled and incorporates a simple water calorimeter to check the heat-flow rate during calibration measurements. A typical heat-flow meter of this type is described later in the chapter.

2.4. Specimen Shields

The upper and lower shields enclosing the central column of the apparatus have an internal diameter of 50 mm, a wall thickness of 5 mm, and measure 70 mm and 120 mm long, respectively. They are attached to each other by a pair of aluminum clamps which locate on flanges having a 5-degree taper. Noninductively wound heaters made with doubled Nichrome wire insulated with Refrasil sleeving are cemented beneath the flanges and held firmly in place using semicircular strips of copper screwed in position. These heaters are used to match the temperature gradients in the shields to those established in the specimen stack. Small holes 0.5 mm diameter by 3 mm deep are provided at intervals along the shields for inserting thermocouples, which are held on an isothermal path around the circumference using pairs of screwed clips. Water cooling is provided at the base of the lower shield to ensure that the temperature at that position matches the temperature at the base of the reference specimen.

2.5. Temperature Measurement

Temperatures in the apparatus are measured using Type E Chromel/Constantan thermocouples made from calibrated 0.2-mm (36 SWG) wire insulated with Refrasil sleeving or PTFE, as appropriate. The reference junctions for the thermocouples are made in an isothermal box containing a large rectangular slab of copper to which are screwed conventional "terminal-block" connectors with the base cut away and reinsulated with mica in the manner shown in Fig. 4. The thermocouple wires are then taken in pairs to one end of these connectors and joined at the other to pairs of screened copper wires attached to a precision switch. In order to correct for the block not being at 273.15 K, it is necessary to correct each thermocouple voltage by adding on a voltage corresponding to the output voltage of the thermocouple at the temperature of the block. This voltage may be measured using an additional thermocouple with its measuring junction attached to the copper block and its reference junctions in an oil-filled tube immersed in a conventional mixture of water and ice at 273.15 K. In practice, we use a commercial temperature-sensing unit (supplied, for example, by Ancom or Cropico Ltd) to monitor the temperature of the block and provide a signal corresponding to a thermocouple at that temperature (see Fig. 5). This signal may then be added to all the thermocouples at the pole of the switch or added numerically in a computer program.

2.6. Instrumentation

In order to avoid undesirable fluctuations in temperature down the specimen column, the power supplies for the specimen heater need to be stable to 1 part in 1000 or preferably better. A constant current/constant voltage supply operating between 0 and 100 V in 0.1 V steps (Kingshill Ltd) has proved particularly suitable for this purpose. Another type of power supply with similar characteristics, but having continuously variable coarse and fine control and a digital display of output voltage and current, is used for some of the

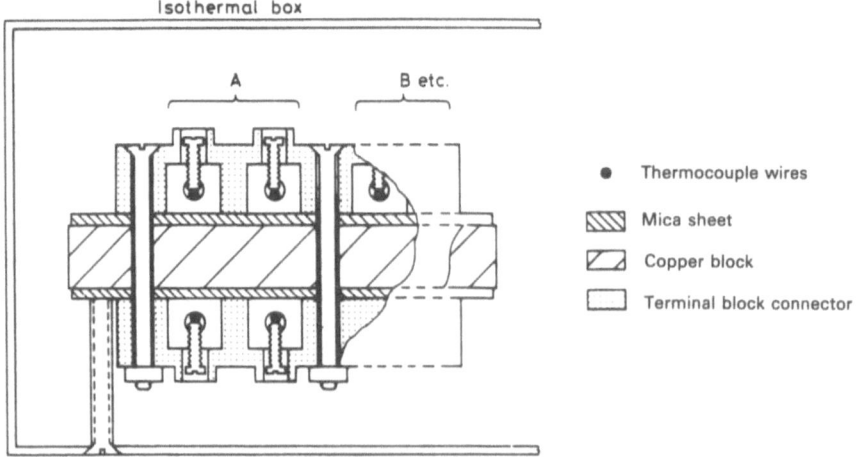

FIGURE 4. Constructional details of the mountings used for thermocouples in the isothermal reference enclosure.

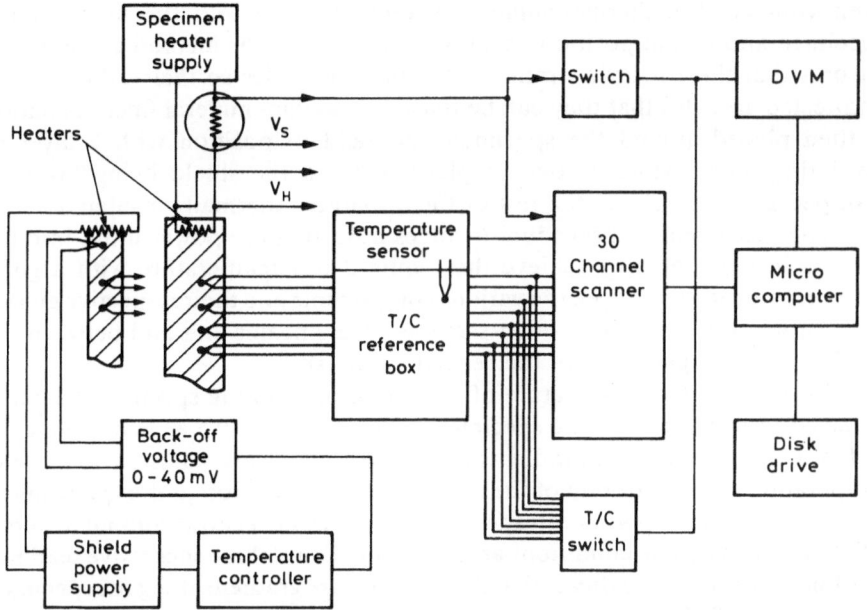

FIGURE 5. Block diagram showing the method used for temperature control, temperature monitoring, and data handling in a thermal conductivity measurement.

other heaters while 0–5 V programmable versions (50 V, 3 A) are also used in conjunction with temperature controllers. A digital voltmeter with a sensitivity of $\pm 1\ \mu$V, six-figure scale, and an accuracy of 0.01% of reading is used for detecting the base-metal thermocouple voltages. A block diagram showing the basis of the circuits used for the thermal conductivity apparatuses described in this chapter is given in Fig. 5. Besides a system of manual switching, which is useful for checking that the apparatus is functioning correctly, it will be noted that there is also a data-logging system. This comprises a low-level scanner operating at the $1\ \mu$V level interconnected to the microcomputer, disk drive unit, and DVM using a standard IEEE interface for control purposes and signal routing. Although the first three units are not essential they greatly facilitate the handling, manipulation, and storage of data. They can also be used to check that the apparatus is approaching equilibrium by comparing results taken at a certain time with results evaluated at a later stage, and so on.

2.7. Experimental Procedure

In order to mount a specimen, the upper shield is raised so that it surrounds the heater unit at the top of the apparatus. The specimen is placed in position and aligned with the clean mating surfaces of the heater and reference specimen before being clamped in position using the compression screw to give firm, even contact. The thermocouples are then inserted into the holes in the specimen and tied in position using Refrasil twine. This method of mounting ensures that there is little strain on the thermocouples so their calibration is unaffected and also that they can be reused. Kaowool mineral fiber insulation is then placed around the specimen and held in position with a layer of insulating paper which is tied in place prior to the shield being lowered, clamped in position, and the rest of the apparatus thermally insulated.

The measurement procedure involves adjusting the power input into the various heaters so as to achieve the required test temperature with a good temperature match between the various components and their respective guards (i.e., between the inner and outer sections of the heater unit and between the shields and the specimens in the central column).

In practice, the temperature of the guard around the specimen heater is adjusted to match that of the specimen heater to within 0.5 K. With reference to Fig. 1, the temperature gradients in the two shields are adjusted so that the temperature at the midpoint of the upper shield and the top of the lower shield match the temperatures at corresponding points in the central column to about 0.5 K. These matching conditions are maintained until three successive readings at 20-minute intervals indicate that the temperature gradient along the specimen is constant to 0.1% or better. At steady state the power input to the specimen heater is measured and all thermocouple voltages recorded. The various voltages are then typed into a microcomputer, which is programmed to convert thermocouple voltages to temperatures, to calculate the radial heat exchange between specimens and their shields as described later, and to determine the thermal conductivity at two different temperatures from the corrected heat-flow rate.

Some results obtained on three NBS reference materials, stainless steel No. 1462, electrolytic iron No. 1464, and arc-cast tungsten No. 1469 are given in Fig. 6 to indicate the performance of the apparatus. Over the temperature range 313 K to 723 K none of the curves deviates by more than 3.1% from the certified values. A fuller discussion is given elsewhere[10] and some additional data on the stainless steel results are provided in Section 4.4.

2.8. Radial Heat-Loss Correction to Conductivity Results

As was noted previously, imperfect temperature matching between the specimen and its surrounding shield gives rise to a small radial component of heat flow, which must be allowed for in deriving the thermal conductivity. Such mismatches occur because the temperature dependence of the thermal conductivity of the shield and of the specimen materials is different, giving rise to nonidentical temperature gradients. Further, the temperature discontinuity at the specimen and shield interfaces will be different because the product of thermal contact resistance and heat-flow rate is different in each case.

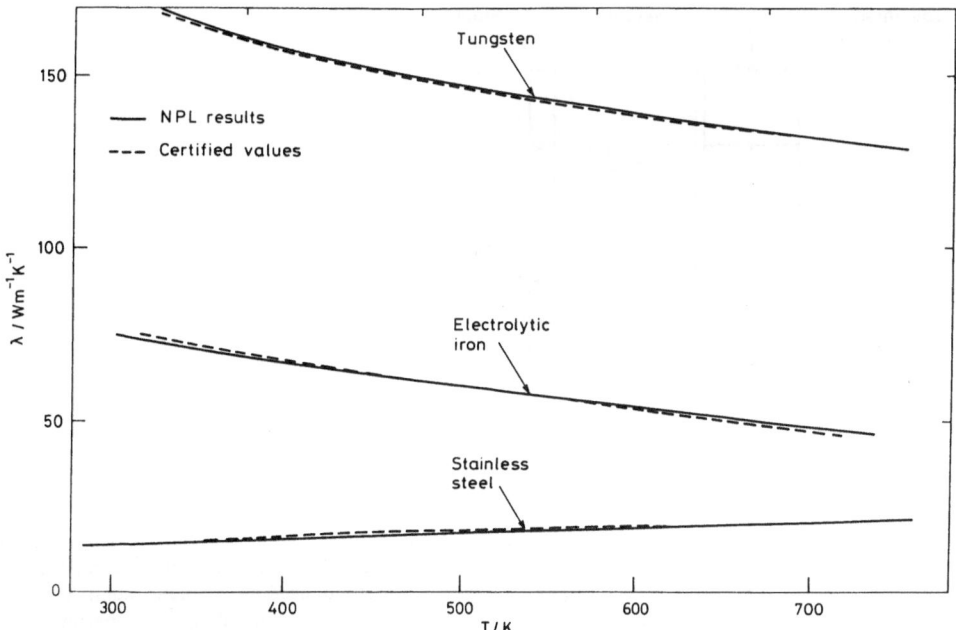

FIGURE 6. Graph comparing NPL results on three NBS reference materials with the certified values.

The heat loss correction may be obtained by assuming the specimen, reference specimen, and their respective shields to be divided into sections whose upper and lower boundaries are defined by horizontal planes passing through the thermocouple locations and the ends of the specimens. A typical section located between two thermocouples on the specimen is shown in Fig. 7. Here $\dot{q}(n)$ and $\dot{q}(n+1)$ represent respectively the axial heat-flow rate into, and out of, the nth section, $\dot{q}(n)$ is the radial loss from the section surface to the shield, and $T(n)$ is the mean temperature of the section.

If $\dot{q}(1)$ is the heat-flow rate from the heater into the first section at the top of the specimen and $\delta\dot{q}(k)$ is the radial heat loss from any section, then the mean axial heat-flow rate through the nth section is given by

$$\dot{q}(n) = \dot{q}(1) - \left[\dot{q}(n)/2 + \sum_{k=1}^{k=n} \delta\dot{q}(k) \right]$$

The terms $\delta\dot{q}(k)$ representing the radial heat loss from specimen to shield are calculated using the formula for radial heat flow between two coaxial cylindrical surfaces assumed to be at mean temperatures $\bar{T}(k)$ and $\bar{T}'(k)$,

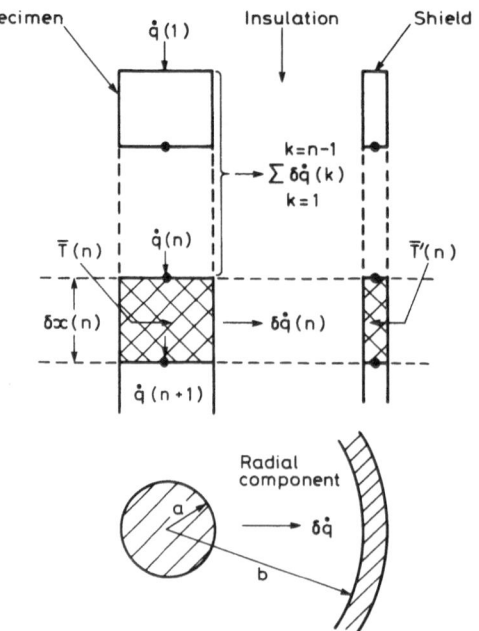

FIGURE 7. Diagram illustrating the method used to evaluate radial heat loss correction along the specimen column.

respectively. Thus

$$\dot{q}(k) = \frac{2\pi\lambda_I[\bar{T}(K) - \bar{T}'(k)]\delta x(k)}{\ln(b/a)}$$

where λ_I is the thermal conductivity of the insulation at the appropriate mean temperature, a and b are the specimen and shield radii, respectively, and $\delta x(k)$ is the length of the kth section.

Calculated values of $\delta\dot{q}(k)$ depend on the effective thermal conductivity of the insulation around the specimen, which may not be the same from experiment to experiment, but since they rarely amount to more than a few percent of the axial heat-flow rate through the specimen even when specimen and shield temperatures are rather poorly matched, large uncertainties in the estimated values of λ_I are relatively unimportant as far as the final thermal conductivity results are concerned.

With the two equations given above, the thermal conductivity $\lambda(n)$ of the nth section can be expressed with the aid of Fourier's equation for linear heat flow as

$$\lambda(n) = \frac{\dot{q}(n)\delta x(n)}{A[T(n) - T(n+1)]}$$

where A is the cross-sectional area of the specimen while $T(n)$ and $T(n + 1)$ are the boundary temperatures of the section.

2.9. Discussion of Uncertainties

The main sources of error in thermal conductivity determinations are associated with unknown heat losses and gains, failure to establish equilibrium, and imprecise measurement of such variables as the temperature, thermocouple spacing, and sample dimensions. Some of these points have already been discussed in Volume 1 of this book,[5] but they will now be considered in the context of the present apparatus. Similar considerations will also apply to the apparatus described later.

A correction for heat transfer in the heater assembly and the associated uncertainty due to imbalance between the temperature of the specimen heater and its guard has been described in Section 2.2 and will not be considered further.

The time taken for the apparatus to reach equilibrium has been assessed by making repeated measurements of the thermal conductivity at a particular temperature while maintaining closely matched temperature conditions between the appropriate parts of the apparatus. It was found that once the initial matching had been achieved by fine adjustment of the power supplies, results taken one hour later agreed to within 0.5% of results taken after a period of 24 hours.

The uncertainty in the determination of the temperature gradient in the specimen depends on the accuracy with which the temperature drop δT between thermocouples can be measured and the precise separation of the thermocouples located within the holes in the specimen. The maximum uncertainty in the measurement of δT using thermocouples giving 60 μV K^{-1} output and a digital voltmeter calibrated to an uncertainty of ± 1 μV is $\pm\frac{1}{30}$ K, corresponding to an error of $\pm 3.3/(\delta T/K)$%. The maximum uncertainty in the determination of the temperature gradient in thermocouple separation is estimated to be 0.55%.

Regarding the measurement of temperature, the Type E thermocouple wire stock was calibrated at NPL over the temperature range 273 K to 723 K. The reproducibility of the calibration for two sample thermocouples prepared from the same wire stock was better than 0.02%, while the overall uncertainty in the calibration was estimated to be about 0.4%.

In order to investigate any possible effect due to nonuniform heat flow at the specimen interfaces, measurements on a stainless-steel specimen were undertaken with the upper and lower thermocouples positioned 5 mm and 15 mm, respectively, from the ends of the specimen. Agreement between this set of measurements and measurements where the thermocouples were positioned symmetrically 10 mm from both ends was better than 0.5%, suggesting that end effects become insignificant beyond 5 mm of an interface, provided the mating surfaces are in good thermal contact.

The estimated maximum uncertainty has been taken as the sum of the maximum uncertainty for the individual sources and amounts to $[3 + 3.3/(\delta T/K)]\%$, where δT was defined above.

3. LOW-TEMPERATURE APPARATUS

3.1. General Description

In this apparatus (see Figs. 8 and 9) the specimen, in the form of a cylindrical solid rod 200 mm long, is heated by a cartridge heater screwed into a recess at the upper end and cooled at the lower end by conduction through

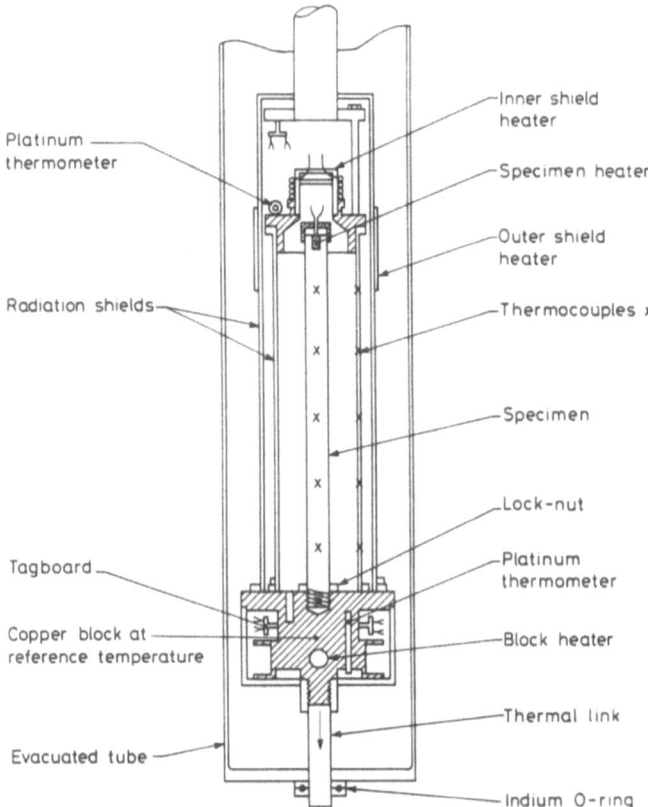

FIGURE 8. Specimen mounting and shield assembly used in the low-temperature thermal conductivity probe.

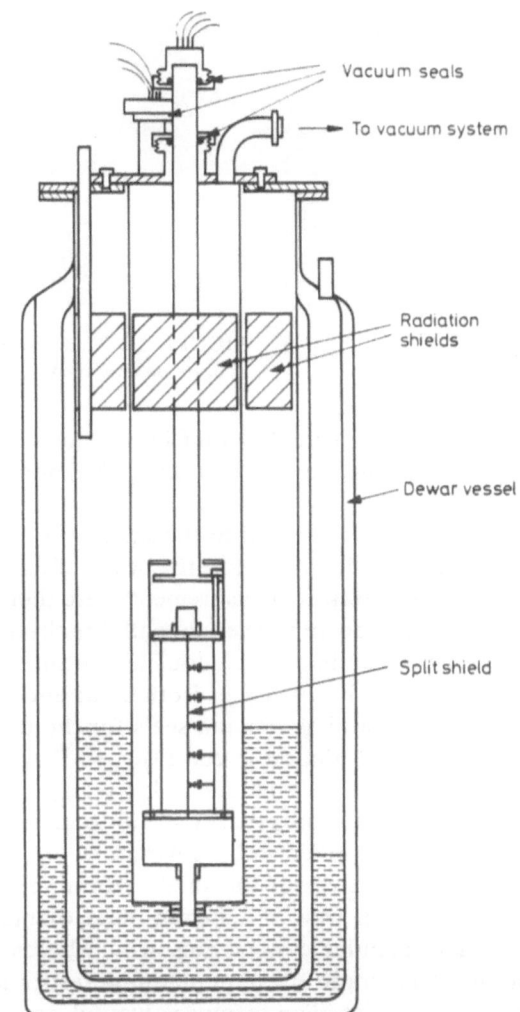

FIGURE 9. General view of the low-temperature apparatus showing vacuum seals and method of refrigeration using a double dewar.

a screwed joint to a copper block maintained at a low temperature. The diameter of the specimen is normally chosen according to the expected thermal conductivity, and ranges from 5 mm for good conductors such as aluminum alloys to 10 mm for poorer conductors like stainless steel. The temperature gradient in the specimen is measured using PTFE-insulated Type E Chromel/Constantan thermocouples (0.2-mm-diameter wire) whose junctions are soldered to small brass clamps (Fig. 10) placed 35 mm apart along the length of the specimen at positions opposite similar thermocouples in the surrounding shield. Each thermocouple on the specimen is brought out at the same level on a spiral path and thermally anchored to the shield by tying it

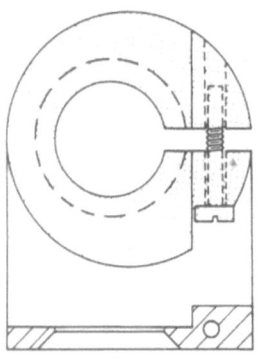

FIGURE 10. Knife-edged brass clamp for mounting thermocouples on specimen.

in position. The shield is split in half lengthwise to allow easy access to the
specimen enclosure for mounting thermocouples and applying glass fiber
insulation.

The edges of the half of the shield permanently fastened to the apparatus
are drilled and slotted to take the beads of the shield thermocouples. To ensure
accurate temperature measurement these thermocouples are taken around the
circumference on an isothermal path for about 50 mm and clamped in position
before being fed in a continuous length through seals in the top flange of the
apparatus (see Fig. 9) to an isothermal enclosure at a known reference temperature. In forming vacuum seals for the thermocouple wires a short length
of the PTFE insulation is cut away, so that when the wires are fed through,
say, either holes in a perspex disk seal (as in the present apparatus) or the
hollow tubes of a proprietary metal-to-glass seal, adhesive or solder, respectively, can be applied to bare wire.

The two halves of the shield (forming a stainless-steel cylinder 180 mm
long by 1.5 mm wall and 46 mm outer diameter) are bolted at their top end
to a large annular copper ring, which holds the shield heater (18 ohm) and
an industrial grade 100 ohm platinum resistance thermometer for temperature
control purposes. The former is wound on an insulated thick-walled copper
former and unscrews from its mounting to permit access to the specimen heater
leads protruding through a copper guard cap over the end of the specimen.

A further stainless-steel shield (260 mm long by 0.7 mm wall by 65 mm
outer diameter) encompasses the split shield and can be moved along the
central supporting tube of the apparatus to provide access to the specimen
chamber when necessary. The heater for this shield is mounted on a 20-mm-
wide copper band whose position can be adjusted empirically to improve the
temperature distribution along the inner shield. A satisfactory position was
obtained with this heater located about 175 mm from the base flange.

The lower ends of both shields are bolted to a copper block in which are
mounted the cartridge heater (25 ohm) and two 100-ohm platinum resistance

thermometers. The block provides a nearly isothermal region, whose temperature is controlled by balancing the power supplied by the various heaters against the heat flux to the refrigerant bath through a thermal link. This link is inserted through the base of the outer vacuum chamber and screwed to the bottom of the block, the necessary vacuum seal being made with a ring of indium wire overlapped and clamped between two smooth-faced flanges. A copper rod 10 mm diameter, with brass fittings brazed in position at either end, is normally used for the link and enables the block to be controlled at a temperature some 15 K above the temperature of the refrigerant bath for a 10-mm-diameter stainless-steel specimen. A simpler alternative might be to join the block to the seal with lengths of thick copper braid and use some form of screw attachment at the inner part of the seal.

3.2. Measurement Method

During measurements the apparatus is evacuated to less than 10^{-1} Pa, and the block temperature is reduced by partly filling the Dewar vessel with liquid nitrogen boiling at 77 K or with a solid CO_2 + alcohol combination giving about 198 K. The temperature of the copper block is then maintained constant at the desired temperature to within about $\frac{1}{100}$ K by using a combination of automatic and manual control. This is achieved by amplifying the out-of-balance signal of a Wheatstone bridge network having a platinum resistance thermometer in one of the arms and controlling the power to the heater in a feedback loop. A degree of manual control is provided by applying a dc voltage of appropriate sign to the summing junction of the second amplifier in the chain. The manual control so provided ensures that the bridge operates close to the balance point and retains its sensitivity. The actual circuit is based on a published design due to Van Hecke and Van Gerven.[14]

The specimen heater is supplied from a constant voltage power supply stable to 1 in 10.000, while the inner shield heater is fed from a conventional temperature controller using the platinum thermometer as sensor. As usual, the power supplied to the specimen is determined from the voltage drop across the specimen heater and the voltage drop across a standard resistance (1 ohm) connected in series with it. Very similar temperature gradients are established in both shields and, once equilibrium has been established, thermocouple and heater voltages are recorded. The temperature gradient in the shield is then altered slightly and, after reestablishment of equilibrium, further readings are taken to obtain the magnitude of the radial heat loss as now described.

3.3. Evaluation of Heat Losses and Results

The heat exchanged between specimen and shield is determined after the method suggested by Watson and Robinson,[15] who assume the heat losses along the length of the specimen to be given by the product of an average heat transfer coefficient f and a temperature difference term S defined by

$$S = \int_0^x [T_{sp}(x) - T_{sh}(x)]\, dx$$

where $T_{sp}(x)$ and $T_{sh}(x)$ are temperatures measured at discrete points along the specimen and shield, respectively. The heat balance equation for a specimen of conductivity λ and cross section A is then given at any point along its length by

$$\dot{Q} - \lambda A\, dT/dx + fS$$

where $\dot{Q}$ is the heat supplied by the specimen heater, dT/dx is the temperature gradient between any pair of thermocouples, and fS is the total heat lost up to the midpoint of the section between the two thermocouples. In order to apply the method the temperature distribution in the specimen and shield is first measured with the temperature gradient in the shield slightly higher than the temperature gradient established in the specimen using a power $\dot{Q}(1)$ applied to the specimen heater. The measurements are then repeated with the temperature gradient in the shield slightly less than the gradient in the specimen, the power supplied to the specimen being increased to a value $\dot{Q}(2)$ to compensate for the extra heat loss from the specimen to the shield. The temperature profiles along the shield and also along the specimen are determined by carrying out low-order polynomial curve-fitting routines in a simple computer program using the thermocouple voltages and thermocouple locations as parameters. The magnitude of the product fS is then determined at points along the specimen by solving, in the present case, four pairs of equations (since five thermocouples are used on the specimen) relating $\dot{Q}(1)$ and $\dot{Q}(2)$ to the temperature gradients established in the two sets of measurements and the calculated values of S. Four values of the thermal conductivity are also obtained at the mean temperature of each individual section between thermocouples when the equations are solved.

4. LONG-BAR APPARATUS

4.1. Description

A general view of the apparatus showing the essential features is given in Fig. 11. These include a water calorimeter arrangement in the stainless-steel

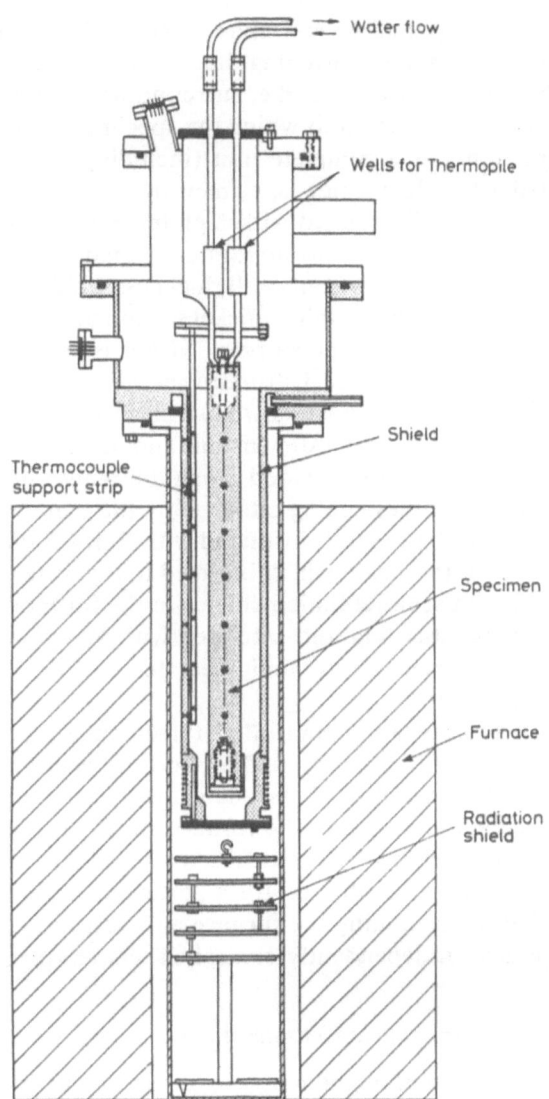

Water flow

Wells for Thermopile

Shield

Thermocouple
support strip

Specimen

Furnace

Radiation
shield

FIGURE 11. High-temperature
apparatus (313 K–1073 K).

tubes supporting the specimen for measuring the heat-flow rate, a vacuum
enclosure for preventing corrosion of materials at high temperature, and a
surrounding furnace to minimize heat losses in the radial direction.

The specimen used in the apparatus is 320 mm long and up to 25 mm
diameter, the ends being recessed to a depth of about 28 mm to take an 18-ohm
heater wound on an insulated nickel former and to mate with the base of the
water calorimeter at the other. It is also provided with 0.5-mm-diameter holes,

3 mm deep, and 35 mm apart along its length to take thermocouples, which are supported by fastening them to a stainless-steel strip in close contact with the inner surface of the surrounding shield. For assembly purposes the calorimeter section to which the specimen is bolted can be removed from the apparatus as a complete unit (including thermocouples and other electrical leads) to allow a new specimen to be mounted, instrumented with thermocouples, and insulated with high-temperature blanket insulation.

The shield in this apparatus is made from a stainless-steel tube, 66 mm outer diameter by 5 mm wall thickness welded to a water-cooled flange at one end and to a thick-walled nickel section at the other containing two axial holes for mounting thermocouples. Nickel is used in this region to give a more uniform temperature distribution at the end of the shield when power is applied to the two shield heaters wound on insulation around the circumference.

Holes of similar dimensions and at similar spacings to those specified for the specimen are provided along the length of the shield for mounting thermocouples, these being attached by the method already described. The main length of the shield is insulated with Refrasil blanket insulation, and enclosed in a vacuum sheath which is sealed to the water-cooled end of the shield using an O-ring seal. Radiation shields in the form of stainless-steel disks separated by metal studding and insulated with Kaowool are placed in the tail of the vacuum enclosure to limit heat losses from the end of the shield to the surroundings.

A temperature-controlled, three-zone furnace of 100 mm bore and operating to 1273 K is positioned around the vacuum enclosure to provide further insulation and to aid the matching of temperature gradients established in specimen and shield. The furnace slides on supporting pillars using counterbalance weights, so its position can easily be adjusted. The most suitable temperature distribution for a particular experiment can then be obtained by suitably positioning the furnace and adjusting the power supplied to each of the three independently controlled temperature zones.

4.2. Thermocouple Considerations

Careful thought must be given to the type of thermocouple used for temperature measurement, since good stability and long life are required. Type K wire, commonly used up to 1273 K, was considered to be unsuitable because large changes in calibration can occur at high temperatures due mainly to oxidization of the electronegative component, while certain phase changes in the electropositive component due to magnetic ordering may also take place at lower temperatures (573 K–673 K) leading to undesirable effects.[16]

Although Pt–Pt/10%Rh thermocouples meet the above requirements, the cost of a large number of thermocouples in this material would be high so it was decided to use the newly developed nicrosil/nisil (Type N) combination, which has been shown to have much higher stability than other base metal thermocouples when exposed to high temperatures for long periods.[17]

Accordingly, thermocouples used in this apparatus were made from lengths of 0.25-mm-diameter nicrosil/nisil wire, insulated with Refrasil sleeving in the heated parts of the apparatus and with PTFE sleeving elsewhere. Although thicker wire would have offered greater stability, any heat leaks along the thermocouple wires would have led to greater uncertainty in the measured temperature. As before, care was taken to ensure near-isothermal conditions in the wires before taking them to the reference junctions through metal/glass seals as described earlier. Useful information on the selection, calibration, and use of thermocouples may be found elsewhere.[18,19]

4.3. Heat-Flow Meter

Although in this apparatus power to the specimen heater may be measured accurately by measuring the voltage drop across the heater electrically, radiative losses at high temperatures due to temperature mismatch between heater and shield can give measurement errors. To avoid this difficulty the heat-flow rate or power received at the cooled end of the specimen is measured using a water calorimeter as shown in Fig. 12. Two partially hollow copper blocks containing thermocouple wells are soldered to stainless-steel tubes through which the cooling water flows, the lower end of each tube being brazed to a hollow cylindrical copper block that locates in the end of the specimen and is screwed in place. In order to obtain good thermal contact indium wire is sandwiched between the base of the copper block and the specimen, and a copper sleeve about 30 mm long is also clamped around the circumference of the specimen and the top of the block once the securing screw for the specimen has been tightened.

The thermopile itself is constructed from Type E (chromel/constantan) 0.1-mm-diameter wires insulated with PTFE. The thermopile is made by connecting alternate wires of the two materials in series by soldering (or spot-welding) to make 20 junctions which are inserted, together with a short length of insulation, into holes in ceramic tubes. These tubes are then placed in the wells and fixed in position using a high-melting-point wax to provide good thermal contact.

In order to have an accurate measure of the power received by the heat-flow meter it is necessary to provide a constant head of water (see Fig. 13). The flow rate is monitored with a commercial turbine flow sensor/flow-meter combination and determined more precisely during a measurement of thermal conductivity by collecting a known mass of water in a given time. In order to calibrate the thermopile a small heater coil is wound temporarily on the section of the calorimeter normally inserted in the specimen and a layer of thermal insulation applied around it. Electrical energy is then supplied to the heater

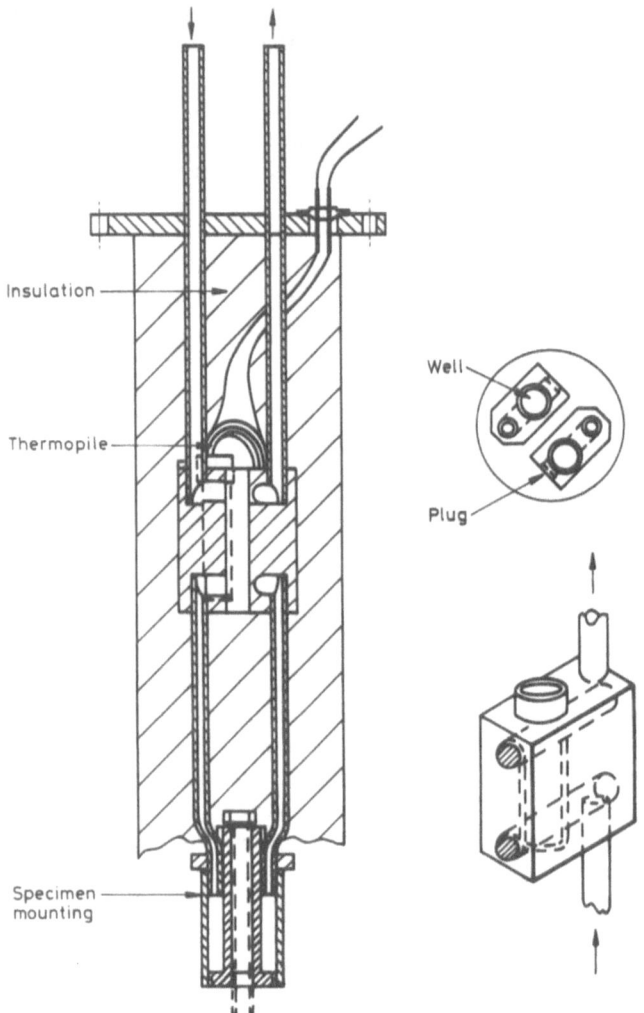

FIGURE 12. Design details of the water calorimeter/heat-flow meter and specimen mounting.

at a rate W to give a small temperature difference indicated by the output voltage E of the thermopile. If m is the mass of water collected in time t, then the sensitivity of the thermopile, S, will be given by

$$S/(\text{V K}^{-1}) = 4.187 mE/(Wt)$$

The present thermopile has a sensitivity of $597 \pm 3 \ \mu\text{V K}^{-1}$ at 293 K, the main uncertainty being caused by small variations in the flow rate of the circulating water.

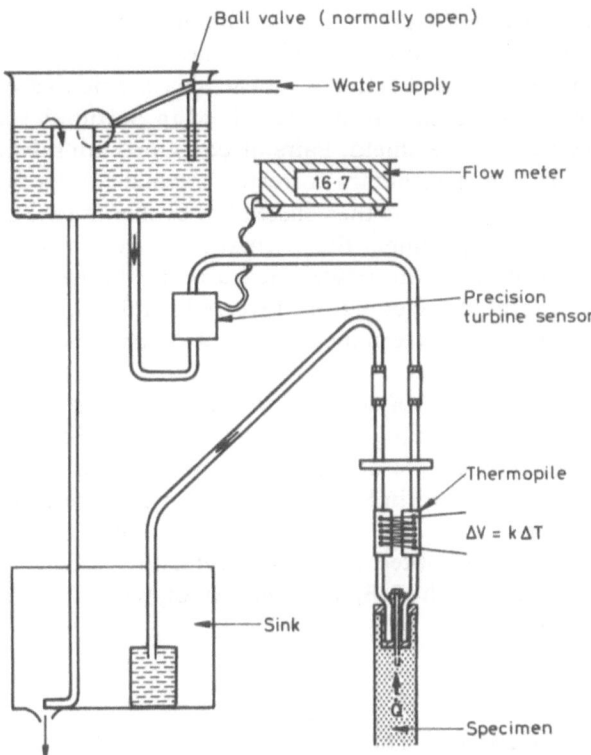

FIGURE 13. Arrangement used to provide a constant flow of water through the calorimeter and to determine the heat-flow rate at the cold end of the specimen.

An uncertainty less than 0.5% between the power measured electrically and the power measured with the heat-flow meter is obtained using the method described.

4.4. Operational Procedure

The specimen to be measured is bolted to the lower end of the water calorimeter and thermocouples and heater are attached. A nickel guard cap is then pushed over the heater end of the specimen and a layer of Refrasil blanket applied along the length of the specimen. The whole assembly is then lowered into the shield, which is normally kept in position inside the furnace. Once the necessary vacuum seals have been made, the apparatus is evacuated (to a pressure <10 Pa) and the heaters are turned on and balanced until equilibrium conditions have been established with the temperature in the shield slightly less than that in the specimen at all points below the heater region. The water flow at the cold end of the specimen and shield can be adjusted independently to ensure a close match of temperatures at the cold end of the specimen. Thermocouple and heater voltages are then recorded together with the heat-flow meter data (see Section 4.3). The measurements are then repeated following a similar procedure to that described in Section 3.3 using a higher temperature gradient in the shield. Pairs of equations are solved as before at points along the specimen to obtain corrected thermal conductivity values, the only significant difference being that, instead of evaluating the equations from the heated end of the specimen, the evaluation is now made from the cold end of the specimen using the power measured at the heat-flow meter with the sign of the fS term in the equations reversed. In this way six values of the thermal conductivity are determined at points along the specimen over a particular temperature range.

Table 1 shows results obtained with this apparatus on NBS stainless steel No. 1462 in the temperature range 323 K to 973 K compared with earlier results obtained on a different sample of the same material measured in the general-purpose apparatus described in Section 2. Agreement of results obtained with the two apparatuses is better than 3%, a proportion of which could be due to material variability. The quoted differences refer to the percent deviation of our measured results from the NBS certified values, which have been smoothed to permit the comparison.

TABLE 1

A Comparison of Thermal Conductivity Results for NBS
Reference Material, Stainless Steel No. 1462, and Their
Deviation from the Certified Values

T (K)	Long-bar apparatus		General-purpose apparatus	
	$W\,m^{-1}\,K^{-1}$	Deviation (%)	$W\,m^{-1}\,K^{-1}$	Deviation (%)
323	14.71	−0.36	14.38	−2.63
373	15.61	−0.79	15.26	−3.09
423	16.47	−0.90	16.13	−3.05
473	17.31	−0.79	16.98	−2.70
523	18.13	−0.54	17.82	−2.16
573	18.91	−0.21	18.65	−1.51
623	19.67	−0.15	19.47	−0.79
673	20.40	0.51	20.27	−0.07
723	21.10	0.85	21.06	0.66
773	21.78	1.14	—	—
823	22.43	1.38	—	—
873	23.05	1.56	—	—
923	23.65	1.68	—	—
973	24.21	1.73	—	—

4.5. Errors and Uncertainties

The most likely cause of error in this apparatus is associated with the heat losses occurring from the surface of the specimen over a rather long path length. The magnitude of this error may be assessed by conducting a series of measurements over a range of temperatures together with some measurements at a relatively low temperature, say 500 K, on a low-conductivity material such as stainless steel ($\lambda \sim 15 \text{ W m}^{-1} \text{ K}^{-1}$). The first set of measurements will show up any inconsistencies in the evaluated thermal conductivity values, because each series of results will not lie on the same smooth curve. The deviation of the respective end points of the curves will therefore give an indication of any outstanding heat-loss error.

Measurements of the second type are particularly significant for low-conductivity specimens, because any radial losses comprise a much greater proportion of the total heat flow through the specimen. At higher temperatures these losses will become relatively less important. In obtaining the results on stainless steel shown in columns 1 to 3 of Table 1, some forty points, fitted to a degree-two polynomial over the temperature range 310 K–973 K, had a standard deviation to the fit of 0.29 W m^{-1} K^{-1} equivalent to about 1.6%. The uncertainty associated with thermal conductivity values obtained for positions away from the ends of the specimen was generally found to be slightly less than for values obtained for the spans between pairs of thermocouples located nearer the hot or cold ends.

The overall uncertainty associated with the apparatus, taking into account factors mentioned in Sections 2.9 and 4.3, is estimated to be less than ±4% over the temprature range 323 K to 973 K.

5. DETERMINATION OF THERMAL CONDUCTIVITY FROM ELECTRICAL RESISTIVITY

5.1. Background of Method

The thermal conductivity of many metals can be predicted to within about 10% from a knowledge of their electrical conductivity and it is recommended that this property be determined as a matter of course whenever thermal conductivity measurements are being undertaken. For pure metals the well-known equation of Wiedemann and Franz

$$\lambda = L_0 \sigma T$$

can be used, where T is the absolute temperature, σ is the electrical conductivity measured in $(\Omega \text{ m})^{-1}$, and L_0 is the Lorenz number equal to $2.39 \times 10^{-8} \text{ (V K}^{-1})^2$. For alloys, more reliable results may be predicted using an

TABLE 2
Values of L and C to Be Used in the Smith and Palmer Equation

Alloy system	$10^8 L$ $(\text{V K}^{-1})^2$	C $(\text{W m}^{-1}\text{K}^{-1})$	Approximate accuracy (%)
Aluminum alloys	2.10	12.6	6
Copper alloys	2.39	7.5	10
Ferrous steels	2.43	9.2	10
Austenitic steels	2.39	4.2	10
Iron–nickel alloys	2.92	3.0	10
Mg and magnesium alloys	2.21	9.6	6
Ni and nickel alloys	2.13	8.4	20
Ni–Cr alloys (Nimonics)	2.20	6.0	5
Ti and titanium alloys	2.62	2.1	10
Zr alloys	2.50	2.2	10

empirical equation of the type proposed by Smith and Palmer,[20]

$$\lambda = L\sigma T + C$$

The parameters L and C are constants in the range 2.1 to 3.0×10^{-8} $(\text{V K}^{-1})^2$ and 2 to $13 \text{ W m}^{-1}\text{ K}^{-1}$, respectively, for many alloys that have been investigated. Typical values of L and C to be used for predicting the thermal conductivity of various alloys are given in Table 2. Fuller details and an analysis of results have been given by Powell.[21]

5.2. Practical Method of Measuring the Electrical Conductivity

The electrical conductivity of a material can be found using the conventional four-wire technique illustrated in Fig. 14. Current is supplied to a specimen of uniform cross section from a constant-current source capable of delivering up to about 20 A, and its magnitude is determined by measuring the voltage drop across a standard resistance of about 0.005 ohm using a digital voltmeter sensitive to 1 μV. The voltage drop along a known length of specimen

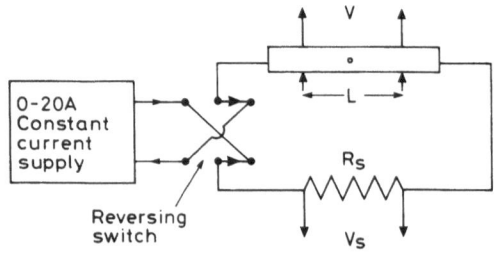

FIGURE 14. Circuit for determining the electrical conductivity of metals.

is also determined, and the measurements are then repeated with the current reversed to eliminate thermal voltages which may arise at the contacts and along the specimen. The electrical conductivity is given by the expression

$$\sigma = V_s L / V R_s A$$

where V is the mean voltage measured over a length L of the specimen, A is the cross-sectional area, and V_s is the mean voltage measured across the standard resistance R_s.

In practice the specimen is clamped against nickel knife edges of known separation that are electrically isolated from their supporting base using ceramic (see Fig. 15), and its temperature is measured with a minerally insulated thermocouple. Current leads are attached to the specimen by screw clamps made from nickel. Copper can also be used, although it oxidizes badly above 650 K. The specimen and support assembly are placed in a low-mass furnace so that the time taken to reach equilibrium at any particular temperature is relatively short. The furnace is set to the temperature required and, once conditions are steady, several readings are taken as outlined above using currents of say 5, 10, and 15 A. The mean voltage measured across the specimen should be proportional to the mean current used if the apparatus is operating satisfactorily.

It should be pointed out that the commercial instruments and products mentioned in this chapter are intended as a guide, and the reader may well

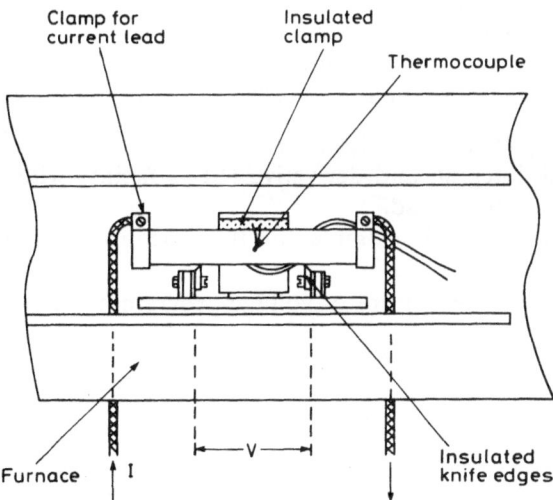

FIGURE 15. Practical arrangement for electrical conductivity measurement at elevated temperatures (schematic).

find equally suitable or better alternatives on the market for any particular application. The following list gives what are considered to be the minimum requirements of equipment used in conjunction with thermal conductivity measurements.

Specification of Equipment

Digital voltmeter—IEEE compatible
Resolution 1 μV on 0.1 V range, 100 V capability.
Accuracy ±0.02% of reading ±3 digits.

Scanner unit
30 channels with noise level about ±1 μV.
IEEE interface for connection to DVM and computer.

Standard resistances
0.1 Ω (2 W) for use with heaters.
0.005 Ω (10 W) for resistivity measurement.
Accuracy ±0.01%.

Electronic reference junction
Stability to random drift $< \pm 1 \mu$V over the period when readings are being taken.

Power supplies
0–50 V, 0–3 A, manual and/or voltage programmable.
Overall stability ±0.1%.
Line and load stability approximately ±0.01%.
0–10 V, 0–30 A (for resistivity measurements).
Overall stability approximately ±0.2%.

Temperature controllers
Proportional, integral, and derivative (PID) facility with approximately 25% proportional band.
Input sensitivity: 1 μV suitable for thermocouple/platinum resistance thermometer.
Output: 0–5 V dc for control of voltage-programmable power supplies.

REFERENCES

1. J.G. Hust, "Thermal Conductivity and Electrical Resistivity, Standard Reference Materials: Tungsten (4 to 3000 K)," *High Temp. High Pressures* **8**, 377–390 (1976).
2. J.G. Hust and P.J. Giarratano, "Standard Reference Materials: Thermal Conductivity and Electrical Resistivity, Standard Reference Materials: Austenitic Stainless Steels, SRM 735 and 798, from 4 to 1200 K," NBS Special Publication 260-45 (1975).

3. J.G. Hurst and P.J. Giarratano, "Thermal Conductivity and Electrical Resistivity, Standard Reference Materials: Electrolytic Iron, SRM 734 and 797 from 4 K to 1000 K," NBS Special Publication 260-50 (1975).
4. R.P. Tye (ed.), *Thermal Conductivity*, Vol. 1, Academic Press, London and New York (1969).
5. K.D. Maglić, A. Cezairliyan, and V.E. Peletsky (eds.), *Compendium of Thermophysical Property Measurement Methods*. Vol. 1, *Survey of Measurement Techniques*, p. 11, Plenum Press, New York and London (1984).
6. M.J. Laubitz, "Measurements of the Thermal Conductivity of Solids at High Temperatures using Steady State Linear and Quasilinear Heat Flow," p. 111, Ref. 4.
7. M.J. Laubitz, "Axial Heat Flow Methods of Measuring Thermal Conductivity," p. 11, Ref. 5.
8. D.L. McElroy and J.P. Moore, "Radial Heat Flow Methods for the Measurement of the Thermal Conductivity of Solids," p. 186, Ref. 3.
9. J.P. Moore, "Analysis of Apparatus with Radial Symmetry for Steady State Measurements of Thermal Conductivity," p. 61, Ref. 5.
10. J.M. Corsan, "A Compact Thermal Conductivity Apparatus for Good Conductors," *J. Phys. E* **17**, 800–807 (1984).
11. R.W. Powell and M.J. Hickman, The Iron and Steel Institute, Second Report of the Alloy Steels Research Committee, Part 3, Special Report 24, Section 9, Part 3, pp. 215-251 (1939).
12. R.W. Powell and R.P. Tye, Thermal and Electrical Conductivities of Nickel-chromium (Nimonic) Alloys," *Engineer* **209**, 729–732 (1960).
13. J.E. Connett and J.M. Corsan, "Thermal Conductivity of a Titanium–Aluminium–Vanadium Alloy," CHEMPOR 85 Conference Proceedings, University of Coimbra, Coimbra, Portugal, April 15–19, Section 31/1 (1985).
14. P. Van Hecke and L. Van Gerven, "A High Performance DC Electronic Regulation System for Temperature Stabilization," *Cryogenics* **10**, 386–388 (1970).
15. T.D.W. Watson and H.E. Robinson, "Thermal Conductivity of some Commercial Iron–Nickel Alloys." *ASME J. Heat Transfer*, 403–408 (1961).
16. N.A. Burley, in: *Temperature: Its Measurement and Control in Science and Industry*, Vol. 4, Part 3, (Ed. H. H. Plumb), pp. 1677-1695, Instrument Society of America, Pittsburgh, PA 15222 (1972).
17. N.A. Burley, R.L. Powell, G.W. Burns, and M.G. Scroger, "The Nicrosil versus Nisil Thermocouple: Properties and Thermoelectric Reference Data." NBS Monograph 161, CODEN:NBSMA6, U.S. Department of Commerce/National Bureau of Standards, Washington, USA (1978).
18. T.J. Quinn, in: *Temperature, Monograph in Physical Measurements* (A.H. Cook, ed.), Academic Press, London and New York (1983).
19. "Manual on the use of Thermocouples in Temperature Measurement," ASTM Special Technical Publication 470B, American Society for Testing and Materials, Publ. Code 04-470020-40 (1981).
20. C.S. Smith and E.W. Palmer, "Thermal and Electrical Conductivities of Copper Alloys," *Trans. Am. Inst. Miner. Metall. Eng.* **117**, 225–243 (1935).
21. R.W. Powell, "Correlation of Metallic Thermal and Electrical Conductivities for both Solid and Liquid Phases," *Int. J. Mass Heat Transfer* **8**, 1033–1045 (1964).

2

Thermal Conductivity of Loose-Fill Materials by a Radial-Heat-Flow Method

DANIEL R. FLYNN

1. INTRODUCTION

Among the several possible methods for measuring thermal conductivities of loose-fill materials (powders, granules, or fibers) at high temperatures, the radial-heat-flow-in-a-cylinder method has the advantage that it inherently reduces the need for thermal guarding or accessory thermal insulation to restrict unwanted heat flow which can cause serious errors in results. This feature is important because, especially at high temperatures, the thermal insulation that may be needed in other methods, such as one using longitudinal heat flow in a cylinder, may be as conductive as the specimen being measured.

Typically, a radial-heat-flow apparatus has a cylindrical heater located along the axis of a cylindrical test specimen of length sufficient to avoid significant errors due to end conditions. Near the mid-length of the cylinder, heat flows radially outward from the axial heater and creates an angularly uniform distribution of temperature in the specimen, the temperature decreasing with increasing radius. After steady temperature conditions are attained, temperatures in the specimen at two different and known radii are measured. The thermal conductivity of the specimen is calculated from these data, and from the measured heat input per unit length of the axial heater.

Such a configuration has the disadvantages that:

- It is difficult to maintain the temperature sensors at precisely known locations within the loose-fill specimen material.
- The temperature sensors may be subject to damage if much force is required to compact the specimen to the desired bulk density.

DANIEL R. FLYNN ● National Institute of Standards and Technology, Gaithersburg, Maryland 20899, USA. Present address: DRF R & D, Millwood, Virginia 22646-0254, USA.

- The temperature sensors may be rather large and may be much more conductive than the specimen material they displace, so that the presence of the sensors may perturb the radial heat flow in the specimen and the effective thermal locations of the sensors may differ from their geometrical locations.
- The longitudinal thermal conductance of the temperature sensors may be sufficiently large that longitudinal heat conduction along the sensors may affect the temperature readings.
- The temperature sensors may be subject to contamination by the specimen material.

The horizontal cross section of a hypothetical radial-heat-flow apparatus is shown in the upper portion of Fig. 1. The specimen is contained between a cylindrical shell, which supports a heater used to achieve the desired elevated temperature, and a coaxial core, which contains a heater used to achieve the desired temperature difference across the specimen. If the temperature sensors

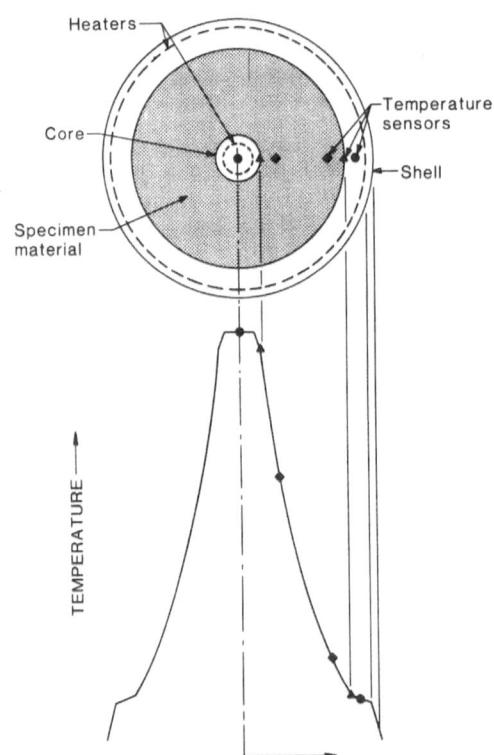

FIGURE 1. Horizontal cross section of a generalized radial-heat-flow apparatus for measuring the thermal conductivity of loose-fill insulations. The lower sketch indicates the radial temperature distribution and alternative locations for temperature sensors.

are located at radial positions within the specimen material, as indicated by the solid diamonds, small uncertainties in their effective locations can result in significant errors in the computed thermal conductivity, because of the large temperature gradients, as shown in the lower portion of Fig. 1, in which the sensors are located.

An alternative approach is to measure the temperatures at the interfaces between the core and the specimen material and between the specimen material and the shell. Temperature sensors at these locations are indicated by the triangles in Fig. 1. This approach is analogous to that used with a guarded hot plate apparatus, where the temperatures of the surfaces of the hot plate and the cold plate are measured and the temperature difference between the two plates is taken equal to the temperature drop across the specimen. For a high-temperature, radial-heat-flow apparatus, particularly if ceramic materials are used, the thermal conductivities of the core and the shell may not greatly exceed the thermal conductivity of the specimen material. This fact, coupled with the inherently large heat flux from the small-diameter core, leads to rather large radial temperature gradients in the shell and, particularly, in the core so that accurate measurement of the surface temperatures presents some difficulties.

For temperature measurements within the specimen material (the solid diamonds in Fig. 1) or for measurements of the temperature of the exterior surface of the core and the temperature of the interior surface of the shell (the triangles in Fig. 1), there is a possibility of angular temperature variations so that temperatures should be measured at several angular positions and averaged.

Inspection of the temperature profile in Fig. 1 indicates that the radial temperature gradient is zero at the center of the core and is rather small between the inner surface of the shell and the radius where the shell heater is located. Thus these locations, indicated by solid circles in Fig. 1, are rather ideal locations for temperature sensors. If a ceramic shell and a ceramic core are used, and the temperature sensors are, for example, thermocouples in ceramic tubing, the temperature sensors can achieve excellent thermal coupling to their environment and will cause very little disturbance to the radial heat flow, particularly in the center of the core where there is no radial heat flow to disturb.

These considerations highlight the potential advantages of high-temperature, radial-heat-flow apparatus, for relatively low conductivity loose-fill specimens, in which temperatures are measured in the isothermal region in the center of the core and in the low-thermal-gradient region within the shell (inside the radius where the shell heater is located). While this approach requires calculating the temperature differences between the temperature sensor locations and the interfaces of the core and of the shell with the specimen, these "adjustments" are small and can be calculated with little uncertainty if the thermal conductivity of the core and that of the shell are well known.

Two different apparatuses based on this approach have been designed and built by the present author.[1-3] The first of these, which was described in 1963,[1] was used to determine the thermal conductivities of loose-fill powders, in air, over the temperature range 100 to 1100 °C. In this apparatus the shell (see Fig. 1) was a furnace that raised the specimen to high temperatures; the core heater produced radial temperature differences of the order of 30 to 60 °C across the specimen.

In 1969,[2,3] a conceptually similar apparatus was designed and built to measure the thermal conductance of soils with the shell held near room temperature and the core raised to hot-side temperatures as high as 1650 °C.

Recently, Moore[4] has analyzed the 1963 apparatus[1] and found that it had certain advantages over many other radial-heat-flow methods for measuring the thermal conductivity of loose-fill materials. Moore[4] stated that "the best features of this apparatus are the small amount of material required and the operation without thermocouples immersed in the specimen." Moore[4] utilized a finite-difference computer program to analyze heat flow in Flynn's[1] 1963 apparatus and confirmed that the influence of deviations from radial heat flow was negligible. In the recommendations at the end of his chapter, Moore stated that:

> "Variations of the apparatus developed by Flynn are probably the best for measuring the thermal conductivity of loose-fill material such as powder. This superiority is based on the lack of temperature sensors within the loose-fill specimen, which could move during thermal cycling and, thus, lead to large measurement errors. Additionally, Flynn's concept can be easily used to measure the thermal conductivity of loose-fill material with different gases at different pressures in the specimen chamber and in the remainder of the system."

These recommendations led the editors to ask that the present chapter be written.

In addition to a description of the 1963 thermal conductivity apparatus,[1] its theory and operation, a description of the 1969 thermal conductance apparatus is included in the present chapter. This later apparatus included changes in the core heater design to enable going to higher temperatures and a more sophisticated analysis of the influence of longitudinal heat flows. Following a discussion of the theory underlying the basic apparatus design and a discussion of the two apparatuses, including theoretical analyses of the effects of longitudinal heat flows, recommendations are given for the design of future apparatuses of this general design.

2. METHOD AND MATHEMATICAL ANALYSIS

A horizontal cross section of the essential elements of the 1963 apparatus[1] is shown in Fig. 2. The specimen was contained within the annular space

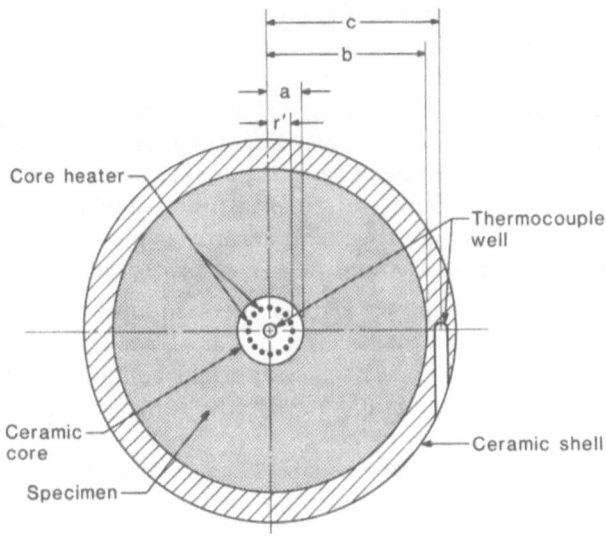

FIGURE 2. Horizontal cross section of the 1963 apparatus.[1]

between the outer radius, a, of the ceramic core and the inner radius, b, of the concentric shell. A known quantity of heat per unit time generated electrically in the ceramic core flowed radially through the specimen to the ceramic shell. The ceramic core had a concentric ring of equally spaced holes at a radius, r', parallel to the axis, each containing a heater wire. Temperatures were measured by an axial thermocouple in the ceramic core and by tangential thermocouples in the ceramic shell at radius c.

For radial heat flow between the cylindrical isothermal surfaces $r = a$ and $r = b$, the heat flow rate, q, through a cylindrical element of unit axial length is

$$q = -2\pi r\lambda(T)\frac{dT}{dr} \tag{1}$$

where temperature is denoted by the symbol T and the temperature-dependent thermal conductivity by $\lambda(T)$. The mean thermal conductivity, $\bar\lambda$, over the temperature range T_b to T_a is given by

$$\bar\lambda = \frac{1}{T_a - T_b}\int_{T_b}^{T_a}\lambda(T)\,dT \tag{2}$$

$$= \frac{q\ln(b/a)}{2\pi(T_a - T_b)} \tag{3}$$

For relatively small temperature differences, $T_a - T_b$, the thermal conductivity of the specimen material can be assumed to vary linearly with temperature and $\bar{\lambda}$ corresponds to the mean temperature $\bar{T} = (T_a + T_b)/2$.

Equation (3) requires a knowledge of the temperatures T_a and T_b at the inner and outer surfaces of the sample. Experimental difficulties discussed in the Introduction (e.g., finite size of temperature sensors and contamination of sensors by the sample) discourage direct measurement of T_a and T_b. Reference to Fig. 2 shows that there are four temperature drops which should be considered in deriving T_a and T_b from the measured temperatures:

- The temperature drop between the thermocouple well in the center of the ceramic core and the surface, $r = a$, of the ceramic core.
- The temperature drop due to the thermal contact resistance between the surface of the ceramic core and the inner surface of the sample, both surfaces being at $r = a$.
- The temperature drop due to the thermal contact resistance between the outer surface of the sample and the inner surface of the shell, both surfaces being at $r = b$.
- The temperature drop between the inner surface, $r = b$, of the shell, and the outer surface, $r = c$, where the temperature is measured.

If the circle of heater wires at $r = r'$ were a continuous cylindrical heat source, the entire region $r < r'$ inside the heater circle would be isothermal. For the case of a finite number of line heat sources at $r = r'$, the temperature measured at the axis is equal to the average temperature, T_0, at the radius of the heater circle.[1,5] Therefore the temperature drop between the thermocouple well in the center of the core and the surface of the core is given by

$$T_0 - T_a' = \frac{q \ln (a/r')}{2\pi\lambda_c} \tag{4}$$

where λ_c is the thermal conductivity of the ceramic core material and T_a' is the temperature at the outer surface, $r = a$, of the ceramic core.

If R_a is designated as the thermal contact resistance per unit area at the core–specimen interface, $r = a$, then the temperature drop across this interface is given by

$$T_a' - T_a = \frac{qR_a}{2\pi a} \tag{5}$$

where T_a is the temperature at the inner surface, also $r = a$, of the sample.

Similarly, the temperature drop at the specimen–shell interface ($r = b$) is given by

$$T_b - T_b' = \frac{qR_b}{2\pi b} \tag{6}$$

where T_b' is the temperature at the inner surface ($r = b$) of the shell, T_b is the temperature at the outer surface (also $r = b$) of the sample, and R_b is the thermal contact resistance per unit area at the specimen–shell interface.

The temperature drop across the shell is

$$T_b' - T_c = \frac{q \ln (c/b)}{2\pi\lambda_s} \tag{7}$$

where T_c is the temperature at $r = c$ in the shell and λ_s is the thermal conductivity of the shell material.

In the apparatus configuration shown in Fig. 2, the temperatures measured correspond to T_0 and T_c as defined above. If equations (3) and (4) through (7) are combined, all of the temperatures involved can be eliminated except T_0 and T_c, yielding

$$\bar{\lambda} = \bar{\lambda}_{app}\left[1 - \frac{\bar{\lambda}_{app}}{\ln (b/a)}\left(\frac{\ln (a/r')}{\lambda_c} + \frac{\ln (c/b)}{\lambda_s} + \frac{R_a}{a} + \frac{R_b}{b}\right)\right]^{-1} \tag{8}$$

where

$$\bar{\lambda}_{app} = \frac{q \ln (b/a)}{2\pi(T_0 - T_c)} \tag{9}$$

is the approximate mean thermal conductivity value obtained by neglecting the correction terms.

Equation (8) is the basic equation to be used for calculating mean thermal conductivity values if the heat is assumed to flow radially. The quantities a, b, q, T_0, and T_c involved in the definition of $\bar{\lambda}_{app}$ are all directly measurable. The quantities λ_c, λ_s, R_a, and R_b (required for the correction terms) must either be known, or the corresponding correction terms shown to be negligible.

For the two apparatuses[1-3] described in this chapter, estimated values were used for λ_c and λ_s while R_a and R_b were assumed to be neglible. For sample material having high thermal conductivity, the error due to neglecting thermal contact resistance may be too large to be acceptable. In that case, it would be necessary to estimate contact resistances from literature data or to determine them experimentally, e.g., by acquiring data with two ceramic cores of different diameters.

3. THERMAL CONDUCTIVITY APPARATUS (1963)

3.1. Apparatus Description

A vertical cross section of the thermal conductivity apparatus is shown in Fig. 3. The specimen was contained in the annular space between the central

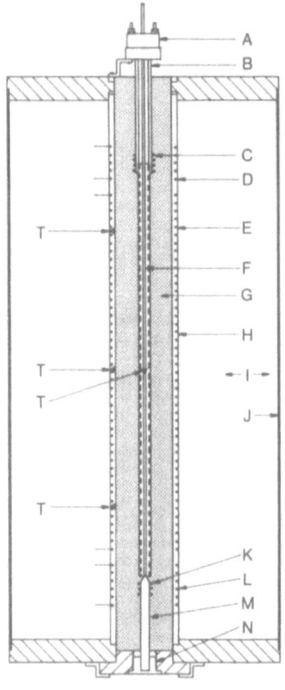

FIGURE 3. Vertical cross section of the 1963 apparatus[1]:
(A) terminal head, (B) ceramic support tube, (C) upper support
heater, (D) upper shell heater, (E) main shell heater,
(F) ceramic core, (G) specimen, (H) ceramic shell, (I) shell
insulation, (J) stainless-steel case, (K) lower support heater,
(L) lower shell heater, (M) ceramic support rod, (N) removable
plug, (T) thermocouple.

ceramic core, which was supported at both ends, and the concentric ceramic
shell. The space between the ceramic shell and the outer stainless-steel case
contained a loose-fill thermal insulation.

The ceramic shell was a hollow mullite (aluminum silicate, $3Al_2O_3 \cdot 2SiO_2$)
cylinder, 61 cm long, with an inner radius of 3.2 cm and an outer radius of
3.8 cm. A helical groove (about 1.6 turns/cm) on the outer surface was wound
with three separate heaters of 1-mm iron–chromium–aluminum alloy heater
wire. The main shell heater was 41 cm in length; the upper and lower shell
heaters were 5 cm in length. A length of 5 cm at each end was not wound. The
upper and lower shell heaters provided extra heat needed to offset end losses,
so that longitudinal temperature differences in the central region could be
minimized.

The central core was an extruded mullite rod, 46 cm long and 1.27 cm in
diameter. An expanded view of the horizontal cross section of this core is
shown in Fig. 4. Sixteen equally spaced holes, 0.8 mm in diameter, extended
the length of the rod. The centers of the holes formed a circle of 0.43 cm
radius. The core heater, which provided the heat flowing radially through the
specimen, consisted of a continuous length of platinum–20% rhodium (0.5 mm
diam) wire threaded back and forth through the 16 holes. Current lead wires

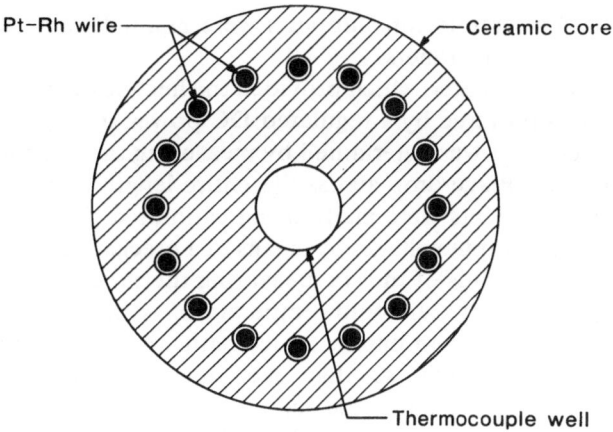

FIGURE 4. Expanded view of ceramic core and heater.

to this heater, and lead wires to two voltage taps located at the upper end of the ceramic core, were brought in through the hollow ceramic upper support for the core. The core was supported below by a ceramic pin. Both the upper and lower supports were provided with small platinum–20% rhodium heaters to minimize longitudinal heat flow from the ends of the ceramic core.

Temperatures in the apparatus were determined by means of six platinum versus platinum–10% rhodium thermocouples, fabricated from calibrated thermocouple wire 0.4 mm in diameter.

Three thermocouples were positioned in the ceramic shell at the midplane of the apparatus. These were inserted into tangential holes (Fig. 2) located 120° apart at a radius of 3.5 cm. The temperature T_c, in equation (9), was obtained by taking the average of the temperatures indicated by these three thermocouples. Two additional tangential thermocouples were located in the ceramic shell, one 15 cm above and one 15 cm below the midplane of the apparatus.

Temperatures in the core were measured by means of a thermocouple which was located in the 0.25-cm-diam axial well and which could be moved vertically by exterior manipulation. A scale fixed above the apparatus indicated the position of the thermocouple junction. The temperature measured at midlength was designated as T_0.

3.2. Instrumentation

The three heaters on the ceramic shell were fed AC current by variable voltage transformers, which in turn were fed by a voltage-regulating transformer. These heaters were individually regulated by on-off controllers actuated by thermocouples located in the ceramic shell adjacent to the heater windings.

The heaters on the ceramic core supports were manually adjusted by means of variable voltage transformers.

The ceramic core heater, which provided the heat flowing radially through the specimen, was powered by a regulated DC power supply. Power input to the ceramic core heater was determined by measuring the DC current through the heater and the voltage drop across the entire length of heater wire in the core. These measurements were made using calibrated shunt and volt boxes and measuring their output voltages by means of a precision DC potentiometer.

The noble metal leads of the six measuring thermocouples were brought to an isothermal zone box at room temperature. A thermocouple with one junction in the zone box and one in an ice bath was placed in series with a double-pole selector switch, so that each measuring thermocouple was automatically referenced against the ice bath. Thermocouple voltages were read to 0.1 μV on the precision potentiometer.

3.3. Test Procedure

The specimen material was placed in the annular space between the ceramic core and the ceramic shell through the opening at the top of the apparatus. The bulk density of the specimen in place was computed from the weight of specimen material and the previously determined volume of the test chamber.

The shell temperature controllers were set at the desired temperature, which for the first test in any series was about 100 °C, and all heaters were turned on. The controllers and power for the heaters at the top and bottom of the ceramic shell were adjusted, if necessary, until the temperatures indicated by the upper and lower tangential thermocouples agreed to within 1 or 2 °C with the average temperature of the three midplane tangential thermocouples. The power to the core heater was adjusted to attain the desired radial temperature drop (typically 30 to 60 °C) across the specimen, and the powers to the heaters on the ceramic core supports were adjusted until a traverse of the thermocouple in the axial hole indicated the desired longitudinal temperature distribution.

When a satisfactory steady-state condition was obtained, the test data were recorded. With the axial thermocouple in the midplane position, three sets of readings (at about 20-min intervals) were taken of all thermocouples and of the output of the volt box and shunt box. At the end of these readings, the axial thermocouple was moved to a position near the top of the ceramic core. After this thermocouple again reached thermal equilibrium (which typically took less than 1 min), its output was read on the potentiometer. Readings were taken at 5-cm intervals over a region 15 cm or more on either side of the midplane. When a complete set of data had been taken, the apparatus was adjusted to a new shell temperature, as desired for the next test.

3.4. Calculation of Results

Calculations of thermal conductivity values by means of equation (8) were performed by means of a digital computer. The data input to the computer consisted of the average emf readings for each thermocouple and the average readings from the volt and shunt boxes. Equations for the estimated thermal conductivities of the ceramic core and shell as functions of temperature, and for temperatures as a function of thermocouple emfs, were contained in the computer program. For the low-conductivity powders, the interface thermal resistances [R_a and R_b in equation (8)] were assumed to be negligible.

3.5. Test Results

Data were reported[1] for a specimen of diatomaceous earth (bulk density of 150 kg m^{-3}) at mean temperatures to 950 °C and for two specimens of finely powdered aluminum oxide powder (400 and 440 kg m^{-3}) at mean temperatures to 1100 °C. (These materials were selected because they were used as insulation in various NBS apparatuses for measuring the thermal conductivity of solids.) The test results for the alumina powder are shown in Fig. 5, along with data obtained near room temperature in an NBS guarded hot plate apparatus.

The data obtained for both the diatomaceous earth and the alumina powder plotted very smoothly versus temperature, with the standard deviation of a test point, estimated from the residuals about least-mean-squares fitted quadratic equations, of about 1% at 500 °C.

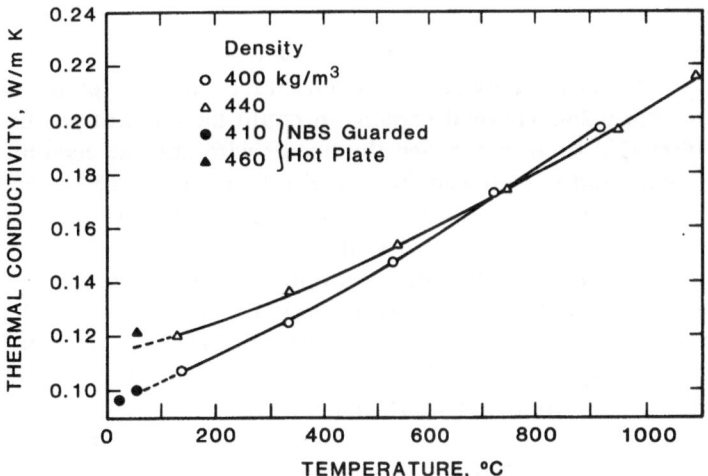

FIGURE 5. Thermal conductivity of alumina powder versus temperature, as measured in 1963 apparatus.[1]

3.6. Uncertainties

3.6.1. Temperature Difference Across the Specimen

The uncertainty (author's estimate of 95% confidence limits) in determining the temperature difference between the axial thermocouple position in the core and the three thermocouple positions in the shell was estimated to be less than 1%. Near 1000 °C, the computed temperature drops in the core ($T_0 - T_a$) and in the shell ($T_b - T_c$) were less than 1.5% and 0.4%, respectively, of the temperature drop through the specimen for the results presented elsewhere,[1] using the best available values for the thermal conductivities of the mullite core and shell, which may have been uncertain by as much as 30%. Near 100 °C, these temperature drops were less than 0.7% and 0.2%, respectively, of the temperature drop through the specimen.

The effective radius, r', of the virtual heating surface in the core was uncertain by not more than the radius of the heater holes, i.e., 10% of r'. The effective radius, c, to the tangential thermocouple wells was uncertain by not more than the radius of the wells. i.e., 4% of c. For fine powders of low thermal conductivity, such as those tested, errors due to thermal contact resistance between the specimen and the surfaces of the core and shell were believed to be negligible. Using standard propagation-of-errors formulas, the temperature drops across the specimen (the expression in braces in the denominator of equation (8)] are conservatively considered to have been known with an uncertainty of 1.5%.

3.6.2. Heat Flux

The total power generated in the core heater was known to within 0.1%. The length of wire heater between the potential taps at room temperature was known to within 0.2%. Thermal expansion could have increased this length by 1% at 1000 °C if friction between the heater wire and the ceramic core at points of contact did not prevent free longitudinal expansion of the wire in the holes. Because the effect of expansion was less than 1% and the exact value uncertain, no expansion corrections were made.

Equation (8) was derived on the assumption of no longitudinal heat flow. For an apparatus of finite length, having imperfect end guarding, the effects of axial heat flow should be considered. An analysis was presented[1] that allows different temperatures at the two ends of the core, and also takes into account the effect of core temperature distribution on the power measurement if the electrical resistivity of the heater wire varies linearly with temperature.

A simplified cross-sectional view of the apparatus is shown in Fig. 6. It is assumed that the surface $r = b$ is isothermal at temperature T_b and that temperatures are measured in the core at three positions equally spaced along

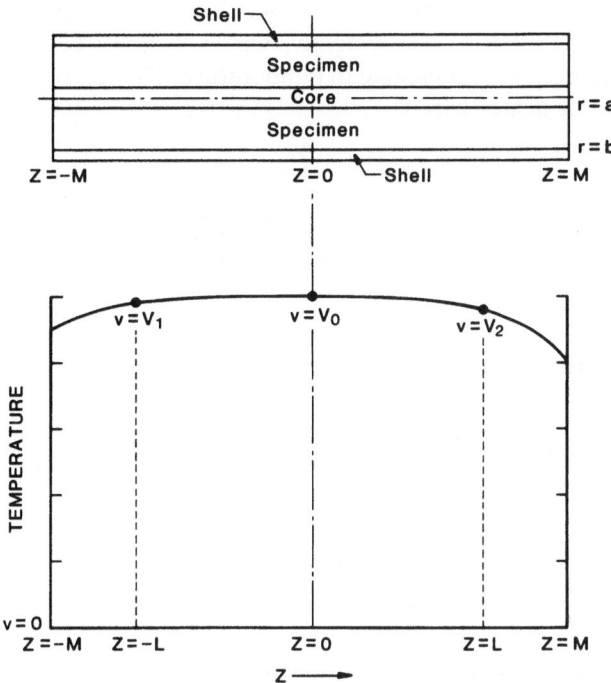

FIGURE 6. General core-temperature profile used in analyzing effects of longitudinal heat flow.[1]

the axis. It is assumed that, for the purpose of analyzing longitudinal heat flow, a transverse cross section of the core can be treated as being isothermal.

By neglecting longitudinal heat flow in the specimen, the differential equation for heat flow along the core can be written as

$$\frac{d^2v}{dz^2} - \frac{2\pi\bar{\lambda}v}{C\ln(b/a)} = \frac{mq'}{C} \tag{10}$$

where $v = T - T_b$, T being the temperature at any point along the core, q' is the power generated per unit length of heater wire, m is the number of heater wires, and C is the longitudinal thermal conductance per unit length of the core, including heater wires and holes.

In general, the rate of heat generation, q', per unit length of heater wire will vary with z if the electrical resistivity of the heater wire varies appreciably with temperature. For reasonable temperature variations, the electrical resistance per unit length of heater wire can be assumed to be a linear function of temperature,

$$\beta = \beta_0(1 + \gamma v) \tag{11}$$

where β and β_0 are the electrical resistances of a unit length of heater wire at temperatures T and T_b, respectively, and γ is the fractional change in electrical resistance per unit change in temperature. The rate of heat generation per unit length of heater wire is

$$q' = I^2\beta = I^2\beta_0(1 + \gamma v) \tag{12}$$

where I is the electrical current flowing through the heater.

The conditions which are to be satisfied are

$$v = V_1 \text{ at } z = -L, \qquad v = V_0 \text{ at } z = 0, \qquad v = V_2 \text{ at } z = L \tag{13}$$

Substitution of these conditions into the solution of equation (10) yields

$$v = V_0 + \left(\frac{V_1 + V_2 - 2V_0}{2}\right)\frac{\cosh \mu z - 1}{\cosh \mu L - 1} + \left(\frac{V_2 - V_1}{2}\right)\frac{\sinh \mu z}{\sinh \mu L} \tag{14}$$

in which

$$\mu^2 = \frac{1}{C}\left[\frac{2\pi\bar\lambda}{\ln(b/a)} - \gamma m I^2\beta_0\right] \tag{15}$$

The thermal conductivity of the specimen is given by

$$\bar\lambda = \frac{mI^2\beta_0 \ln(b/a)}{2\pi V_0}\left[1 + \gamma V_0 + \frac{V_1 + V_2 - 2V_0}{2V_0 \cosh \mu L - (V_1 + V_2)}\right] \tag{16}$$

The total power delivered by the core heater is determined by the current, I, through the heater and the voltage drop, E, across the total length of heater wire in the core. The total resistance of the heater is given by

$$R = E/I = \beta_0(1 + \gamma\bar V)S \tag{17}$$

where S is the total length of heater wire and $\bar V$ is the average temperature of the core heater, obtained by integration of equation (14) over the entire length, $2M$, of the core:

$$\bar V = \frac{1}{2M}\int_{-M}^{M} v\, dz = V_0 + \left(\frac{V_1 + V_2 - 2V_0}{2}\right)\frac{\sinh \mu M}{\mu M(\cosh \mu L - 1)} \tag{18}$$

Equation (16) can be rewritten as

$$\bar\lambda = \bar\lambda_{app}\left\{1 + \frac{V_1 + V_2 - 2V_0}{1 + \gamma\bar V}\left[\frac{1}{2V_0 \cosh \mu L - (V_1 + V_2)} - \frac{\gamma(\sinh \mu M - \mu M)}{2\mu M(\cosh \mu L - 1)}\right]\right\} \tag{19}$$

where

$$\bar{\lambda}_{\text{app}} = \frac{mEI \ln (b/a)}{2\pi V_0 S} \tag{20}$$

is the approximate value of thermal conductivity obtained if no corrections are made for the effects of axial heat flow. Since $\bar{\lambda}$ is defined in terms of μ and μ in terms of $\bar{\lambda}$, iterations are necessary in order to obtain $\bar{\lambda}$. A first approximation to $\bar{\lambda}$ is given by $\bar{\lambda}_{\text{app}}$; substitution of $\bar{\lambda}$ into equation (15) yields a value of μ which can be used in equation (19). The resultant $\bar{\lambda}$ can be used to obtain a better value of μ, etc. The convergence is quite rapid in all practical cases. Equation (20) corresponds to equation (9), the equation that was used for purposes of calculation, with EI/S corresponding to q'. The first expression in the square brackets in equation (19) represents the correction for axial heat flow; the second expression represents the correction in power measurement to take account of the variation of electrical resistance with temperature. For a positive value of γ, these corrections are of opposite sign.

For the alumina powder material reported in 1963,[1] experimental data were obtained in which the longitudinal core temperature imbalance, defined as $(V_1 + V_2)/2V_0$, for $L = 15$ cm, varied by $\pm 15\%$ about unity. The corresponding differences in the measured thermal conductivity values with no correction for longitudinal heat flow were less than $\mp 0.4\%$, respectively. By correcting for longitudinal heat flow using equation (19) and the estimated longitudinal conductance, C, of the ceramic core and heater wires, values of thermal conductivity were obtained for the alumina powder that were not significantly different (less than 0.1%) from the values obtained with $(V_1 + V_2)/2V_0$ equal to unity.

Although no corrections were made routinely for longitudinal heat flow, this could easily have been done, with resultant uncertainties of less than 0.1%. Thus considering the uncertainties in the power input, the length of heater wire, and the effects of longitudinal heat flow, the overall uncertainty in the heat flux, if longitudinal heat flow corrections had been made, would have been 0.3% at 100 °C, increasing (as a result of the uncertainty in heater wire expansion) to 1.3% at 1000 °C.

3.6.3. Eccentricity

The effect of eccentricity between the outer surface of the ceramic core and the inner surface of the ceramic shell was considered in 1963,[1] an equation for the influence of this effect being given, as well as a plot of error versus eccentricity. For the apparatus described, the uncertainty due to eccentricity was less than 0.1%.

3.6.4. Overall Precision and Accuracy

The precision of measurements in the 1963 paper[1] is indicated by the standard deviations (estimated) of the data, which are of the order of 1%.

It was estimated that temperature drops across the specimen were determined with an uncertainty of 1.5%. The heat flux was known to 0.3% at 100 °C and 1.3% at 1000 °C. Uncertainties in the ratio of the shell radius to that of the core did not introduce more than 0.2% uncertainty in the results. The error due to possible eccentricity of the ceramic core in the ceramic shell did not exceed 0.1%. Combining all of the uncertainties discussed, uncertainties in the thermal conductivity results presented in the 1963 paper were estimated to be less than 2% at 100 °C and, because of heater wire expansion, 3% at 1000 °C.

4. IMPROVED LONGITUDINAL-HEAT-FLOW CORRECTION

For the 1969 thermal conductance apparatus,[2,3] described below in Section 5, the apparatus was of an overall configuration similar to that of the 1963 apparatus but the ceramic core and its heater were supported at both ends by fixtures that were at room temperature. It was not practical to provide auxiliary heaters to minimize longitudinal heat flows. The longitudinal thermal conductance of the alumina core used in the 1969 apparatus was considerably higher than the conductance of the mullite core used in the 1963 apparatus, particularly at lower temperatures. These factors made it necessary to derive a more sophisticated analysis of the influence of longitudinal heat flow in the ceramic core and in the specimen material.

Peavy[5] considered the effect of longitudinal heat flow in apparatuses such as those used in 1963[1] and 1969.[2,3] His analysis is exact for the case in which both the core and the specimen have constant thermal conductivity (i.e., independent of temperature) and the heat generation per unit length does not vary along the core. The analysis from my paper,[1] also given above in Section 3.6.2, considers the effect of longitudinal heat flow along the core for the case in which the heat generation per unit length is a linear function of temperature. This analysis neglects longitudinal heat flow in the specimen and assumes a constant thermal conductivity for the core.

In the 1969 investigation, the very large temperature differences within the apparatus made it imperative that the temperature dependence of the thermal conductivity of the core, the temperature dependence of the thermal conductivity of the specimen, and the temperature dependence of the heat generation per unit length all be taken into account. Peavy's analysis[5] would be very difficult to adapt to the case in which all these parameters have large temperature dependences. The analysis in Section 3.6.2 could easily be modified to include the effect of the temperature dependence of the thermal conductivity of the core and, with a little more effort, could be extended to include a more complicated (than linear) temperature dependence for the heat generation per unit length in the core. However, this analysis is inherently incapable of dealing with the effects of longitudinal heat flow in the specimen.

Since the apparatus design[1-3] permitted direct experimental measurement of the temperature distribution along the axis of the core, more information is available than is the case in boundary value problems in which only the conditions on the outside of the system are prescribed. This additional information enables computation of the effects of longitudinal heat flow in a rather straightforward manner with only minor approximations being required. If necessary, even these approximations could be circumvented by a simple process of iteration.

If the situation in the ceramic core is examined first, the thermal conductivity of high-density, high-purity alumina decreases by a factor of more than six between room temperature and 1600 °C. The electrical resistivity of the platinum–rhodium alloy used for the heater winding in the 1969 apparatus increases by a factor of three between room temperature and 1600 °C. Both the thermal conductivity of the core and the electrical resistivity of the heater (and hence the heater generation per unit length) are explicit functions of temperature. Since the temperature distribution along the axis of the core was measured and since the temperature at the axis of the core was essentially the same as the temperature of the heater winding, the thermal conductivity of the core (inside the circle of heater wires) and the heat generation per unit length could be determined as functions of position. For a given test both of these quantities can be treated as if they were explicit functions of position.

Let us consider an element of the core of length dz. The heat generated in this element is $mI^2\beta(z)\,dz$, where m is the number of heater wires, I the electric current flow through the heater winding, and $\beta(z)$ the electrical resistance per unit length of heater wire. The radial heat flow from the convex surface of this element into the specimen is $q(z)\,dz$, where $q(z)$ varies with longitudinal position. The heat flowing into this element in the positive z-direction is $-C(z)(d\bar{T}/dz)$, where $C(z)$ is the ("position-dependent") thermal conductance of the core, including heater wires, and $\bar{T} = \bar{T}(z)$ is the average temperature of the core at longitudinal position z. Under steady-state conditions, the heat flux into the specimen at any longitudinal position is

$$q(z) = mI^2\beta(z) + \frac{d}{dz}\left[C(z)\frac{d\bar{T}(z)}{dz}\right] \tag{21}$$

where the average core temperature at any longitudinal position is given by

$$\bar{T}(z) = \frac{2}{a^2}\int_0^a rT(r, z)\,dr \tag{22}$$

Except very near the ends of the apparatus, most of the heat flow in the core is radial rather than longitudinal. For the relatively small radial temperature differences between $r = r'$ and $r = a$, one can justifiably neglect the temperature dependence of λ_c and, as shown elsewhere,[3] obtain

$$\bar{T}(z) = T(0, z) - \frac{mI^2\beta(z)}{4\pi\lambda_c(z)}\left(2\ln\frac{a}{r'} - 1 + \frac{r'^2}{a^2}\right) \tag{23}$$

where $T(0, z) = T_0(z)$ is the experimentally determined temperature distribution along the axis of the core. By writing equation (21) in terms of the average core temperature, $\bar{T}(z)$, as given by equation (23), the heat flow, $q(z)$, into the inner surface of the specimen can be determined. From this, the analysis needed to permit computation of the thermal conductivity of the specimen follows.

We now consider an apparatus having the horizontal cross section shown in Fig. 1, which is of finite length, extending from $z = 0$ to $z = l$, with the flat ends cooled. Neglecting contact resistances, the boundary conditions for the specimen are

$$
\begin{array}{cccc}
a < r \leqslant b & z = 0 \text{ and } z = l & & T = T_b \\[2mm]
r = b & 0 \leqslant z \leqslant l & & T = T_b \\[2mm]
r = a & 0 < z < l & -2\pi a\lambda(T)\dfrac{\partial T}{\partial r} = q(z)
\end{array} \tag{24}
$$

In order to simplify the boundary condition at the surface of the core, $r = a$, a new variable is introduced, defined by

$$y = \frac{1}{\lambda_0}\int_{T_b}^{T}\lambda(T)\,dT = \frac{\bar{\lambda}(T_b, T)}{\lambda_0}(T - T_b) \tag{25}$$

where $\lambda_0 = \lambda(T = T_b)$. In terms of y, the boundary conditions in equation (24) can be written as

$$
\begin{array}{cccc}
a < r \leqslant b & z = 0 \text{ and } z = l & & y = 0 \\[2mm]
r = b & 0 \leqslant z \leqslant l & & y = 0 \\[2mm]
r = a & 0 < z < l & -2\pi a\lambda_0\dfrac{\partial y}{\partial r} = q(z)
\end{array} \tag{26}
$$

With these boundary conditions, the solution of the Laplacian equation for the potential, y, is given by Carslaw and Jaeger[6] as

$$y(r, z) = \frac{-1}{\lambda_0} \sum_{n=1}^{\infty} \frac{l A_n F_0(n\pi r/l; n\pi b/l)}{n\pi F_1(n\pi a/l; n\pi b/l)} \sin \frac{n\pi z}{l} \qquad (27)$$

where

$$A_n = \frac{1}{\pi a l} \int_0^l q(z') \sin \frac{n\pi z'}{l} dz' \qquad (28)$$

$F_0(x; y) = I_0(x)K_0(y) - I_0(y)K_0(x)$, $F_1(x; y) = I_1(x)K_0(y) + I_0(y)K_0(x)$, and $I_n(x)$ and $K_n(x)$ are modified Bessel functions of the first and second kind, respectively, of order n.

Evaluation of equation (27) at $r = a$ and comparison with equation (25) enables one to write

$$\bar{\lambda}(v) = \frac{q(z) \ln (b/a)}{2\pi[T_a(z) - T_b]}[1 - U(z)] \qquad (29)$$

where $U(z)$, a correction term for the effects of longitudinal heat flow in the specimen, is given by

$$U(z) = \frac{2\pi a}{q(z)} \sum_{n=1}^{\infty} A_n \left[1 - \frac{1}{\ln (b/a)} \cdot \frac{l F_0(n\pi a/l; n\pi b/l)}{n\pi a F_1(n\pi a/l; n\pi b/l)} \right] \sin \frac{n\pi z}{l} \qquad (30)$$

In deriving equation (29), the influence of thermal contact resistances was neglected. Since $U(z)$ is a small correction, it is assumed that it is still valid even if thermal contact resistances are taken into account. By comparing equations (8), (9), and (29), one can write

$$\bar{\lambda}(T_b, T_a) = \frac{\bar{\lambda}_{app}[1 - U(z)]}{\left[1 - \frac{\bar{\lambda}_{app}}{\ln (b/a)} \left(\frac{\ln (a/r')}{\lambda_c} + \frac{\ln (c/b)}{\lambda_s} + \frac{R_a}{a} + \frac{R_b}{b} \right) \right]} \qquad (31)$$

where $\bar{\lambda}_{app}$ now has the more general definition

$$\bar{\lambda}_{app} = \frac{q(z) \ln (b/a)}{2\pi[T_0(z) - T_c]} \tag{32}$$

and $q(z)$ is given by equation (22). Equation (31) is the equation used in the 1969 investigation to compute $\bar{\lambda}$.

5. THERMAL CONDUCTANCE APPARATUS (1969)

In conjunction with safety evaluation of space nuclear power systems, it was desired to be able to predict the maximum steady-state temperature which would be reached by a given nuclear power supply after reentry impact burial in earth. In order to do this it was necessary to have information regarding the thermal conductance of soils under conditions in which heat would flow from a source that might be as hot as 1700 °C to a sink at ambient temperature. It was decided to build an apparatus based on the same principle of operation as that which had been used for the 1963 thermal conductivity apparatus,[1] but to replace the ceramic shell with a metal shell maintained at room temperature. In this way the average thermal conductivity [$\bar{\lambda}$ in equation (2)], or, equivalently, the thermal conductance from room temperature to a given hot side temperature, could be determined. The specimen size was selected large enough that the heat flow path would be much larger than the average particle size of the soil. For the expected maximum particle diameter of 0.2 cm, it seemed appropriate for the sample to be at least 2 to 3 cm thick from the hot surface to the cold surface. Similarly, it was decided that the heat source in contact with the specimen should be large in comparison with the average particle size of the soil.

5.1. Apparatus Description

A horizontal cross section of the apparatus that was designed and built is shown in Fig. 7. The specimen was contained within the annular space between the outer radius, a, of a ceramic core and the inner radius, b, of a brass shell. A measured quantity of heat per unit time, generated electrically in the ceramic core, flowed radially through the specimen to the inner concentric water-cooled brass shell. The ceramic core had a concentric ring of equally spaced holes, parallel to the axis, at a radius, r', each containing a heater wire. Temperatures were measured by an axial thermocouple in the ceramic core and by thermocouples attached to the outer surface of the inner brass shell at radius c. A vertical cross section of the apparatus is shown in Fig. 8.

The central core was an extruded dense alumina rod, 46 cm long and 1.25 cm in diameter. Sixteen equally-spaced holes, 0.9 mm in diameter, exten-

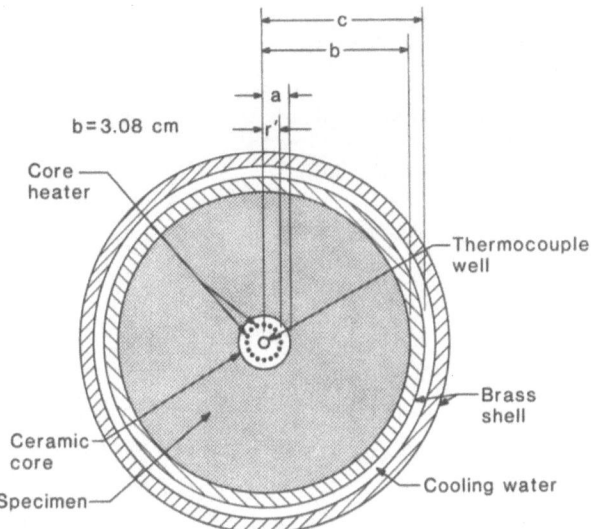

FIGURE 7. Horizontal cross section of the apparatus for measuring the average thermal conductivity of soils with the cold side held near room temperature.[2,3]

ded the length of the rod. The centers of these holes formed a circle of 0.44 cm radius. The core heater, which provided the heat flowing radially through the specimen, consisted of lengths of platinum-40% rhodium (0.6 mm diameter) wire threaded back and forth through the sixteen holes. The core was held from beneath by a ceramic insulating support designed to permit free expansion of the heater wire. The upper end of the core passed through a hole in the removable flange at the top of the apparatus; this hole was sufficiently loose to permit free expansion of the core. Current leads and voltage taps were attached to the heater winding at the upper end of the core.

For the first tests conducted, heaters were fabricated by connecting all sixteen heater wires in series, as shown in Fig. 9a. With this wiring arrangement, the current leads to and from the heater came to adjacent holes in the ceramic core. This resulted in large voltage gradients within the alumina core with consequent dielectric breakdown of the alumina at core temperatures around 1400 °C. The heater wiring was subsequently changed to a series–parallel arrangement in which the heater was wired as two sets of eight series-connected wires with these two sets connected in parallel, as shown in Fig. 9b. This greatly reduced the voltage gradients in the ceramic core and there was no further trouble with dielectric breakdown.

In the center of the alumina core there was an axial hole 0.25 cm in diameter. The hole accommodated a thermocouple which could be moved vertically by exterior manipulation. This thermocouple was fabricated from

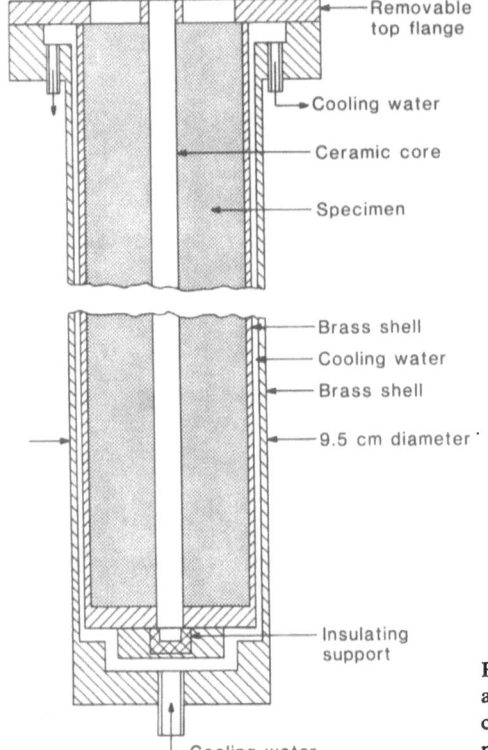

FIGURE 8. Vertical cross section of the apparatus for measuring the average thermal conductivity of soils with the cold side held near room temperature.[2,3]

0.4-mm-diameter platinum and platinum–10% rhodium wire, contained in double-bore alumina tubing which was a slip fit in the thermocouple well. The temperature measured at the midlength of the apparatus was designated T_0. This thermocouple could also be utilized to obtain the longitudinal temperature distribution along the core; the latter information was used to make corrections for longitudinal heat losses.

With the exception of the ceramic core, all surfaces in contact with the specimen were water-cooled to maintain them at room temperature. As shown in Fig. 8, the cooling water entered the center of the bottom of the apparatus, passed upward in the annulus between the inner and outer brass shells, and exited at the top of the apparatus.

Two copper-versus-constantan thermocouples (0.25 mm diameter) were attached to the outer surface of the inner brass shell at the midplane of the apparatus. The junctions of these thermocouples were thermally insulated from the cooling water by plastic electrical tape. The average of these two temperatures was designated T_c. Additional copper-versus-constantan ther-

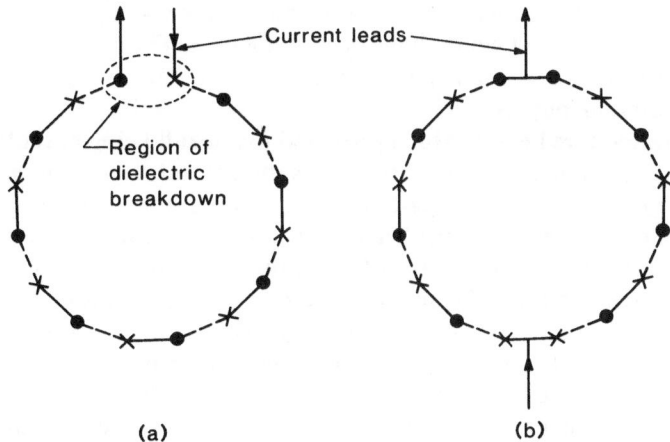

(a)   (b)

FIGURE 9. Two different heater winding procedures. The crosses (×) represent wires going down a hole; the solid circles represent wires coming up a hole. The solid lines indicate wires crossing from hole to hole at the top of the heater; the dashed lines indicate crossovers at the bottom of the heater. With the straight series arrangement on the left, there was high-temperature dielectric breakdown in the region shown. The series–parallel arrangement on the right worked very well to temperatures above 1600 °C.

mocouples were used to monitor temperatures at different locations in the circulating system.

5.2. Instrumentation

Individual cold junctions in an ice bath were used with the platinum versus platinum–10% rhodium thermocouple. The constantan leads of the copper-versus-constantan thermocouples were brought to an isothermal zone box at room temperature. A thermouple with one junction in the zone box and one in an ice bath was placed in series with a double-pole selector switch, so that each measuring thermocouple was automatically referenced against the ice bath.

Thermocouple voltages were read, to the nearest microvolt, using an integrating digital DC voltmeter having 140-dB common-mode rejection. This instrument was able to read accurately the thermocouple in the ceramic core even though, at the higher temperatures, the electrical resistivity of the alumina core became low enough that there was considerable AC leakage from the AC-powered core heater to the axial thermocouple. An alternative procedure would have been to use metal-sheathed thermocouples fabricated by swaging techniques. The temperatures involved would have required the use of ther-mocouples with noble-metal sheaths, thus greatly increasing the cost of the thermocouples. Since it was desired to use a new thermocouple for each specimen (the high temperatures could cause thermocouples to lose their calibration), the total cost would have been quite high. The use of an integrating digital voltmeter, rather than a DC potentiometer, made it possible to use unshielded thermocouples.

The ceramic core heater, which provided the heat flowing radially through a specimen, was fed AC power from a saturable core reactor which was controlled by a current-adjusting-type proportional controller incorporating automatic rate and reset action. This controller was modified to operate in either of two modes. It could adjust the power to the core heater so as to maintain the temperature, T_0, at the center of the core at a constant value. Alternatively, it would control the saturable core reactor so as to provide constant current to the core heater. The power dissipated in the core heater was determined by measuring the current through the heater and the voltage drop across the heater utilizing electrodynamic voltmeters and ammeters having an accuracy of one-quarter percent of full scale.

5.3. Test Procedure

The procedures used for compacting the soil specimens to the desired bulk density are described elsewhere.[2,3] Each specimen was compacted into the inner brass shell (see Figs. 7 and 8), with a core heater in place.

The specimen container, with the ceramic core heater and the compacted specimen installed, was placed in the outer brass shell as shown in Fig. 8. Rubber O-rings, not shown, provided water-tight seals at the top of the apparatus. The current leads and voltage taps were attached to the heater leads extending from the top of the ceramic core. A thermocouple was installed in the axial hole in the ceramic core so that the hot junction was at the mid-height of the apparatus. The upper end of this thermocouple was attached to a height gage which permitted moving the thermocouple vertically by known distances. The water-circulating system was activated to maintain the outside of the specimen container at the desired temperature.

The specimen heater was energized with the proportional controller operating in a constant-temperature mode so as to bring and maintain the temperature at the core thermocouple to the desired value. This procedure effectively prevented significant temperature overshoot, allowing monotonic heating of the hot face of the sample but yet permitting rapid equilibrium to be attained. When the temperature indicated by the core thermocouple was steady and when the electrical power required to maintain this temperature had also leveled out to a steady value, the proportional controller was switched to a constant-current mode so as to maintain the current to the heater at the value it had before switching over.

After any small thermal disturbances due to changing the mode of control had damped out, several successive readings were taken of the current through the heater, the voltage drop across the heater, the emf of the platinum versus platinum–10% rhodium thermocouple (at the mid-height of the apparatus) in the axial hole in the ceramic core, and the emfs of the two copper-versus-constantan thermocouples on the outside of the specimen container (i.e., the inner brass shell). After satisfactory data were obtained with the core thermocouple at the mid-height of the apparatus, the thermocouple junction was moved to a position approximately level with the bottom of the specimen. The emf of the core thermocouple was read at this position and then at approximately 2.5-cm intervals along the entire length of the specimen, time being allowed at each position for the thermocouple junction to reach thermal equilibrium with the ceramic core. The core thermocouple junction was then returned to the mid-height of the apparatus and readings were again taken of the heater current, heater voltage, and thermocouple emfs.

Following completion of one set of data, the heater controller was switched to constant-temperature mode and the heater and both controllers were adjusted to bring the specimen to the next condition desired. Data were normally taken at 200-°C intervals to about 1100 °C (hot side) and then at 100 °C intervals until either the heater burned out or the core thermocouple ceased to function properly.

5.4. Calculation of Results

A computer program was used to calculate mean thermal conductivity values. The input data consisted of the voltmeter and ammeter readings (for the core heater) and the thermocouple positions and emfs for each test. The calculation procedure was as follows:

1. The ammeter and voltmeter readings were corrected when necessary using subprograms which contained the meter calibrations.
2. Temperatures were computed from the thermocouple emfs using subprograms which contained the thermocouple calibrations.
3. The heat flux in the absence of longitudinal heat losses was computed for each position along the ceramic core using the leading term in equation (21). The electrical resistance per unit length, $\beta(z)$, of the heater wire was computed using an equation representing NBS-obtained data on the temperature-dependence of the resistance of platinum–40% rhodium alloy. It was assumed that the heater wire was at the temperature $T_0(z)$ measured at the center of the core.
4. The average core temperature at each longitudinal position was computed using equation (23).
5. A fourth-order polynomial was fitted (by least squares) to the (usually 19) values of $\bar{T}(z)$. From this equation and an equation which was used to compute C, the longitudinal thermal conductance of the core, the second term in equation (21) was evaluated.

6. From the resultant values of $q(z)$, the mean thermal conductivity, $\bar{\lambda}(T_b, T_a)$, was computed at each position using equation (31). The computer program contained a value for λ_s, the thermal conductivity of the brass shell, and an equation for computing λ_c, the thermal conductivity of the ceramic core, as a function of temperature. Since no knowledge was available regarding the thermal contact resistances, R_a and R_b were set equal to zero; thus the $\bar{\lambda}$ values obtained were apparent mean thermal conductivity values. A separate subprogram was used to compute $U(z)$, the correction for longitudinal heat flow in the specimen.

7. An equation of the form $\bar{\lambda}(T_b, T_a) = a + bT_a + cT_a^2 + dT_a^n$ (with $n \geqslant 3$) was fitted (least squares) to the computed $\bar{\lambda}(T_b, T_a)$ values corresponding to the mid-height of the apparatus, the $\bar{\lambda}$ values being adjusted slightly so that they all corresponded to a cold-side temperature of $T_b = 25\,°C$. These smoothed mean thermal-conductivity values were reported elsewhere.[2,3]

8. An estimate of the apparent hot-side thermal conductivity of each sample was computed from this equation using the relation

$$\lambda(T_a) = \bar{\lambda}(T_b, T_a) + (T_a - T_b)\frac{\partial \bar{\lambda}(T_b, T_a)}{\partial T_a}$$

which is obtained by differentiating equation (2) with respect to the hot-side temperature.

5.5. Test Results

In order to illustrate the type of data obtained and the relative magnitudes of the various corrections involved, a set of data and a typical calculation for one of the soil specimens for a hot-side temperature near $1100\,°C$ is presented.

The upper curve in Fig. 10 illustrates the measured values of $T_0(z)$, the temperature along the axis of the ceramic core. The dashed curve, labeled T_a', is the computed longitudinal temperature distribution at the surface, $r = a$, of the core. The electrical resistance per unit length of the heater wire, corresponding to the measured values of T_0, was used to compute the power generated per unit length, q', which is the leading term in equation (21) and is shown in Fig. 10. The average core temperature at any longitudinal position, computed by equation (23), was used in equation (21) to correct q' for the effects of longitudinal heat flow in the core and thus obtain the heat flux, q, into the specimen at any longitudinal position; this quantity is also shown in Fig. 10. Using this curve for $q(z)$, $\bar{\lambda}(T_b, T_a)$ values were computed at a number of longitudinal positions using equation (31). Only the thermal conductivity values corresponding to the mid-height position of the apparatus were used for the curves shown by Flynn and Watson.[2,3]

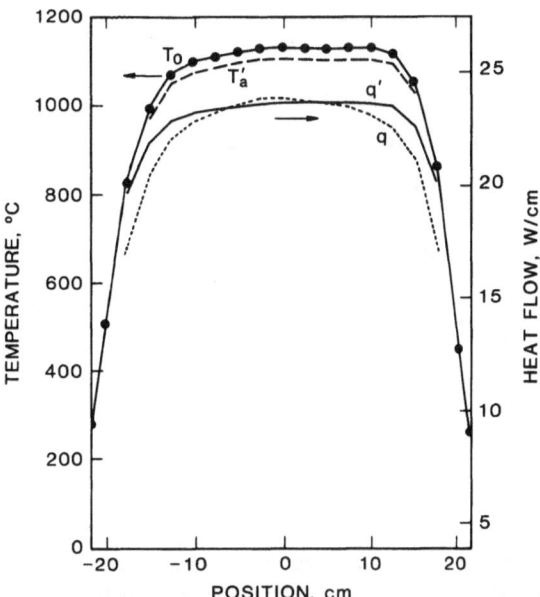

FIGURE 10. Temperature and heat flux distributions for a specimen of soil at hot-side temperatures of about 1100 °C.[2,3] The four curves are identified in the text.

For this test, the different parameters defined in Section 4 had the following values (at the mid-height of the apparatus):

$$T_0 = 1131.1\ °C \qquad T_b = T_b' = 36.8\ °C \qquad q' = 237 \times 10^3\ \text{W m}^{-1}$$

$$\bar{T} = 1124.7\ °C \qquad T_c = 36.5\ °C \qquad q = 239 \times 10^3\ \text{W m}^{-1}$$

$$T_a = T_a' = 1108.3\ °C \qquad\qquad\qquad U = -0.007$$

$$\bar{\lambda}(36.8, 1108.3) = 0.646\ \text{W m}^{-1}\ \text{K}^{-1}$$

For these data the correction from T_0 to T_a was 2.1% of the temperature difference across the sample while the correction for the temperature drop across the brass shell was only 0.03% of $(T_a - T_b)$. At the mid-height of the apparatus, even with the ends near room temperature, the correction for longitudinal heat flow was only 0.7%. It is noteworthy that the longitudinal heat flow present near the mid-height of the apparatus was mainly due to a slight skewness in the temperature distribution, and hence in the heat input from the core. Such a skewness occurred for many of the soil samples, probably because of density variation with height in the specimen.

For two specimens of this same soil sample, the thermal conductivities obtained are shown in Fig. 11, where $\bar{\lambda}$ represents the mean thermal conductivity, corrected to a cold-side temperature of 25 °C, plotted against hot-side

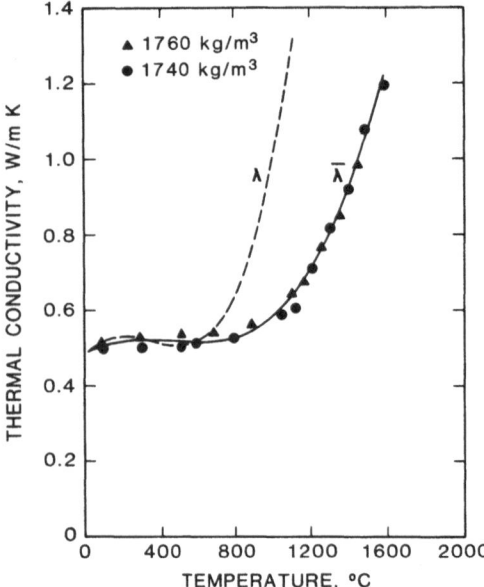

FIGURE 11. Thermal conductivity of a particular soil at two different bulk densities.[2,3] The solid curve represents a cubic fitted to the mean thermal conductivity data (represented by the solid symbols) plotted versus mean temperature. The dashed curve represents the apparent hot-side thermal conductivity, obtained by differentiating the solid curve, plotted versus hot-side temperature.

temperature, T_a, and λ represents the apparent hot-side thermal conductivity obtained by differentiating the curve that had been fitted to the $\bar{\lambda}$ data.

Prior to undertaking measurements of the thermal conductivity of soil samples, measurements of mean thermal conductivity were made on a specimen of diatomaceous earth from the same lot of material that had been tested using the 1963 thermal conductivity apparatus. Measurement of this lot of diatomaceous earth in the 1969 apparatus provided a check on the corrections for longitudinal heat flow, which were rather large because of the low thermal conductivity of diatomaceous earth.

The solid triangles shown in Fig. 12 represent the $\bar{\lambda}(25, T_a)$ data points plotted versus T_a, the hot-side temperature. The solid curve is the cubic of a least-squares fit to these data. The dashed curve represents the $\lambda(T_a)$ values obtained by differentiating that cubic equation. The open circles represent the smoothed thermal conductivity values reported in 1963[1] on diatomaceous earth from the same lot of material. The agreement is good, except at the higher temperatures where the result of differentiating a fitted curve is very sensitive to even small errors in the fitted curve of $\bar{\lambda}$ versus T_a. In addition, in the 1963 investigation, the diatomaceous earth was observed to sinter at high temperatures.

With a dense alumina core, for good insulators such as diatomaceous earth, the correction, $T_0 - T_a$, for the temperature drop in the core was only a few tenths of a percent of $T_a - T_b$, the total temperature drop across the

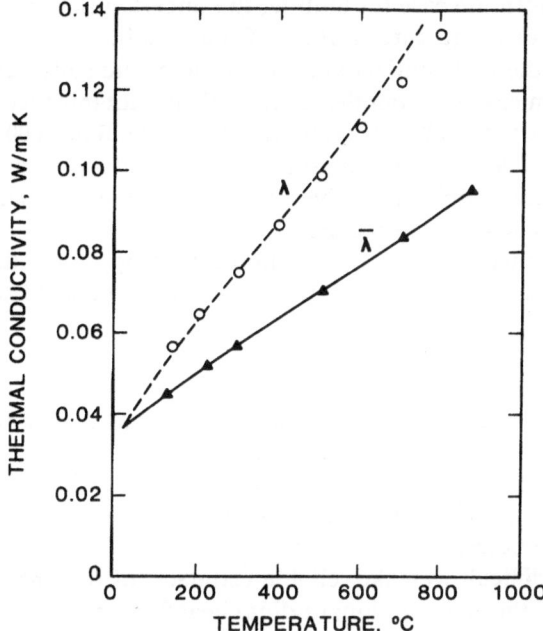

FIGURE 12. Thermal conductivity of diatomaceous earth. The solid line represents a cubic fitted to the mean thermal conductivity data (solid triangles) plotted versus mean temperature, as measured in the 1969 apparatus with the cold side near room temperature.[2,3] The dashed line represents the apparent hot-side thermal conductivity, obtained by differentiating the solid curve, plotted versus hot-side temperature. The open circles represent the smoothed data obtained on the same material with the 1963 apparatus.[1]

specimen. For such low-conductivity specimens, however, with the ends of the apparatus near room temperature the corrections for longitudinal heat flow in the rather conductive alumina core were several percent.

5.6. Uncertainties

5.6.1. Temperature Difference Across the Specimen

The uncertainty in determining the temperature difference between the axial thermocouple in the core and the two thermocouples on the exterior of the brass shell was estimated to be less than 1% for temperatures up to 1100 °C, increasing to perhaps 2% at 1500 °C. It was not possible to make reliable estimates of the uncertainties in measuring axial temperatures above about 1500 °C, due to thermocouple instabilities.

Because the thermal conductivity of the alumina core used for the 1969 apparatus was much higher than that of the mullite core used for the 1963 apparatus, and because the thermal conductivity of alumina was known much better than had been the case for mullite, the uncertainties in the temperature drops from the axis to the outer surface of the alumina core were quite small, even though the thermal conductivities of some of the soils were rather high. For the soil specimens, uncertainties in $T_0 - T_a$ did not introduce uncertainties in $T_a - T_b$ greater than 0.2% for temperatures up to about 600 °C, increasing to about 0.6% at 1500 °C. The brass shell was so conductive that uncertainties in the temperature drop across the shell had negligible influence on the temperature drop across the specimen.

The overall uncertainty in the temperature drop across the sample was less than 1.2% at low temperatures, increasing to perhaps 2.6% at 1500 °C.

5.6.2. Heat Flux

The total power generated in the core heater was known to within 0.4%. The length of heater wire between the potential taps at room temperature was known to within 0.2%. Thermal expansion could have increased this length by about 1.7% at 1500 °C if friction did not prevent free longitudinal expansion of the wire in the holes.

It was very difficult to make reliable estimates of the uncertainties in the determinations of the effects of longitudinal heat flow, particularly considering the nature of the soil specimens, which not only had density variations but which, at higher temperatures, were subject to sintering and partial melting. However, it is estimated that for a well-behaved, homogeneous specimen, having a thermal conductivity similar to those of the soils tested, the uncertainties in heat flow due to uncertainties in longitudinal-heat-flow corrections would not have exceeded 0.2% at 100 °C and 0.5% at 1500 °C. (For specimens with very low thermal conductivity, the longitudinal-heat-flow corrections would be larger and, for this apparatus with the ends of the specimen near room temperature, uncertainties in heat flow of a few percent could occur.)

5.6.3. Eccentricity

The uncertainty due to eccentricity was less than 0.1% (see Section 3.6.3).

5.6.4. Overall Precision and Accuracy

The temperature drop across the specimen was uncertain by 1.2% at 100 °C, increasing to 2.6% at 1500 °C. The heat flux was uncertain by about 0.7% at 100 °C and 2.1% at 1500 °C. Uncertainties in the ratio of the shell radius to that of the core did not introduce more than 0.1% uncertainty, and the effect of eccentricity was less than 0.1%. Combining all of the uncertainties discussed, uncertainties in the mean thermal conductivities presented for the 1969 apparatus were estimated to be less than 2% at 100 °C and 4% at 1500 °C for a well-behaved, homogeneous specimen having a thermal conductivity similar to those of the soils tested.

The uncertainties in the thermal conductivities at the hot-side temperatures, obtained by differentiation of the curves fitted to the mean thermal conductivities, were much larger, depending upon the shape of the curve of $\bar{\lambda}$ versus temperature.

6. RECOMMENDATIONS FOR FUTURE DESIGNS

This section conveys the author's thoughts concerning the "best" way to build an apparatus of the type described in this chapter. It is assumed that the actual thermal conductivity, rather than a mean value over a large temperature range, is desired, so the recommendations below pertain mainly to apparatuses in which relatively small temperature differences would be used. The recommendations do, however, include some features derived from the 1969 apparatus in which the outside of the specimen was held near room temperature.

6.1. Apparatus Design

6.1.1. Overall Configuration

If the apparatus is only going to be used over a temperature range where liquid coolants can be used, the temperature of the shell should be controlled by the circulation of an appropriate heat transfer fluid around the exterior of the shell or by immersion of the shell into a temperature-controlled liquid bath. Otherwise, the design can be similar to the apparatus described below.

For high-temperature use, it is recommended that the apparatus be built so that the test section, consisting of the shell with the specimen and core installed in it, can be placed in a vertical tube furnace that will provide the heat to raise the mean temperature to the desired value. This approach is somewhat in contrast to the 1963 design, described in Section 3.1, where the shell also was the main furnace heater. It is believed that it is better not to have a heater in the central portion of the shell since this could lead to local temperature inhomogeneities and, at very high temperatures, to electrical leakage from the shell heater to the temperature sensors. The tube furnace can be of any type that provides the necessary space, upper temperature limit, and spatial and temporal uniformity. An alternative approach to a furnace would be to immerse the test section in a heated fluidized bed.

A further advantage of having a test section that can be separated from the furnace is that installation of the specimen can be greatly facilitated. For example, the test section can be vibrated in order to promote homogeneity and to attain the desired bulk density for the test sample.

Figure 13 is a schematic vertical cross section of a test-section design that uses a minimal amount of specimen material. The core is held concentric within the shell by centering fixtures that are well inside the tube furnace (not shown in Fig. 13). If the quantity of specimen material that can be used is quite limited, or if the tube furnace is not very long, this design allows the core and shell to be quite short (e.g., of the order of five times the inner diameter of the shell), since end heaters for both the shell and core are provided to promote a uniform longitudinal temperature distribution and to reduce the effects of longitudinal heat flow on the test results. Depending on the temperature range of interest, these high-temperature centering fixtures can be metal "spiders" or can be fabricated from insulating fire brick. The upper and lower supports that position the test section in the furnace can be thin-walled tubes, individual support rods, or even wires. In general, it is preferable to make these supports sufficiently conductive to carry away the excess heat from the core and guard heaters, so that the ends of the shell will not be hotter

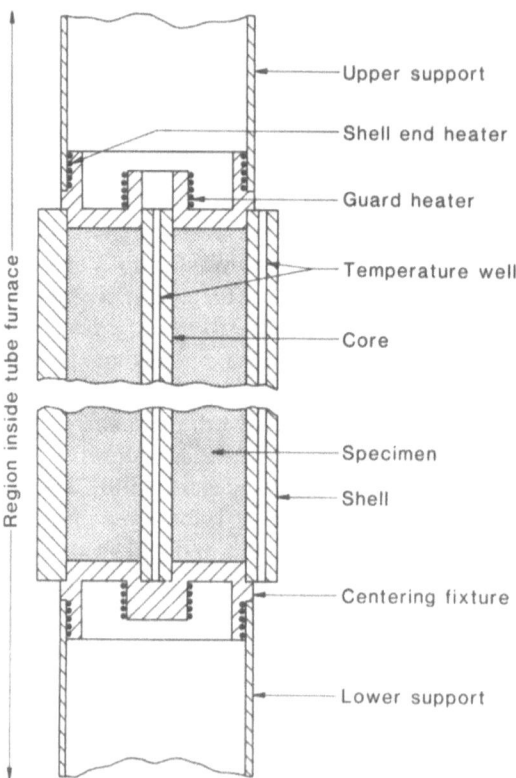

FIGURE 13. Schematic vertical cross section of a recommended apparatus design that uses high-temperature centering fixtures. This design requires very little specimen material.

than the midplane of the shell and so that some heat from the shell end heaters will be required to make the shell essentially isothermal.

If there are no severe limitations on the quantity of specimen material and if the tube furnace is long, the design shown in Fig. 13 can be simplified by using a longer core and a longer shell, omitting the guard heaters and the shell end heaters, and letting the furnace provide a sufficiently long isothermal region so that the effects of end heat losses are negligible. If these simplifications are desired and the quantity of specimen material is limited, it would be possible to fill only the center section of the apparatus with the test material of interest and to use some other type of insulating material near either end of the apparatus, although it should be noted that experimental uncertainty due to end effects may increase rapidly with the difference between the specimen and the extension material conductivity.

In general, for the design shown in Fig. 13, it would not be practical to make the specimen chamber gas tight. Thus if it were desired to make thermal conductivity measurements in different gases or at different gas pressures, the entire work zone of the tube furnace would have to be filled with the particular gas of interest, at the desired pressure. Figure 14 shows an alternative design for the test section, in which the shell extends out of the tube furnace at each end so that flanges with O-rings (not shown) can be used to provide gas-tight or vacuum seals at each end. Depending upon how far the ends of the shell extend from the furnace, it could be necessary to provide water cooling for the flanges. Shell end heaters are shown; these could be dispensed with if the test section and the tube furnace were sufficiently long that an adequate isothermal region can be provided near the midplane of the test section. Figure 14 shows the core being held between two coaxial support tubes, of the same diameter as the core. Again, if the core is long enough, the guard heaters that are shown could be eliminated. Alternatively, the core support tubes could be omitted and the core could extend all the way to the room-temperature centering fixtures, as was done in the apparatus described in Section 5.1. Elimination of either the shell heaters or the core support tubes and the guard heaters would require corrections to be made for longitudinal heat flow, unless the test section is extremely long.

At the top of Fig. 14, a tube is shown that is labeled "to atmospheric control system." This tube can be made sufficiently long, and preferably out of glass, to allow space for the core-axis thermocouple to be manipulated (e.g., magnetically) to obtain the longitudinal temperature distribution along the center of the core.

6.1.2. Shell Design

Figure 15 is a horizontal cross section, at the midplane, of the apparatus designs shown in Figs. 13 and 14. Three temperature wells are provided in the

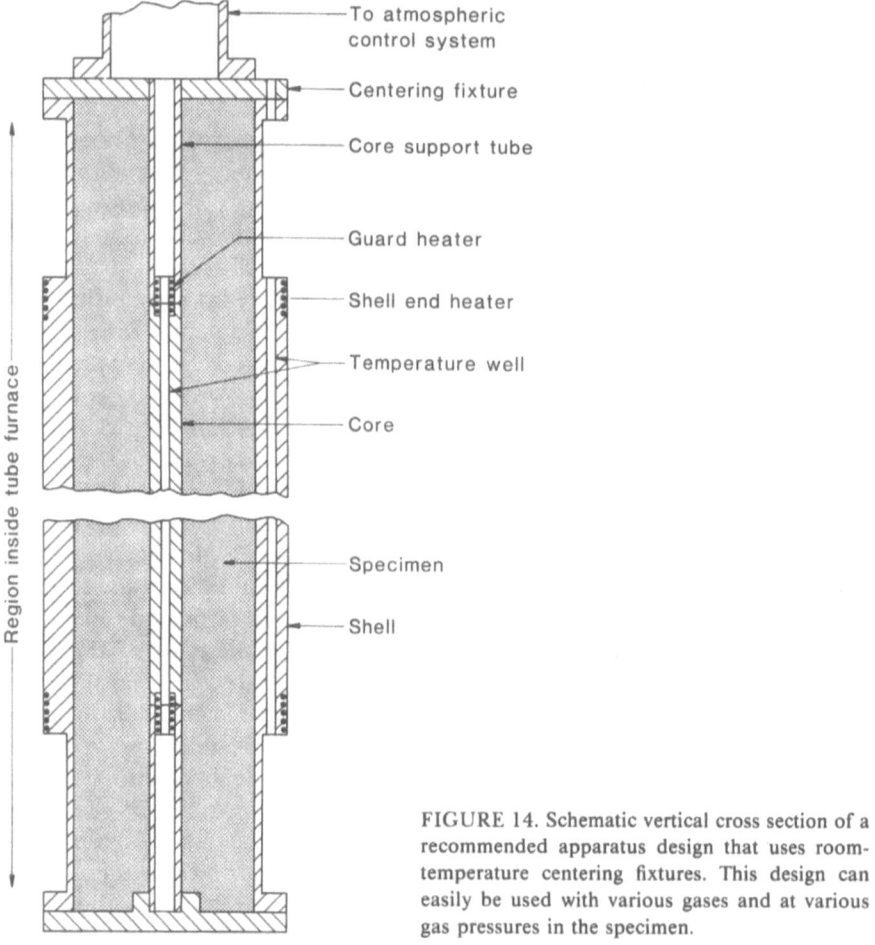

FIGURE 14. Schematic vertical cross section of a recommended apparatus design that uses room-temperature centering fixtures. This design can easily be used with various gases and at various gas pressures in the specimen.

shell, parallel to the axis of the core and spaced 120° apart. Reference back to Figs. 13 and 14 indicates that in these recommended designs the three shell thermocouples can be manipulated, in a manner similar to that used for the core axis thermocouple, to allow measurement of the longitudinal temperature profiles along the shell. In general, it is recommended that this be done, even if the shell is so long that the effects of longitudinal heat flow are negligible, because the recommended designs allow for replacement of thermocouples while the apparatus is in operation. Furthermore, these designs keep the thermocouples protected from mechanical or chemical interaction with the inner liner of the tube furnace. However, if it is not planned to operate the apparatus at sufficiently high temperature that the thermocouples might drift from calibration, fixed thermocouples could be installed in the shell.

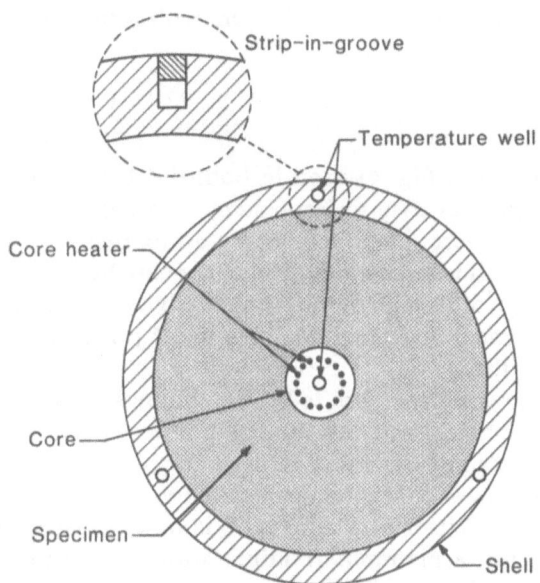

FIGURE 15. Schematic horizontal cross section of the recommended designs that use three longitudinal temperature wells in the shell. These designs allow easy determination of the longitudinal temperature distribution in the shell and also allow all temperature sensors to be removed and replaced while the apparatus is in operation.

The shell design indicated in Fig. 13 could be fabricated either from ceramic or from metal. A ceramic tube of the proper size could be extruded with the temperature wells in place and then fired. Alternatively, the shell could be cast with organic rods in it that would burn off during firing, leaving the desired temperature wells, or these holes could be drilled into the ceramic in the green or in the partially fired state. It would not be practical to drill holes so long in a metal shell. Rather, it would be preferable to mill or saw longitudinal grooves in the outside surface of the shell and weld strips of metal into the grooves so as to leave a temperature well, as shown in the inset in Fig. 15.

The shell design shown in Fig. 14 could be fabricated from a single tube of metal, with the wall made thinner near the ends to reduce the longitudinal thermal conductance, with grooves for temperature wells fabricated as described in the previous paragraph. This design could also be fabricated from dense ceramic, with or without making the wall thinner near the ends. The shell design shown in Fig. 14 is, aside from the longitudinal temperature wells, very similar to the guard design described by Flynn and O'Hagan[7] in which the "shell" was fabricated from molybdenum and the heaters (corresponding to the shell end heaters) were swaged heaters with a platinum–10% rhodium sheath.

6.1.3. Core Design

The core design (see Fig. 4) used in both the 1963 and 1969 apparatus worked very well and was easy to implement. The cores that were procured for these apparatuses were manufactured by an extrusion process and were relatively inexpensive. It is recommended that the core be made of high-density alumina, both because of its high thermal conductivity and because the thermal conductivity of alumina is reasonably well known. A possible alternative would be high-density beryllia.

The best way to wind the core is to use unannealed wire, which is easy to thread through the holes, and only anneal it, if necessary, at the ends where the wire comes out of one hole and doubles back into the adjacent hole; this is easily done as the heater wire is being installed in the core. For temperatures to, say, 1000 °C, nickel–chromium alloy wire is quite satisfactory. For higher temperature use in air, a platinum–rhodium alloy is recommended as being easy to work with, particularly for welding of current leads and voltage taps. For very high temperatures in an inert or vacuum atmosphere, a refractory metal such as molybdenum, tantalum, or tungsten could be used; in some cases it might be necessary to use individual wires, rather than a single wire threaded back and forth, and weld the end of each wire to the adjacent wire.

The use of a ceramic core has the particular advantages that the ceramic material will usually be rather inert with respect to the sample materials and that the heater wires are electrically well insulated from the specimen material so that the electrical conductivity of the specimen material is of no consequence. If, however, the specimen material is known to be a reasonably good electrical insulator that will not react chemically with a particular metal over the temperature range of interest, the core can be fabricated as a metal (or electrically conductive refractory material such as silicon carbide or, at high temperatures, zirconia) tube through which electrical current passes directly. Three designs for directly-heated cores are shown in Fig. 16. The first of these, on the left, is simply an electrically-conductive tube through which the current passes. In order to measure the temperature along the axis, the core thermocouple could be of a rather large diameter, along its entire length, so as to almost fill the hole in the heater tube, or a short metal or ceramic cylinder could be used around the thermocouple junction to make it nearly as large as the hole in the metal heater tube. This design has the disadvantage of allowing considerable radiation heat transfer along the interior of the heater tube. Whether the entire thermocouple is made large or whether just the junction is enlarged, it would not be practical to interchange thermocouples between the wells in the shell and the well in the core (see Section 6.3).

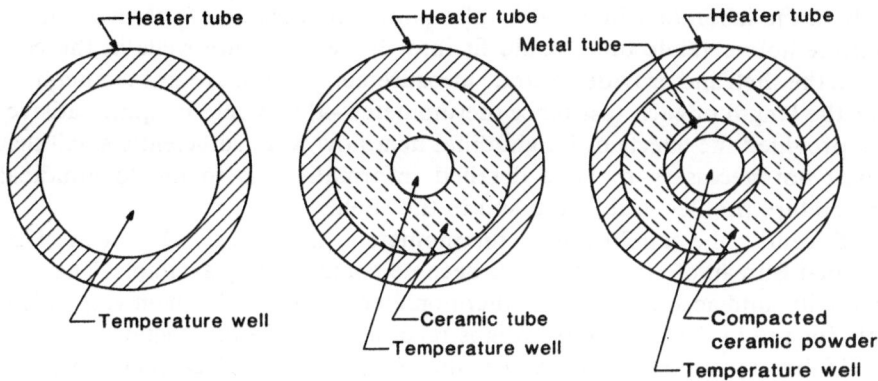

FIGURE 16. Three possible core designs in which the heater is simply an electrically conductive tube through which electric current passes. The advantages and disadvantages of these alternatives are discussed in the text.

It is recommended that either the design in the middle or that on the right of Fig. 16 be used. In the middle design, a ceramic tube, with a center hole properly sized to accommodate the thermocouple, is slipped inside the heater tube. A more complicated, but still practical approach would be that on the right of Fig. 16, where the core is made as an assembly with a metal temperature well inside a swaged metal heater tube, with magnesia or alumina insulation in between. Regardless of what core design is selected, the outside diameter of the core should be large compared to the average particle size of the specimen under test.

For a hollow heater tube that is heated directly by passage of electrical current, the temperature drop between the temperature well in the center of the core and the surface of the core [cf. equation (4)] is given by

$$T_0 - T_a' = \frac{A[a^2 - c^2 - 2c^2 \ln (a/c)]}{4\lambda_c} \tag{33}$$

where a, c, and λ_c are, respectively, the outer radius, the inner radius, and the thermal conductivity of the heater tube, and A is the power generated per unit volume in the tube.

6.1.4. Temperature Sensors

For most applications, thermocouples are probably the temperature sensors of choice. However, for the apparatus designs shown in Figs. 13 and 14, slender thermistors or resistance thermometers could be used over the temperature ranges where they are suitable. Platinum resistance thermometers, made in the laboratory by winding platinum wire back and forth in a slender multiple-hole ceramic tube, could fit into the temperature wells in the core and in the shell and provide accurate measurements of the average temperature over the length of each resistance thermometer—this would be quite satisfactory provided the effects of longitudinal heat flow were sufficiently small that it was not necessary to have detailed information as to the longitudinal temperature profiles in the core and in the shell.

Recently, the sapphire fiber thermometers invented by Dils at the U.S. National Bureau of Standards have become commercially available. The size, sensitivity, and accuracy of these thermometers would make them very acceptable for use in the apparatus designs recommended in this section.

If thermocouples are selected as the temperature choices, they can either be fabricated in the laboratory from bare wire inserted into high-purity, high-density alumina tubing or commercial sheathed thermocouples can be used. For temperatures up to, say, 1100-1400 °C in a clean, nonreducing atmosphere, platinum–10% rhodium versus platinum thermocouples are recommended. For temperatures approaching 1700 °C, platinum–30% rhodium versus platinum–6% rhodium should be used. Appropriate cleanliness should be observed in fabricating platinum–rhodium alloy thermocouples and proper annealing is important. For accurate thermal conductivity measurements, the thermocouple wire should be calibrated relative to the International Practical Temperature Scale.

6.2. Instrumentation

6.2.1. Temperature Control

In general it is recommended that all heaters, other than the core heater that provides the radial temperature drop across the specimen, be powered with AC, rather than DC, current. Not only is it easier to work with AC but there is less likelihood of leakage currents adversely affecting the temperature readings from the thermocouples. In general, the use of isolation transformers on each heater is recommended to further reduce the possibility of leakage currents. Grounded shielding, for example along the inside liner of the tube furnace, will further reduce problems due to leakage currents. Wherever possible, heaters should be wound to be noninductive.

Either proportional controllers or "on/off" controllers should be satisfactory for the type of apparatus under discussion here. There is some advantage in using on/off controllers, either with the current switched completely on and off or with a power resistor intermittently placed in series with the heater so as to change the current by, say, 10%. If there are problems with leakage currents or induced voltages causing errors in the thermocouple emfs, such problems are more likely to be obvious when the heater power is switched from one value to another rather than being continuously modulated. A controller should be selected that has a very stable setpoint temperature so that temperatures in the test section can be maintained constant to a few hundredths of a degree.

It is recommended that a controller be selected whose setpoint is adjustable under computer control so that the entire operation of the apparatus can be fully automated. Some modern controllers are computer addressable. Alternatively, a digital-to-analog converter can be used to generate the setpoint voltage to which the control thermocouple emf is compared.

Ideally, a separate temperature controller would be used for each heater in the apparatus. If this is not possible, the tube furnace can be provided with a good temperature controller and the currents to the guard heaters and shell end heaters can be adjusted to fixed values using variable voltage transformers. However, this approach would make full automation of the apparatus difficult.

6.2.2. Power Measurement

For temperatures up to perhaps 1400 °C, the core heater can be powered by DC current from a regulated power supply (preferably a constant-current supply or a constant-voltage supply with remote sense leads to control the voltage at the voltage taps on the high-temperature portion of the heater). The current through the heater can be obtained by measuring the voltage drop across a standard resistor in series with the heater and the voltage drop across the heater can be measured directly. It is recommended that a calibrated precision integrating digital voltmeter, interfaced to a computer, be used for these measurements.

For temperatures above 1400 °C, there is a strong likelihood that leakage currents from the core heater can affect thermocouple emfs unless special precautions are taken. If the thermocouples are sheathed, the core heater can be powered with DC current to quite high temperatures. If the thermocouples are not sheathed, it is recommended that the core heater be powered with AC current so that the common-mode and normal-mode rejection capabilities of modern integrating digital voltmeters can be used to read the DC thermocouple voltages even if large AC leakage currents are present. If AC current is used for the core heater, the ideal way to measure the power would be to use a digital AC wattmeter, interfaced to the computer. However, the core heater designs that are recommended above should have rather low inductance at power-line frequencies so that a calibrated, high-accuracy digital AC voltmeter could be used to measure, separately, the current through the heater and the voltage drop across it. If this is done, the power factor of the circuit should be checked to determine whether or not the heater current and voltage may be assumed to be in phase.

6.2.3. Temperature Measurement

It is recommended that a precision integrating digital DC voltmeter, with good normal-mode and common-mode rejection at line frequencies, be used for reading thermocouple voltages. The normal-mode and common-mode rejection capabilities are particularly important for operation of the apparatus at very high temperatures. Since temperature differences of 10 °C or more can easily be used with loose-fill thermal conductivity specimens, it normally would be adequate for the digital voltmeter to have a resolution of 0.1 μV.

If a computer-operated multiplexer is used to switch among the thermocouples being read by the digital voltmeter, care should be taken to select a break-before-make multiplexer with high, low, and guard lines, and with very low thermal offset voltages.

In general, it is recommended to use a constant-temperature block for the thermocouple cold junctions and to measure the temperature of that reference block directly, rather than to rely upon commercial compensating cold junctions that insert a voltage into the measurement circuit.

6.3. Test Procedure

6.3.1. Specimen Installation

For low-density insulations, the specimen installation procedure is relatively straightforward. For high-temperature measurements, it is generally desirable to oven-dry the sample material at, say, 110 °C to drive off any free moisture that may be present. The desired weight of material should be put into the test section with vibration or compaction as necessary to attain the desired bulk density. It often will be necessary to try this several times in order to attain the desired final bulk density and to have a uniform density throughout the test specimen.

The additional procedures that were used for soil samples where considerable compaction was needed in order to achieve the desired densities are available elsewhere.[2,3]

6.3.2. Testing Sequence

It is strongly recommended that an "isothermal correction," as discussed by Moore[4] (see also Flynn and O'Hagan,[7] Laubitz and McElroy,[8] and Laubitz[9]), always be used. This procedure compensates for differences among the several temperature sensors and, to a great extent, for the effects of longitudinal heat flows. The particular operational sequence will depend somewhat upon the particular apparatus design—for the following discussion it is assumed that there are shell end heaters and core guard heaters.

Isothermal Test. Using the tube furnace, bring the shell temperature to the desired value, taking care not to overshoot the desired temperature during the heating process (so as to avoid possible changes to the specimen material, for example due to moisture release or sintering). Adjust the shell end heaters so that the shell is as close to isothermal as is practical. Without supplying any power to the main core heater, adjust the core guard heaters so that the core is isothermal at very nearly the same temperature as that of the shell. Measure the temperatures at the midplane of the apparatus, using the temperature sensor in the core and the three temperature sensors in the shell. If calculations or experience indicate that a correction for longitudinal heat flow may be necessary, measure the longitudinal temperature distribution for each temperature well.

Gradient Test. Keeping the shell temperature constant, provide sufficient power to the main core heater to attain the desired temperature difference (e.g., 10 to 20 °C) between the temperature sensor in the core and those in the shell. Adjust the core guard heaters so as to achieve isothermal conditions along the axis of the core. Measure the power input to the main core heater. Measure the temperatures at the midplane of the apparatus and, if necessary, measure the longitudinal temperature distributions.

The thermal conductivity can be computed from the data acquired during the isothermal test and during the gradient test, using the procedure given in Section 6.4. If data are being taken at temperatures high enough to cause drift in the temperature sensors, it is recommended to carry out an additional isothermal test, after the gradient test, and use the average of the data from the two isothermal tests, in conjunction with the data from the gradient test, to compute the thermal conductivity.

For both the isothermal and gradient tests, the test section should be allowed to attain thermal equilibrium so that there are no monotonic drifts in temperature large enough to influence the final test results. For an automated apparatus, where data are being acquired under computer control, thermal conductivity values should be computed as equilibrium is approached and criteria established for convergence to acceptably stable values. As suggested by Laubitz,[9] the use of computers to both acquire data and compute results may allow more sophisticated analyses that do not place such stringent requirements on thermal equilibrium.

For measurements at very high temperatures, where drift of the temperature sensors is likely, the apparatus designs recommended above allow the temperature sensors to be removed from the apparatus and new, or recalibrated, sensors inserted into the hot regions of the apparatus only for short periods of time after thermal equilibrium is essentially established.

At high temperatures where leakage currents may cause a problem, checks should be made, preferably after each test, for extraneous emfs that cause errors in the thermocouple voltages. An easy way to do this is to read the thermocouples and then quickly shut off all heaters and see if the apparent thermocouple voltages change suddenly. (The apparatus will have enough heat capacity to keep the actual temperatures from changing very rapidly.)

6.4. Calculations

The computer should be provided with instrumentation and temperature sensor calibrations, subprograms for computing temperatures from the sensor voltages, the apparatus geometry, the thermal conductivities of the core and shell materials, and, if longitudinal heat flow corrections will be needed, the longitudinal thermal conductance of the core and the temperature dependence of the electrical resistance of the core heater.

For radial heat flow, with *neglible* effects due to longitudinal heat flow, the mean thermal conductivity is given by [cf. equation (3)]

$$\bar{\lambda} = \frac{q \ln (b/a)}{2\pi[(T_a - T_b) - (T_{a0} - T_{b0})]} \tag{34}$$

where $T_{a0} - T_{b0}$ is the apparent temperature difference between the axis of the core and the temperature wells in the shell as measured during the "isothermal test," and the other quantities are as defined in Section 2. The proper calculation procedure for the case where longitudinal heat flows cannot be negelcted can be derived without great difficulty (see Moore,[4] Flynn and O'Hagan,[7] Laubitz and McElroy,[8] and Laubitz[9]). The particular procedure for dealing with longitudinal heat flows will depend upon the specific apparatus design. Variations of the analyses in Section 3.6.2 and in Section 4 could be used or a finite-element code, such as used by Moore,[4] could be incorporated into the routine data-analysis procedure.

REFERENCES

1. D.R. Flynn, *J. Res. Natl. Bur. Stand.* **67C**, 129 (1963).
2. D.R. Flynn and T.W. Watson, "Measurement of the Thermal Conductivity of Soils to High Temperature—Final Report," Aerospace Nuclear Safety Rept. SC-CR-69-3059, Sandia Laboratories (April, 1969).
3. D.R. Flynn and T.W. Watson, in: *Thermal Conductivity*, Proc. Eighth Conf. (C.Y. Ho and R.E. Taylor, eds.), p. 913, Plenum Press, New York (1969).
4. J.P. Moore, in: *Compendium of Thermophysical Property Measurement Methods*, Vol. 1, *Survey of Measurement Techniques* (K.D. Maglić, A. Cezairliyan, and V.E. Peletsky, eds.), p. 61, Plenum Press, New York and London (1984).
5. B.A. Peavy, *J. Res. Natl. Bur. Stand.* **67C**, 119 (1963).

6. H.S. Carslaw and J.C. Jeager, *Conduction of Heat in Solids*, 2nd ed., p. 221, Oxford University Press, London (1959).
7. D.R. Flynn and M.E. O'Hagan, *J. Res. Natl. Bur. Stand.* **71C**, 255 (1967).
8. M.J. Laubitz and D.L. McElroy, *Metrologia* **7**, 1 (1971).
9. M.J. Laubitz, in: *Compendium of Thermophysical Property Measurement Methods*, Vol. 1, *Survey of Measurement Techniques* (K.D. Maglić, A. Cezairliyan, and V.E. Peletsky, eds.), p. 11, Plenum Press, New York and London (1984).

3

The Measurement of Thermal Conductivity by the Comparative Method

RONALD P. TYE

1. INTRODUCTION

A variety of so-called "secondary" steady-state measurement techniques utilizing "comparison" of parameters of an unknown specimen with those of one or more specimens of known thermal properties exists. In general, when the term comparative method or "comparator" is used, it is understood that it applies to the basic method whereby a specimen is joined directly to, or sandwiched between, one or two reference materials and surrounded by a heated guard cylinder. A temperature gradient is established along the test stack and longitudinal heat flow is assured or maximized by adjustment of either the temperature gradient in, or the isothermal temperature of, the guard cylinder.

Other techniques involving similar principles, but utilizing heat flux transducers and calibration specimens or transfer standards to calibrate the measurement system, are also used. In particular, they are especially useful for measurements on thermal insulations[1,2] and, with modification, of materials of low and intermediate thermal conductivity.[3,4] These methods are not addressed directly in the present chapter, since their applications are more restricted and specific than the general comparative method.

The comparative method has been described as the "workhorse of the thermal conductivity field" and with good reason. Essentially, it can be used to evaluate the properties of homogeneous, heterogeneous, and composite solid materials which have an intrinsic, apparent, or effective thermal conductivity, λ, in the approximate range of $0.15 < \lambda < 150\ \mathrm{W\,m^{-1}\,K^{-1}}$. The overall temperature range of use is generally 100 to 1300 K.

RONALD P. TYE • Consultant, Sinku Riko Inc., Cohasset, Massachusetts 02025, USA.

This very wide range of λ is between that of the true thermal insulation for which absolute guarded hot plate[5,6] or heat flow meter methods[1,2] utilizing large flat slab specimens are required, and the metallic and other highly conducting materials where long, thin, rod-type specimens are necessary.[7,8] With suitable modifications and careful experimental techniques[9] the λ range can be extended at each end of the spectrum.

Because of these factors, it is a compromise blend of techniques of somewhat lower but acceptable accuracy than the other primary methods. However, it fills the void where such methods are not truly practical and it is also in the range of many engineering materials and components. In addition, it is a relatively simple concept in both design and practice and, thereby, has a number of significant advantages over radial flow techniques,[10] which are more complicated in configuration and significantly more time consuming in operation.

The comparative method of measurement is especially useful for engineering materials including ceramics, polymers, metals and alloys, refractories, carbons and graphites, including combinations, and other composite forms of each. It is capable of generating relatively accurate data on specimens which are too small (or possibly have an unusual shape) to accept heater and heat sink designs required by absolute methods.

The method can also be used for materials in other forms such as powders and particulates,[11] for molten materials,[12,13] and for a variety of contact resistance and conductance applications.[14,15] These include contact resistances between dissimilar materials and thermal conductances of composites including layered, graded, and coated specimens. A major advantage here is the requirement for smaller, more simple specimen configurations than those required by other absolute methods.

Proper design of a guarded-longitudinal system is difficult. It is not practical to try to establish details of construction and procedures to cover all contingencies that might offer difficulties to a person without technical knowledge concerning theory of heat flow, temperature measurements, and general testing practices. This chapter provides the basic information necessary for undertaking reliable measurements. Standardization of this method is necessary and has been undertaken,[16] but it is not intended to restrict in any way the future development by research workers of new or improved methods or improved procedures. However, new or improved techniques must be thoroughly tested and requirements for qualifying an apparatus are also outlined.

Overall, therefore, the method has a major advantage in that it is versatile, flexible, highly adaptable, and especially useful for thermal conductivity ranges between those of absolute methods and for materials which are available in relatively small quantities. With smaller configurations, it is somewhat more rapid in operation than the absolute methods, especially when automated. In recent years, real-time control of experiments and their automation have made comparative techniques more productive and competitive with some non-stationary state methods where λ is obtained from thermal diffusivity and specific heat measurements.

2. TERMINOLOGY

2.1. Definitions

2.1.1. Thermal Conductivity, λ

In an optically thick solid with constant properties, the quantity of heat flow per unit area at steady state, q', is given by

$$q' = \lambda \nabla T$$

where λ is the thermal conductivity. For one-dimensional heat flow in the Z direction, this reduces to

$$q' = \lambda \frac{dT}{dZ}$$

where dT/dZ is the temperature gradient. When only two sensors are used, the differential must be replaced with $\Delta T/\Delta Z$, the temperature difference between the two sensors.

2.1.2. Apparent Thermal Conductivity, λ_A

In some specimens the quantity of heat flow will be a function of additional system parameters such as the emittances of bounding surfaces and the specimen size. In these cases, the comparator determines only an apparent thermal conductivity, λ_A, defined as

$$\lambda_A = -q' \frac{\Delta Z}{\Delta T}$$

3. APPARATUS

3.1. Basic Considerations

In general, a material or system is considered homogeneous if the apparent thermal conductivity of the specimen, λ_A, does not vary with changes of thickness or cross-sectional area by more than ±5%. For composites or heterogeneous systems consisting of slabs or plates bonded together, the specimen should be more than 20 units wide and 20 units thick, respectively, where a unit is the thickness of the thickest slab or plate, so that diameter or length changes of one-half unit will affect the apparent λ by less than ±5%.

For systems which are nonopaque or partially transparent in the infrared, the combined error due to inhomogeneity and photon transmission should be less than ±5%. However, measurements on highly transparent solids should be accompanied with infrared absorption coefficient information *or* the results must always be considered as an apparent or effective thermal performance property. By choosing the comparative longitudinal heat flow technique, the experimentalist is provided with the opportunity of measuring not only the intrinsic property, but also the various performance parameters of representative specimens of the appropriate material or system.

Since the method is a compromise, the specimen configuration lies between those required for the absolute methods. Thus, for the range $0.15 < \lambda < 2$ W m^{-1} K^{-1}, the "flat slab" configuration[17] is to be recommended, while for $\lambda > 2$ W m^{-1} K^{-1}, the "cut bar" configuration[18] is to be preferred. Such limits are not rigid in application nor are the configurations for the respective ranges. Modified comparative rod methods[19,20] have been used successfully, especially for $\lambda > 10$ W m^{-1} K^{-1}, but, in such cases, much larger and longer specimens must be used.

Specimen configurations for the "flat slab" are usually some 50 to 100 mm in diameter or square, and for the "cut bar" they are cylindrical rods of diameter between 17 and 37 mm. Smaller sizes can be used where material availability and form present particular problems. However, specimen dimensions smaller than 17 mm diameter are not to be recommended unless the material is at a premium or a lower measurement accuracy is acceptable. Less common and more complex configurations can be used, providing suitable reference materials can be made available in appropriate sizes.

The major restriction to its use outside 100 to 1300 K is the unavailability of a range of reference materials having sufficient and widely different characterized thermal properties at or outside the above range. A very limited number of pure metals have been characterized at high temperature but these are all of high thermal conductivity and, in some cases, are not compatible with all other materials. In addition, some require special environments, especially high vacuum for protection purposes and, as will be discussed later, this factor can influence the use of this method for many applications. Many more materials require characterization for use as reference materials, not only to extend the present temperature range but also to increase both the flexibility of the method and reduce the uncertainty of current measurements.

3.2. General Features

The general features of the guarded longitudinal heat flow technique are shown schematically in Figs. 1a and 1b and a typical cut-bar apparatus and test stack in Figs. 2a and 2b, respectively. A specimen of unknown conductivity, λ_s, but having an estimated thermal conductance of λ_s/l_s, is mounted between two reference materials or meter bars of known thermal conductivity, λ_m, ideally of the same cross section and length and similar thermal conductance, λ_m/l_m. A more complex but suitable arrangement is a column consisting of a

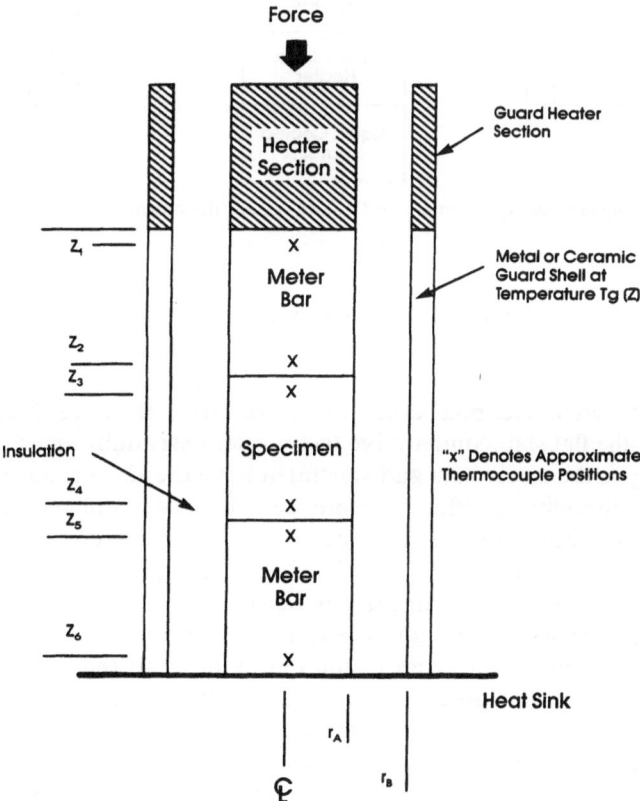

FIGURE 1a. Schematic of a comparative guarded longitudinal heat flow system showing possible locations of temperature sensors.

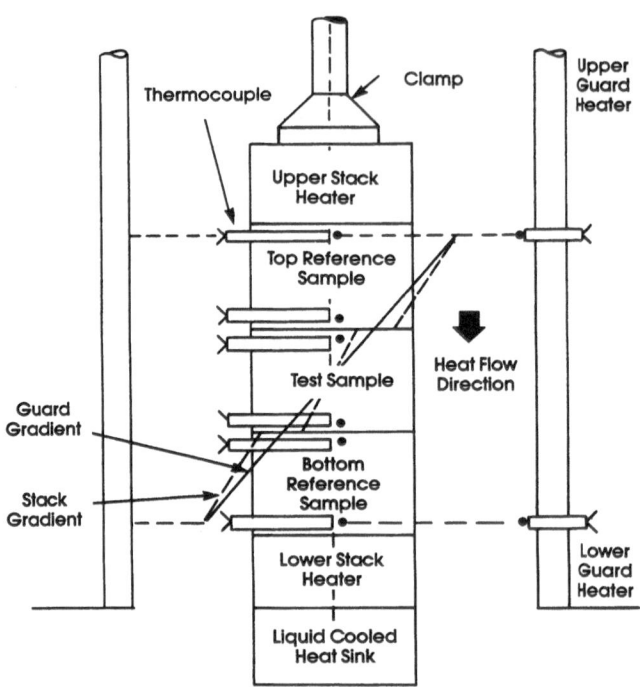

FIGURE 1b. Schematic of typical stack and guard system illustrating matching of temperature gradients.

disk heater with a specimen and a meter bar on each side between heater and heat sink. In this case, approximately one-half of the power flows through each specimen.

When the cross sectional dimensions are larger than the thickness, it is described as the flat slab comparative technique. Essentially any shape can be used, as long as the meter bars and specimen have the same conduction areas. Due to the availability of different reference materials in different thicknesses, it is possible to obtain a wide variation in thermal conductance of unknown specimens. It is not essential for λ_m to be similar to λ_s, although it is preferable for the overall λ values and dimensions not to be too dissimilar.

A force, F, provided to retain the test stack in position, is applied to the column that is surrounded by an insulation material of thermal conductivity λ_I. The insulation is enclosed in a guard shell with radius r_B, and the shell is held at temperature $T_g(z)$. A temperature gradient is imposed on the column by maintaining the top at a temperature T_T and the bottom at temperature T_B. Function $T_g(z)$ is usually a linear temperature gradient situation with the gradient matched approximately to that established in the test stack. However, an isothermal guard with $T_g(z)$ equal to the average temperature of the specimen may also be used. An unguarded system is not recommended due to the potential very large errors, particularly at elevated temperatures. At steady state, the temperature gradients along the sections are calculated from measured temperatures along the two meter bars and the specimen. The value of λ_s (λ_s as uncorrected for heat shunting) can then be determined using

$$\lambda_s^1 = \frac{Z_4 - Z_3}{T_4 - T_3}\left[\frac{\lambda_m^1}{2}\left(\frac{T_2 - T_1}{Z_2 - Z_1}\right) + \frac{\lambda_m^2}{2}\left(\frac{T_6 - T_5}{Z_6 - Z_5}\right)\right] \tag{1}$$

where the notation is shown in Fig. 1a.

This is a highly idealized situation, however, since it assumes no heat exchange between the column and insulation at any position and uniform heat transfer at each meter bar–specimen interface. The errors caused by these two assumptions vary widely and are discussed in a later section. Because of these two effects, restrictions are necessary if the desired accuracy is to be achieved.

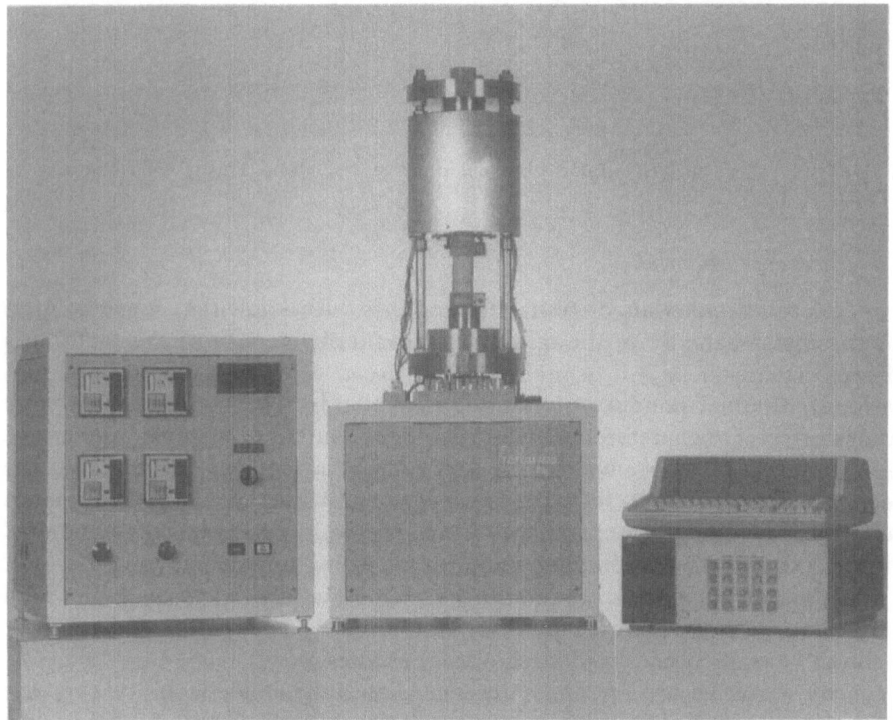

FIGURE 2a. A guarded longitudinal heat flow comparative test apparatus.

FIGURE 2b. Close-up of typical cut-bar comparative test stack (Photographs Courtesy of Holometrix, Inc.).

3.3. Reference Materials

Reference materials or transfer standards with known λ_m must be used for the meter bars. Since the minimum measurement error of the method is the uncertainty in λ_m, it is preferable to use a certified national (or international) thermal conductivity standard available from national standards laboratories. Other reference materials are available, because numerous measurements of λ have been made and general acceptance of the values has been obtained. Table 1 lists the currently available, recognized reference materials, including those available from the National Bureau of Standards. Figure 3 shows the approximate variation of λ_m with temperature.

Table 1 is not exhaustive and other stable materials can be used, provided they have been evaluated separately and reliably by absolute methods or through "round-robin" or interlaboratory studies.

The requirements for any reference material include stability over the temperature range of operation, compatibility with other system components, reasonable cost, ease of thermocouple attachment, and an accurately known

TABLE 1
Reference Materials for Use as Meter Bars

Material	Temperature range (K)	Percent uncertainty in λ (±%)	$\lambda_M(\mathrm{W\,m^{-1}\,K^{-1}})$	Material source
Electrolytic Iron SRM734	To 1000	2	NBS Special Publ. 260-90	NBS[a]
Tungsten SRM730	4–300 300–2000	2 2–5	NBS Special Publ. 260-90	NBS[a]
Austenitic stainless SRM735	4–1200	<5%	NBS Special Publ. 260-90 $\lambda = 1.22\,T^{0.432}$ $T > 200$ K	NBS[a]
Graphite RM8424, 25, & 26	<300 300–2500	2 10 max	NBS Special Publ. 260-89	NBS[a]
Armco iron	80–1200	2	λ_M should be calculated (b, c) from measured values	—
OFHC copper	90–1250	<2	$\lambda_M = 416.3 - 0.05904\,T$ $+7.087 \times 10^7/T^3$ [d]	Manufacturer
Pyroceram code 9606	90–1200	6	e, f	Corning glass
Fused silica[b]	300–1300 Up to 900 K	<8	$\lambda_M = (84.7/T) + 1.484$ $+ 4.94 \times 10^{-4}\,T$ $+ 9.6 \times 10^{-13}\,T^4$ [g,h]	Corning glass
Pyrex 7740	90–600	6	e, f	Corning glass

[a] Office of Standard Reference Materials, National Bureau of Standards, Washington, DC 20899.
[b] W. Fulkerson et al., Phys. Rev. **167**, 765 (1968).
[c] C.F. Lucks, J. Test Eval. **1** (5), 422 (1973).
[d] J.P. Moore, R.S. Graves, and D.L. McElroy, Can. J. Phys. **45**, 3849 (1967).
[e] "Thermal Conductivity of Selected Materials," Report NSRDS-NBS 8, National Bureau of Standards (1966).
[f] L.C. Hulstrom, R.P. Tye, and S.E. Smith, in: Thermal Conductivity 19 (D.W. Yarbrough, ed.) pp. 199–211, Plenum Press, New York (1988). [See also High Temp. High Pressures **17**, 707 (1985).]
[g] Above 700 K a large fraction of heat conduction in fused silica will occur by radiation and the actual effective values may depend on the emittances of bounding surfaces and meter-bar size.
[h] Recommended values from Table 3017 A-R-2 of the Thermophysical Properties Research Center Data Book, Vol. 3, Nonmetallic Elements, Compounds, and Mixtures, Purdue University, Lafayette, Indiana.

thermal conductivity. Since heat shunting errors for a specific value of λ_I increase as λ_m/λ_s varies from unity,[21] the reference which has a λ_m value nearest to λ_s should be used for the meter bars unless values of λ_s/l_s preclude this arrangement.

If a sample has a λ_s value between two reference materials, the reference with the higher λ_m should be used to reduce the total temperature drop along the column.

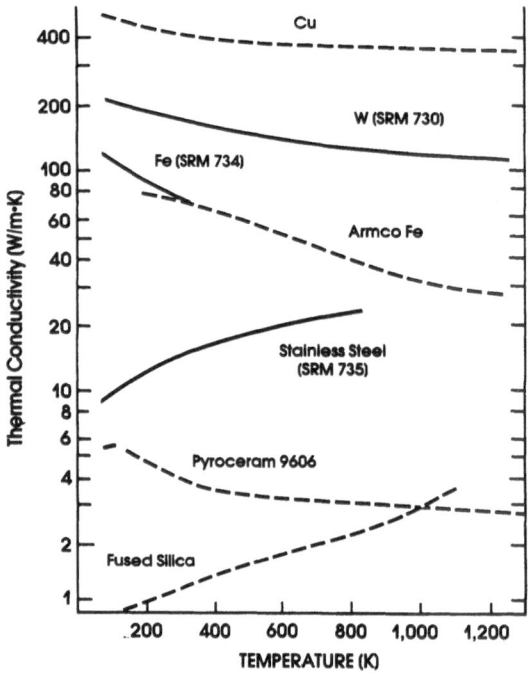

FIGURE 3. Approximate values for the thermal conductivity of several possible reference materials for meter bars. Where possible the material selected should have a thermal conductivity as near as possible to that of the unknown.

3.4. Insulation Materials

A large variety of powder, particulate, and fiber materials exist for reducing both radial heat flow in the column-guard annulus and surrounds and heat shunting along the column. Several factors must be considered during selection of the most appropriate insulation. The insulation must be stable over the anticipated temperature range, have a low λ_I value, should transmit minimal energy by radiation, and be easy to handle. In addition, the insulation should not contaminate system components such as the temperature sensors, it must have low toxicity, and it should not conduct electricity. In general, powders and particulates are used since they pack readily. However, low-density fiber blankets can also be used, depending on the temperature range. Table 2 contains a list of commonly used insulation materials.

3.5. Temperature Sensors

Reliable sensor attachment and temperature measurements are critical for this method[22] since the whole technique is based upon comparison of temperature gradients. A minimum of two temperature sensors is necessary on each meter bar and two on the specimen, unless the thickness of the test

TABLE 2
Suitable Thermal Insulation Materials[a]

	Typical thermal conductivity ($W m^{-1} K^{-1}$)		
	300 K	800 K	1300 K
Poured powders			
Diatomaceous earth	0.053	0.10	0.154
Bubbled alumina	0.21	0.37	0.41
Bubbled zirconia	0.19	0.33	0.37
Vermiculite	0.07	0.16	—
Perlite	0.050	0.17	—
Blankets and felts			
Aluminosilicate 60–120 kg m^{-3}	0.044	0.13	0.33
Zirconia 60–90 kg m^{-3}	0.039	0.09	0.25

[a]All materials listed can be used up to the 1300 K limit of the comparative longitudinal except where noted.

specimen precludes attachment. Whenever possible, the meter bars and specimen should each contain three sensors. The extra sensors are useful in confirming linearity of temperature versus distance along the column or indicating an error due to a temperature sensor decalibration. However, in many cases, specimens of thickness less than the order of 6 mm have to be evaluated and attachment of temperature sensors is not suitable due to potential distortion of unidirectional heat flux.

In this case, extrapolation of temperatures elsewhere in the test stack is undertaken to determine the temperature of the specimen. Great care is required in such cases to ensure that contact resistances between specimen and references are eliminated, minimized, or are included in the system calibration measurements. This is necessary to ensure that the extrapolated temperatures are truly representative of those at the appropriate positions.

The type of temperature sensor depends on the system size, temperature range, and the system environment as controlled by the insulation, meter bars, specimen, and gas within the system. Any sensor possessing adequate accuracy may be used for temperature measurement and be used in large systems where heat flow perturbation by the temperature sensors would be negligible. Thermocouples are normally employed; their small size, usually wires 0.1 mm or less in diameter, and the ease of attachment are distinct advantages.

When thermocouples are employed they should be fabricated from small-diameter wires of high-grade material, either calibrated thermocouple wire or wire that has been certified by the supplier to conform with standard specifications such as ASTM E-230[23] to within the given special limits of error. Such specifications should be consulted for information regarding thermocouple selection and calibration. A constant temperature reference is necessary for all cold junctions. This reference can be an ice–water slurry, a constant temperature zone box, or an electronic ice point reference.

Thermocouple attachment is very important in this technique so as to ensure that reliable temperature measurements are made at specific points. The various techniques are illustrated in Fig. 4. Intrinsic junctions can be obtained with metals and alloys by welding individual thermocouple wires to the surfaces (Fig. 4a). Butt or bead welded thermocouple junctions can be rigidly attached by peening, cementing, or welding in fine grooves or small holes (Fig. 4b–d).

In Fig. 4b the thermocouple resides in a radial slot, and in Fig. 4c the thermocouple is pulled through a radial hole in the material. When a sheathed thermocouple or a thermocouple with both thermoelements in a two-hole electrical insulator is used, the thermocouple attachment shown in Fig. 4d can be used. In the latter three cases, the thermocouple should be thermally connected to the solid surface using a suitable glue or high-temperature cement.

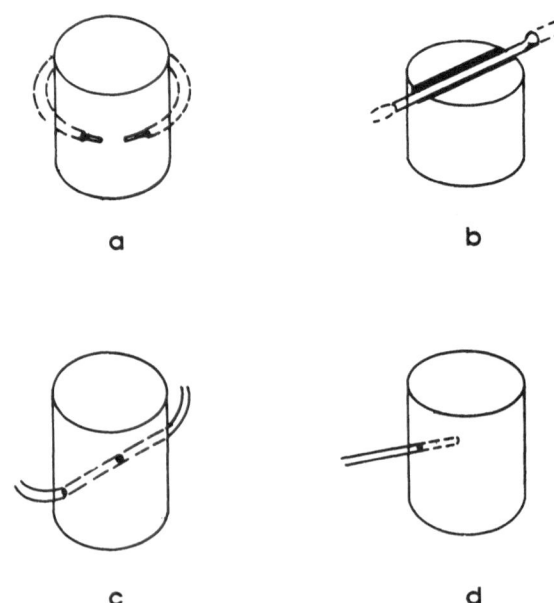

FIGURE 4. Thermocouple attachments. In all cases, the thermoelements should be thermally tempered and/or thermally grounded on the guard to minimize temperature measurement errors due to heat flow into or out of the hot junction.

All four procedures shown in Fig. 4 should include wire tempering on the surfaces, wire loops in isothermal zones, thermal wire grounds on the guard, or a combination of all three.[24]

Since uncertainty in temperature-sensor location leads to large errors, special care must be taken to determine the correct distance between sensors to calculate the possible error resulting from any uncertainty and to be sure that extrapolation of any temperatures is based on good experimental practices which ensure that the measurements represent the temperatures of the parts of the object required.

3.6. Reduction of Contact Resistance

The test method requires uniform heat transfer at the meter bar–specimen interfaces, particularly when the temperature sensors are within a distance equal to R_A from an interface.[21] This requirement necessitates a uniform contact resistance across the adjoining areas of meter bars and specimens. Ideally, intimate contact by soldering, brazing, etc. is desirable. However, due to the wide variety of materials, material types and dissimilarities, and the overall temperature range to be covered, this is often impractical and other means are required to obtain the desired uniformity. This is normally achieved by use of an applied axial load in conjunction with a conducting medium at the interfaces.

Measurements in a vacuum environment are not recommended, unless the vacuum is required for the purposes of protection or to study behavior of materials or contact resistances in such environments. This is due to the fact that the gaseous fluid at the interface has a relatively high thermal conductivity compared to that of a vacuum and its removal can contribute to significant increase in contact resistances.

For the relatively thin specimens normally used for materials having a low thermal conductivity, temperature sensors must be mounted close to the surface and, in consequence, the uniformity of contact resistance is critical. In such cases, a very thin layer of a compatible highly conductive fluid, paste, soft metal foil, or screen is normally introduced at the interfaces.

Some means is normally provided for imposing a reproducible and constant load along the column with the primary purpose of minimizing interfacial resistances at meter bar–specimen interfaces. Since the force applied to the column usually affects the contact resistance, it is desirable that this force be variable to ensure that the value of λ_s does not change due to force variation during the measurements. This force can be applied either pneumatically, hydraulically, or by putting a dead weight on the column. The above load mechanisms have the advantage of remaining constant with change in column temperature. In some cases, the compressive strength of the specimen might be so low that the applied force must be limited to the dead weight of the upper reference bar. In such cases, special care should be taken to limit errors resulting from poor contact by judicious positioning of temperature sensors away from any heat flow perturbation at the interfaces.

3.7. Guard Cylinder

The specimen–meter bar column is enclosed by a guard cylinder normally of right-circular symmetry. This guard cylinder can be either a metal or a ceramic, but its inside radius should be such that the ratio r_B/r_A is between 2.0 and 3.5.[21] The guard cylinder contains at least two heaters for controlling the temperature profile along the guard, or one heater to maintain an isothermal guard.

The guard is constructed and operated such that the temperature of the guard surface is either isothermal and equal to the approximate mean temperature of the specimen, or preferably it has an approximately linear profile with the top and bottom ends of the guard matched to corresponding positions along the column. In each case, at least three temperature sensors are normally attached to the guard at known positions to measure the temperature profile.

3.8. System Instrumentation

The combination of temperature sensor and the instrument used for measuring the sensor output should be adequate to ensure a temperature measurement precision of ± 0.05 K and an absolute error less than $\pm 0.5\%$.

Instrumentation for this technique should be adequate to maintain the required temperature control and measure all pertinent output voltages with accuracy commensurate with the system capability. Although control can be manual, a technique of this general description can be automated so that a computer carries out all the control functions, acquires all pertinent voltages, and calculates the thermal conductivity.

3.9. Test Specimen

As discussed earlier, the method is not restricted to a particular geometry. General practice is to use cylindrical or square cross sections. The conduction areas of the specimen and reference samples should be the same to within 1%. In some cases this requirement is not necessary. For example, some apparatus might consist of meter bars and specimen with high values of λ_M and λ_S, so that thermal shunting errors would be small for long sections. These sections might be long enough to permit temperature-sensor attachment far enough away from the interfaces that heat flow is uniform. However, the contact areas of mating surfaces must still be within the previously stated criteria.

Any difference in area is normally taken into account in the calculations of the result. For the cylindrical configuration, the radii of the specimen and meter bars should agree to within ±1% and the specimen radius, r_A, must be such that r_B/r_A is between 2.0 and 3.5. Each flat surface of the specimen and reference must be flat with a surface finish equal to or better than 32√ with the normal to each end parallel with the specimen axis to within ±10 minutes.

Where possible, the specimen length should be selected based on considerations of radius and thermal conductivity. For example, when λ_M is higher than the thermal conductivity of SRM 735 (stainless steel), long specimens with length/$r_A \gg 1$ can be used. These long specimens permit the use of large distances between temperature sensors and this reduces the percent error from uncertainty in sensor position. When the value of λ_M is lower than the thermal conductivity of SRM 735, the length must be reduced because uncertainty from heat shunting becomes too large.

4. PROCEDURE

Where possible and practical, reference specimens are selected such that the thermal conductance is of the same order as that expected for the test specimen. After instrumenting and installing the references, the specimen is instrumented similarly and inserted into the test stack such that it is aligned between the meter bars with at least 99% of each specimen surface in contact with the adjacent meter bar.

Soft foil or other contacting medium, if any, is used to reduce interfacial resistance. A large variety of contact media have been used. The choice depends upon the temperature range to be covered and the materials under evaluation. Thin (<0.001 mm) foils of indium, lead, copper, platinum, and platinum alloys and graphite and graphoil have been used particularly for higher temperatures of operation and numerous greases, especially metal and graphite particle filled used for moderate and lower temperatures.

If the system must be protected from oxidation during the test or if operation requires a particular gas or gas pressure to control the value of λ_I, the system should be pumped and purged, and the operating gas and pressure established. A predetermined force required for reducing the effects of nonuniform interfacial resistance is applied to the load column.

Heaters either attached to the ends of the meter bars or attached to a structure adjacent to the meter bars are energized with AC or DC power. The power to these heaters should be steady enough to maintain short-term T fluctuations less than 0.1 K on the meter-bar thermocouple nearest the heater. The two heaters, in conjunction with the guard shell heater and the system coolant, should maintain long-term temperature drift less than ±0.1 K h^{-1}. After being energized, the power to the various heaters is adjusted either manually or automatically until the T differences between positions Z_1 and Z_2, Z_3 and Z_4, and Z_5 and Z_6 are between 200 times the imprecision of the ΔT measurements and approximately 30 K, with the specimen at the average temperature desired for the measurement. Although the exact temperature profile along the guard is not important for $r_B/r_A \geqslant 3$, the power to the guard heaters should be adjusted until the temperature profile along the guard, $T_g(z)$, is constant with respect to time to within ±0.1 K and either:

1. Approximately linear so that $T_g(z)$ coincides with the temperature along the sample column at a minimum of three places, including the temperature at the top sensor on the top meter bar, the bottom sensor on the bottom bar, and the specimen midplane.
   or
2. Constant with respect to z to within ±5 K and matched to the average temperature of the test specimen.

When the system has reached steady state (ΔT drift in each section of the stack $<0.1\ \mathrm{K\,h^{-1}}$), and the guard gradient is matched to the test stack, the output of all temperature sensors is measured and the temperatures obtained from the appropriate conversion tables.

5. CALCULATION

5.1. Approximate λ_s Value

Following the conversion of voltages to temperature, the apparent heat flow per unit area (q') in the meter bars is calculated using

$$q'_T = \lambda_M \frac{T_2 - T_1}{Z_2 - Z_1} \qquad\qquad \text{top bar}$$

$$q'_B = \lambda_M \frac{T_6 - T_5}{Z_6 - Z_5} \qquad\qquad \text{bottom bar}$$

In each equation, the λ_M value is obtained from the information for the average meter-bar value of T. This calculation procedure actually requires only two T sensors on each column section. In this case, the third sensor on each section serves as a test for consistency of the other two. Some calculation procedures require more than the two sensors to obtain more knowledge about dT/dZ.

Measurements on materials where the ratios λ_m/λ_I and λ_s/λ_I do not fall within these boundaries should be accompanied by corrections for extraneous heat flow. These corrections may be determined in three different ways:

1. Use of analytical techniques such as those described by Didion[21] and Flynn.[25]
2. Using calculations from finite-difference or finite-element heat conduction codes such as those of Sweet and colleagues.[26,27]
3. Determined experimentally under identical conditions by using several reference materials or transfer standards of different thermal resistance as system calibration specimens. The procedure must be used cautiously, since all such specimens should have the same size as the specimen with an unknown thermal conductivity and have the same surface finish.

6. CALIBRATION AND VERIFICATION

There are many situations which call for equipment tests before operations on unknown materials can be successfully accomplished. These include:

1. After initial equipment construction.
2. When the ratio of λ_M to λ_S is less than 0.3 or greater than 3 and it is not possible to match thermal conductance values.
3. When the specimen shape is complex or the specimen is inordinately small.
4. When changes have been made in the system geometry.
5. When meter-bar or insulation material other than those listed are considered for use.
6. When the apparatus has reached a high enough temperature to change the properties of a component, such as thermocouple sensitivity.

A recommended procedure in such cases[14] is to run tests by comparing at least two reference materials in the following manner:

1. A reference material which has the closest λ_m value to the estimated specimen of a λ_s value should be machined according to recommended guidelines.
2. The λ value of this specimen should then be measured as described using meter bars fabricated from another reference material which has the closest λ value to that of the specimen. For example, technique tests might be made by measuring on Pyroceram 9606 using SRM 735 stainless-steel reference. If the measured thermal conductivity of the specimen disagrees with the value from Table 1 after all corrections for heat exchange are applied, additional effort is required to discover the error source(s).

7. PRECISION AND BIAS

7.1. Determinate Errors

The determinate error of equation (1) is

$$
\left|\frac{\delta\lambda'_S}{\lambda_S}\right| = \left|\frac{\delta\lambda_M}{\lambda_M}\right| + \left|\frac{\delta(Z_4 - Z_3)}{Z_4 - Z_3}\right| + \left|\frac{\delta(T_4 - T_3)}{T_4 - T_3}\right|
$$

$$
+ \frac{\delta(T_2 - T_1)}{(Z_2 - Z_1)C} + \frac{\delta(T_2 - T_1)\delta(Z_2 - Z_1)}{(Z_2 - Z_1)^2 C}
$$

$$
+ \frac{\delta(T_6 - T_5)}{(Z_6 - Z_5)C} + \frac{\delta(T_6 - T_5)\delta(Z_6 - Z_5)}{(Z_6 - Z_5)^2 C}
$$

where

$$
C = \frac{T_2 - T_1}{Z_2 - Z_1} + \frac{T_6 - T_5}{Z_6 - Z_5}
$$

Usually the heat flows and temperature-sensor spacings are essentially the same in the meter bars, so that C approaches a numerical value of 2.

7.2. Example of Determinate Error

We assume a system where both meter bars and the specimen are of equal length, and the sensor spacings are all 13 mm and $\lambda_M \approx \lambda_S$:

$$
\left|\frac{\delta\lambda_M}{\lambda_M}\right| = 0.003
$$

$$
Z_2 - Z_1 \sim Z_4 - Z_3 \sim Z_6 - Z_5 = 13 \text{ mm}
$$

$$
T_2 - T_1 \sim T_4 - T_3 \sim T_6 - T_5 = 10 \text{ K}
$$

$$
\delta(Z_2 - Z_1) \sim \delta(Z_4 - Z_3) \sim \delta(Z_6 - Z_5) = 0.2 \text{ mm}
$$

$$
\delta(T_2 - T_1) \sim \delta(T_4 - T_3) \sim \delta(T_6 - T_5) = 0.04 \text{ K}
$$

The maximum value of $\delta(Z_2 - Z_1)$ etc. was approximated by assuming an uncertainty of $\pm 1/2$ (sensor diameter) at each temperature measurement position. Therefore, if the diameter of each sensor is 0.2 mm, the uncertainty in the difference would be ± 0.2 mm. The number for $\delta(T_2 - T_1)$ etc. was calculated based on the sensor absolute accuracy.

With these values the fractional uncertainty in λ_S', namely $\delta\lambda_S'/\lambda_S'$, will be 0.069, or $\pm 6.9\%$.

7.3. Indeterminate Errors

There are at least three other errors that can contribute to the total system error; these are (1) nonuniform interfacial resistance, (2) heat exchange between the column and guard, and (3) heat shunting through the insulation around the column. These three errors must be minimized or appropriate corrections applied to the data if the desired accuracy is to be obtained.

The contributions from the last two errors can be determined approximately using results from appropriate experiments carried out at different levels of guard temperature to specimen stack temperature out of balance.

7.4. Overall

A recent international, interlaboratory round-robin study also involving absolute methods[28] has shown that a precision of $\pm 6.8\%$ can be attained over the temperature range 300–600 K. Although no definite bias could be established, there were indications that the values were an order of 2% lower than those obtained by absolute methods.

NOTATION

$\lambda_m(T)$	Thermal conductivity of meter bars (reference materials) as a function of temperature (W m^{-1} K^{-1})
λ_m^1	Thermal conductivity of top meter bar (W m^{-1} K^{-1})
λ_m^2	Thermal conductivity of bottom meter bar (W m^{-1} K^{-1})
$\lambda_S(T)$	Thermal conductivity of specimen corrected for heat exchange where necessary (W m^{-1} K^{-1})
$\lambda_S'(T)$	Thermal conductivity of specimen calculated by ignoring heat exchange correction (W m^{-1} K^{-1})
$\lambda_I(T)$	Thermal conductivity of insulation as a function of temperature (W m^{-1} K^{-1})
T	Absolute temperature in degrees Kelvin (K)
z	Position as measured from the upper end of the column (m)
l	Specimen length (m)
T_i	Temperature at $Z_i(\lambda)$
q'	Heat flow per unit area
$\delta\lambda, \delta T$, etc.	Uncertainty in λ, T, etc.
r_A	Specimen radius (m)
r_B	Guard cylinder inner radius (m)
$T_g(z)$	Guard temperature as a function of position, z, λ
F	Compressive force on column (Pa)

REFERENCES

1. ASTM C518 "Test Method for Steady-State Thermal Transmission Properties by Means of the Heat Flow Meter," ASTM Stand., Philadelphia, PA, Latest Revision (1990).
2. R.P. Tye, K.G. Coumou, A.O. Desjarlais, and D.M. Haines, "Historical Development of Large Heat Flow Meter Apparatus for Measurements of Thermal Resistance of Insulations," in: *Thermal Insulation: Materials and Systems*, ASTM STP 922 (F.J. Powell and S.M. Mathews, eds.), pp. 656–664, ASTM, Philadelphia, PA (1987).
3. K.G. Coumou and R.P. Tye, "A Laboratory Instrument for Rapid Determination of Thermal Conductivities in the Range 0.4–5 W/mK," *High Temp. High Pressures* 13, 695 (1981).
4. Y. Agari, M. Tanaka, and S. Nagai, "Thermal Conductivity of Carbon Fiber Filled Polymer Composites," in: *Thermophysical Properties 8*, Japanese Society of Thermophysical Properties, Tokyo (1987).
5. ASTM C177 "Test Method for Steady-State Thermal Transmission Properties by Means of the Guarded Hot Plate," ASTM Stand., Philadelphia, PA, Latest Revision (1990).
6. S. Klarsfeld, "Guarded Hot Plate Method for Thermal Conductivity Measurements" in: *Compendium of Thermophysical Property Measurement Methods*, 1 (K.D. Maglić, A. Cezairliyan, and V.E. Peletsky, eds.), Plenum Press, New York and London (1984).
7. M.J. Laubitz, "Axial Heat Flow Methods of Measuring Thermal Conductivity," in: *Compendium of Thermophysical Property Measurement Methods*, 1 (K.D. Maglić, A. Cezairliyan, and V.E. Peletsky, eds.), Plenum Press, New York and London (1984).
8. R.E. Taylor, "Thermophysical Property Determinations Using Direct Heating Methods," in: *Compendium of Thermophysical Property Measurement Methods*, 1 (K.D. Maglić, A. Cezairliyan, and V.E. Peletsky, eds.), Plenum Press, New York and London (1984).
9. R.P. Tye, "The Thermal and Electrical Conductivities of Porous Copper and Stainless Steel to Elevated Temperatures," Paper 3-HT-46, Proceedings of ASME–AICHE Heat Transfer Conference, Atlanta, GA (1973).
10. J.P. Moore, "Analysis of Apparatus with Radial Symmetry for Steady-State Measurements of Thermal Conductivity," in: *Compendium of Thermophysical Property Measurement Methods*, 1 (K.D. Maglić, A. Cezairliyan, and V.E. Peletsky, eds.), Plenum Press, New York and London (1984).
11. E. Huebner and W. Neumann, "Measurement of Thermal Conductivity by Means of a Comparative Longitudinal Method," *Bulletin of Austrian Society for Nuclear Studies*, SGAE, Ber. No. 2837 (November, 1977).
12. R.P. Tye, A.O. Desjarlais, and J.G. Bourne, "Thermophysical Property Measurements on Thermal Energy Storage Materials," in: *Proc. Seventh Symposium on Thermophysical Properties*, pp. 189–197, A.S.M.E., New York (1977).
13. R.W. Powell and R.P. Tye, "The Thermal and Electrical Conductivity of Liquid Mercury," *Int. Developments and Heat Transfer* 4, 856 (1961).
14. E. Fried, "Thermal Conduction Contribution to Heat Transfer at Contacts," in: *Thermal Conductivity, Vol. II* (R.P. Tye, ed.), Academic Press, London (1969).
15. S. Begej, J.E. Garnier, R.P. Tye, and A.O. Desjarlais, "Ex-Reactor Determination of Thermal Contact Conductance Between Uranium Dioxide and Zircaloy 4 Interfaces," in: *Thermal Conductivity*, 16 (D.C. Larsen, ed.), pp. 221–232, Plenum Press, New York (1983).
16. ASTM E 1225 "Test Method for Thermal Conductivity of Solid by Means of the Guarded Comparative Longitudinal Heat-Flow Techniques," ASTM Stand., Philadelphia, PA (1987).
17. R.P. Tye and A.O. Desjarlais, "The Thermophysical Properties of Coated Reinforced Carbon-Carbon Composites," in: *Thermal Conductivity*, 17 (J.G. Hust, ed.), pp. 711–725, Plenum Press, New York (1983).
18. R.P. Tye and E.J. Clougherty, "The Thermal and Electrical Conductivities of a Series of Electrically Conducting Compounds," in: *Proc. Fifth Symposium in Thermophysical Properties*, pp. 396–403, ASME, New York (1970).

19. R.W. Powell, "The Thermal and Electrical Conductivity of Armco Iron," *Proc. Phys. Soc.* **46**, 659 (1934).

20. R.W. Powell and R.P. Tye, "Thermal and Electrical Conductivities of Nickel–Chromium (Nimonic) Alloys," *The Engineer* **209**, 729 (1960).

21. D.A. Didion, "An Analysis and Design of a Linear Guarded Cut-Bar Apparatus for Thermal Conductivity Measurements," AD-665789 (January, 1968). Available from the National Technical Information Service, Springfield, VA.

22. R.P. Tye and J.R. Hurley, "The Measurement of Thermal Conductivity with Particular Reference to Temperature Measurement," *Proc. Fifth Temperature Measurement Society*, II C-1 (1967).

23. ASTM E 230 "Temperature Electromotive Force (EMF) Tables for Standardized Thermocouples," ASTM Stand., Philadelphia, PA, Latest Revision (1989).

24. R.L. Anderson and T.G. Kollie, "Problems in High Temperature Thermometry," July, pp. 171–221, 1976.

25. D.R. Flynn, "Thermal Conductivity of Ceramics," in: *Mechanical and Thermal Properties of Ceramics*, Special Publication 303, National Bureau of Standards (1969).

26. M. Moss, J.A. Koski, and G.M. Haseman, "Measurement of Thermal Conductivity by the Comparative Method," U.S. Govt. Report SAND82-0109 UC 1970 (March, 1982). Available from NTIS, U.S. Department of Commerce, Springfield, VA.

27. J.N. Sweet, M. Moss, and C.E. Sisson, "The Use of Numerical Heat Transfer Techniques to Analyze Thermal Comparator Conductivity Measurements," in: *Thermal Conductivity 18* (T. Ashworth and D.R. Smith, eds.), pp. 43–59, Plenum Press, New York (1985).

28. L.C. Hulstrom, R.P. Tye, and S.E. Smith, "Round-Robin Testing of Thermal Conductivity Reference Materials," in: *Thermal Conductivity 19* (D.W. Yarborough, ed.), pp. 199–211, Plenum Press, New York (1988). See also *High Temp. High Pressures* **17**, 707 (1985).

4

Reference Guarded Hot Plate Apparatus for the Determination of Steady-State Thermal Transmission Properties

FRANCESCO DE PONTE, CATHERINE LANGLAIS, and SORIN KLARSFELD

1. INTRODUCTION

The new international standard ISO 8302 is written for both designers and users of the guarded hot plate apparatus. With this aim, following a very detailed analysis, the standard gives:

- The requirements for design and construction of the apparatus as a function of desired performance.
- The method to experimentally evaluate the actual performance.

The purpose of this chapter is to give a practical example of how to use the recommendations of this standard to design a new apparatus, strictly following the given guidelines.

The procedure for achieving accurate measurements of the thermal conductivity of insulating materials involves:

- The design of a "reference" absolute apparatus, following unanimously accepted rules.
- The use of one or more national or international Standard Reference Materials (SRMs) to check the performances of this apparatus.

In Europe the Bureau Communautaire de Reference (BCR; Commission of the European Communities) supplies a high-density resin-bonded glass fiber board as SRM.[1] In the USA high-density and low-density glass-fiber

FRANCESCO DE PONTE ● Istituto di Fisica Tecnica, Università di Padova, 35131 Padova, Italy. CATHERINE LANGLAIS ● Isover Saint-Gobain, Centre de Recherches Industrielles, 60290 Rantigny, France. SORIN KLARSFELD ● Saint-Gobain Recherche, 93304 Aubervilliers, France.

boards are available from the National Bureau of Standards. This paper describes the design of a "reference" guarded hot plate apparatus. The purpose here is not to present an original design but to give an example of how to follow the procedure originating from the work of an International Working Group (7 countries) and described in the document ISO DIS 8302[2] in order to achieve a high-performance apparatus at a relatively moderate cost.

2. CHOICE OF TECHNICAL CHARACTERISTICS AND PERFORMANCE SPECIFICATIONS

The ISO document[2] does not specify any particular type of guarded hot plate apparatus: "Considerable latitude both in the temperature range and in the geometry of the apparatus is given to the designer of a new equipment since various forms have been found to give comparable results."

The designer is advised to start by taking preliminary decisions on the choice of several design parameters which are listed in Table 1. The parameters of Table 1 define just the minimum and maximum expected testing conditions. From these, overall apparatus sizes can be derived. The ISO document suggests starting the design with a hot plate side four times the maximum expected specimen thickness and guard external side eight times the maximum expected specimen thickness. In the present case this results in a hot plate side of 0.6 m and guard side of 1.2 m. The nearest specimen size suggested in the ISO document is 1.0 m. In Europe and elsewhere a large number of 0.5 m × 0.5 m and 0.6 m × 0.6 m guarded hot plate apparatus exist. In addition many 0.61 m × 0.61 m heat flowmeters are used for quality control in laboratories. The ideal apparatus and plants should be compatible with both sets of apparatus. In addition the use of a secondary guard is strongly supported. From these considerations emerged the decision of trying the design of a double-guarded hot plate with a main guard 0.5 m wide and secondary guard 0.61 m wide.

The ISO document also states limits for flatness (0.025%, i.e., 6×10^{-5} m over 0.25 m), hot plate temperature stability (0.3% of the temperature difference $T_1 - T_2$ through the specimen), cold plate stability (2% of $T_1 - T_2$), plate temperature uniformity (2% of $T_1 - T_2$), imbalance and edge heat loss error (0.5%), and accuracy in thickness transducers (0.5%). By referring to the two worst-case testing conditions given in Table 1, the sizes and performance requirements of the current hot plate are summarized in Table 2.

The next design step is the evaluation of both the imbalance and edge heat loss errors and the plate temperature uniformity. These require a detailed definition of such elements as metal plate thickness, mechanical connections between central section and guard ring of the hot plate, layout of the balancing system, and so on. In the actual design process this happens step by step, discarding some solutions due to the error analysis. Throughout this chapter,

TABLE 1

Design Parameters

Parameters		Present choice	Units
Minimum specimen thickness	d_m	0.020	m
Maximum specimen thickness	d_M	0.150	m
Minimum specimen resistance	R_m	0.1	m^2 K W^{-1}
Maximum specimen resistance	R_M	5	m^2 K W^{-1}
Minimum temperature difference across the specimen	$T_1 - T_2$	5	K
Maximum temperature difference across the specimen	$T_1 - T_2$	50	K
Sensitivity of the balancing system for the guard ring section	V	± 5	μV
Minimum cooling unit temperature	T_{cm}	210^a	K
		240^a	K
		260^a	K
Maximum heating unit temperature	T_{HM}	345	K
Noise and drifts		0.5%	
Reproducibility on the same specimen removed and mounted again		0.5%	
Overall apparatus accuracy as maximum acceptable error in measured property in defined worst-case conditions (mean test temperature $T = 297$ K and $d = 0.150$ m; $R_M = 5$ m^2 K W^{-1}; $T_1 - T_2 = 20$ K or $d = 0.02$ m; $R_m = 0.1$ m^2 K W^{-1}; $T_1 - T_2 = 5$ K)		$\pm 1.5\%$	
Surrounding environment temperature	$T_A = (T_c + T_H)/2$	255 to 375	K

^{a}Depending on chilling unit power.

for the sake of an easier understanding of the text, the resulting apparatus will be described first and then the error analysis supporting the choices will follow.

3. APPARATUS DESCRIPTION

The present guarded hot plate is a conventional "two specimen apparatus." The assembly is placed in a temperature and humidity controlled environment, with surrounding air flows along a "cold" and "hot" source in a closed test chamber, as shown schematically in Fig. 1.

The hot plate is made of photoetched heaters sandwiched between 10-mm aluminum plates. It is positioned horizontally by a system of crossbars fixed at the bottom of the test chamber and lies on four points on teflon supports

TABLE 2
Apparatus Sizes and Performance Requirements

Apparatus Sizes	
Metering area	$0.25 \times 0.25 \text{ m}^2$
First guard ring	$0.50 \times 0.50 \text{ m}^2$
(external-internal)	$0.25 \times 0.25 \text{ m}^2$
Second guard ring	$0.61 \times 0.61 \text{ m}^2$
(external-internal)	$0.50 \times 0.50 \text{ m}^2$
Cold plate	$0.61 \times 0.61 \text{ m}^2$

Performance Requirements	
Flatness	0.025%
Parallelism	10^{-4} m
Hot plate temperature stability	$\pm 0.015 \text{ K}$
Cold plate temperature stability	$\pm 0.1 \text{ K}$
Temperature uniformity	$\pm 0.1 \text{ K}$
Sum of imbalance and edge losses	0.5%
Accuracy of thickness measurement	$\pm 10^{-4} \text{ m}$

of small cross section (10^{-4} m^2). The two cold plates are mobile with a symmetrical displacement with respect to the hot plate, carried out by means of a mechanical device operated from the outside of the test chamber with a hand wheel. A rectilinear potentiometer allows for the electrical measurement of the displacement. The metering area is surrounded by two guards with only eight epoxy mechanical connections embedded within the thickness of the two repartition plates; the reason for this choice will be explained later. The apparatus and hot plate structures are described on Figs. 2-7. The temperature measurements are achieved with type K thermocouples, positioned as indicated on Fig. 3. Platinum resistance thermometers are used as temperature-control sensors. The imbalance between the metering area and the first guard, as well as between the first and the second guard, is controlled by thermopiles with 100 and 200 thermocouple junctions, respectively.

The principle of the electrical circuit is given on Fig. 8. All thermocouple outputs and voltages are connected to a data logger (Hewlett Packard, model HP 34977 A) driven by a supervisory computer (model HP 9915). The resolution of the built-in digital voltmeter is 1 μV, as most of today's digital voltmeters. This datum has to be compared with the ISO document that requires for the instrumentation a resolution and accuracy better than $0.002(T_1 - T_2)$. To meet these requirements, the data-logger measurement routines must take care of instrumentation drifts and must use averaging techniques. In such a way a temperature difference $T_1 - T_2 = 20 \text{ K}$ can easily be achieved, while a difference of 5 K is possible but rather critical. If desired, the plate temperature

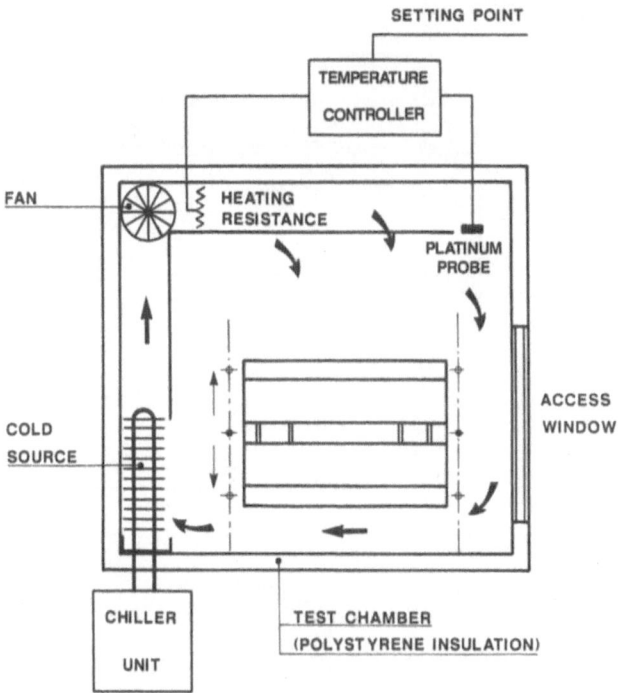

FIGURE 1. Principle of the temperature control in the test chamber.

controllers (Eurotherm, model 820) can be monitored by the computer with a continuously-monitoring program establishing equilibrium criteria and recording all the data on magnetic tape cassettes.

The minimum cold plate temperature is limited only by the present power of the chilling units and can be lowered to 210 K in the future. The maximum hot plate temperature is limited to 340 K due to the maximum service temperature for the epoxy used in the mechanical junctions between the guard and the metering area. Other mechanical junctions withstanding higher temperatures could have been used, but a substantial limit only a few tens of Kelvin higher comes from the use of photoetched heaters. If this type of heater is abandoned, the achievement of required hot plate flatnesss becomes much more critical.

4. APPARATUS DESIGN

Once the central section and guard ring dimensions are defined, the edge heat loss error can be evaluated. The worst-case condition is for maximum

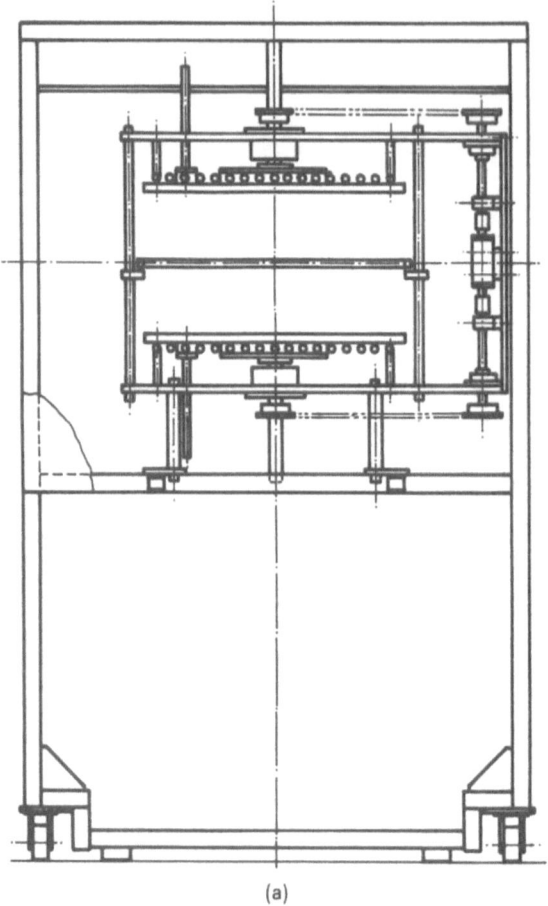

(a)

FIGURE 2. Apparatus general view.

specimen thickness and maximum specimen thermal resistance. For the present apparatus the specimen thickness is $d = 0.15$ m and the specimen side is the external side 0.61 m of the secondary guard. If the specimen edge is kept at a uniform temperature matching the mean test temperature to less than 5% of the temperature difference $T_1 - T_2$ (match to 1 K for $T_1 - T_2 = 20$ K), an edge heat loss error should be smaller than 0.4%. For the case of temperature matching to 2.5% (or 0.5 K in the above hypothesis) the corresponding error should be reduced to 0.2%. A performance check had shown that this limit can be achieved by the air temperature controller for the cabinet enclosing the apparatus.

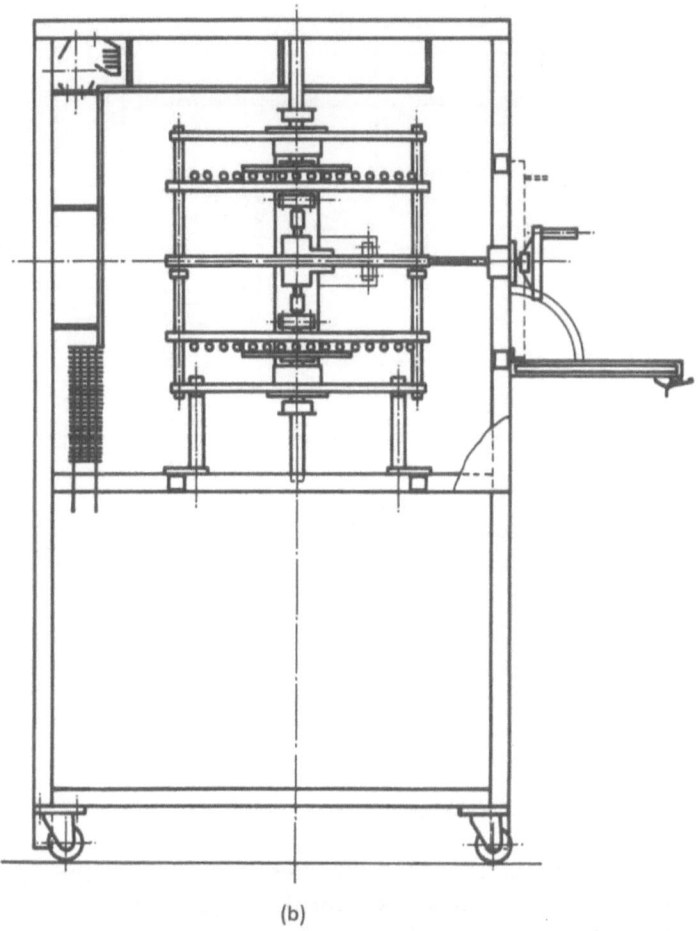

(b)

FIGURE 2. (*Continued*).

The gap width can then be chosen to evaluate imbalance error and to compare it with the edge heat loss error. The gap width should be as large as possible, provided the error due to the uncertainty in the definition of the metering area is acceptable. The ISO document fixes the maximum for the area of the gap at 5% of the metering area. Hence, if an uncertainty not greater than 5% of the gap width is assumed in the definition of the edge of the metering area, the resultant error in the definition of the metering area should not exceed 0.25%. This corresponds to the 3.1×10^{-3} m gap width retained in the apparatus. The gap width affects the minimum specimen thickness, since this should be some 5 to 20 times larger than the gap width (the ISO document

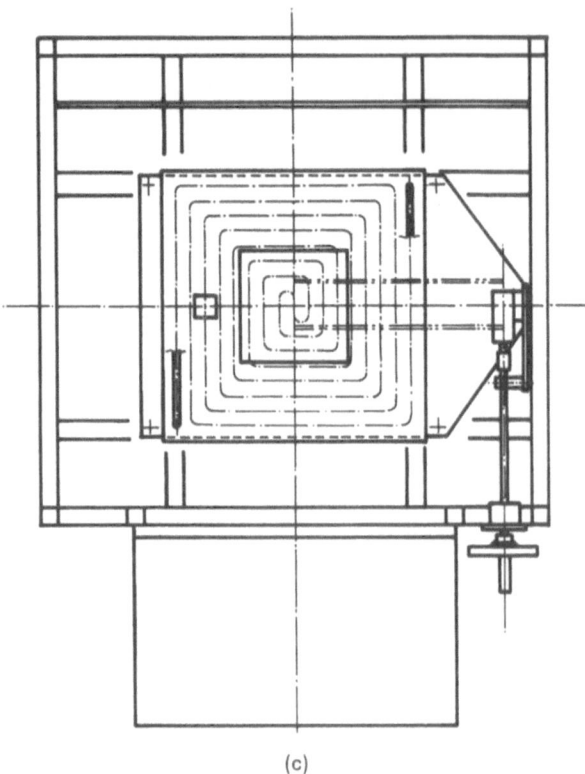

(c)

FIGURE 2. (*Continued*).

suggests 10, i.e., a minimum specimen thickness of 0.03 m for this apparatus; however, there is no theoretical support for this suggestion).

As the imbalance error depends not only on the gap size but also on the gap design, the balancing system design, and the sensitivity of the instrumentation used to detect and control imbalance, a choice for the imbalance-detection system has to be made. This is a most critical point in the development of the apparatus. Error analysis showed first that a mechanical connection between the central section and the guard ring consisting of metal bridges or even a plastic sheet supporting the electric heaters had a thermal resistance which was too low. Consequently the level of acceptable imbalance would have been extremely low. These considerations resulted in the choice of the araldite bridges and in selecting the number of junctions for the balancing thermopile to meet the 5 μV imbalance limit stated in Table 1. This limit, which applies to the lowest density of heat flow rate, is rather conservative from an electrical viewpoint, so that imbalance error could be reduced further if better instrumentation were used. However, the main limit for imbalance detection

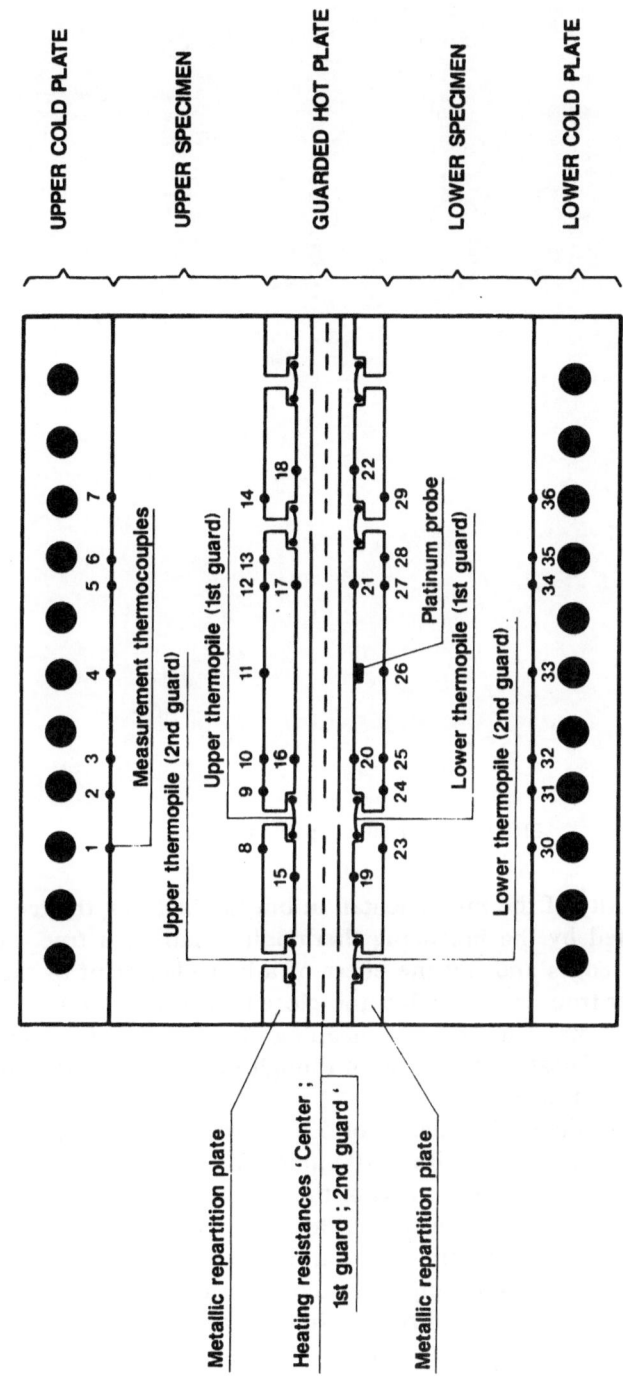

FIGURE 3. Guarded hot plate apparatus structure and sensor positions.

FIGURE 4. Metallic repartition plate.

is not the sensitivity of the instrumentation but the fact that the temperature difference detected by the balancing thermopile shall be a true imbalance through the gap edges and not the effect of a bad placement of thermopile junctions (too far from the gap edge, too close to a heater, etc.). In turn, the most severe problems in detecting a meaningful imbalance are found at the highest densities of heat flow rate, even though in this situation a somewhat lower sensitivity is needed.

The stated maximum imbalance of 5 μV corresponds to approximately 2.5 mK at room temperature with a 100-junction thermopile. A check must then be made when testing specimens of the minimum expected resistance to see how this level is meaningful at the highest densities of heat flow rate. Because of the heater power limit, it is testing practice to run such tests at low temperature differences; for example, 5 K corresponds, in the present case, to 50 W m^{-2}. This density of heat flow rate implies a temperature difference of 3 mK through the 10-mm aluminum plate of the hot plate. If balancing sensors

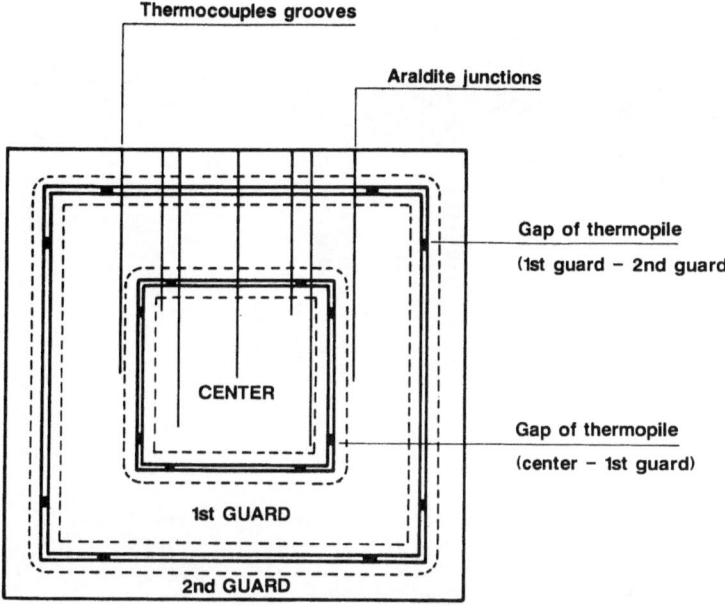

FIGURE 5. Hot plate top view.

are placed in grooves as shown in Fig. 9—top, the maximum uncertainty in the temperature they detect is well below the 3 mK temperature difference through the aluminum plate.

This solution, even though very satisfactory from a thermal viewpoint, can create problems while assembling the apparatus. An alternative solution, that of embedding balancing junctions in sheets placed either between the metal plate and the heater or between the metal plate and the specimen (see Fig. 9—middle), introduces severe errors because the position of the junction has small displacements within the insulating sheet and because an additional nonuniform resistance is added due to the contact between the sheet and the metal plates. For example, at room temperature, a 10^{-4} m air gap between a sheet not glued to the metal plate and the metal plate itself adds a temperature error of 15 mK even with the highest-resistance specimen under a temperature difference of 20 K. The compromise of the present design between ease of assembling and low resistances is shown in Fig. 9—bottom. Even under the highest densities of heat flow rate (that create problems in detecting a meaningful imbalance but do not require the highest sensitivity) the uncertainty in imbalance has been limited to 4 mK.

If we refer to the balancing system just explained and call Φ the heat flow rate through the metering area of the two specimens, a theoretical

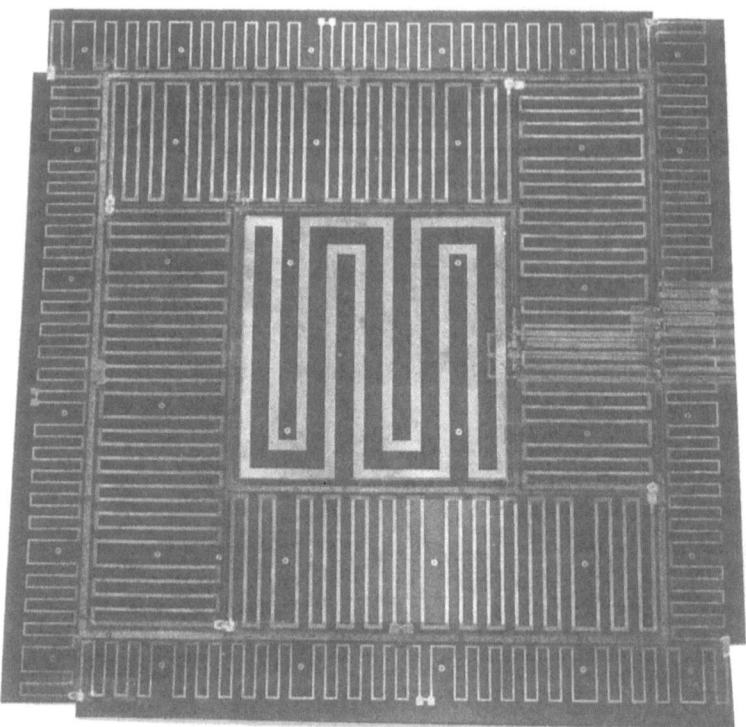

FIGURE 6. Photoetched heaters.

evaluation of imbalance errors, E_g, defined by

$$E_g = \Phi_g/\Phi \tag{1}$$

is possible by starting from the equations derived by Woodside[3]:

$$\Phi_g = (\Phi_0 + \lambda c)\Delta T_g \tag{2}$$

$$c = \frac{16}{\pi} l \ln \frac{4}{1 - e^{-(2\pi g/d)}} \tag{3}$$

and by computing the parameter Φ_0 as the following sum:

$$\Phi_0 = \Phi_a + \Phi_r + \Phi_p + \Phi_w \tag{4}$$

In these equations the symbols have the following meaning:

c an apparatus parameter that can be derived either experimentally or that can be theoretically predicted by expression (3),

d specimen thickness,

FIGURE 7. View of the test chamber.

$2g$ the gap width,

l half side of the metering section,

Φ_a heat flow rate by conduction through the air gap for a 1 K imbalance,

λc heat flow rate through the two specimens of thermal conductivity, λ, for a 1 K imbalance,

Φ_w heat flow rate through power leads and thermocouple wires for a 1 K imbalance,

Φ_g heat flow rate between the metering and guard sections of the hot plate for a temperature imbalance ΔT_g,

Φ_0 heat flow rate across the gap for a 1 K imbalance,

Φ_p heat flow rate by conduction through the junctions between the metering and guard sections for a 1 K imbalance,

Φ_r heat flow rate by radiation through the gap for a 1 K imbalance.

For the present plate at 307 K, assuming an emittance of 0.9 for the gap surfaces and with the following numerical values: $d = 0.150$ m, $2g = 0.003$ m,

$h = 2 \times 0.01$ (gap height), $l = 0.125$ m, and $\lambda = 0.030$ W m^{-1} K^{-1}, we obtain

$$\Phi_a = 0.173 \text{ W K}^{-1}, \qquad \Phi_w = 0.326 \text{ W K}^{-1}, \qquad \Phi_0 = 0.631 \text{ W K}^{-1}$$

$$\Phi_p = 0.025 \text{ W K}^{-1}, \qquad \Phi_r = 0.107 \text{ W K}^{-1}, \qquad c = 2.66 \text{ m}$$

and finally $\Phi_g = 0.711 \Delta T_g$.

To achieve $\Phi_g \leqslant 0.003 \ \Phi$ for $R = 5$ m^2 K W^{-1} and $T_1 - T_2 = 20$ K, $\Delta T_G \leqslant \pm 2.1$ mK must be ensured. This means that the emf of the thermopile (50 thermocouples; 40 μV K^{-1}) must be controlled to 0 within $\pm 4.3 \ \mu$V, close to 5 μV, as stated in Table 1. The requirement of keeping the sum of imbalance

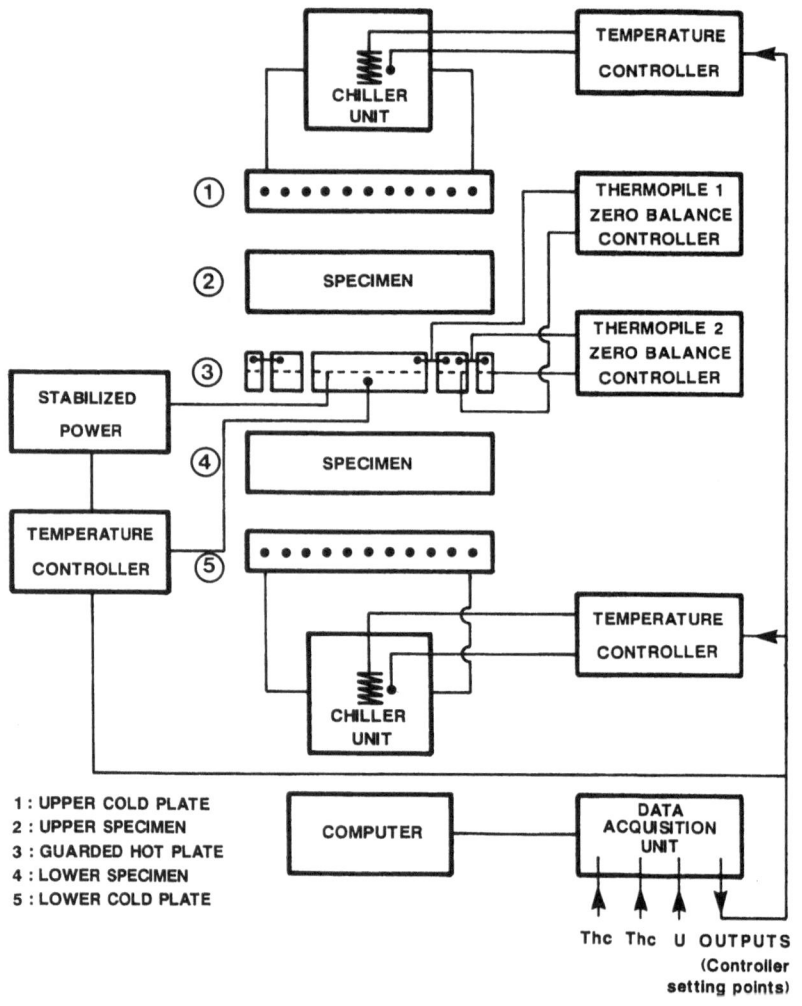

1 : UPPER COLD PLATE
2 : UPPER SPECIMEN
3 : GUARDED HOT PLATE
4 : LOWER SPECIMEN
5 : LOWER COLD PLATE

Thc Thc U OUTPUTS
(Controller
setting points)

FIGURE 8. Principle of the electric circuit.

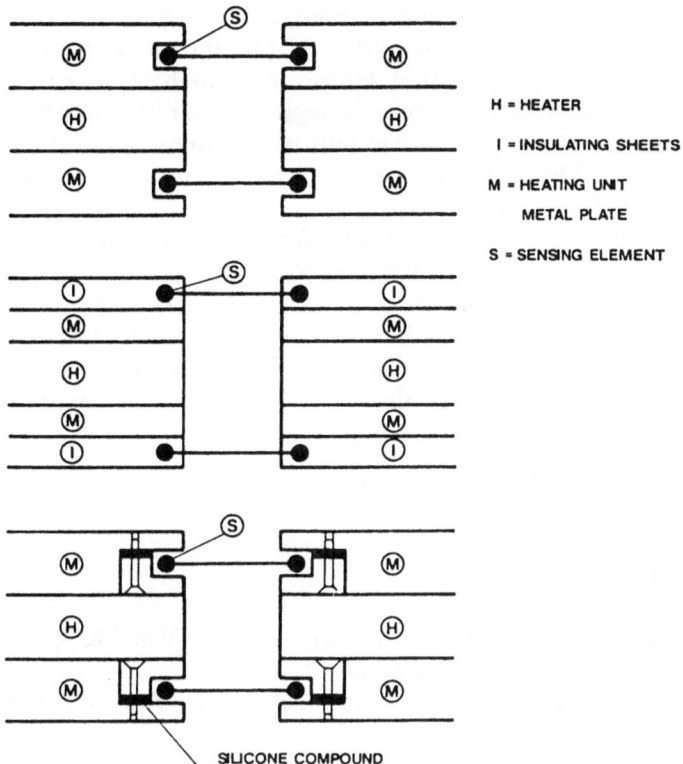

FIGURE 9. Sensing elements to detect imbalance: (top) junctions in a groove within the metal plate; (middle) junctions embedded in insulating sheets; (bottom) junction groove created in a screw-fastened metal bar.

and edge heat loss error within 0.5% is thus fulfilled. It is also pointed out that the introduction of a secondary guard together with tight control of the air temperature within the cabinet allows a significant size reduction with reference to the recommended dimensions to start a guarded hot plate design.

The next check to be performed is the uncertainty in the detection of surface temperatures in the apparatus when the sensing elements are placed into grooves in the metal plates. An approximate estimate is to evaluate the temperature drop in a layer of the cement, used to embed the thermocouples in grooves, a few tenths of a millimetre thick. This is found to be some hundredths of K at the highest heat flow rates, which correspond to an error in the detection of the apparatus-surface temperatures usually below 0.5% of $T_1 - T_2$.

A further check to be performed is of the temperature uniformity of the metal hot plates. This must take into account the fact that lateral losses both through the specimen and through the edge of the hot plate create a heat conduction parallel to the metal plates. At present no accurate and simple procedure exists for this evaluation. However, a rough approach is to assume that the total lateral losses will flow through the guard plate from the center to its edge for a length equal to one half of the guard width, i.e., 0.0625 m. This evaluation yields a temperature uniformity up to 1% of the temperature difference through the specimen.

The final check is the temperature uniformity of the cold plates. A double spiral configuration was chosen for the cold plates (see Fig. 2c). Here again a step-by-step process is necessary to find a satisfactory compromise between the chilling-fluid mass-flow rate, the thickness of the metal plates, the pipe-to-pipe distance, and the thermal resistance between the pipe and metal plate. Assuming water as chilling liquid, the final cold plate parameters are as follows:

Water mass flow rate, $w = 0.13$ kg s^{-1},
Water specific heat, $c = 4180$ J kg^{-1} K^{-1},
Pipe diameter 0.012 m,
Pipe length, $21_t = 10.4$ m,
Pipe pitch, $d_t = 0.036$ m,
Cold plate side, $l_p = 0.61$ m,
Metal plate conductivity (aluminum), $\lambda_p = 165$ W m^{-1} K^{-1},
Metal plate thickness, $h_p = 0.02$ m,
Resistance between 1 m of pipe and metal plate, $R'_m = 0.016$ m K W^{-1}.

From the aforementioned data the following dimensionless parameters can be derived:

$$H_1 = l_t/d_t = 145 \tag{5}$$

$$H_2 = \lambda_p h_p / cw = 5.9 \times 10^{-3} \tag{6}$$

$$H_3 = R'_m \lambda_p h_p \frac{\sqrt{2}}{l_p} = 0.12 \tag{7}$$

According to the theory of De Ponte and Di Filippo[4] this should result in a temperature nonuniformity smaller than 0.25 of the temperature difference of the chilling water between inlet and outlet on the cold plate. At the highest load (200 W, due to a large amount of heat transfer on the surfaces of the cold plates not in contact with the specimen) the above difference should be 0.36 K. In consequence the cold plate uniformity should be within 0.09 K. Insulation of the side of the cold plates not in contact with the specimen will reduce this value. This computation is provided as a guideline on the order of magnitude of the temperature uniformity, since the model used corresponds only partly to the actual plate layout.

With all of the aforementioned considerations combined with the instrumentation specifications an overall error estimate is as follows:

- Electrical power supplied to the metering section: 0.1%.
- Temperature difference through the specimens (due to thermocouple mounting): 0.25% to 1%.
- Temperature difference through the specimens (due to thermocouple calibration): 0.5%.
- Specimen thickness: 0.1% to 0.5%.
- Definition of the metering area: 0.25%.
- Sum of imbalance and edge heat loss error: 0.1% to 0.5%.

The sum of the maximum values of each systematic error is then close to 3%, but systematic errors do not reach their maximum value all together in the same test and are not all of the same sign. A 0.4% reproducibility when mounting and dismounting the same specimen shall also be added to systematic errors, while the effect of noise and drifts can be almost eliminated by averaging enough subsequent measurements. Then it is possible to foresee an overall accuracy between 1% and 2%, dependent upon the specimen type. Subsequent measurements on an SRM fell well within its 1.5% confidence level, indicating that this error estimate was conservative.

5. MEASUREMENT RESULTS AND PERFORMANCE CHECK

The current apparatus characteristics are summarized in Table 3. All the measurements reported hereafter have been performed on the resin-bonded glass-fiber board reference material from the BCR, with the following characteristics: $d = 0.035$ m; $\rho = 88$ kg m^{-3}; most frequent fiber diameter = 4.1×10^{-6} m; resin content (mass fraction) = 16%.

5.1. Influence of the Temperature Difference

Measurements were conducted initially with 3 different temperature-difference levels between the faces of the specimens ($10 \leqslant T_1 - T_2 \leqslant 30$ K), at the same mean temperature ($T = 297$ K). A plot of the electrical power, P, dissipated in the metering section versus the temperature difference, $T_1 - T_2$, should be a straight line passing through the origin, since a positive or negative intercept indicates the presence of heat losses or gains. The present results are shown in Fig. 10.

5.2. Influence of Air Temperature in the Test Chamber

A series of tests was performed to evaluate the influence of the ambient air temperature, T_a (surrounding the plate assembly), on the apparatus (temperature uniformity, thermocouple calibration, imbalance detection, etc.).

TABLE 3
Measured Technical Characteristics

Plates	
Flatness	$\pm 10^{-4}$ m
Parallelism	$\pm 10^{-4}$ m

Temperature Stabilities	
Hot plate	± 0.015 K
Cold plate	± 0.05 K
Air temperature in the test chamber	± 0.4 K

Power Stability	
Heating resistance voltage $(\Delta U/U)$	$\pm$ 0.4%

Temperature Uniformity	
Upper hot plate	± 0.025 K
Upper cold plate	± 0.1 K
Lower hot plate	± 0.035 K
Lower cold plate	± 0.05 K

Thermopile Imbalance Control	
Thermophile output V	$0 \leqslant V \leqslant 5\,\mu$V

Range of Operating Temperatures	
Maximum hot plate temperature	345 K
Minimum cold plate temperature	260 K

Ambient temperature was varied between 289.5 and 304 K, for a mean temperature of 297 K, all other measurement conditions being constant. Due to the low specimen thickness ($d = 0.035$ m), edge heat losses through the specimen should be negligible. Results plotted in Fig. 11 indicate the very small influence of the ambient temperature:

$$\Delta\lambda = 0.33 \times 10^{-3} \text{ W m}^{-1}\text{ K}^{-1} \text{ for } \Delta T_{a} = 14.5 \text{ K}$$

These results show that the apparatus is not temperature-sensitive and that measured ambient-temperature variations (± 0.4 K, see Table 3) have a very small influence ($\Delta\lambda/\lambda \leqslant \pm 0.05\%$) on the value of the measured thermal conductivity.

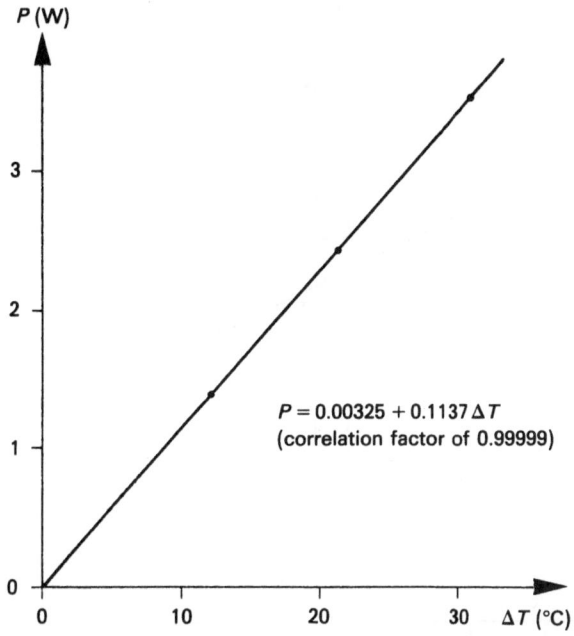

FIGURE 10. Dissipated power versus temperature difference (mean test temperature of 297 K).

5.3. Influence of Guard Imbalance

Several measurements were made for variable thermopile voltages ($V \neq 0$), acting on the set-point of the zero-balance controller, in order to determine the influence of the guard imbalance (thermopile No. 1 on Fig. 8) experimentally. A plot of the measured thermal conductivity, λ, in SI units versus the thermopile voltage, V, in μV (Fig. 12) gives the following relationship:

$$\lambda = 0.03176 - 3.947 \times 10^{-6} V \tag{8}$$

Taking into account the thermopile sensitivity, it is possible to derive the imbalance heat flow, Φ_g, as a function of temperature imbalance ΔT_g:

$$\Phi_g = 0.620 \Delta T_g \text{ (W K}^{-1}) \tag{9}$$

The experimental value of 0.620 is even lower than the theoretical value 0.688 that would pertain to the BCR reference specimen 0.035-m thick. It should be pointed out that the theoretical computations are pessimistic and that the specified sensitivity of the balancing system (± 5 μV; see Table 1) allows the condition $\Phi_g \leq 0.3\%$ of the main heat flow in any testing condition to be fulfilled.

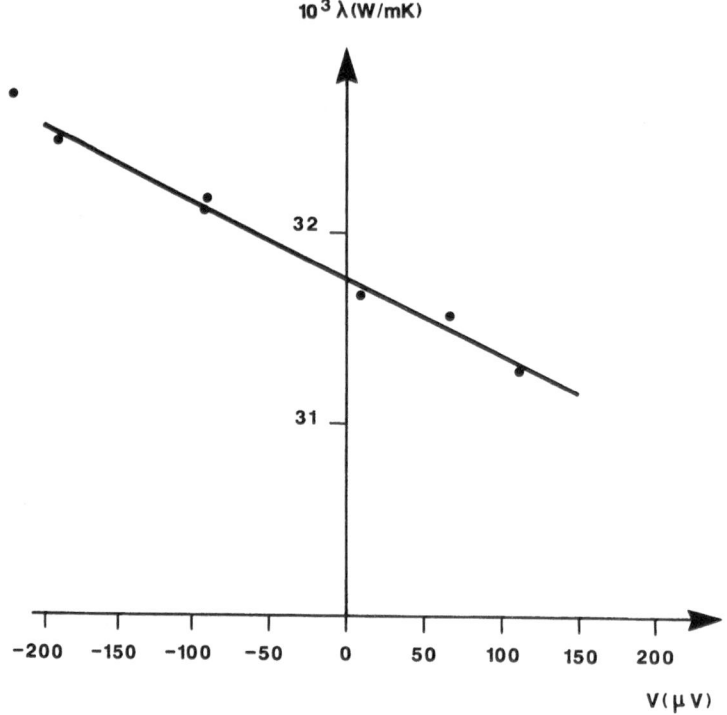

FIGURE 11. Influence on measured thermal conductivity of air temperature variations in the test chamber.

5.4. *Influence of Edge Heat Losses*

If $\Delta T_g = 0$ is assumed, the edge heat loss error, E_e, is defined as equal to the ratio of the lateral loss heat flow, Φ_e (transferred from the metering section to the guard section of the specimen, or *vice versa*), to the unidirectional heat flow, Φ, such as it would be for no lateral losses:

$$E_e = \frac{\Phi_e}{\Phi} = \frac{\lambda_m - \lambda}{\lambda} \tag{10}$$

where λ is the actual thermal conductivity of the specimen and λ_m its measured thermal conductivity in the above conditions.

According to Woodside,[6] the error E_e is expressed by the relation

$$E_e = \left\{ \frac{d}{\pi l} \left[e \ln \frac{\cosh\{\pi[(b+l)/d]\} + 1}{\cosh\{\pi(b/d)\} + 1} \right.\right.$$
$$\left.\left. + (1-e) \ln \frac{\cosh\{\pi[(b+l)/d]\} - 1}{\cosh\{\pi(b/d)\} - 1} \right] \right\}^2 - 1 \tag{11}$$

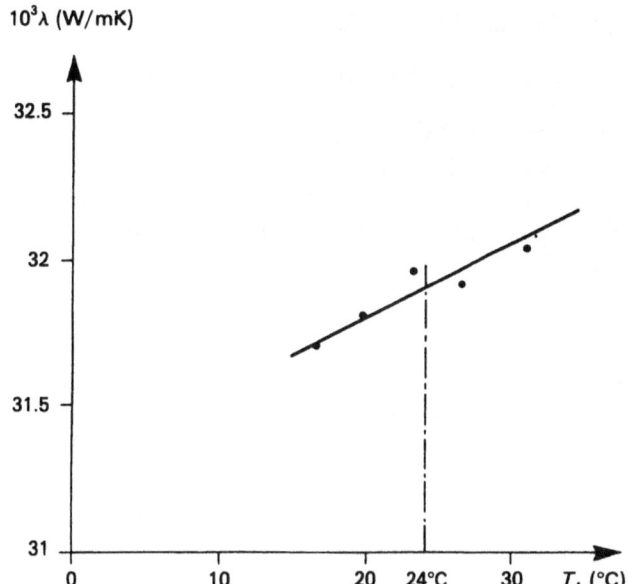

$10^3 \lambda$ (W/mK)

FIGURE 12. Variation of thermal conductivity versus thermopile imbalance.

The relationship was derived assuming the two surfaces of the specimens isothermal at the temperatures T_1 and $T_2 = 0$, and the edge surfaces at the temperature $T_a = eT_1$ with $0 \leq e \leq 1$ in normal test practice; b is the guard width.

A plot of the error, E_e, as a function of e is given on Fig. 13, for a specimen with $d = 0.150$ m (maximum specimen thickness for the present apparatus). For $T = 297.2$ K, $T_1 - T_2 = 20$ K and $\lambda = 0.0335$ W m^{-1} K^{-1}; T_a was measured successively from 288.6 K, 296.6 K, 302.1 K, and 307.7 K with E_e respectively 4.1%, 0.03%, -2.35%, and -3.28%.

Good agreement can be seen between measured edge heat losses and theoretical values computed through equation (11). Thus it can be concluded that the edge heat loss relationship given by equation (11) and used in the design of the apparatus is valid. Consequently, in normal conditions of use, $T_a = (T_1 + T_2)/2 \pm 0.4$ K; the edge heat loss error never exceeds ± 0.2%.

5.5. Comparison with the BCR Certified Curve

The measured values and the BCR reference curve are compared in Fig. 14. For mean test temperatures, T, ranging from 275 to 335 K, the following regression was obtained for λ, expressed in W m^{-1} K^{-1}:

$$\lambda = 0.000305 + 0.106113 \times 10^{-3} T$$

It can be compared to the BCR regression

$$\lambda = 0.00014160 + 0.10285 \times 10^{-3} T \pm 0.00045$$

A difference of $-0.22 \times 10^{-3} \, \text{W m}^{-1} \, \text{K}^{-1}$, i.e., 0.75% on the intercept at $T = 273$ K, and +3.2% on the slope, i.e., less than 1% over the full temperature range for the apparatus, can be seen. These differences are well within the $\pm 0.00045 \, \text{W m}^{-1} \, \text{K}^{-1}$ confidence level for the value of the BCR Standard Reference Material.[1]

5.6. Reproducibility

Four tests have been performed to evaluate the reproducibility of the apparatus, taking the specimens in and out three times. The results indicated a relative variation in thermal conductivity of $\pm 0.45\%$, within the expected value of $\pm 0.5\%$.

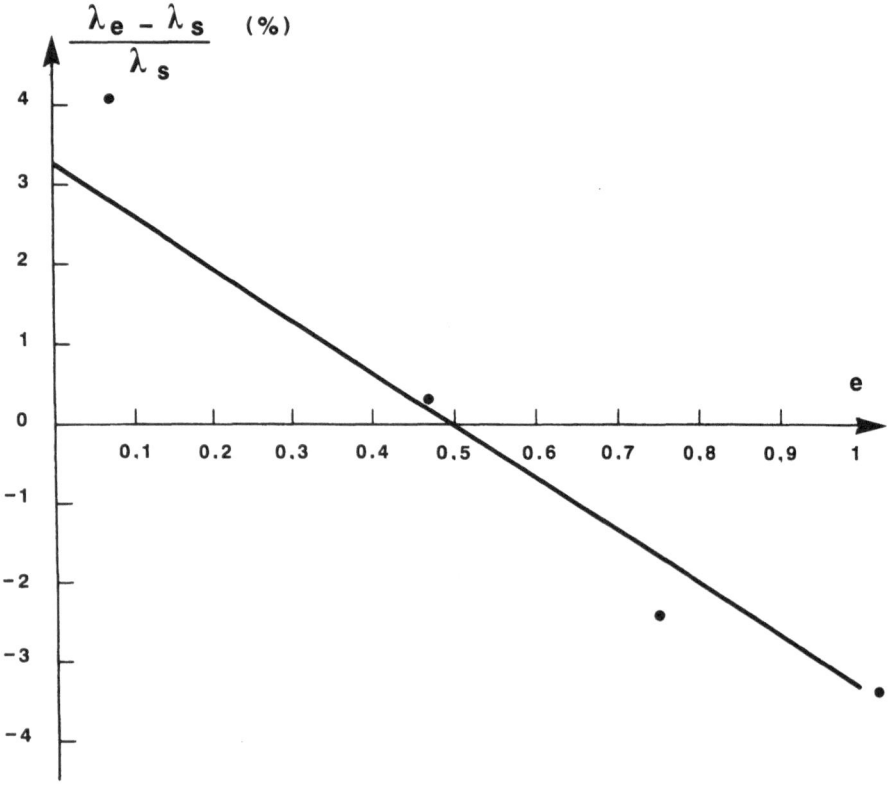

FIGURE 13. Comparison between theoretical and measured edge heat losses.

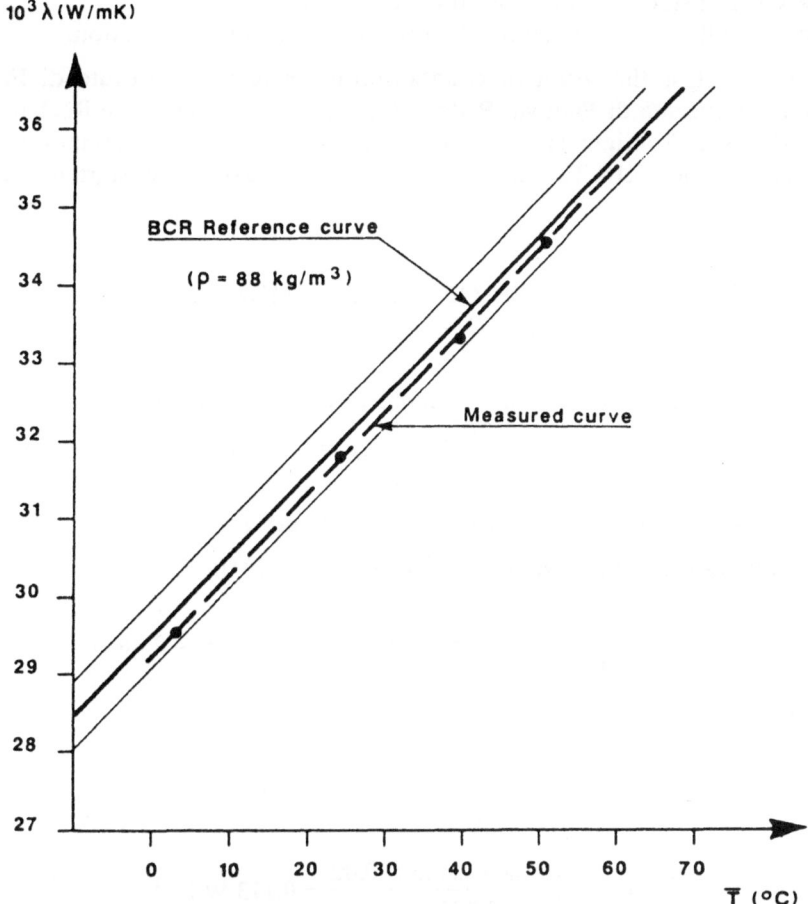

FIGURE 14. Comparison of reference and measured conductivities.

6. CONCLUSIONS

The aim of this chapter has been to illustrate the procedure proposed in ISO DIS 8302 to design a guarded hot plate apparatus. Based on these proposals an apparatus has been designed and built. Performance checks on a European Standard Reference Material (high-density resin-bonded glass-fiber board) have shown that:

- Errors predicted by simple theoretical computations have not been exceeded.
- Agreement against a Standard Reference Material has been better than its claimed confidence level.

- Overall stability and reproducibility is better than 1%.
- Overall accuracy is better than the 1.5% design specification.

This work is the result of collaboration between the Istituto di Fisica Tecnica, Università di Padova, Padova (Italy) and the Centre de Recherches Industrielles de Rantigny (France) of Isover Saint-Gobain. The apparatus was completed in the CRIR Thermal Laboratory and measurement started in June 1985.

APPENDIX A. INFLUENCE OF GUARD IMBALANCE: THEORETICAL EVALUATION

Implementation of Theoretical Calculations in the Present Apparatus

Apparatus dimensions: $l = 0.125$ m, $2g = 0.003$ m, and gap height $h = 2 \times 0.010$ m.

Specimen properties: $\lambda = 0.030$ W m^{-1} K^{-1} and $d = 0.150$ m.

Computation of the parameter c [see equation (3)]:

$$c = 0.125 \frac{16}{3.14} \ln \frac{4}{1 - \exp(-3.14 \times 0.003/0.150)} = 2.66 \text{ m}$$

Computation of Φ_0: Conduction through the air in the gap,

$$\Phi_a = \frac{\lambda_{air} \text{ (lateral surface of the gap)}}{\text{gap width}}$$

$$= \frac{0.026 \times 8 \times 0.125 \times 0.02}{0.003} = 0.173 \text{ W K}^{-1}$$

Radiation at $T_1 = 307$ K with $\varepsilon = 0.9$ ($\sigma_n = 5.67 \ 10^{-8}$ W m^{-2} K^{-4}),

$$\Phi_r = \frac{4\sigma_n T_1^3 \text{ (lateral surface of the gap)}}{(2/\varepsilon - 1)}$$

$$= \frac{4 \times 307^3 \times 5.67 \times 10^{-8} \times 0.020}{(2/0.9 - 1)} = 0.107 \text{ W K}^{-1}$$

Conduction through 16 epoxy connections (front section 0.005 m $\times$ 0.004 m; length 0.003 m; epoxy conductivity $\lambda = 0.232$ W m^{-1} K^{-1}),

$$\Phi_p = \frac{16 \times (0.005 \times 0.004) \times 0.232}{0.003} = 0.025 \text{ W K}^{-1}$$

Conduction through the wires

- 2 power ribbons (copper; 0.006 m × 0.0001 m),
- balancing thermopile consisting of 100 junctions of Ni and NiCr wires, diameter 0.0002 m,
- 16 temperature-measurement Ni–Ni Cr thermocouples, diameter 0.0002 m,
- platinum resistance thermometer for the hot plate temperature control connected through 4 copper wires, diameter 0.0005 m.

The total copper cross section, S_1, is then given by

$$S_1 = 2 \times (0.006 \times 0.0001) + 4 \times (3.14 \times 0.0005^2/4) = 1.985 \times 10^{-6} \text{ m}^2$$

The total Ni and NiCr cross section, S_2, is then given by

$$S_2 = 100 \times (3.14 \times 0.0002^2/4) + 16 \times 2 \times (3.14 \times 0.0002^2/4) = 4.147 \times 10^{-6} \text{ m}^2$$

Assuming $\lambda = 389 \text{ W m}^{-1} \text{ K}^{-1}$ for the copper and $\lambda = 50 \text{ W m}^{-1} \text{ K}^{-1}$ for the thermocouple wires, we then obtain

$$\Phi_w = \frac{S_1 \times 389 + S_2 \times 50}{2g} = \frac{(1.985 \times 389 + 4.147 \times 50) \times 10^{-6}}{0.003} = 0.326 \text{ W K}^{-1}$$

Computation of Φ_0 [see equation (4)]:

$$\Phi_o = \Phi_a + \Phi_r + \Phi_p + \Phi_w = 0.173 + 0.107 + 0.025 + 0.326 = 0.631 \text{ W K}^{-1}$$

Computation of Φ_g [see equation (2)]:

$$\Phi_g = (\Phi_o + \lambda c)\Delta T_g = (0.631 + 0.030 \times 2.66)\Delta T_g = 0.711\Delta T_g$$

Computation of ΔT_g: By imposing $E_{g\,max} = \pm 0.003$ we get from equations (1) and (2) that

$$\Delta T_g = \frac{0.003\Phi}{0.711}$$

From the specimen properties and testing conditions

$$\Phi = \frac{2 \times (2l)^2 \times \lambda \times (T_1 - T_2)}{d} = \frac{2 \times 0.25^2 \times 0.030 \times 20}{0.150} = 0.5 \text{ W}$$

so that

$$\Delta T_g \leqslant \pm 2.1 \text{ mK}$$

As the balancing thermopile consists of 100 junctions, i.e., 50 Ni–Ni Cr thermocouples having an approximate sensitivity of 40 μV K^{-1}, this means that imbalance control is better than

$$V = \pm 2.1 \times 10^{-3} \times 40 \times 10^{-6} \times 50 = \pm 4.2 \ \mu\text{V}$$

APPENDIX B. INFLUENCE OF GUARD IMBALANCE: INTERPRETATION OF EXPERIMENTAL RESULTS

Having assumed that the imbalance between the central section and the main guard ring will be kept within 5 μV (see Table 1), several measurements were conducted by imposing known voltage offsets to the zero-balance controller in order to determine the actual influence of a guard imbalance. Reference testing conditions, already described in Section 5.3, were $(T_1 + T_2)/2 = 297$ K, $(T_1 - T_2) = 21$ K, and $d = 0.035$ m. Test results are summarized in Fig. 12, which is a plot of the measured thermal conductivity, λ, in SI units versus the thermopile voltage, V, in μV. Least-squares analysis applied to those data gives the following relationship:

$$\lambda = 0.03176 - 3.947 \times 10^{-6} V \tag{8}$$

The imbalance heat flow, Φ_g, can be computed from equation (8) if the error component $3.947 \times 10^{-6} V$ of the measured thermal conductivity is multiplied by $2 \times (2l)^2 (T_1 - T_2)/d = 2 \times 0.25^2 \times 21/0.035 = 75$ m K. We then obtain

$$\Phi_g = 3.947 \times 10^{-6} \times 75 = 2.960 \times 10^{-4} V \text{ (W)}$$

The balancing thermopile has 50 junctions with a sensitivity of 42 μV K^{-1} at the test conditions. Therefore $V = 42 \times 50 \Delta T_g$ and hence

$$\Phi_g = 2.960 \times 10^{-4} \times 2100 \Delta T_g = 0.62 \Delta T_g \text{ (W)}$$

This is the value applied by equation (9) in Section 5.3.

APPENDIX C. MAXIMUM SPECIMEN THICKNESS

C.1. This appendix provides a program for the theoretical estimate of the maximum thickness of specimens that can be used in a conventional guarded hot plate apparatus. A sample run is given in Table 4. The reader can easily construct similar tables for different values. The calculations are based on the formula compiled by De Ponte and Di Filippo.[5] A number of assumptions have been made in the calculations. These are tabulated for convenience as follows:

1. two copper leads 0.129 mm^2 (0.000 2 in.^{2} or 26 Awg) cross the gap (see statement 6 of the program given below);
2. the number of thermocouples crossing the gap for sensing the metering section temperature is $10\sqrt{\text{area of plate (m}^2)}$ or 2, whichever is greater;
3. four differential thermocouples are used and are of the same type as those in (2) above (see statement 6 of the program);

TABLE 4

Maximum Thickness of Specimens Used in Conventional Guarded Hot Plate Apparatus

Sample Run, Dimensions in Centimeters, Inch Dimensions in Parentheses

$E = 0.50$, RATIO $T = 0.0010,$[a] RATIO $K = 30,000.0,$[a]

RATIO $L = 1.05$, GAP $= 0.317$

			MAX THICKNESS					
PLATE	GUARD	%ERR	0.1	0.2	0.5	1.0	2.0	5.0
10.16	1.69		0.2	0.4	1.0	1.8	2.5	3.5
(4.00)	2.54		0.1	0.2	0.6	1.2	2.2	3.9
	3.39		0.1	0.1	0.3	0.5	1.0	2.6
20.32	3.39		0.8	1.5	3.3	4.5	5.6	7.3
(8.00)	5.08		0.4	0.9	2.2	4.1	6.3	8.8
	6.77		0.2	0.4	1.0	2.0	3.9	8.2
30.48	5.08		1.6	3.1	5.6	7.1	8.7	11.2
(12.00)	7.62		1.0	1.9	4.5	7.6	10.2	13.6
	10.16		0.4	0.9	2.2	4.2	8.1	13.8
45.72	7.62		3.1	5.7	9.0	11.0	13.2	16.9
(18.00)	11.43		2.1	3.8	8.6	12.7	16.1	20.9
	15.24		1.0	1.9	4.5	8.6	15.0	22.2
60.96	10.16		4.8	8.4	12.3	14.9	17.8	22.7
(24.00)	15.24		3.1	5.9	12.8	17.7	21.9	28.1
	20.32		1.6	3.1	7.2	13.7	21.7	30.6
91.44								
(36.00)	15.24		8.0	13.3	18.7	22.5	26.8	34.1
	22.86		5.4	10.0	20.5	27.2	33.3	42.5
	30.48		3.1	5.9	13.5	24.4	35.0	47.2
121.92	20.32		11.1	18.1	25.1	30.1	35.8	45.5
(48.00)	30.48		8.0	14.7	28.6	36.9	44.8	56.9
	40.64		4.9	9.0	20.3	35.2	48.0	63.8

[a]See Appendix C for RATIO T and RATIO K.

4. guard and metering section temperatures are both uniform in temperature, but may be at different levels;
5. the edge temperature of the specimens is uniform.

A number of parameters are used as follows:

E	is the amount the edge temperature is above the cooling unit temperature as a fraction of the difference in temperature between the heating- and cooling-unit plates:

$$E = [T(\text{edge}) - T(\text{cold})]/[T(\text{warm}) - T(\text{cold})]$$

RATIO T	is the ratio of the temperature difference across the gap to the temperature difference between the plates
RATIO K	is the ratio of the conductivity of the power leads to the conductivity of the specimen
RATIO L	is the ratio of the conductivity of the thermocouples to the conductivity of the copper wires (variable KCU)
GAP	is the gap width, in centimeters
PLATE	is the overall heating-unit dimension, in centimeters (variable WPLATE)
GUARD	is the guard-section width, in centimeters
% ERR	is the total theoretical error in the measurement
MAX THICKNESS	is the maximum thickness of the specimens that produce the designated total error

C.2. After constructing a set of tables, use them as follows. For a given apparatus, determine RATIO K, RATIO L, and GAP. If these correspond at least approximately to the values in the tables, then the tables apply. If the values differ widely, then consult the references. Calculate RATIO T for normal test conditions. Select the table that corresponds and find the maximum allowable specimen thickness for the heating-unit size and guard-section size of the apparatus. Interpolate between tables, if necessary. Note that the value in the table applies only for $E = 0.5$, which is an ideal condition. When other values of E are to be used, consult the design criteria of Ponte and Di Filippo[5] to carry out computations, or modify the given program.

```
C                                                                                1
C                                                                                2
C       PROGRAM TO CALCULATE MAXIMUM SPECIMEN THICKNESS ALLOWED FOR VARIOUS      3
C       PERCENTAGE ERRORS (0.1,0.2,0.5,1.0,2.0,5.0)                              4
C                                                                                5
C       MAXIMUM THICKNESSES ARE CALCULATED FOR THREE GUARD WIDTHS: 1/6 PLATE     6
C       WIDTH, 1/4 PLATE WIDTH, 1/3 PLATE WIDTH.                                 7
C                                                                                8
C       INPUT (CONSISTS OF: E = FRACTION OF TEMPERATURE DIFFERENCE BETWEEN PLATES 9
C                       AT SPECIMEN EDGE.                                        10
C                           RATIOT = RATIO OF GAP TEMPERATURE DIFFERENCE TO      11
C                       SPECIMEN TEMPERATURE DIFFERENCE.                         12
C                           RATIOK = RATIO OF THERMAL CONDUCTIVITY OF            13
C                       LEADS ACROSS GAP TO THERMAL CONDUCTIVITY OF SPECIMEN.    14
C                           KCU = THERMAL CONDUCTIVITY OF T/C ACROSS GAP         15
C                       RELATIVE TO THERMAL CONDUCTIVITY OF COPPER.              16
C                           D = GAP WIDTH.                                       17
C                           WPLATE = PLATE WIDTH.                                18
C                                                                                19
C       NO. OF T/C ACROSS GAP IS BASED ON ASTM FORMULA (NO. = 1/8*SQRT (AREA PLATE)) 20
C       WITH A MINIMUM OF 2 BEING ASSUMED IF NO. = 2 OR LESS.                    21
C       ONE COPPER LEAD (0.0002 SQ. IN.) WAS ALSO ASSUMED ACROSS THE GAP         22
C       AND FOUR DIFFERENTIAL T/C, OF THE SAME MATERIAL AS THE TEST AREA T/C,    23
C       WERE ASSUMED ACROSS THE GAP.                                            24
C                                                                                25
        DIMENSION G(5),ER(10),ANS(10),GM(10),ANSN(10)                           26
        REAL L,KCU                                                               27
        INTEGER CARD                                                            28
        COMMON WPLATE,E,RATIOT,RATIOK,AC,WGUARD,ERR,D,L                         29
        CARD = 1                                                                30
        LP = 3                                                                  31
        K = 1                                                                   32
        ER(1) = 0.1                                                             33
        ER(2) = 0.2                                                             34
        ER(3) = 0.5                                                             35
        ER(4) = 1.0                                                             36
        ER(5) = 2.0                                                             37
        ER(6) = 5.0                                                             38
        EPS = 0.00001                                                           39
        IEND = 20                                                               40
C                                                                                41
1000    READ (CARD,200) WPLATE                                                  42
        IF (WPLATE) 1,99,2                                                      43
1       READ (CARD,100) E,RATIOT,RATIOK,KCU,D                                  44
        DM = D/0.39370079                                                       45
        WPLATE = −WPLATE                                                        46
C                                                                                47
        IF (K) 73,72,73                                                        48
72      WRITE (LP,900)                                                          49
        WRITE (LP,901)                                                          50
73      WRITE (LP,700) E,RATIOT,RATIOK,KCU,DM                                  51
        WRITE (LP,500)                                                          52
        WRITE (LP,600) (ER(I),I = 1,6)                                         53
```

```
C                                                                              54
2       G(1) = WPLATE/6.                                                       55
        G(2) = WPLATE/4.                                                       56
        G(3) = WPLATE/3.                                                       57
        K = 0                                                                  58
C                                                                              59
        WPLATM = WPLATE/0.39370079                                            60
        WRITE (LP,401) WPLATM                                                  61
        DO 40 I = 1,3                                                          62
        L-WPLATE/2.-G(I)                                                       63
        WGUARD = G(I)                                                          64
        GM(I) = G(I)/0.39370079                                               65
C                                                                              66
        X = L/4.                                                               67
        IX = X                                                                 68
        IF (X – IX) 4,4,3                                                      69
3       IX = IX + 1                                                            70
4       IF (IX – 1) 5,5,6                                                      71
5       IX = 2                                                                 72
6       AC = (IX + 4.)*0.00007894*KCU + 0.0002                                73
C                                                                              74
        H1 = 0.0001*WPLATE                                                    75
        DO 30 J = 1,6                                                          76
        ERR = ER(J)                                                            77
7       H2 = 2.*H1                                                             78
        IF (FCT(H1)*FCT(H2)) 9,9,8                                            79
8       H1 = H2                                                                80
        GO TO 7                                                                81
9       CALL RTMIX(ANSH,ANSE,FCT,H1,H2,EPS,IEND,IER)                          82
        IF (IER) 10,20,10                                                     83
10      ANS(J) = 0.0                                                           84
        GO TO 30                                                               85
20      ANS(J) = ANSH                                                          86
        ANSM(J) = ANS(J)/0.39370079                                          87
30      CONTINUE                                                               88
        IF (I – 2) 70,71,70                                                   89
71      WRITE (LP,400) WPLATE                                                 90
70      WRITE (LP,301) GM(I),(ANSM(J),J = 1,6)                               91
40      CONTINUE                                                               92
        WRITE (LP,200)                                                         93
        GO TO 1000                                                            94
99      WRITE (LP,900)                                                         95
        WRITE (LP,901)                                                         96
        WRITE (LP,999)                                                         97
        STOP                                                                   98
100     FORMAT (F4.2,F6.4,F8.6,F4.2,F6.4)                                     99
200     FORMAT (F6.2)                                                         100
300     FORMAT (32X,'0',F5.2,')',8X,6('(',F6.3,')'))                         101
301     FORMAT (33X,F5.2,7X,6(1X,F7.1))                                      102
400     FORMAT ('-',16X,'(',F5.2,')')                                        103
401     FORMAT ('-',16X,F6.2)                                                104
500     FORMAT (18X,'PLATE',10X,'GUARD',26X,'max. thickness')               105
600     FORMAT (44X,'%ERR',2X,F3.1,5(5X,F3.1)/)                             106
700     FORMAT ('1',19X,'E = ',F4.2,4X,'RATIO T = ',F6.4,4X,'RATIO K = ',F7.1,4X   107
     1, 'T/C K = ',F4.2,4X,'GAP = ',F5.3,//)                                 108
```

```
900    FORMAT (/,32X,'NOTE: 1) TERMS NOT BRACKETED MEASURED IN CENTIMETER      109
       1S.')                                                                    110
901    FORMAT (38X,') BRACKETED TERMS MEASURED IN INCHES.')                     111
999    FORMAT ('1')                                                             112
       END                                                                      113

       SUBROUTINE RTMI(X,F,FCT,XLI,XRI,EPS,IEND,IER)                             1

       IER = 0                                                                  2
       XL = XLI                                                                 3
       XR = XRI                                                                 4
       X = XL                                                                   5
       TOL = X                                                                  6
       F = FCT(TOL)                                                             7
       IF(F)1,16,1                                                              8
1      FL = F                                                                   9
       X = XR                                                                   10
       TOL = X                                                                  11
       F = FCT(TOL)                                                             12
       IF(F)2,16,2                                                              13
2      FR = F                                                                   14
       IF(FL*FR)3,25,25                                                         15
C                                                                               16
3      I = 0                                                                    17
       TOLF = 100.*EPS                                                          18
C                                                                               19
4      I = I + 1                                                                20
C                                                                               21
       DO 13 K = 1, IEND                                                        22
       X = .5*(XL + XR)                                                         23
       TOL = X                                                                  24
       F = FCT(TOL)                                                             25
       IF(F)5,16,5                                                              26
5      IF(F*FR)6,7,7                                                            27
C                                                                               28
6      TOL = XL                                                                 29
       XL = XR                                                                  30
       XR = TOL                                                                 31
       TOL = FL                                                                 32
       FL = FR                                                                  33
       FR = TOL                                                                 34
7      TOL = F - FL                                                             35
       A = F*TOL                                                                36
       A = A + A                                                                37
       IF(A - FR*(FR - FL))8,9,9                                                38
8      IF(I - IEND)17,17,9                                                      39
9      XR = X                                                                   40
       FR = F                                                                   41
C                                                                               42
       TOL = EPS                                                                43
       A = ABS(XR)                                                              44
       IF(A - 1.)11,11,10                                                       45
10     TOL = TOL*A                                                              46
11     IF(ABS(XR - XL) - TOL)12,12,13                                           47
12     IF(ABS(FR - FL) - TOLF)14,14,13                                          48
13     CONTINUE                                                                 49
```

```
C                                                                       50
      IER = 1                                                           51
14    IF(ABS(FR) − ABS(FL))16,16,15                                     52
15    X = XL                                                            53
      F = FL                                                            54
16    RETURN                                                            55
C                                                                       56
17    A = FR − F                                                        57
      DX = (X − XL)*FL*(1. + F*(A − TOL)/(A*(FR − FL)))/TOL             58
      XM = X                                                            59
      FM = F                                                            60
      X = XL − DX                                                       61
      TOL = X                                                           62
      F = FCT(TOL)                                                      63
      IF(F)18,16,18                                                     64
C                                                                       65
18    TOL = EPS                                                         66
      A = ABS(X)                                                        67
      IF(A − 1.)20,20,19                                                68
19    TOL = TOL*A                                                       69
20    IF(ABS(DX) − TOL)21,21,22                                         70
21    If(ABS(F) − TOLF)16,16,22                                         71
C                                                                       72
22    IF(F*FL)23,23,24                                                  73
23    XR = X                                                            74
      FR = F                                                            75
      GO TO 4                                                           76
24    XL = X                                                            77
      FL = F                                                            78
      XR = XM                                                           79
      FR = FM                                                           80
      GO TO 4                                                           81
C                                                                       82
25    IER = 2                                                           83
      RETURN                                                            84
      END                                                               85


      FUNCTION FCT(H)                                                    1
      REAL NUM, L,LC                                                     2
      COMMON WPLATE,E,RATIOT,RATIOK,AC,WGUARD,ERR,D,L                    3
      PI = 3.1415926535                                                  4
C                                                                        5
      CONST = 16.*ALOG(4.)/PI                                            6
C                                                                        7
      LC = D                                                             8
C                                                                        9
      VARIA1 = 1. − EXP(−2.*PI*D/H)                                     10
      TERM1 = RATIOK*AC/(LC*L)                                          11
      TERM3 = 16.*ALOG(VARIA1)/PI                                       12
      EPSIG = H*RATIOT*(TERM1 + CONST − TERM3)/(8.*L)                   13
C                                                                       14
      TERM4 = COSH(PI*(WGUARD + L)/H)                                   15
      TERM5 = COSH(PI*WGUARD/H)                                         16
      TERM6 = ALOG((TERM4 + 1.)/(TERM5 + 1.))                          17
      TERM7 = ALOG((TERM4 − 1.)/(TERM5 − 1.))                          18
      NUM = E*TERM6 + (1. − E)*TERM7                                    19
      EPSIL = ((NUM*H)/(PI*L))**2 − 1.                                 20
```

```
C                                                               21
      FCT = ERR/100. – ABS(EPSIG + EPSIL)                       22
C                                                               23
      RETURN                                                    24
      END                                                       25
```

REFERENCES

1. H. Ziebland, "Certification Report on a Reference Material for the Thermal Conductivity of Insulating Materials between 170 K and 370 K," Resin-Bonded Glass Fiber Board BCR, No. 64, Commission of the European Communities EUR 7677 EN (1982).
2. ISO DIS 8302, "Determination of Steady-state Thermal Resistance and Related Properties—Guarded Hot Plate Apparatus."
3. W. Woodside and A.G. Wilson, "Unbalance Errors in Guarded Hot Plate Measurements," in *ASTM STP* No. 217, Philadelphia, pp. 32–48 (1957).
4. F. De Ponte and P. Di Filippo, "Some Remarks on the Design of Isothermal Plates," *Bull. IIF*, Annex 1973-4, Commission B1, Zurich, pp. 145–155 (1973).
5. F. De Ponte and P. Di Filippo, "Design Criteria for Guarded Hot Plate Apparatus," in *ASTM STP* No. 544, Philadelphia, pp. 97–117 (1974).
6. W. Woodside, "Analysis of Errors Due to Edge Heat Loss in Guarded Hot Plates," in *ASTM STP* No. 217, Philadelphia, pp. 49–64 (1957).
7. D. Fournier and S. Klarsfeld, "Mesures de conductivité thermique des materiaux isolants par un appareil orientable à placque chaude bi-guardée," *Bull. IIF*, Annex 1969-7, pp. 321–331.
8. B. Rennex, "Error Analysis for the National Bureau of Standards 1016 mm Guarded Hot Plate," *J. Therm. Insulation* 7, 18–51 (1984).
9. D. Fournier and F. De Ponte, "Orientations prises dans la mise au point de methodes d'essais normalisées par l'ISO pour la mesure des propriétés de transport thermique en régime stationnaire sur des materiaux isolants: historique et structures," 16e Congrès de l'IIF, Paris, pp. 453–460 (1983).
10. F. De Ponte and S. Klarsfeld, "What Property Do We Measure? Consideration on a Decade of ISO/TC 163," *J. Therm. Insulation* 13, 160–190 (1990).

5

Apparatus for Testing High-Temperature Thermal-Conductivity Standard Reference Materials with Conductivities Above $1\,W\,m^{-1}\,K^{-1}$ in the Temperature Range 400 to 2500 K

V.E. PELETSKY

1. INTRODUCTION

The reliable measurement of thermal conductivity is a problem that has not lost its urgency, this being particularly true for substances possessing a complex electronic structure and constituting in the solid state the backbone of modern structural materials. The achievements of solid-state physics notwithstanding, determination of the numerical values of phenomenological kinetic factors will long remain a problem to be solved experimentally.

Solving the problem in question presumes constant development and refinement of appropriate instrumentation and broadening the range of experimental techniques embodied in the instruments being developed. The latter aspect is of special importance, as a wide variation in the physical properties of substances and the temperature ranges of their application make it impracticable to perform a valid investigation of the thermophysical characteristics of these substances with just a single instrument or setup.

What is suited for work with metals may be totally useless in connection with an investigation of the properties of a dielectric, while a technique adapted for a given temperature range is likely to fail over a different temperature range. Moreover, for one and the same object of investigation it is essential that the instrumentation available be based on different methodological principles.

V.E. PELETSKY • Institute for High Temperatures, USSR Academy of Sciences, Moscow 127412, USSR.

The history of metrology shows the approach described above is mandatory in cases in which one wishes to ascertain the existing sources of systematic errors inherent in any measurement technique. An objective metrological evaluation of the level of a given measuring device therefore necessitates its testing relative to previously investigated specimens of materials, the so-called reference specimens. Broadening the range of such specimens and their careful testing are a major element of the program for the metrological support of experimental studies.

This chapter describes apparatus developed at the Institute for High Temperatures of the USSR Academy of Sciences to investigate the thermal conductivity of materials having a relatively high conduction level. The apparatus in question was employed for developing the thermal conductivity reference material Molybdenum SOTM-2150 GSO 2317-82. What follows is a brief description of the design features of the apparatus, and a discussion regarding the limits of its optimum employment and the main sources of measurement errors.

2. SCHEMATIC OUTLINE OF THE METHOD AND THE BASIC MATHEMATICAL MODEL

In selecting a method for the measurement of some property, one should proceed from the geometry and dimensions of the test specimen submitted by the customer. The specimen may be a foil strip, wire, casting, bar, or a rod, and each case calls for finding a specific solution.

The variant described below pertains to the situation where the requirement to be adequate in structure to real matter necessitates test specimen fabrication in the shape of a rod from a few millimeters to 10–15 mm in diameter and up to tens of millimeters long. The employment of such test specimens makes it possible to investigate single and polycrystals, cermets, and porous materials, as well as modern composite materials with reinforcing inclusions in the form of spherical particles or fibers.

Specimens with this geometry find application in the apparatus operating in the so-called longitudinal (axial) heat flow mode. In this case the specimen is secured between a heater and a cooler, while the lateral surface of the specimen is surrounded by heat insulation intended to ensure heat flow constancy in all specimen sections.

The magnitude of the heat flow is measured either calorimetrically in the cooler or determined on the basis of the heater power, provided the heater is guarded by a system of compensating heaters. Temperature distribution measured from the readings of two or more temperature sensors located in different specimen sections makes it possible to calculate the thermal conductivity.

The success of measurements or, to be more exact, their error depends primarily on how far the heat flow has been preserved or the losses thereof taken into account. The higher the temperature, the more difficult the attainment of these aims, and traditional approaches have been found to be operative at temperatures not exceeding 1000–1200 K.

It turned out that the temperature limit of the longitudinal heat flow method can be raised significantly by designing the measuring device so as to enable it to find the local heat flow passing through a preselected cross section of the specimen. Then the principal design formula of the method will be the Biot–Fourier relation, wherein use is made of the thermal conductivity as the proportionality factor between the heat flow density q and the temperature gradient:

$$q = -\lambda \text{ grad } T \tag{1}$$

In resorting to this approach, we dispense with the thermal insulation of the lateral surface of the rod and, accordingly, the heat flow from the heater will be dissipated from this lateral surface, resulting in the formation of a two-dimensional temperature field. Hence, apart from the useful longitudinal temperature gradients, in the specimen radial gradients will be operative, the magnitude of the latter being the higher, the greater the degree of specimen overheating relative to the ambient medium.

Stated differently, on the lateral surface the boundary condition should be

$$-\lambda \frac{\partial T}{\partial r}\bigg|_{r=R} = q_s \tag{2}$$

where r is the radial coordinate, R the external radius, and q_s the heat flow density lost from the lateral surface.

To assess the extent of interference exerted by this two-dimensional pattern on measurements, the analysis of the corresponding boundary problem is pertinent. Let us formulate the problem as follows:

The heat flow q_0 is introduced via the specimen end face (plane $z = 0$) through an area having radius r_1, i.e.,

$$z = 0, \quad r \leq r_1, \quad -\lambda \frac{\partial T}{\partial z} = q_0 \tag{3}$$

the remaining area of the end face $z = 0$ being insulated, i.e.,

$$r_1 < r < R, \quad -\lambda \frac{\partial T}{\partial z} = 0$$

Heat release into the ambient space having zero temperature through both the lateral surface and the "cold" end obeys the following law:

$$z = L, \qquad -\frac{\partial T}{\partial z} = hT \tag{4}$$

$$r = R, \qquad -\frac{\partial T}{\partial r} = hT, \qquad h = \alpha/\lambda \tag{5}$$

where α is the heat exchange coefficient.

The solution of the problem formulated in this manner may be written in the form

$$T = \frac{2q_0 r_1}{\lambda} \sum_{k=1}^{\infty} \frac{J_1[n_k(r_1/R)]J_0[n_k(r/R)]}{J_0^2(n_k)[n_k^2 + (hR)^2]} \varphi_k(z) \tag{6}$$

where n_k denotes the roots of the equation

$$nJ_1(n) = hRJ_0(n) \tag{7}$$

and

$$\varphi_k(z) = \frac{\operatorname{sh}\left[\{(L-z)/R\}n_k\right] + \dfrac{n_k}{hR} \operatorname{ch}\left[\{(L-z)/R\}n_k\right]}{\operatorname{ch}\left[(L/R)n_k\right] + (n_k/hR)\operatorname{sh}\left[(L/R)n_k\right]} \tag{8}$$

Analysis of this solution carried out elsewhere showed that, at some distance from the heated end face of the specimen, the temperature field ceases to be dependent on the profile of the heat flow density introduced into the specimen. Heat flow inhomogeneity and the associated temperature profile in the plane of the hot end face are not "memorized" by the temperature field, so that it would be reasonable to assume the existence of a temperature field ordering zone in the rod being heated, wherein the extent of the zone equals 1–$1.5R$, where R is the specimen radius. The radial temperature distribution $T(z)$ in cross sections located outside the ordering zone are described practically by the first term of series (6).

Accordingly, quite simple relationships can be obtained for assessing the role of field two-dimensionality. If it is assumed that the total heat flow in the cross section z could have been measured, then it is related to the field of the longitudinal gradients in this cross section as follows:

$$Q_z = -2\pi\lambda \int_0^R \frac{\partial T(r, z)}{\partial z} r \, dr \tag{9}$$

For calculating the thermal conductivity it is pertinent to know the quantity

$$\overline{\text{grad}}_z T = \frac{2}{R^2} \int_0^R \frac{\partial T(r, z)}{\partial z} r \, dr \tag{10}$$

In practice, temperature measurements are generally localized and yield temperature values at selected points of the specimen.

It is feasible, for example, to find the longitudinal temperature gradient at $r = R$ by welding thermocouples to the specimen surface. It is also practicable to obtain another value of the longitudinal gradient by providing radial channels in which temperature sensors are inserted as far as the specimen axis $r = 0$. In either case, substituting in equation (9) the local longitudinal gradient for the quantity (10) would result in the appearance of an error

$$\delta\lambda = \frac{\lambda - \lambda_c}{\lambda} - 1 - \frac{\overline{\text{grad}}_z T}{(\text{grad}_z T)_r} = 1 - c(r) \tag{11}$$

where λ is the true value of thermal conductivity, and λ_c is the thermal conductivity value calculated from the local gradient value.

The correction $c(r)$ in the case $r = 0$ will be

$$c_{r=0}(r) = \left\{ \sum \frac{2hR\psi(z)}{n_k^2[n_k^2 + (hR)^2]} \right\} \cdot \left\{ \sum \frac{\psi(z)}{J_0(n_k)[n_k^2 + (hR)^2]} \right\}^{-1} \tag{12}$$

For the zone of a stabilized temperature profile, provided we confine ourselves to the first terms in the series, the correction will be

$$c_{r=0}(r) = \frac{2hR}{n_1^2} J_0(n_1) \tag{13}$$

The result obtained is of vital importance. Indeed, in the zone of a stabilized temperature profile, measurements of the two-dimensional temperature distribution are not mandatory, and it would be sufficient for a selected radius (one and the same in different specimen cross sections) to measure the temperatures in order to find the magnitude of the longitudinal temperature gradient. Gradient deviations from the requisite average longitudinal gradient can be readily calculated if the Biot parameter is available.

In practice, however, the accuracy of the known radiative heat-exchange coefficient value is very low. The condition of the constancy of h over the entire specimen surface employed in deriving the solution is likewise not observed, so that it is essential to realize what are the corrections in question.

For subsequent solutions, the range of variations displayed by thermal parameters in the equations obtained is of great importance.

Let us assume the test specimen has the following properties: thermal conductivity $\lambda = 10 \text{ W m}^{-1} \text{ K}^{-1}$, total hemispherical emittance $\varepsilon = 0.5$, and radius $R = 0.6 \text{ cm}$. Then the equation $\alpha(T - T_0) = \varepsilon\sigma(T^4 - T_0^4)$ can be used to calculate the heat transfer coefficient, where $\sigma = 5.6687 \times 10^{-12} \text{ W cm}^{-2} \text{ K}^{-4}$ is the Stefan–Boltzmann constant. The following dependence of the Biot parameter on temperature is obtained:

T (K)	500	1000	1500	2000	2500
$\text{Bi} = hR$	5×10^{-3}	0.02	0.07	0.16	0.3
$\delta\lambda(\%)$	0.12	0.5	1.7	3.8	7.0

For these Biot parameter values, the Bessel functions in equation (7) can be replaced by the first two terms of their expansion, thereby making it possible to establish the relation of the first root of equation (7) to the Biot parameter, viz.,

$$n_1 = \left(\frac{8hR}{4 + hR}\right)^{1/2} \tag{14}$$

Thus the exact value of the root of equation (7) for $hR = 1.0$ will be $n_1 = 1.2558$, while using expression (14) yields $n_1^* = 1.2649$, so the deviation from the exact value is only 0.7%. For smaller Biot parameters, deviations will be even lower. Correction for two-dimensionality should accordingly be expressed as follows:

$$c(r = 0) = 1 - 0.25hR + \frac{(hR)^2}{4(4 + hR)} \tag{15}$$

The above table presents the values of errors in thermal conductivity determination calculated for a hypothetical specimen. It is apparent that introducing the correction for two-dimensionality in the case under consideration would be expedient at temperatures above 1500 K.

The thermal conduction of refractory metals and alloys is for the most part from two to three times as great as that of the hypothetical specimen, and many such materials do not require the aforementioned correction for two-dimensionality over the entire temperature range studied. Specimen radius reduction provides an additional possibility of minimizing the role of this correction.

At high temperatures, heat exchange on the specimen surface makes it imperative to correct the reference temperature in view of the fact that in the net (working) section there is a radial temperature drop.

In the stabilized temperature profile zone, the relative drop magnitude may be evaluated using the formula

$$\frac{T(r = 0) - T(r = R)}{T(r = 0)} = \frac{2hR}{4 + hR} + \frac{(hR)^2}{(4 + hR)^2} \tag{16}$$

At a temperature of 2500 K, in the hypothetical specimen under consideration the temperature drop in the cross section would equal 14% of the temperature at the bar axis, a significant quantity which makes refinement of the reference temperature T_λ an important problem.

If the thermal conductivity–temperature relationship is assumed to be linear, then T_λ should be found from

$$T_\lambda = \left[\int_0^R rT(r)g(r)\, dr \right] \bigg/ \left[\int_0^R rg(r)\, dr \right] \tag{17}$$

Here $T(r)$ is the function describing the variation of temperature with radius in the selected specimen section, and $g(r)$ pertains to the temperature gradient.

For the stabilized temperature profile zone

$$T_\lambda = T_{r=R}\left(0.5hR + \frac{n_1^2}{2hR} \right) \approx T_{r=R}(1 + 0.25hR) \tag{18}$$

or, in terms of the axial temperature,

$$T_\lambda \approx (1 - 0.25hR)\,T_{r=0}$$

The aforementioned relationships constitute the mathematical basis of the high-temperature version of the longitudinal heat flow method, which comprises heating one end face of a rod specimen with allowed heat exchange on the surface, measuring the heat flow passing through the selected specimen cross section designated as the net section, measuring on the selected radius the temperatures at several points located in the vicinity of the net section, and calculating the thermal conductivity using the Fourier relation.

The success of implementing this method is governed by the accuracy of measuring the thermal flow in a given cross section.

3. EXPERIMENTAL SETUP

3.1. Measuring Cell Design

The problem of determining the local heat flow lends itself to solution by various approaches. If, for example, the total hemispherical emittance of the test material surface, ε, is well known and the true temperature distribution $T(z)$ is measured experimentally, then the local heat flow can be calculated. Thus, for a freely radiating surface in a finite rod heated from one end (Fig. 1), the thermal balance equation will be valid:

$$Q_z = 2\pi R \varepsilon \sigma \int_{z_m}^{L} (T_R^4(z) - T_0^4)\, dz + 2\pi \int_{0}^{R} (T_L^4(r) - T_0^4) r\, dr \qquad (19)$$

The total thermal flow that has passed through the selected cross section z_m of the specimen is emitted into the ambient space by radiation.

The computational procedure permits the determination of Q_z with high accuracy. The above method found application in a study of refractory metals described elsewhere.[2]

Subsequent investigations showed the emittance to be subject to marked variations depending on experimental conditions used and the specimen's prehistory, these variations being particularly significant at temperatures below 1300–1600 K where, for many metals *in vacuo*, there commences active sublimation of oxide films from the specimen surface.

Under steady-state conditions when the temperature field $T(z)$ has ceased to vary in the course of time, there appears to be a possibility of forming the $\varepsilon(z)$ field, for which *a priori* valuations appear to present difficulties. In view of this situation, in subsequent work the computational method of determining Q_z using expression (19) was abandoned, although a variation of this method disclosed elsewhere[3] contemplated carrying out on the same specimen a direct determination of the temperature dependence $\varepsilon(T)$.

The employment of a calorimetric device operating on the principle illustrated in Fig. 2 proved to be a more reliable way of heat flow measurement.

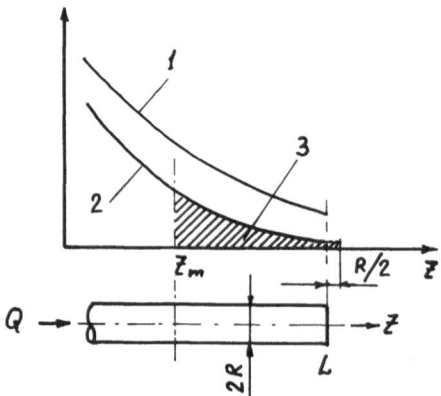

FIGURE 1. Distribution of temperature and heat losses along a freely radiating specimen: 1—$T(z)$; 2—$q_r(z)$; 3—Q_z.

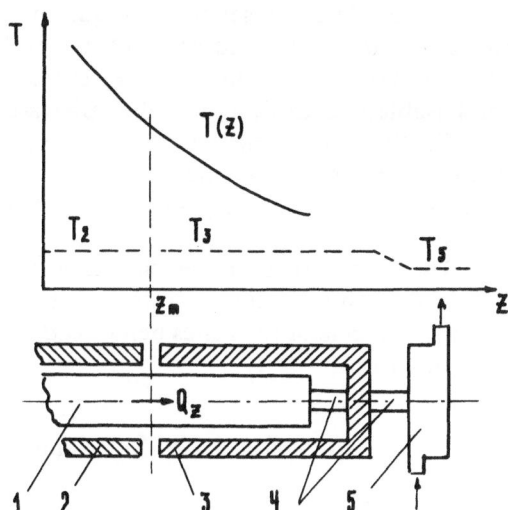

FIGURE 2. Diagram of heat flow measurement in a given cross section of the specimen: 1—specimen; 2—massive shield; 3—heat absorber; 4—thermal resistance inserts; 5—thermostat.

Here specimen 1 is placed in a cylindrical cavity formed by the cylinder of heat absorber 3 and the cylinder of shield 2. Elements 2 and 3 are separated from each other by a small gap (0.2-0.5 mm), and in the plane of this gap with the coordinate z_m there lies the specimen cross section designated below as the net cross section.

The number 4 denotes a thermal resistance of the element that links the specimen and the heat absorber, while number 5 pertains to a thermostat intended to absorb heat and assign the initial temperature level to elements 3 and 2. Over the entire range of working temperatures, elements 2 and 3 have temperatures close to that of the thermostat temperature.

If element 3 is a calorimeter, then it is useful for measuring the following thermal power:

$$Q_3 = Q_z + Q_{13}(z < z_m) - Q_{12}(z > z_m) + Q_{23} \qquad (20)$$

Heat losses from the outer surface of the calorimeter to the environment are negligible due to the low temperature of this member. In equation (20) $Q_{13}(z < z_m)$ denotes the heat flow irradiated from the specimen surface to the left of the working cross section and entering the calorimeter through the annular gap between the heat absorber and the specimen; Q_{12} denotes heat losses from the heat absorber zone toward element 2 via the annular gap, and Q_{23} is radiative heat exchange between elements 2 and 3.

The discussion here is concerned with radiative heat exchange only, insofar as the entire measuring cell is assumed to be *in vacuo*. Intuitively, it is apparent that Q_3 should be closer to the required Q_z, the smaller the radial gaps and the closer the temperatures of elements 2 and 3 are to each other. In the absence, however, of quantitative estimates of this heat transfer it would be inadvisable to start designing the apparatus.

For the sake of calculations, let us assume the internal surfaces of elements 2 and 3 are absolutely black and possess temperatures significantly below that of the specimen (it should be noted in passing that this is the worst case).

The next step comprises breaking the specimen surface to the right and left of the cross section z_m into annular elements each having length l, the ordinal numbers of the zones being read off in both directions from z_m. Hence we can write

$$Q_{13} = \sum_i \varphi_{i3} F_i \varepsilon_i \sigma T_{i2}^4 \quad \text{and} \quad Q_{12} = \sum_i \varphi_{i2} F_i \varepsilon_i \sigma T_{i3}^4 \tag{21}$$

where φ_{i3} and φ_{i2} are the "viewing factors" of the ith zone and the element formed by the annular-gap cross section between the specimen and the heat absorber in the plane z_m. It follows from symmetry considerations that for identical ordinal numbers to the left and right of the plane z_m these coefficients should be equal. Hence

$$Q_{13} - Q_{12} = \sigma F_i \sum_i \varphi_i \varepsilon_i (T_{i2}^4 - T_{i3}^4)$$

As regards the linear distribution of temperatures in the vicinity of z_m, the temperature difference between the elements having identical subscripts may be expressed in terms of the temperature gradient G as follows:

$$T_{i2} - T_{i3} = G[z_{i2} - z_{i3}] = G(2i - 1)l$$

If, for evaluation purposes, we assume that $\varepsilon_i = \text{const} = \varepsilon(z_m)$, then

$$Q_{13} - Q_{12} = 4\varepsilon(z_m)\sigma\pi D l^2 T^3(z_m) G(z_m) \sum_{i=1}^{k} \varphi_i (2i - 1)$$

$$= 4\sigma\varepsilon(z_m) T^3(z_m) G(z_m)\Gamma \tag{22}$$

where Γ is the factor governed by a particular geometry of the system, and σ is the Stefan–Boltzmann constant.

Dividing the quantity $Q_{13} - Q_{12}$ by the heat flow in the net cross section yields the following expression for the relative magnitude of the correction in heat conduction for radiative heat transfer in the annular gap:

$$\frac{\Delta\lambda}{\lambda} = \frac{16\sigma\varepsilon T^3(z_m) \cdot D}{\lambda} \frac{l^2}{D^2} \sum_{i=1}^{k} \varphi_i(2i-1) \tag{23}$$

In the linear approximation of the temperature field, the correction is therefore independent of the gradient, to the first approximation.

The values of the "view factors," which are of interest in the case under consideration, are cited by Aleksandrov.[4] For example, when the specimen diameter $D = 1.2$ cm, the heat-absorber diameter equals 1.4 cm, and $\lambda = 0.2$ W cm^{-1} K^{-1}, $\varepsilon = 0.3$, $l = 0.5$, $\varphi_1 = 0.08$, and at the temperature of 2000 K in the net cross section we obtain $\Delta\lambda/\lambda = 2.6 \times 10^{-2}$, i.e., approximately 3%. It is therefore desirable to make the gap as small as possible, but even for a gap size of 1 mm the correction is fairly small and decreases rapidly as the temperature drops.

In the variant under consideration, other aspects of heat transfer are fundamentally insignificant.

It can be seen from the foregoing that the schematic diagram shown in Fig. 2 provides the possibility of acceptable measurements of the thermal flow.

Let us consider the design of the measuring cell used for carrying out all measurements in question and, for the sake of brevity, designated subsequently as the calorimeter, since the cell is intended for thermal flow determination. Figure 3 shows that all calorimeter components are mounted on ring 1 furnished with two vertical uprights 2 fabricated from stainless steel. Provision is made for flange 3, capable of being displaced along the uprights and having guide bushings centered relative to uprights 2 by means of teflon inserts.

Secured onto flange 3 is a massive copper shield 12 with a copper tubing soldered to its bottom, intended for cooling water. A teflon ring 4 is secured on flange 3 coaxially with the shield, and carries the upper block of the calorimeter comprising the heat conduit cylinder 5 and heat absorber 6 with lids 8 and 9.

The heat absorber 6, made of copper, has in the top a well which accommodates test specimen 11 secured in place with slot washer 7, while lid 8 serves to clamp tightly the specimen in the heat absorber well.

On the external surface of the heat conduit there is located a pile of differential (copper–constantan) thermocouples 10, the thermocouple junctions being cemented into special wells bored in the upper and lower portions of the heat conduit. As in the case of the shield, a copper tube soldered to the heat absorber bottom allows the passage of cooling water.

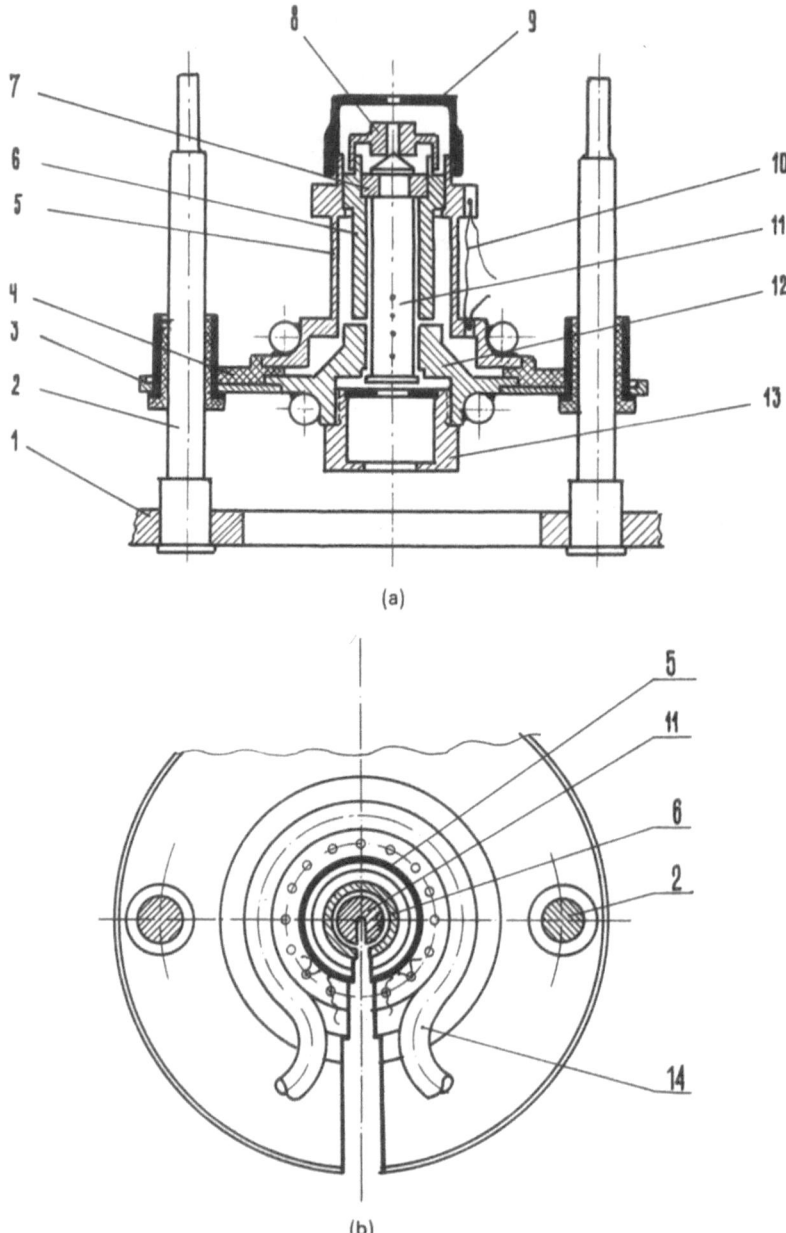

(a)

(b)

FIGURE 3. (a) Calorimeter: 1—base ring; 2—uprights; 3—flange; 4—fluoroplastic ring; 5—heat conduit; 6—heat absorber; 7—washer; 8, 9—lids; 10—differential thermocouple pile; 11—specimen; 12—shield; 13—diaphragm; 14—cooling coil. (b) Calorimeter cross section in the plane of the first pyrometric channel.

The thickness of the teflon ring is selected so as to obtain a gap of 0.5 mm between the end faces of heat absorber 6 and shield 12. The plane of this gap forms the net cross section shown earlier in Fig. 2, temperature measurements being made in the neighborhood of this cross section. To do so, four radial pyrometric channels are bored in the specimen (boring depth $0.7-0.8D$, where D is the specimen diameter).

All calorimeter components have a longitudinal slot ca 2 mm wide (cf. Fig. 3b) intended for pyrometer sighting, as well as for thermocouple laying (at low temperatures).

The calorimeter components function as follows. A heat flow (in the form of an energy beam) falls onto the bottom end face of the specimen via a limiting diaphragm (a tantalum disk in copper cylinder 13). Shaping the regular temperature profile produced by the heat loss as a result of radiation from the specimen surface occurs in the specimen length from the bottom end face to the first pyrometric hole. Two bottom pyrometric holes are located in the zone of shield 12.

A fraction of the input power is transmitted radiatively, via the gap, from the specimen to shield 12 and flows through to the cooling water, while the remaining heat flow is introduced across the net cross section into the zone of heat absorber 6. The heat absorber accumulates the heat flow radiated from the specimen surface and conveys the collected heat flow to the heat conduit.

The passage of the heat through heat conduit 5 to the cooling water produces a temperature gradient, measured by differential thermocouple 10. The signal generated by the thermocouple is the principal measured value of the calorimeter and serves to determine the magnitude of the heat flow.

It goes without saying that here steady-state temperature conditions are implicit (stationary problem).

The dimensions of all elements listed above should be selected so as to observe the requirement of minimal temperature lag for predetermined resolution of heat flow measurements.

In order to monitor the calorimeter temperature pattern, provision is made for thermocouples embedded in calorimeter components: a thermocouple in the neighborhood of the bottom end face of the cylinder of heat absorber 6, a thermocouple in the vicinity of the upper end face of shield 12, and differential thermocouples on the inside of teflon ring 4 (for controlling the temperature difference between the shield and heat conduit coolers). The readings of these temperature sensors are not included in the design formulas, but made it possible to monitor the working condition of the apparatus.

It is further essential to emphasize the advantages inherent in the measuring cell used. To begin, the measuring cell can be used in combination with any means of specimen heating, be it a laser beam, imaging furnace, or electron bombardment, let alone the conventional technique involving the use of a heater winding around the end of the rod specimen. The latter technique is, however, unsuitable for work at temperatures above 1000 K and is also disadvantageous in that it increases by a large factor the temperature lag and hence prolongs the measurement period.

Second, varying the materials of washers 7 and the method of fixing the specimen in the heat absorber hole markedly change the thermal resistance in the specimen–calorimeter system, thereby enabling reasonable temperature gradients to be selected for the predetermined level of the mean specimen temperature.

Third, the measuring cell is adapted for work not only with metals, but with a multitude of other materials as well. The selected basic diameter of the specimen, 12 mm, makes it also representative in the case of investigating various porous materials, such as pseudoalloys and composites of different structure. Where it is desired to study specimens of smaller diameter, suitable inserts with a longitudinal groove should be inserted into the heat absorber and shield cylinders.

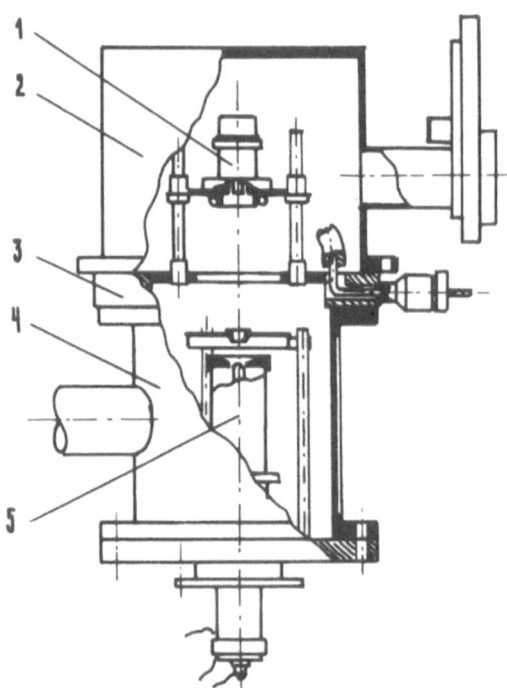

FIGURE 4. Vacuum chamber of the apparatus: 1—calorimeter; 2—upper hood; 3—calorimeter flange; 4—heating source chamber; 5—electron gun.

3.2. Description of Apparatus

The principal assemblies that form the vacuum chamber are shown in Fig. 4. The measuring cell 1 described above is secured on flange 3 that accommodates hermetic lead-ins of all electric circuits and the inlet of cooling water for calorimeter thermostating. Provision is made for thermally insulating the inlet of water flowing to the heat conduit coil from flange 3 in order to minimize the effect of temperature fluctuations of the flange on the heat conduit temperature. Flexible vacuum tubing (15/4 diameter) is used to connect the cooling coils of the heat conduit and shield to the water inlet and outlet pipes.

This arrangement enables easy travel of the calorimeter 1 on its uprights during system adjustment without breaking the hermetic seal. The vacuum tubing is also instrumental in providing electrical insulation between the calorimeter and the housing.

Flange 3 is mounted on the upper flange of heating source chamber 4 which serves as a base for installing the apparatus and is furnished with a branch pipe for connecting it to the air evacuation system. The flange of electron gun 5 is attached to the bottom flange of the heating source chamber.

After placing the test specimen in the calorimeter, hood 2 is lowered onto flange 3, provision being made in the hood for a branch pipe with a window pyrometry and a system of glass protection. An arrangement for the protection of the window against the molecular beam sublimated from the hot specimen is placed in the same branch pipe.

All flange joints contain rubber seals and are vacuum-tight. The evacuation system is designed to obtain a vacuum in the chamber of 10^{-3}–10^{-2} Pa or better.

All elements are water cooled in the vacuum chamber. Cooling of the top hood ensures the thermostating of surfaces around the calorimeter in order to conserve calorimeter calibration conditions.

Cooling of flange 3 and chamber 4 removes the heat evolved by the electron gun (several tens of watts). Poor electron gun adjustment may cause a fraction of the electron beam to fall onto the anode, thereby increasing heat evolution in chamber 4 and necessitating the removal of this heat by the water that cools the chamber.

Selecting the velocity of water flow in the heat conduit coil is as far as its function is concerned; it is essential here to minimize as much as possible the wall-to-liquid temperature drop and to maintain its steady minimal value for each operating mode. This suggests the importance of maintaining the constancy of water flow rate at a level providing a fully developed turbulent flow.

The requirements to be met by the electron gun as a source of the operating heat flow are as follows:

1. In order to reduce the intensity of X-ray radiation, it is desirable that the accelerating voltages used be as low as possible. In the present

apparatus recourse is made to accelerating voltages in the 1.0 to 10 kV range, depending upon the specimen temperature.

2. The electron beam should preferably ensure a uniform thermal load over the entire end-face area of the specimen. The greater the failure to observe this requirement, the lower the temperatures at which reliable thermal conductivity measurements are practicable.

3. Since the power required for specimen heating varies over a broad range, the beam intensity should be controlled both by the magnitude of the emission current and by the level of the accelerating voltage.

4. The system of automatically maintaining the preset operating conditions should be based on holding the desired temperature in the topmost, hottest pyrometric channel.

3.3. Control and Measuring Circuits

Figure 5 shows the diagram of the heat flow source, i.e., an electron gun with a direct filament cathode made of a tantalum strip, wherein the filament voltage is supplied from the secondary winding of the isolation transformer T_1. In the transformer, use is made of reinforced insulation between the windings, which is rated to withstand the total accelerating voltage (up to 15 kV).

The voltage of the primary winding is set by the autotransformer AT_1 (coarse adjustment) and, via the adjusting transformer T_2, by the autotransfor-

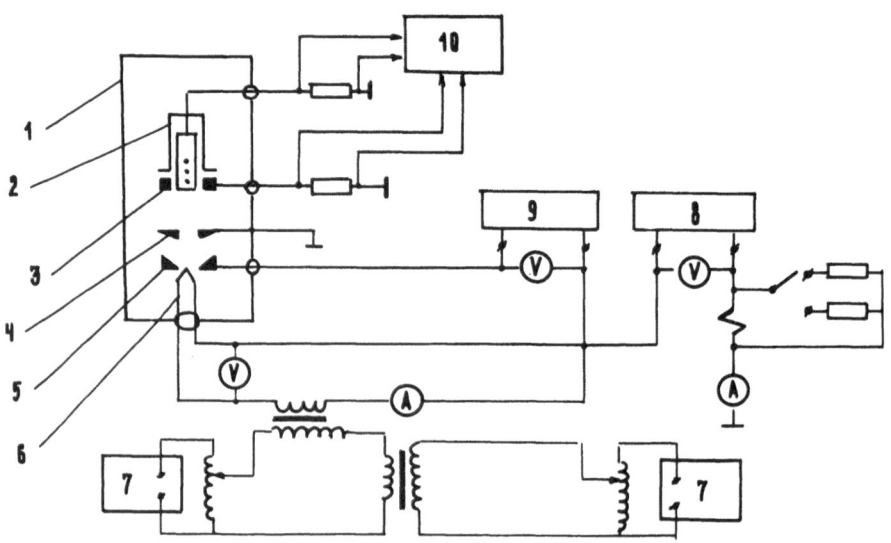

FIGURE 5. Circuit diagram of the working heat flow source.

mer AT_2 (fine adjustment). Where recourse was made to automatic adjustment, the autotransformer AT_2 should be replaced by a controlled thyristor.

The rectifier 8 (rating up to 5 kW) serves as a source of regulated high voltage. Its positive output is earthed via the emission current control ammeter A_1 and the protective relay winding W_r, which can be shunted by the resistance R_{sh} in order to set the requisite magnitude of operating current.

The normally closed contacts of this relay are incorporated in the rectifier protective circuit and serve to cut off the rectifier 8 once there occurs a current surge. The auxiliary electrode potential is preset by the control votage source 9 (this potential generally varies from -100 to -200 V relative to the cathode).

To control the distribution of electron beam current between the specimen and the calorimeter shield, these elements should be earthed outside the vacuum chamber via the reference coils of the resistance R_0, voltage drop across these coils being monitored by means of a digital voltmeter.

The measuring circuits of thermocouples are conventional and need not be described here.

4. PRE-STARTING AND MEASUREMENT PROCEDURES

4.1. Calorimeter Calibration

As pointed out earlier, the calorimeter output signal is the emf of the differential thermopile mounted on the heat conduit. The thermal flow in the heat conduit can, in principle, be calculated, provided the thermal conductivity and geometry of the heat conduit and the characteristics of the thermocouples are known.

More accurate results, however, are obtainable by the direct determination, in a suitable experiment, of the thermopile signal E dependence on the heat flow Q_z supplied into the calorimeter. In this experiment, a special-type heater is placed in the calorimeter instead of the specimen, and the electric power evolved under steady conditions by the heater is conveyed via the heat conduit to the cooler, whereby the thermopile signal is formed.

It is convenient to introduce the quantity $R = E/W$ (mV W^{-1}), which characterizes the sensitivity of the converter used. Having measured E and W for several power values, one can find, by the method of least squares, an approximating expression for the relationship. It is noteworthy that such an experiment automatically accounts for heat losses Q_l from the calorimeter surface into the ambient medium. Hence, strictly speaking,

$$W = Q_l + kE$$

where k is the proportionality factor which takes into consideration thermopile sensitivity, the thermal conductivity of the heat conduit material, and heat conduit geometry, so that

$$R = \frac{W - Q_l}{kW} = \frac{1}{k}\left(1 - \frac{Q}{W}\right) \tag{24}$$

For a given flow rate and temperature of the cooling water all terms on the right-hand side of the reduced equation are the single-valued function of the power introduced or of the quantity E interrelated therewith.

The value of $R(E)$ turned out to be variable, due to the dependence of the thermal conductivity of the heat conduit material upon temperatures and because of the effect of heat losses. For instance, in the case of the copper-constantan junctions of the new thermopile employed in the apparatus under consideration, calibration yields the following expression for $R(E)$ in mV W^{-1} units:

$$R = 0.1939 + 1.57 \times 10^{-3} E$$

With a voltmeter resolving power of 10 μV, heat flow sensitivity equals 0.05 W. Since the magnitude of thermal flows is at least 10 W, the error of thermal flow measurement will be 0.5% maximum.

4.2. Auxiliary Procedures Required for Calibration Accuracy Enhancement

As noted above, the calibrating heater is introduced into the heat absorber 6 in Fig. 3. With a view to preventing heat losses by radiation from the heating element toward shield 12, the outlet opening of the heat absorber cylinder is plugged with a double copper shield in contact with the walls. A thin tube made of copper foil is inserted into the longitudinal groove in the wall of cylinder 6 so as to contact the walls. Electric leads (two current and two potential leads), on being previously placed in an annular groove between lids 8 and 9, are brought out through a slit in the top of cylinder 6. All these precautions are taken in order to prevent uncontrollable heat losses from the calorimeter zone in the course of calorimeter calibration.

Figure 6 presents the design of a calibrating heater wherein the heating element comprises three (or four) parallel-connected tungsten coils 2. At one end the spirals are connected to copper cylinder 3 clamped in the well for fastening the specimen, so that the calorimeter copper casing functions as a current lead. This arrangement minimizes overheating in the heater in relation to the calorimeter at a given heater rating and reduces the thermal lag of the system. The other ends of the coils are connected to the central rod made of stainless steel, the current lead-out being attached to the upper part thereof.

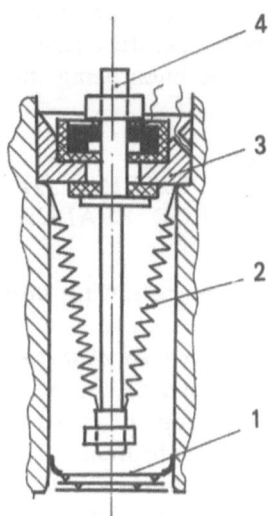

FIGURE 6. Calibrating heater: 1—pack of shields; 2—heating winding; 3—casing; 4—central electrode.

The heater is energized by DC current from a regulated rectifier (rated power up to 300 W), and the useful power is measured by the four-probe technique. Calibration is carried in a vacuum identical with that employed in the experiments proper, the same being true of the temperature of water used for cooling the chamber walls.

4.3. Distinctive Features of Measurements

The measurement procedure is contained in the description of the technique used. After mounting the specimen, the vacuum chamber is pumped down, and in the cold apparatus check measurements are made of the electric parameters of all sensors (electrical resistances and generated thermoelectromotive forces).

Next, the cooling system is set in operation and adjusted to attain the preset temperature and flow rate of water passing through the calorimeter. Check measurements of relevant parameters in all sensors should again be carried out.

A voltage is then applied to the electron gun cathode, and the systems for recording the signals from sensors are switched in, followed after an interval of 3 to 5 minutes by turning on the electron gun high voltage and bringing cathode incandescence to the level adequate for obtaining the requisite emission current. Steady-state conditions having been attained, all relevant parameters should be measured and, in compliance with the test program, electron gun power should be adjusted for further operating conditions.

The test program is to involve periodic return to one of the previous low-temperature tests in order to detect possible changes in the properties of the specimen being studied.

5. PROCESSING OF EXPERIMENTAL RESULTS AND ERROR EVALUATION

The following information is available for thermal conductivity calculation.

Specimen characteristics prior to test. Dimensions: diameter, D; cross-sectional area, F; the coordinates of pyrometric channel centers, z_1, z_2, z_3, z_4. Dimensions of radial pyrometric channels: diameter, d; depth, l_h. Net cross-section coordinate: z_m, $z_2 < z_m < z_3$.

Calibration result. $R = f(E)$.

Results of measurements. The emf of the thermocouples on a specimen, $U_T(z_i)$, or current strength in the pyrometer tube, $i_p(z_i)$. Thermocouple sensor signal, E_i. Signals of thermocouples defining the state of the calorimeter and its environment.

The algorithm of operating parameter processing is as follows:

1. Approximation of relationship $U_T(z)$ by four points for a suitable analytical function.
2. Calculation of $U_T(z = z_m)$ and derivative $\partial U_T(z)/\partial z$ at the point $z = z_m$.
3. Calculation of the measured value of the longitudinal temperature gradient,

$$G = \left(\frac{\partial T}{\partial z}\right)_{z=z_m} = \left(\frac{\partial U_T(z)}{\partial z}\right)\left(\frac{\partial T}{\partial U}\right)_{U=U(z_m)}$$

   where $(\partial T/\partial U)_{U=U(z_m)}$ denotes the thermocouple sensitivity for the value of the emf at the point z_m as determined during thermocouple calibration.
4. Calculation of the thermal flow $Q_z = E_i/R_{E=E_i}$.
5. Calculation of the measured value of the thermal conductivity,

$$\lambda_m = Q_z/(FG)$$

6. Determination of the reference temperature,

$$T_m = f(U_{z_m})$$

5.1. Calculation of Corrections

The value of the thermal conductivity found at stage 5 of the latter algorithm should be corrected due to the deviation of actual experimental conditions from the mathematical model. Taking into consideration the principal factors, one can write

$$\lambda = \lambda_{\mathrm{m}}\left(1 + \sum_i c_i\right) \tag{25}$$

where c_i denotes corrections.

1. Correction c_1 for thermal expansion. Heating the specimen changes the cross-sectional area and the distance between pyrometric channels. Allowing for this circumstance necessitates the introduction of a correction for thermal expansion calculated using the formula

$$c_1 = -\alpha(T_z - T_0)$$

where α is the average linear thermal expansion coefficient in the T_z-T_0 temperature range, and T_0 is the temperature at which the specimen geometrical dimensions were measured. Allowance for this correction reduces the required value of the thermal conductivity.

2. Correction c_2 for radiative heat exchange in the gaps of the apparatus structure is calculated using the expression $c_2 = \Delta Q_r / Q_z$ where ΔQ_r is the resultant radiative thermal flow entering the heat absorber and recorded together with the principal thermal flow Q_z.

The main component of this correction was introduced in the course of substantiating the present method [cf. equation (23)]. Additional contributions to the correction value are likely to appear, depending on the specimen geometry as well as the number and dimensions of design slits in the calorimeter. Allowance for this correction reduces the sought thermal conductivity value.

3. Correction c_3 for the effect of pyrometric channels is introduced when the method of measurement involves cavity-type blackbody models in the form of blind radial borings $2b$ in diameter and having depth h. Such channels enhance the effective thermal resistance of the specimen. The theory underlying the introduction of this correction is described elsewhere,[5] the numerical value being found from the equation

$$c_3 = \frac{2}{\sqrt{\pi}} \frac{h}{D} \frac{\eta_0}{1 + \eta_0} \tag{26}$$

where η_0 can be found from the plot in Fig. 7 as a function of the expression $\pi b^2/(aL) = 2\sqrt{\pi}b^2/(DL)$, L being the distance between the nearest adjacent channels.

Taking into account this correction enhances the required value of the thermal conductivity.

4. Correction c_4 for temperature field two-dimensionality. If use is made of the local value of the longitudinal temperature gradient in calculating the value of λ_m [cf. equation (25)], then there arises the necessity of introducing a correction to take account of the deviation in this gradient from the average value across the specimen cross section. On taking into consideration the results of the theoretical analysis represented by equations (12), (13), and (14), the correction should be written in the following form:

$$c_4 = 1 - c(r)$$

Specifically, in the case of temperature measurement on the rod axis $\rho = 0$

$$c_4 = 0.25hR - \frac{(hR)^2}{4(4 + hR)}$$

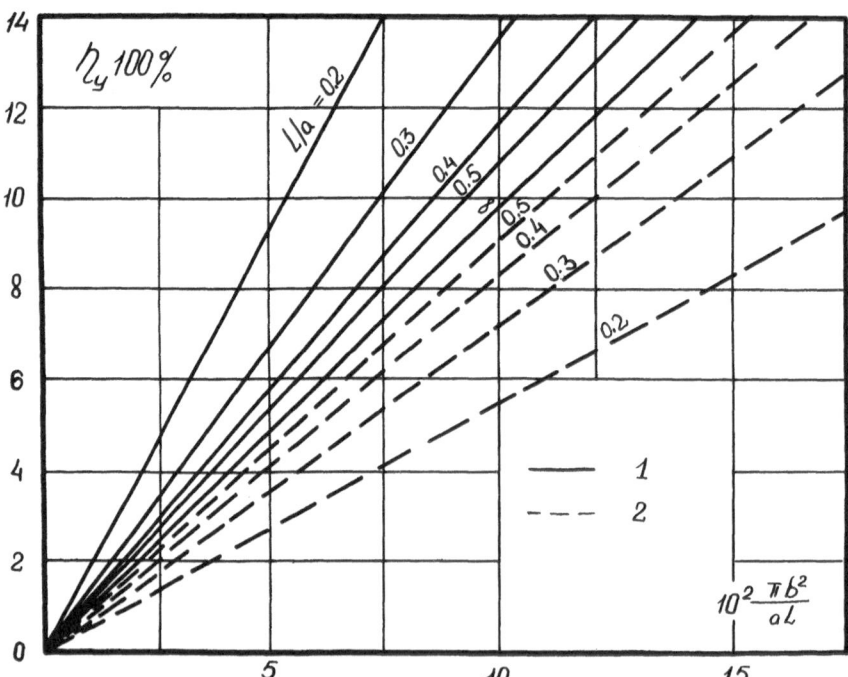

FIGURE 7. Correction for the effect of pyrometric channels with the probes placed at the following specimen points: (1) $y = 0$ (specimen axis); (2) $y = R\sqrt{\pi}$.

The introduction of this correction increases the sought value of the thermal conductivity.

It is noteworthy that the accuracy of determining this error is not high, particularly so in the case of temperature measurements by an optical pyrometer wherein use is made of radial pyrometric channels for absolute blackbody simulation. The effective radiation in the channel is associated with the temperature distribution in an intricate manner and, accordingly, correlating the measured temperature with the local one presents serious difficulties.

Here, the practical procedure comprises increasing the channel depth to *ca* 1.5*R*, where *R* denotes the specimen radius, under the assumption that the sought mean gradient is being measured in the experiment. Judicious selection of the specimen geometry permits decreasing the possible error to tolerable values.

If, for example, the thermal conductivity of the specimen at 2500 K equals 50 W m^{-1} K^{-1}, the emittance being 0.25, the possible error due to neglect of this correction would be 0.7% maximum. This correction can therefore be disregarded in a number of instances.

When temperature measurements are performed on the specimen surface, the sign of this correction is reversed, and in this case the mean gradient is greater than the longitudinal gradient for $r = R$.

5.2. *Estimate of Measurement Error*

The resultant relative systematic error of thermal conductivity measurements may be assessed using the expression

$$\delta_\lambda = \pm 1.1\sqrt{\delta_Q^2 + \delta_G^2 + \delta_F^2 + \delta_M^2} \qquad (27)$$

where δ_Q is the residual systematic error not eliminated in thermal flow determination, δ_G is the same in temperature gradient determination, δ_F the same in specimen cross-section determination, while δ_M denotes the possible effect of other factors and inaccuracies in reference temperature determination. Each of the error species depends on specimen properties and specific conditions of the test.

The value of δ_Q, with allowance for the inaccuracy of taking into account radiative transfer over the slits [cf. equation (23)], may vary from about 0.5% at 1300 K to 1.0–1.5% at ~2500 K.

The error δ_G will depend on the temperature measurement technique used. For thermocouples, it will not exceed 1.5–2.0%, while in the case of a disappearing-filament optical pyrometer the error may be as high as 2.5–4.5%.

The error δ_F, with allowance for the inaccuracy of taking into account the thermal expansion and for the pyrometric channel role correction, is likely to reach 0.2–0.3%.

The overall error calculated by equation (27) may vary from 2.5% when $T < 1300$ K (temperature measurement by means of thermocouples) to 5.5% when $T \sim 2500$ K (temperature measurement using the optical pyrometer).

The error may be reduced, depending on the specific experimental conditions and the type of instruments used. In particular, the employment of photoelectric pyrometers with a sensitivity nearly an order of magnitude higher than that of disappearing-filament optical pyrometers results in a substantial reduction in the error thanks to the enhanced accuracy of temperature gradient determination.

6. FUNCTIONAL POTENTIALITIES AND LIMITATIONS OF THE APPARATUS

Range of test substances. The apparatus is primarily intended for investigating the properties of metals and alloys. It is, however, adapted for the study of other materials such as carbides, nitrides, carbon-graphite composites, etc.

Range of thermal conductivity values. The higher the conduction, the more effective the present technique. For values ~ 10 W m^{-1} K^{-1}, measurements can be carried out using the specimens possessing the geometry described above. At lower conduction values, it is feasible to resort to composite specimens furnished with an extension made of a metal having high thermal conductivity.

Temperature range. The specimen end face can be brought to the melting point. Then, lowering the maximum working temperature, at which thermal conduction values are determined, would be equal to the temperature drop in the shield zone. This temperature drop is governed by the specimen properties, by the manner in which the specimen is secured in the heat absorber, and also by the temperature level.

For practical purposes in the case of metals, the maximum temperatures should equal 0.8-0.9 of the melting point. This temperature drop can be reduced, provided the electron source used, namely, the electron gun, ensures uniform distribution of the heat load over the plane of the specimen end face. In this situation, the hot zone of temperature profile stabilization can be shortened or, stated differently, the net cross section can be brought closer to the heating zone.

Certain limitations imposed on the temperature are related to the specimen vapor pressure. Active sublimation of the specimen material affects adversely the stability of the electron gun operation, the practical upper limit being the temperature at which the pressure of specimen saturated vapors attains 10^{-3} to 10^{-2} mm Hg.

The lower temperature limit is defined by the coolant temperature. In the experiments under consideration, use is made of water having working temperature 300 K. The lower operating temperature of the specimen in this situation might be equal to 320–350 K.

In conclusion, it appears appropriate to note that the presence of an intense heating source and a means of heat flow measurement enables some additional characteristics of metals to be measured.[6] Thus, securing a test substance tablet in the shield zone makes it possible to carry out an experimental determination of the integral emissivity of the metal of interest. Heating a tablet disposed on the upper edge of the heat absorber enables the heat capacity of a test substance to be determined.

7. EXAMPLES OF APPLICATION

In order to check the apparatus one needs a reference material with well-established properties. At moderate temperatures austenitic stainless steel (12 Ch 18 N 10 T) can be used as such a reference material.

Experimental data available for this steel were analyzed by Sergeev.[7] The most reliable values of the thermal conductivity for this substance were approximated by the relation

$$\lambda = 10.15 + 0.01607T \ (W \, m^{-1} \, K^{-1}) \tag{28}$$

which refers to the temperature range of 273–1100 K. The error of this recommendation is equal to ±4% and included the possible effect of the steel composition variations.

The specimen of this steel studied in this work was fabricated by casting in the following composition (in mass percent): Cr—17.28; Ni—10.05; Ti—0.6; C—0.11; Mn—1.82; Si—0.41; S—0.018; P—0.08, the rest being Fe. The density of the metal was equal to 7.70 g cm^{-3}; the electrical resistivity at 295 K was 79.5 μohm cm.

The diameter of the specimen was 12 mm. Four cylindrical cavities of 6-mm depth and 1-mm diameter were drilled in the specimen with a pith of 4 mm in the net section vicinity.

The twisting of the chromel–alumel thermocouple wires was inserted into the cavity and welded there on the wall of the cavity. The test specimen with the fasted thermocouples was fixed in the seat of the calorimeter with the aid of a slatted washer manufactured from stainless steel. Measurements were carried out at a pressure of approximately 10^{-2} Pa inside the vacuum chamber. No heat treatment of the specimen was applied before the testings.

TABLE 1
Thermal Conductivity Data of Stainless Steel 12 Ch 18 N 10 T

T (K)	471	540	597	681	738	808	880
λ(W m^{-1} K^{-1})	17.54	18.52	19.69	21.64	22.22	23.28	24.43
T		897	1030	922	873	795	541
λ		24.72	26.62	24.90	24.24	22.79	18.02

The results in the order they were obtained are presented in Table 1. All necessary corrections, including the effect of thermal expansion, have been made to obtain the data.

The least-squares method was employed in order to approximate the data by the following relationship:

$$\lambda = 9.55 + 0.01682\,T \ (\text{W m}^{-1}\,\text{K}^{-1}) \tag{29}$$

which is in good accord with the above-mentioned recommendation. The deviation of one mean curve from another does not exceed ±2%.

In order to prove the setup at higher temperatures the experiments were performed on molybdenum specimens. The metal was 99.9% pure. Its density was equal to 10.2 g cm^{-3}, and the specific electrical resistivity at 300 K was 5.86×10^{-6} ohm cm.

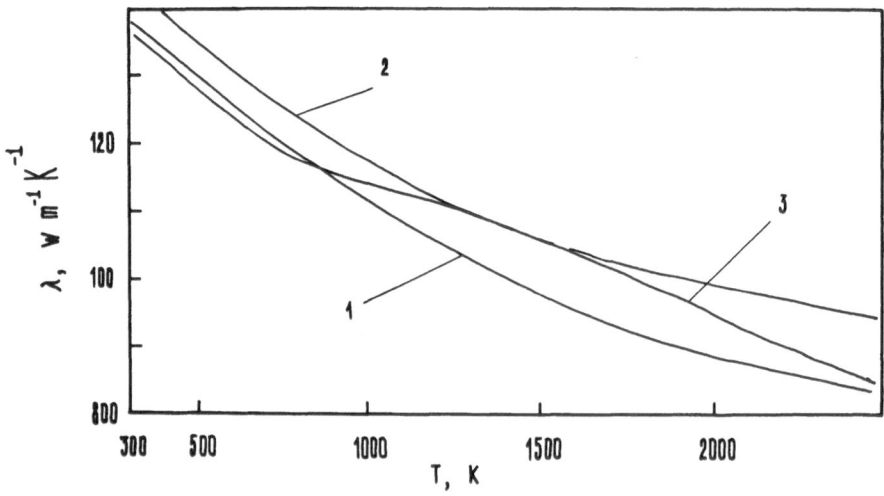

FIGURE 8. Thermal conductivity of molybdenum: (1) published data[8]; (2) published data[9]; (3) new data.

The chromel–alumel thermocouples were used to measure the temperature distribution along the specimen at temperatures lower than 1300 K. Higher temperatures were indicated by an optical pyrometer. The blind holes in the specimen acted as blackbody models.

The results obtained for this specimen are plotted in Fig. 8 alongside the generalized curve taken from a well-known reference book.[8] The uncertainty of the curve at high temperatures is ±(4–10)%. Taking into account that this curve relates to 99.95% pure molybdenum one can compare it with our data without any corrections for the difference in the composition.

With provision for the inaccuracy in our results (±5–6%) their agreement with the recommended values is good enough over the temperature range investigated (350–2500 K).

The above results have confirmed the serviceability of the described setup and substantiated the legitimacy of the principles underlying the technique. The most important of them should be emphasized once more. If one can implement direct measurements of the total heat flux penetrating a cross section of the investigated specimen, there is no longer any necessity to thermally isolate its lateral surface. The refusal to create a thermal isolation system makes it possible to extend the temperature range of the longitudinal heat flow technique without any marked loss in accuracy.

REFERENCES

1. V.E. Peletsky, "On the Role of Temperature Field Two-Dimensionality in Thermal Conductivity Studies" (in Russian), *Teplofizika Visokikh Temperatur* 6(1), 133–138 (1968).
2. D.L. Timrot and V.E. Peletsky, "Application of Electronic Heating in Thermal Conductivity Investigation" (in Russian), *Teplofizika Visokikh Temperatur* 1(2), 168–172 (1963).
3. D.L. Timrot and V.E. Peletsky, "Investigation of Integral Emissivity and Thermal Conductivity of Zirconium" (in Russian), *Teplofizika Visokikh Temperatur* 3(2), 223–227 (1965).
4. V.T. Aleksandrov, "On the Determination of Angular Emissivity for a System of Two Coaxially Disposed Cylindrical Bodies" (in Russian), *Inzhenerno-Fizicheskii Zhurnal* 8(5), 609–613 (1965).
5. V.E. Peletsky, "The Effect of Pyrometric Channels on the Results of Measuring Thermal and Electrical Conductivities in Solid Bodies" (in Russian), *Teplofizika Visokikh Temperatur* 12(2), 359–362 (1974).
6. V.E. Peletsky, *Studies of Thermophysical Properties of Materials under Electronic Heating Conditions* (in Russian), P.H. Nauka, Moscow (1983).
7. O.A. Sergeev, in: *Teplofizicheskie Svoistva Veshchestv i Materialov*, No. 13, p. 133, GSSD, Izdatelstvo Standartov, Moscow (1979).
8. Y.S. Toulokian (ed.), *Thermophysical Properties of Matter*, Vol. 1. *Thermal Conductivity*, 1469 pp., IFI/Plenum, New York (1970).
9. "Molybdenum. Thermal Conductivity at Temperatures from 200 to 2600 K" (in Russian), GSSD 39-82, Izdatelstvo Standartov, Moscow (1983).

6

The Probe Method for Measurement of Thermal Conductivity

ALFRED E. WECHSLER

1. INTRODUCTION

The line heat source method and the probe method for measurement of thermal conductivity are based on the same (or a very similar) theory. Both methods have been used to measure thermal conductivity of insulations, soils, biological materials, liquids, rocks, ceramics, and glass over a wide range of temperatures and other environmental conditions.

The line heat source method was suggested by Schleiermacher[1] in 1888 and later by Stalhane and Pyk[2] in 1931. The first practical use of the method was in 1949 by Van der Held and Van Drunen,[3] who measured the thermal conductivity of liquids. Although more simple in analytical and experimental nature than the thermal conductivity probe, the line heat source equipment is relatively fragile, is not useful for severe environmental conditions or field measurements, and is best suited to insulations, fine granular materials, and liquids.

The first probe utilizing the principle of the line heat source was described in 1950 by Hooper and Lepper,[4] who measured the thermal conductivity of soils. In the two decades which followed, many investigators analyzed the measurement theory of finite-dimension probes and an equal number experimented successfully with practical application of probes, incorporating designs of different geometries, materials, and physical characteristics, and measuring the thermal properties of soils, foods, granular materials, and solids. Today, probes are used for most materials (except perhaps for liquids) because of their ease of handling and ruggedness.

ALFRED E. WECHSLER • Arthur D. Little, Inc., Cambridge, Massachusetts 02140, USA.

An important attribute of both the line heat source and the probe methods is their ability, under most practical conditions, to measure thermal conductivity directly, even though the methods are "transient." Under certain conditions, the line heat source and probe methods can be used to measure both thermal conductivity and thermal diffusivity independently. Another important attribute of the probe is its ability to measure thermal conductivity "in situ."

The focus of this chapter will be on thermal conductivity probes; however, background and descriptions will be given of the line heat source method to show the differences and practical uses. Following a presentation of measurement theory and analysis, a description of apparatus, measurement methods, and instrumentation will be given. Practical information on the use of probes—measurement problems and possible solutions—will conclude this chapter.

2. MEASUREMENT THEORY AND ANALYSES

2.1. The Line Heat Source

The constant production of heat by a linear source of heat in an infinite volume of homogeneous material produces a cylindrical temperature field. The temperature rise at any point in the material depends upon the thermal conductivity. For an infinitely long linear heat source of strength q per unit length, the temperature, θ, at a distance, r, from the source is given by Carslaw and Jaeger[5] as

$$\theta = -\frac{q}{4\pi\lambda} \text{Ei}\left(-\frac{r^2}{4\alpha t}\right) \tag{1}$$

where

$$-\text{Ei}(-x) = \int_x^\infty \frac{e^{-x}}{x} dx \tag{2}$$

while α is the thermal diffusivity of the material, λ the thermal conductivity, and t time. [Ei(x) is an exponential integral of the form

$$\text{Ei}(x) = -\delta - \ln x - \frac{x^2}{2 \cdot 2!} + \frac{x^3}{3 \cdot 3!} - \frac{x^4}{4 \cdot 4!} + \cdots$$

where δ = Euler's constant = 0.577216.] The boundary conditions above are: $t = 0$, $r \neq 0$, $\theta = 0$; $t > 0$, $r = \infty$, $\theta = 0$; and $t > 0$, $r \to 0$, q = constant = $-2\pi r\lambda \, d\theta/dr$.

For values of $r^2/4\alpha t$, small compared to unity, corresponding to small radii and/or large values of time, equation (1) reduces to

$$\theta = \frac{q}{4\pi\lambda}\left(-0.5772 - \ln\frac{r^2}{4\alpha t}\right) \tag{3}$$

Then, for a fixed radius, the temperature rise θ_2 and θ_1 at times t_2 and t_1 are related by

$$\theta_2 - \theta_1 = \frac{q}{4\pi\lambda}\ln\frac{t_2}{t_1} \tag{4}$$

or the thermal conductivity is given as

$$\lambda = \frac{q}{4\pi}\left(\frac{\ln t_2/t_1}{\theta_2 - \theta_1}\right) \tag{5}$$

Thus, from the temperature rise at two different times and from the strength of the heat source, the thermal conductivity can be calculated. Alternatively, from the slope of a plot of temperature rise versus logarithm of time, the conductivity can be determined if the strength of the heat source is known. We note that it is not necessary to specify the radius at which the temperature is measured, provided the value of $r^2/4\alpha t$ is small.

2.2. The Finite-Dimension Probe

Any experimental device embodying the principles of the line heat source has finite dimensions and is usually constructed of materials with thermal properties different from those of the materials being tested. Therefore an extension of the line heat source theory is needed to account for finite dimensions and practical materials of construction.

Theoretical interpretations of probes for thermal conductivity measurements have been given by Carslaw and Jaeger,[5] Blackwell,[6] Jaeger,[7] and de Vries.[8] These investigators considered the thermal behavior of heated, infinitely long, solid or hollow cylinders, of finite or infinite thermal conductivity, with zero or finite heat capacities, and zero or finite contact resistance between the cylinder and the surrounding medium.

A useful analysis by Jaeger,[7] shows that the temperature rise of a cylindrical probe of infinite conductivity, heat capacity per unit length of $M_1 C_1$, and contact resistance of $1/H$, with a radius of b, is given as a function of time and other parameters as

$$\theta = \frac{q}{\lambda}G(\lambda/bH, \beta, \tau) \tag{6}$$

where $G(\lambda/bH, \beta, \tau)$ is an integral function of the three parameters which give the influence of the contact resistance, the mass, and specific heat of the probe, and time after heating; $\tau = \alpha t/b^2$ and $\beta = 2\pi b^2 \rho C/M_1 C_1$, where b^2/α is a characteristic time, τ is a dimensionless parameter, ρ and c are respectively the density and heat capacity of the medium, and β is a dimensionless parameter equal to twice the ratio of the volumetric specific heat of the medium to that of the probe.

For large values of time, the temperature rise of the probe is given by

$$\theta = \frac{q}{4\pi\lambda}\left[\frac{2\lambda}{bH} + (\ln 4\tau - \gamma) - \left(\frac{4\lambda/bH - \beta}{2\beta\tau}\right) + \frac{\beta - 2}{2\beta\tau}(\ln 4\tau - \gamma)\right] \quad (7)$$

It is seen from equation (7) that at large values of time (and large values of τ) the temperature rise becomes directly proportional to the logarithm of time. [Equation (7) reduces to equation (3), with the radius of the probe b replacing r and the inclusion of a term for contact resistance at the probe/medium interface.]

Blackwell[6] has analyzed a hollow cylindrical probe of infinite conducting material with heat applied at the probe surface, in a manner similar to Jaeger's analysis, and derived approximate solutions for the temperature of the probe at small and large times. The small-time approximate solution is of limited significance to actual probe measurement; the long-time results are identical to those of Jaeger.[7]

Blackwell[6] also considered a hollow cylindrical probe with finite thermal conductivity (a reasonable representation of most thermal conductivity probes) and obtained the following equation for the temperature rise at the inner probe radius:

$$\theta_a = \frac{q}{4\pi\lambda}\left\{\ln 4\tau - \gamma + \frac{2\lambda}{bH}\right.$$

$$+ \frac{1}{2\tau}\left[\ln 4\tau - \gamma + 1 - \frac{2}{\beta}\left(\ln 4\tau - \gamma + \frac{2\lambda}{bH}\right)\right.$$

$$\left.\left. - \frac{2\alpha}{b^2}(\Delta_1 + \Delta_2)\right] + O\left(\frac{1}{\tau^2}\right)\right\} \quad (8)$$

where a is the inner probe radius while Δ_1 and Δ_2 are functions of the thermal diffusivity of the medium to be measured and of the inner and outer radii of the probe, as follows:

$$\Delta_1 = \frac{1}{\alpha}\left[\frac{b^2}{8} - \frac{3a^2}{8} + \frac{1}{2}\ln\left(\frac{b}{a}\right)\left(\frac{a^4}{b^2 - a^2}\right)\right] \tag{9}$$

$$\Delta_2 = \frac{1}{\alpha}\left[\frac{b^2}{8} - \frac{a^2}{8} - \frac{1}{2}\ln\left(\frac{b}{a}\right)\left(\frac{a^2 b^2}{b^2 - a^2}\right)\right] \tag{10}$$

The contribution of Δ_1 and Δ_2 to the temperature rise of the probe is often negligible, provided the probe wall is relatively thin and the time is long.

A similar analysis of a hollow probe of finite conductivity, with a line heat source at its center (another good representation of an actual probe), was carried out by de Vries and Peck.[9] The results of these authors are the same as those of Jaeger[7] and Blackwell[6] for corresponding conditions.

Examination of equations (7) and (8) shows that the terms involving contact resistance, finite probe dimensions, conductivities, and heat capacities become small at large values of time or τ. Then the equation for temperature rise of the probe reduces to

$$\theta = \frac{q}{4\pi\lambda}\left(\ln\frac{4\alpha t}{b^2} - \gamma + \frac{2\lambda}{bH}\right) \tag{11}$$

The temperature rise at long times is proportional to the logarithm of time, the same as the line heat source, with an additional "constant" term associated with the contact resistance of the probe/medium interface.

Thus in most practical implementations of either the probe or line heat source, the same general method can be used to determine thermal conductivity, *provided* appropriate probe dimensions and materials are selected, "long times" ($t \gg b^2/\alpha$) are used for measurement, and other experimental errors discussed below are not large.

2.3. Analysis of Errors and Corrections

2.3.1. Line Heat Source

Although most investigators attempt to use equipment for line heat source measurements which approximate the theoretical conditions described in Section 2.1, several potential errors associated with practical systems must be considered and evaluated.

The principal sources of error in using equation (5) to evaluate experimental data are: (a) finite diameter and length of heater wire (heat source); (b) finite dimensions of the temperature sensor; (c) finite size of the sample; (d) variations in experimental conditions (temperature and source strength);

(e) variations in the sample (inhomogeneity or anisotropy); (f) contact resistances at the source/sample and sample/temperature sensor interfaces; and (g) improper timing in the experiment, e.g., improper use of the long-term solution of equation (1).

The effects of most of these sources of error have been examined in the literature (Arthur D. Little, Inc.,[10] Prelovsek and Uran[11]). The effects of finite sizes of heat source and temperature sensor are usually minimized by maintaining the heat source and temperature sensor parallel and making each with as small a radius as possible (consistent with strength and stability) and having the length of the source and thermal sensor many times (30–50) its radius and/or the distance between the source and sensor. Some practical guidance for dimensions is given by Prelovsek and Uran,[11] who suggest source and sensor diameters less than 0.1 mm and distances from the source to sensor of 1.5 to 3 mm. Similar dimensions are recommended by Wechsler and Black.[12] A further examination of the effects of source and sensor dimensions is given below when evaluating errors associated with finite probe size.

The size of the sample is relatively easy to control. In order for the external boundary of the sample (of radius R) not to significantly influence the temperature rise at the sensor (an accuracy of $<1\%$), Prelovsek and Uran[11] suggest that the maximum time for measurement be limited to

$$t_{\max} < 0.1 R^2/\alpha$$

One must make sure that $t_{\max}$ is sufficient to use the long-term solution to equation (1), so that $t_{\max}$ must also satisfy the following:

$$t_{\max} \gg r^2/4\alpha$$

where r is the distance from the source to the sensor.

Asher et al.,[13] in referring to Jaeger's work, suggest that for a probe of radius b the sample radius R needed to yield less than 10% error in probe temperature is given by

$$R/b \gg 3\left(\frac{\alpha t_{\max}}{b^2}\right)^{1/2}$$

which is essentially identical to the expression given earlier. In practice, if the sample radius is greater than 10 to 20 times the distance from the source to the sensor, the effect of sample size is negligible.

Variations in sample conditions and homogeneity are controllable by sample specification and will not be discussed here. (Prelovsek and Uran[11] describe a line heat source used between rectangular solid samples of different conductivities and suggest the errors associated with inhomogeneous samples.)

The error associated with contact resistance should generally be small, at least if any such resistance remains constant during the measurements and is small compared to the thermal resistance between the source and sensor. The effect of contact resistance on measurements of granular samples is described later for probe measurements.

The general conclusion is that, by proper experimental design, namely, long, thin heat sources and temperature sensors, properly spaced and within a large enough sample, the errors described above can be maintained on the order of 1% or less.

2.3.2. Thermal Conductivity Probe

The errors involved in the thermal conductivity probe method include those associated with (a) the finite dimensions and thermal properties of the probe, (b) axial heat flow in the probe, (c) finite specimen size, (d) contact resistance with the specimen, (e) inhomogeneity and anisotropy of the specimen, (f) variation in measurement parameters, including source strength and temperature of specimen, and (g) improper timing in the experiment. Depending upon the dimensions and materials of construction of the probes, these parameters can have a significant effect on measurement accuracy.

The effects of the finite radial dimensions and thermal properties of probes have been discussed earlier in the analysis of Jaeger[7] and Blackwell.[6] Thomas and Ewen,[14] using Blackwell's solution for the long-term response of a probe, show the fractional error in measurement of thermal conductivity as a function of thermal diffusivity of the sample, probe radius, heat capacities of the probe relative to the sample, ratio of outer to inner radius of the probe, and contact resistance.

Their equation for fractional error in $1/\lambda$ is

$$\text{fractional error} = \frac{b^2}{2\alpha t}\left[(1-\phi)\left(\delta - \ln\frac{4\alpha t}{b^2}\right) - \phi + \frac{2\phi\lambda}{bH}\right] \qquad (12)$$

where $\phi = (\rho C/\rho_1 C_1)[1 - (a/b)^2]$. For perfect contact, Thomas and Ewen[14] show that for values of ϕ between 0.75 and 2, the fractional error is less than 5% for $\alpha t/b^2 > 20$, and decreases to $\pm 2\%$ for $\alpha t/b^2 > 100$. To reduce the error in measurement (or the time at which the transient heating of the probe affects the results) it is important (in general) to minimize the thickness of the probe wall or to provide for increased values of the probe's heat capacity relative to that of the sample for values of ϕ up to 1.25. The authors suggest that for values of $\phi \simeq 1.25$ to 1.5, these initial errors should be small. The effect of contact resistance is also shown to decrease at $\alpha t/b^2$ increases, with fractional errors less than about 5% for values of $\alpha t/b^2 > 20$, $\phi = 2.0$, and $bH/\lambda > 10$.

The finite dimensions of the probe, the heat capacity of the materials, and the contact resistance generally have the effect of extending the initial time of measurement before which the temperature rise of the probe becomes proportional to the logarithm of time. Wechsler[15] estimated initial errors for probes used in several sample materials. Typical results suggest that for a typical stainless-steel probe of outer radius $(b) = 0.08$ cm, inner radius $(a) = 0.054$ cm, and length 15.6 cm, the initial errors would be less than 1.5% in soils $(\lambda > 0.0016$ W cm^{-1} K$^{-1})$ after about 1000-s operation, but up to 25% in low-density polystyrene insulations $(\lambda \simeq 3 \times 10^{-4}$ W cm^{-1} K$^{-1})$ at the same measurement time. [At 1000 s, $\alpha t/b^2 \simeq 240$ for dry soils, showing the consistency of these results with those of Thomas and Ewen.[14]]

Another error associated with probe dimensions is the axial heat flow in the probe. Blackwell[16] analyzed a cylindrical probe which is infinitely long but heated for only a finite length. He states that the actual axial heat loss for a finite probe should be less than that indicated by his analysis. He concludes that the fractional error in calculation of thermal conductivity, due to axial heat flow, is a complex function of time, probe parameters, and sample parameters:

$$\text{relative error} = \pi^{1/2} \exp\left(\frac{-Z^2}{4\tau}\right)\left[\frac{(4\tau)^{1/2}}{Z} + 2\sigma Z(\varepsilon - \eta)(4\tau)^{-3/2}S\right] \quad (13)$$

where $\sigma = A_C/\pi b^2$ for hollow probes, 1 for solid probes; $Z = L/b$ (ratio of probe heated length to radius of probe); $\varepsilon = $ ratio of thermal conductivity of probe material to that of the sample; $\eta = $ ratio of volumetric heat capacity of probe material to that of the sample; and $S = \ln 4\lambda - \delta + 2\lambda/bH$.

We note that the error for axial heat flow increases with time, so that for probes of finite dimensions there will be a measurement time after which axial heat-flow errors are considerable, a plot of temperature rise versus logarithm of time is no longer linear, and the simplified calculations and analysis for the probe are no longer valid. Blackwell[16] considers large ($\simeq 25$ cm) hollow probes and concludes that values of $Z > 25$ are sufficient for axial flow less than 1%. Wechsler[15] evaluated the axial heat-flow error using Blackwell's approach and concluded that for the same stainless-steel probe as described above ($b = 0.08$ cm, $a = 0.054$ cm, $L = 15.6$ cm) the errors for axial heat flow in dry soils would be less than 1% for times of about 3000 s. Errors could reach 5% for higher materials of conductivity in 1500 s, and such a probe would not be very suitable for materials of very low density and low conductivity.

In a study of a needle probe for measurement of thermal conductivities of liquids, Asher et al.[13] use a probe of length 3.5 cm and radius 0.045 cm ($L/r = 78$) and conclude, using the work of Kierkus et al.,[17] that the measurement error was less than 2%.

Thus it is important to consider both initial errors, which get smaller as the measurement interval increases, and axial heat-flow errors, which get larger as the measurement interval increases. By proper choice of probe dimensions and construction materials these errors can be made small enough, so there is a time interval suitable for experimental measurements in which simple probe theory can be used for calculations.

In general, the larger the ratio of probe length to radius, the smaller the probe wall thickness, and the lower the conductivity of the probe materials, the longer the time for onset of axial heat-flow errors. Similarly, the smaller the wall thickness and diameter of the probe, the shorter will be the time during which effects of probe properties are important.

The effect of finite sample size is the same for that of a line heat source. Generally, when $t < 0.1R^2/\alpha$ (where R is the sample radius) the effect of sample boundary conditions will be small.

The effect of contact resistance of the probe has already been mentioned, as it increases the time of the initial transient effects. The effects of contact resistance on the nonsteady-state probe method was studied by Bruijn et al.[18] These authors have expanded the original Jaeger model of a probe with contact resistance at the surface to include a four-region model of a probe, with a central heating wire source surrounded by an insulating material, an outer shell or jacket (probe), and a temperature sensor between the outer shell and the heater. This model is an excellent representation of a probe, but is very difficult to use in practice.

Inaba[19] examined the effect of contact resistance by considering a probe used for measurement of materials made up of spherical particles. Through both analysis and measurements using a thermal conductivity probe ($L/r = 200$, $r = 0.075$ cm to $r = 0.5$ cm), Inaba concludes that the contact resistance is sufficiently small that measurement accuracies of several percent can be obtained if the ratio of probe radius to sample particle radius is about unity or larger. In most experiments, the contact resistance is small because of the fluid in the sample or the force between the sample and probe is about the same as among the particles comprising the sample.

As in the case of the line heat source, errors associated with inhomogeneity and anisotropy of the sample, and measurement parameters such as temperature and heat source stability, must be controlled by the investigator.

The conclusions from the error analysis are that under carefully controlled experimental conditions, with careful a priori evaluation of probe design parameters, probes can be used to measure thermal conductivity of a wide variety of materials with measurement errors less than 5%. To use simple probe theory for measurement, care must be taken to insure an appropriate "measurement time," when initial transient errors have become small and axial heat flow and sample size errors are not yet large.

3. DESCRIPTION OF APPARATUS AND INSTRUMENTATION

3.1. Line Heat Source

3.1.1. Conceptual Designs

The conceptual design of a line heat source apparatus is very simple, namely, a linear source of heat which has a minimal radial dimension and a large longitudinal dimension, immersed in a homogeneous sample of large dimensions. The heat source should be easily controllable, i.e., it should provide a stable source of energy of uniform strength along the length of the source. In principle, either the temperature of the linear heat source or the temperature at a suitable position in the sample can be measured.

The sample should be uniform, isotropic, and extend to sufficient radial and longitudinal dimensions to simulate an "infinite sample" during the period of measurement. The sample container should provide for uniform control of temperature of the sample and permit maintenance of other appropriate environmental conditions such as gas pressure.

3.1.2. Practical Design Considerations

In practice, the line heat source apparatus normally consists of a long, thin, electrically heated resistance wire and a similar temperature sensor.

The heater wire is often made of Nichrome, constantan, or other alloy which exhibits relatively small changes in resistance with temperature. The wire is usually suspended between electrically and thermally insulated supports and stretched tightly. Typical dimensions range from about 0.02 mm to 0.2 mm in diameter and from 15 to 30 cm in length. Thin ribbon heaters have also been used (widths of 1 to 2 mm), and are often more practical when measuring the thermal conductivity of solid materials such as ceramics or insulation boards, since they often have greater strength and can be placed between two flat samples. In apparatus for measuring the properties of liquids, the heater wire is often used as both heater and temperature sensor (see Fig. 1a). In this case, the resistance of the heater changes with temperature; appropriate instrumentation must be used both to maintain constant heater energy output per unit length as well as to measure the heater wire resistance (temperature). Platinum and other similar heater wires have been used, with bridge circuitry

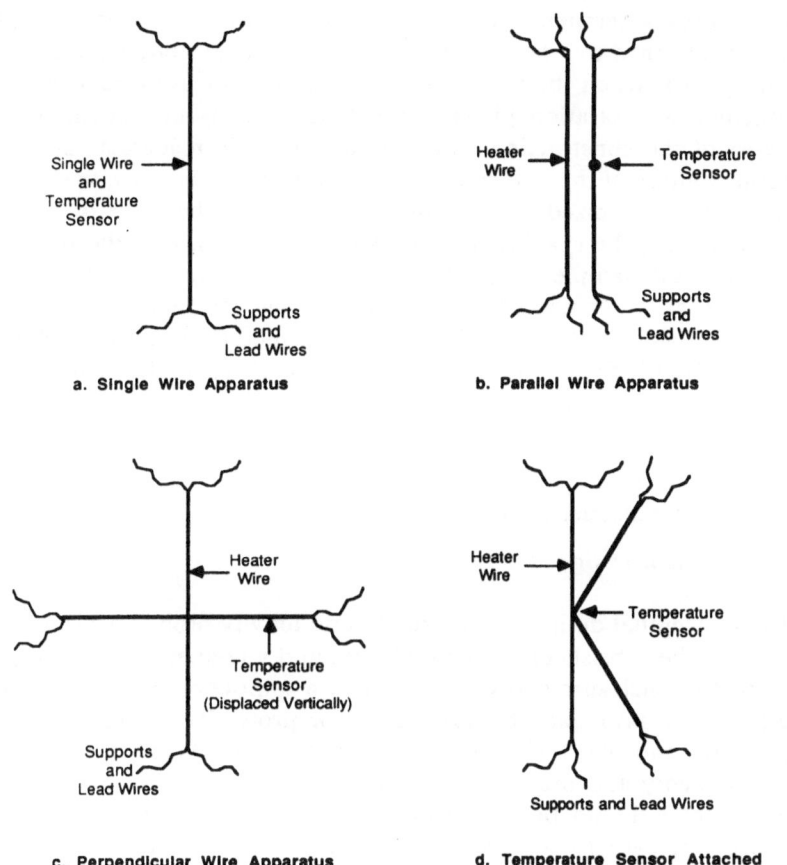

FIGURE 1. Conceptual designs of line heat source apparatus.

used to maintain heater output and to measure temperature. The end effects of such apparatus and compensating approaches are discussed by Knibbe.[20]

The temperature sensors for line heat source measurements are usually thin butt-welded (or welded and trimmed) thermocouples (of materials suitable for the temperature range employed in the experiment) or thermistors. In either case, the thermocouple (thermistor) wire dimensions are similar to those of the heater. The temperature sensor is also stretched between insulating supports. In a practical apparatus, the temperature sensor may be oriented perpendicular to the line heat source (as recommended by the German Standards Committee, 1976 testing of ceramic materials, "determination of thermal conductivity to 1600 °C by the hot-wire method"; see Fig. 1c) or parallel to the heater wire and spaced a small distance from it (see Fig. 1b). Still another apparatus uses a temperature sensor (in the form of a thermocouple or thermistor) attached to a heater wire, but with leads extending parallel to the heater (Fig. 1d). When the heater wire and thermal measurement "wire" are not attached, a separation distance must be chosen—dependent upon the properties of the construction materials and sample materials—so that the long-term solution of the line heat source equation can be used, the axial heat flow losses are minimized, and the measurements can be made in a practical time frame. Also, the spacing must be such that the size of the particles or fibers within the sample is small compared to the spacing of heater and temperature sensor. In practice, spacings of about 2 to 5 mm are appropriate. A good discussion of the generalized hot wire method is given by Prelovsek and Uran,[11] in which materials, variations in sample properties, and apparatus dimensions are discussed.

3.2. Thermal Conductivity Probe

3.2.1. Conceptual Design

The conceptual design of a thermal conductivity probe apparatus is also simple. A probe consists of a source of heat and a temperature sensor placed in a protective enclosure and surrounded by a sample of large size compared to the probe. In principle, the heat source or protective enclosure can be of any geometry and size. Cylindrical and spherical probes are most often considered in theory and practice owing to ease of analysis, construction, and use. The probe should be of uniform geometry, i.e., it should be radially symmetric (for cylindrical and spherical devices) and uniform with length. The components of the probe should be thermally in good contact with themselves and with the protective enclosure, but electrically isolated from each other; and the entire probe should have sufficient structural integrity to prevent being damaged by the sample environment.

3.2.2. Practical Design Considerations

The practical design of a thermal conductivity probe is a compromise among materials which are easy to obtain and use, construction techniques of various difficulty, small radial dimensions, and structural considerations. Most laboratory and field probes are long, thin cylinders, although probes of other shapes for laboratory experiments—large and small cylinders, thin plates, and spheres—are also described.[21] The discussion below focuses on thin cylindrical probes, since these are generally the easiest and most practical to develop, purchase, analyze, and use.

A general schematic of a thermal conductivity probe is shown in Fig. 2. Probes generally consist of the following components: (a) a heater or heat source, (b) a temperature sensor, (c) filler materials which are electrically insulating but thermally conducting, (d) a protective sheath, and (e) a connector or end assembly. Figure 2a represents a simple two-wire probe with a heater wire and a temperature sensor.

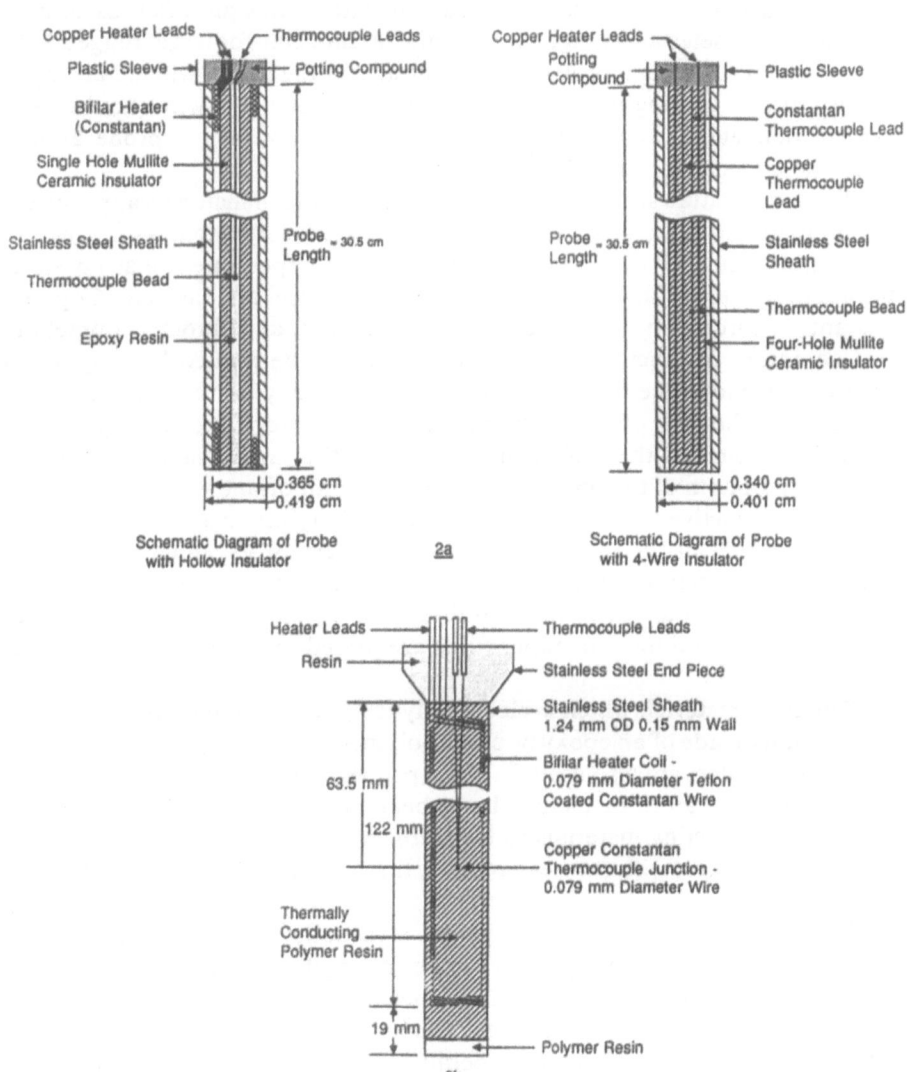

FIGURE 2. Engineering schematic of laboratory type probes.

The heat or heat source is usually a thin, insulated, electrical heater whose resistance is minimally sensitive to change in temperature. Nichrome, constantan, manganin, or other such alloys are used. The wire is often placed in the center of the probe and is doubled back, i.e., it extends from the top of the probe to the bottom and back again. An excellent heater-wire arrangement is a bifilar coil, first used by D'Eustachio and Schreiner.[22] In this application (Fig. 2b) the wire coil extends from the bottom to the top of the probe and is located adjacent to the protective sheath. Heater wires are often as thin as 0.02 mm in diameter and range up to 1 mm in diameter for large, rugged field probes. In constructing a probe, it is essential that the heater wire be arranged so that uniform heating occurs axially along the probe. The wire should be of uniform diameter, extend from the top to the bottom of the probe and, if coiled, be coiled uniformly throughout.

The temperature sensor is usually a thermocouple of materials appropriate for the temperature range of experiments under consideration. Also, since in most experiments it is desirable to use as small a temperature rise as possible, thermocouple materials with high output response are desired. Copper-constantan, chromel p–constantan, iron–constantan, and chromel–alumel are most common. In practice, the thermocouple may be located halfway along the length of the probe from the top connector or may be "butt welded" and extend to the bottom of the probe and back to the top. The latter promotes thermal symmetry of the probe, since there are the same number of wires in the top and bottom of the probe. This is often unimportant when using probes where the protective sheath has a high conductivity and has a wall thickness greater than that of the thermocouple wires. Thermistors which have a high resistance/temperature coefficient are also used. Thermocouple and thermistor wires also range from about 0.02 mm to almost 0.5 mm. In most probes, both the heater wires and thermocouple wires are joined to larger lead wires in the end connector.

The filler material, which is electrically insulating and thermally conducting, is usually made of an epoxy or other polymeric resin. Magnesia or alumina insulating powders are also used. In some probes, standard two- and four-wire thermocouple ceramic insulators have been used. It is important for the insulating/conducting materials to be uniform along the length of the probe, and in good contact with the heater wires, thermocouple wires, and protective sheath.

The protective sheath of the probe is usually made from thin-walled stainless-steel tubing, although other materials such as rigid plastic and glass have been used. Stainless steel is ideal because of its strength, availability, and chemical and physical properties. The bottom end of the probes are often welded tight, plugged with high temperature cement or insulating material, or filled with a polymer.

The connector or end assembly serves to hold the probe, acts as a connector for heater and sensor wires, and supports the connection of the heater or sensor wires to thicker and stronger lead wires. The end assembly may be a molded plastic or potting compound, a larger stainless-steel tubing, or other material fixed to the upper end of the probe. The connector or end assembly often provides for flexibility of the lead wires.

Thermal conductivity probes have three practical size ranges—miniature, small, and large or field probes. Miniature probes are often built to fit within a hypodermic needle. A typical example, described by Asher, Sloan and Graboski,[13] consists of a thin heater wire and thermistor emplaced in a 20-gauge hypodermic needle, of length 35 mm and diameter 0.9 mm, with epoxy cement filling. Sweat and Haugh[23] described a similar probe using an insulated constantan heater of 0.051-mm diameter and an insulated chromel-constantan thermocouple of 0.051-mm diameter emplaced in a 23-gauge hypodermic needle, 0.66-mm diameter and 39-mm long. These probes are particularly suited to measurements on small samples, biological materials, foodstuffs, etc. Caution must be followed in analyzing the results of measurements, since the ratios of length to diameter for these probes are only about 40 to 60 and axial heat flow may distort measurements.

A more common dimension probe is similar to that described by Wechsler[15] and D'Eustachio and Schreiner.[22] The typical probe used by Wechsler is stainless steel and has a diameter of about 1 to 1.5 mm and a length of 140 to 170 mm, and a wall thickness ranging from 0.15 to 0.25 mm. Heater and thermocouple wires are typically Teflon-insulated 0.062-mm constantan or copper–constantan. The probe used by d'Eustachio and Schreiner[22] (and commercially available) has a diameter of 0.5 mm. It uses a fine chromel-P thermocouple of 0.013-mm diameter, and a heater wire coil of 0.025-mm constantan. These probes have a ratio of length to diameter of 100 to 400. The probe of D'Eustachio and Schreiner[22] is more flexible, but more fragile, than the ones described by Wechsler.[15]

Probes for field or *in situ* use in soils, sands, etc., can be 3 to 6 mm in diameter, 400 to 600 mm in length, and use thermocouple and heater wires of greater diameter than those for laboratory probes. Again caution must be used with these larger probes to avoid axial heat flow losses and to reduce undesirable effects of the dimensions and thermal properties of the probe.

General guidelines for construction of probes are:

1. Use a metal sheath having a diameter and wall thickness as thin as is consistent with the flexibility and strength desired from the probe.
2. Use a length-to-diameter ratio of 100 or more if possible.
3. Position the temperature sensor at the mid-point of the probe.
4. Use insulated sensor and heater wires of as small diameter as practical; provide for constant heater wire geometry over the length of the probe.
5. Fix the heater and sensor wires in place using an insulation or potting compound.
6. Try to achieve radial symmetry in the probe.

7. Seal the top and bottom; for the heater and probe use lead wires of sufficient strength and flexibility.

3.2.3. Instrumentation for Probes (and Line Heat Source Apparatus)

The instrumentation required for operation of the probe and line heat source is the same. The instruments may be manually operated electrical instruments or automated electronic equipment operated by computers.

Two types of instrumentation are required—one for the heater and one for the temperature sensor (see Fig. 3). A stabilized, regulated DC power supply is essential for measurements. (Some experimenters have used the same wires for heating and temperature measurement with an AC power supply for the heater, reading the filtered DC output of the thermocouple—but this is not recommended.) Some means for measuring the heater power—typically a voltmeter and ammeter—are required.

Instrumentation for measurement of temperature is also relatively standard. A cold or reference junction is appropriate if a thermocouple is used as the temperature sensor. A potentiometer/galvanometer can be used to measure the output from the temperature sensor or a recorder can be used directly to obtain a plot of output versus time. If a thermistor is used as a temperature

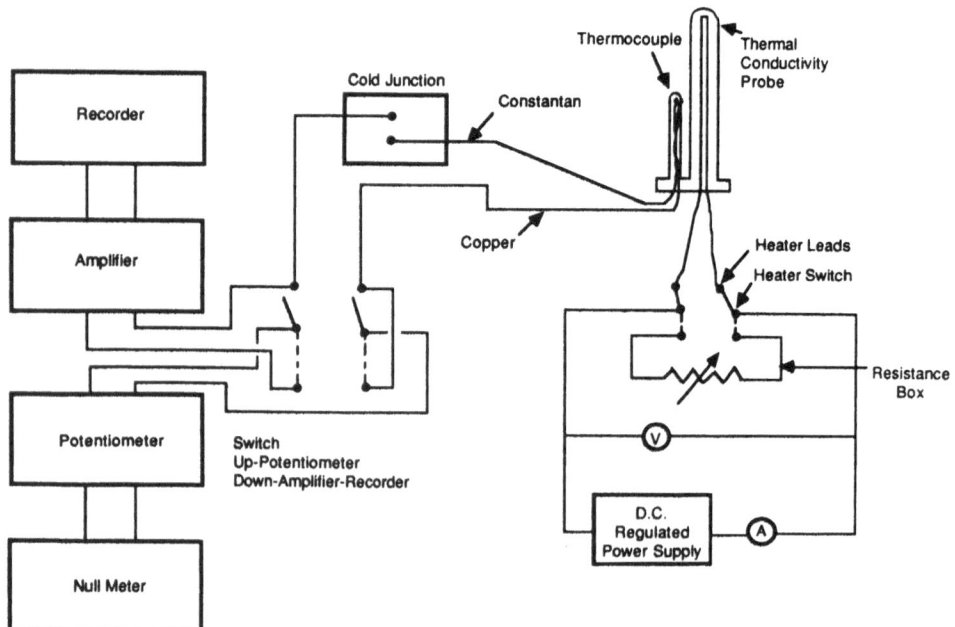

FIGURE 3. Schematic diagram of instrumentation for probe or line heat source apparatus.

sensor, instrumentation is needed to measure the resistance of the thermistor bead so that its temperature can be determined.

Manually operated high-quality laboratory instrumentation is sufficient for most measurements. More complex—and expensive—computer compatible devices can be used to automatically measure and record heater power and thermocouple or thermistor output and to calculate the results of the experimental measurements.

The power supply should produce an output constant to 0.5% or better. Since it is customary to measure temperature changes of only several degrees in probe or line heat source measurements, the instrumentation for temperature measurement should permit a precision of 0.001 to 0.10 °C over time.

3.2.4. Commercially Available Equipment

A probe similar to that used originally by D'Eustachio and Schreiner[22] is available from Custom Scientific Instruments, Inc., P.O. Box A, Whippany, New Jersey, 07981, U.S.A.

According to Asher et al.,[13] probes similar to those used by Von Herzen and Maxwell[24] are available from Fluid Dynamics Corporation, Boulder, Colorado. [The author has not been able to locate a source of these probes.]

Instrumentation for probe and line heat source measurements is readily available from commercial sources.

3.3. Peripheral Equipment

Peripheral equipment is required for providing environmental control for the samples and probes during measurement. This equipment depends upon the range of environmental conditions considered during the experiments—namely, temperature and atmospheric pressure. Since the probes and line heat source apparatus are generally small and self-contained, it is customary to place the entire apparatus and sample in a "constant temperature" chamber which can be cooled or heated to appropriate levels and stabilized during the experiment. Since the temperature changes during thermal conductivity measurements with a probe or line heat source are often as small as 0.5 to 3 °C, it is important that the sample chamber temperature be carefully stabilized. A period of constant temperature (±0.01 °C or better) is needed prior to measurements so that the measurements of temperature versus time reflect those associated with the probe or line source apparatus, and not those associated with the sample chamber or environmental conditions. The size of the chamber needed depends upon the size of the probe or line heat source and the sample which surrounds it. Because of the relatively small size of the probe and sample, a constant-temperature chamber having dimensions as small as 25 cm × 25 cm × 25 cm is often adequate for measurements.

Some investigators desire to change the gaseous environment surrounding the sample—vacuum, pressure, or gases other than air. Therefore, the sample chamber should be designed to be placed in a vacuum or pressurized system as appropriate. A standard combination of vacuum table and bell jar often suffices, provided there are appropriate feed-through devices for heater power, temperature measurement, and provision of temperature control of the sample holder.

4. MEASUREMENT METHODS

4.1. Operating Procedures

4.1.1. Assembling the Probe and Sample

The first step in the probe method is to carefully fix the probe within the sample to be measured. A clean, straight probe (the dimensions, materials, thermal and electrical characteristics having been previously established and recorded) is easily inserted in a granular, fibrous, or foam material. Care should be taken to assure good contact between the sample and the probe. This may be accomplished by vibrating a granular sample or by slightly compressing a foam or fibrous sample, and avoiding opening a hole larger than the probe in fibrous or foam material. In a cylindrical sample the probe should be placed along the axis. In a rectangular sample, the probe should be placed near the center with the probe aligned along the longest dimension. For relatively hard samples where it is not possible (or appropriate) to drill a hole of the appropriate dimensions for a probe, a groove may be prepared in each of two flat samples, and the probe "sandwiched" between the samples. Care must be taken to avoid gaps in the sample or other conditions which lead to heat leaks or anisotropy of the sample. For some sample materials, it may be appropriate to use a contacting medium—a fluid, conducting grease, soft metal, etc.—to assure good contact with a solid sample.

Probes are often used *in situ* for insulations and soils. The procedure is the same—either the probe is inserted directly into the sample or a hole of appropriate diameter is made in the sample and the probe inserted.

4.1.2. Stabilization

After the probe and sample have been assembled, electrical connections made for providing heater power and temperature measurement, and the sample placed in an appropriate sample chamber, the environment of the sample and probe must be stabilized. The time required varies with the sizes of the sample and probe, previous conditions, and the experimental conditions to be used. The most direct way to test for stabilization is to monitor the temperature of the probe over time. The degree of stabilization required will depend upon the properties of the probe, the sample, the probe temperature rise expected during measurement, and the duration of the measurement. Practical experience in probe measurement helps the investigator determine the required stabilization of the sample temperature. For example, if the temperature rise of the probe during measurements is 3 or 4 °C during a 5- to 10-min measurement period, the probe and sample should be stabilized to within 0.01 °C per minute redundant prior to measurement. Applying more heater power to the probe can reduce the requirement for stabilization. However, it must be remembered that there will be a greater probe (and sample) temperature rise during the measurement period and the measurement can be in error if the thermal properties of the sample vary considerably with temperature.

In *in situ* measurements, reaching stabilization may be difficult and it is sometimes appropriate to try to isolate the portion of the sample used in measurements from its surroundings through insulation or other means.

In probe measurements with moist or fluid-containing samples or in samples which have large pore spaces, it is usually desirable to have only small temperature rises during measurement to reduce moisture transport and/or convection. In these situations greater thermal stabilization may be required.

Some investigators also monitor the temperature near the outer boundary of the sample during the stabilization period and later during experiments. This provides another check on stability, permits stability monitoring during the measurement, and is a useful indicator of the limits to the time of measurement. For example, when the boundary temperature begins to rise significantly, the useful measurement period is over.

4.1.3. Probe Operation and Measurements

After stabilization, power to the heater is applied and the probe temperature monitored over time. Temperature measurement may be continuous or periodic and is usually accomplished with an analog or digital temperature–time recording system. Electrical power to the heater should be measured and monitored periodically to assure stability. Measurements continue for as long a period as necessary. The time required for accurate measurements can be estimated from the equations given in Section 2, by plotting the temperature rise versus the logarithm of time elapsed since measurement initiation and noting the linear portion, or by experience. When the electrical power is turned off, temperature recording may continue since some investigators use the "cooling" portion of the measurement cycle as a check on the measurement. By simple theory, the cooling curve should be the inverse or mirror image of the heating curve, in the absence of effects of probe thermal mass or axial heat flow and boundaries. After passage of time sufficient for stabilization, measurements can be repeated.

4.2. Analysis of Data

Depending upon the probe, samples used, and measurement conditions, analysis of data can be either simple or complex. If the probe has been properly designed and is of a size, shape, and materials appropriate for the experimental sample, if temperature stability has been achieved in a large enough sample, if heater power is stable, and if power and temperature measurements are accurate and precise over a sufficient period, then data analysis is simple. Typically, a plot of temperature rise of the probe versus logarithm of elapsed time is made (see Fig. 4). Such a plot usually has an initial curved portion, a result of the finite dimensions and mass of the probe and the time region, where $\alpha t/r^2$ is relatively small; a linear portion, where simple probe theory is valid; and a final curved portion, where sample size and boundaries or axial heat losses are important. From the power per unit length and the slope of the temperature/time curve, the thermal conductivity can be calculated according to equation (5).

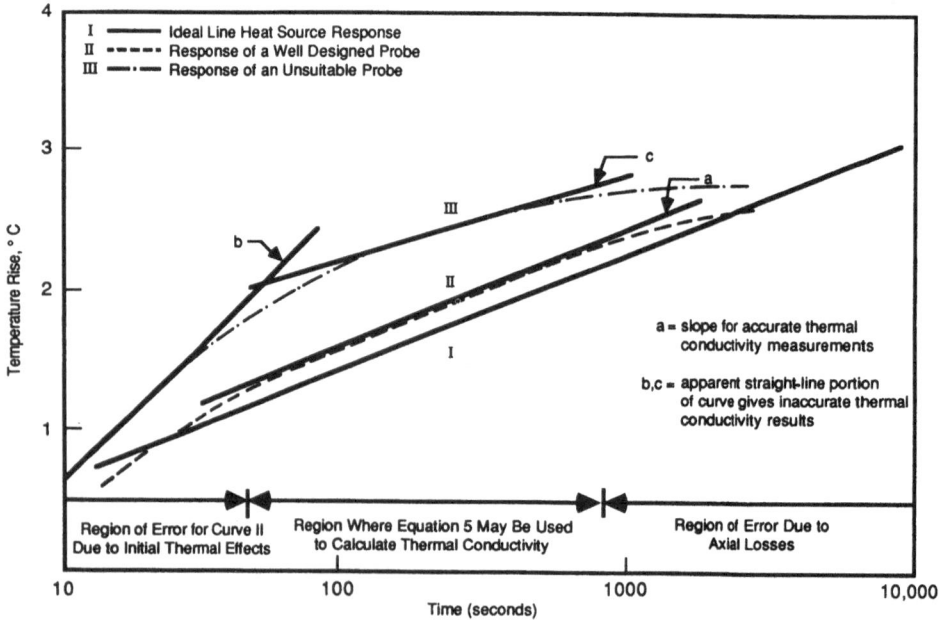

FIGURE 4. Typical temperature–time profiles for line heat source and probe.

Most investigators prefer to design probes so that they are compatible with the simplified temperature-versus-logarithm-time approach to data reduction, although this is not always possible. From probe dimensions and properties, the equations presented in Section 2 can be used to establish the measurement time during which a linear temperature-versus-logarithm-time relationship should occur.

Numerical methods can also be used to calculate the thermal conductivity from the early time period if the probe dimensions and physical properties are known. An example of a parameter estimation technique that uses both early and late times is given by Koski and McVey.[25] The approach permits estimates of probe contact resistance and can shorten the experimental times required for measurements of low-conductivity materials.

For line heat source measurements, the simple approach is almost always used. A plot of temperature rise versus logarithm of time yields a linear relationship when $r^2/4\alpha t$ is small, and the thermal conductivity is calculated directly from the heater power per unit length and the slope of the temperature/time curve.

A "curve matching" technique for analysis of data from line heat source measurements was also presented by Wechsler and Black.[12] A plot of logarithm of temperature rise versus logarithm time is "matched" with a plot of $\ln\left[-Ei(-1/x)\right]$ vs $\ln x$. From the temperature rise corresponding to a value of $-Ei(-1/x)$ of unity (called T^*) thermal conductivity is calculated as $q/4\pi T^*$.

4.3. Error Analysis

A discussion of errors in applying simplified formulas to data reduction has already been given in Section 2. In general, for line heat source measurements, errors of 5% or more occur if the values of $\alpha t/r^2$ used in calculations are smaller than 5, but errors of 1% to 2% are more typical of most measurements.

For measurements with thermal conductivity probes careful attention to design can limit the errors to several percent. The normal sources of error—temperature and power measurement, sample thermal stability, and moisture migration—are often of more significance than errors attributed to probe construction, axial heat flow, or simplified data-reduction techniques.

4.4. Standardization and Calibration

The probe and line heat source methods are both "absolute" methods in that thermal conductivity is directly estimated from measured parameters. Neither calibration nor standardization is necessary. Investigators find it useful to make comparisons of conductivity measured with probe or line heat source methods and those measured using the guarded hot plate approach. Comparisons are generally within several percent, and may be attributed to variation in sample condition rather than errors inherent in the methods. Glass beads, well-established insulation materials, glass, or liquids are sometimes used as reference materials for probe and line source methods.

5. PRACTICAL CONSIDERATIONS IN THE USE OF THE PROBE AND LINE HEAT SOURCE METHODS

5.1. Limitations of the Methods

Both the line heat source and the thermal conductivity probe methods have some inherent limitations. It must be easy and practical to place the probe (or line heat source) within the sample for measurement with sufficient sample surrounding the device for proper measurement. Thus, the probe and line heat source are not particularly suited to very hard or stiff sample materials, thin, or very small samples. Because the heat flow is radial, the instruments may give unreliable results in samples that are not homogeneous or radially symmetric, for example, layered samples that show considerable anisotropy, or grossly heterogeneous samples with large differences in properties between the components. The probe and line heat source are not desirable for very high conductivity materials (e.g., metals) or very low conductivity materials (e.g., evacuated multilayer insulations) although a properly designed line heat source may be used for low-conductivity powdered insulations. Probes and line heat sources are useful for soils or particulate or fibrous materials if the particle or fiber sizes are generally smaller than the probe dimensions or smaller than the spacing between the heater and temperature measurement wires of the line heat source. The methods are not appropriate in fluid or fluid/solid systems where convection is an important heat transfer mechanism compared with conduction. Because of the nature of the construction materials (e.g., polymer coated wires, potting compounds) probes may not be suitable for use at temperatures greater than 200 °C unless special precautions are taken in design and construction.

5.2. Suggestions for Design and Construction

The general principle to be followed in the design construction of cylindrical probes is to use as thin and long a probe as possible, consistent with the rigidity or flexibility requirements of the samples and measurement conditions. Ratios of length to diameter for cylindrical probes should be greater than 100. A construction using thin walls is generally better than one using thick walls. Radial symmetry should be preserved in a probe; heating along the probe length should be uniform. Temperature sensors should be firmly positioned within the probe and have a sensitivity as high as practicable.

Prior to designing a probe, an analysis should be undertaken—using the methods described in Section 2—to estimate the errors associated with axial heat flow, finite probe size, and sample size for typical samples. Then one can choose construction materials that will help minimize these errors. Probes should be constructed of materials of known thermophysical properties so that errors associated with deviations from line source theory can be estimated and corrections made to measurements, if necessary.

Similar suggestions apply to the design and construction of a line heat source—thin wires, appropriately spaced, with a large ratio of wire length to spacing between heater and temperature sensor, and careful *a priori* analysis to determine if accurate measurements can be made.

5.3. Suggestions for Operation and Analysis

Prior to initiating measurements, make sure that the probe is firmly implaced in the sample and in good contact with it. Also be certain that the sample materials will not shift redundant, or be compacted during measurement. It is essential to stabilize the temperature and gas pressure of the sample environment. Temperature fluctuations of the sample and sample container should be small compared to the rise of the sample temperature during measurement. Assure that the power to the probe or heater wire is constant during measurements. Limit the power applied (and thereby the rise of temperature) to only that required for precise measurement of the rate of rise of temperature. This is especially important if the sample is moist, contains fluids which can convect, or has a thermal conductivity which varies significantly with temperature. Make sure that measurements are made for a long enough redundant period (determined from analysis).

When analyzing the data, carefully plot the temperature rise versus the logarithm of elapsed time. Do not necessarily interpret an early steep rise in temperature as the linear portion of the temperature-log time plot. A linear rise may also occur during the initial heating of the probe, especially in materials of low conductivity. Use an analysis of the probe to estimate when the linear portion of the temperature rise–logarithm time relationship should begin and use the subsequent time period for estimating thermal conductivity. If appropriate, check the measurements using probes of different dimensions or "calibrate" the probe for its behavior with a similar sample of known thermal properties.

Careful application of the principles mentioned in the early sections of this paper should provide the investigator with a relatively simple and accurate method of measurement of thermal conductivity for a variety of materials.

ACKNOWLEDGMENT. The author gratefully acknowledges the constructive review of the manuscript by Dr. David R. Smith, Center for Chemical Engineering, National Bureau of Standards, Boulder, Colorado.

NOTATION

a	Inner radius of hollow probe (m)
A_C	Cross-sectional area of probe (m^2)
b	Radius of probe (m)
C	Heat capacity of sample (W s kg^{-1} K^{-1})
C_1	Heat capacity of probe (W s kg^{-1} K^{-1})
Ei (x)	Exponential integral [see equation (2)]
G	Function of parameters given by Jaeger[7]
H	Conductance at the probe/sample interface (W m^{-2} K^{-1})
M_1	Mass per unit length of probe (kg)
q	Strength of heat source per unit length (W m^{-1})
r	Distance from source of heat (m)
R	Radius of sample (m)
S	Parameter defined by Blackwell;[16] see equation (13)
t	Time since initiation of experiment (s)
t_{max}	Maximum measurement time (s)
Z	Ratio of length of heated portion of probe to radius
α	Thermal diffusivity (m^2 s^{-1})
β	Defined as $2\pi b^2 \rho C / M_1 C_1$
$\Delta_1 \Delta_2$	Functions of diffusivity of sample and probe dimensions [see equation (8)]
δ	Euler's constant = 0.577216
ϕ	Parameter defined by equation (12)
λ	Thermal conductivity (W m^{-1} K^{-1})
ε	Ratio of thermal conductivity of probe to that of sample
η	Ratio of volumetric heat capacity of probe material to that of sample
θ	Temperature (°C or K)
ρ	Density of sample (kg m^{-3})
ρ_1	Density of probe (kg m^{-3})
σ	Parameter defined by Blackwell[16]
τ	Defined as $\alpha t / b^2$

REFERENCES

1. A.L.E.F. Schleiermacher, "Uber die Waermeleitung der Gase," *Ann. Phys. Chem.* **34**, 623 (1888).
2. B. Stalhane and S. Pyk, "New Method for Determining the Coefficients of Thermal Conductivity," *Tek. Tidskr.* **61**, 389 (1931).
3. E.M.F. Van der Held and F.G. Van Drunen, "A Method of Measuring the Thermal Conductivity of Liquids," *Physics* **15**, 865 (1949).
4. F.C. Hooper and F.R. Lepper, "Transient Heat Flow Apparatus for the Determination of Thermal Conductivities," *ASHVE Trans.* **56**, 309 (1950).
5. H.S. Carslaw and J.C. Jaeger, *Conduction of Heat in Solids*, Oxford Univ. Press. London (1959).
6. J.H. Blackwell, "A Transient-Flow Method for Determination of Thermal Constants of Insulating Materials in Bulk," *J. Appl. Phys.* **25**, 137 (1954).

7. J.C. Jaeger, "Conduction of Heat in an Infinite Region Bounded Internally by a Circular Cylinder of a Perfect Conductor," *Aust. J. Phys.* **9**, 167 (1956).

8. D.A. de Vries, "A Non-Stationary Method for Determining Thermal Conductivity of Soil in Situ," *Soil Sci.* **73**, 83 (1952).

9. D.A. de Vries and A.J. Peck, "On the Cylindrical Probe Method of Measuring Thermal Conductivity with Special Reference to Soils," *Aust. J. Phys.* **11**, 255 (1958).

10. Arthur D. Little, Inc., "Studies of the Characteristics of Probable Lunar Surface Materials," Final Report, prepared for Air Force Cambridge Research Laboratories under Contract AF 19(628)-421 (1962).

11. P. Prelovsek and B. Uran, "Generalised Hot Wire Method for Thermal Conductivity Measurements," *The Institute of Physics*, 0022-3735/84/080674, p. 674 (1984).

12. A.E. Wechsler and I.A. Black, "Design, Development and Fabrication of Thermal Measuring Systems," prepared for National Aeronautics and Space Administration, George C. Marshall Space Flight Center, Huntsville, Alabama, Contract No. NAS 8-11708 (25 June 1964–25 March 1965).

13. G.B. Asher, E.D. Sloan, and M.S. Graboski, "A Computer-Controlled Transient Needle-Probe Thermal Conductivity Instrument for Liquids," *Int. J. Thermophys.* **7**(2), 285–294 (1986).

14. H.R. Thomas and J. Ewen, "A Reappraisal of Measurement Errors Rising From the Use of a Thermal Conductivity Probe," *J. Heat Transfer* **108**, 705 (1986).

15. A.E. Wechsler, "Development of Thermal Conductivity Probes for Soils and Insulations," prepared for U.S. Army Cold Regions Research and Engineering Lab., Hanover, N.H., Contract No. DA 27-021-AMC-25(X) (Sept. 1965).

16. J.H. Blackwell, "The Axial-Flow Error in the Thermal-Conductivity Probe," *Can. J. Phys.* **34**, 412 (1956).

17. W.T. Kierkus, N. Mani, and J.E.S. Venart, "Radial–Axial Transient Heat Conduction in a Region Bounded Internally by a Circular Cylinder of Finite Length and Appreciable Heat Capacity," *Can. J. Phys.* **51**, 1182 (1973).

18. Pieter J. Bruijn, Izaak A. Haneghem, and Jacob Schenk, "An Improved Nonsteady-State Probe Method for Measurements in Granular Materials: Part I: Theory," *High Temp. High Pressures* **15**, 359–366 (1983).

19. H. Inaba, "Measurement of the Effective Thermal Conductivity of Agricultural Products," *Int. J. Thermophys.* **7**(4), 773–787 (1986).

20. P.G. Knibbe, "The End-Effect Error in the Determination of Thermal Conductivity Using a Hot-Wire Apparatus," *Int. J. Heat Mass Transfer* **29**(3), 463–473 (1986).

21. H.J. Goldsmid, K.E. Davies, and V. Papazian, "Probes for Measuring the Thermal Conductivity of Granular Materials," *J. Phys. E.* **14**, 1149–1152 (1981).

22. D. D'Eustachio and R.E. Schreiner, "A Study of a Transient Method for Measuring Thermal Conductivity," *ASHVE Trans.* **58**, 331 (1952).

23. V.E. Sweat and C.G. Haugh, "A Thermal Conductivity Probe for Small Food Samples," *Trans. ASAE*, 56 (1974).

24. R. Von Herzen and A.E. Maxwell, "The Measurement of Thermal Conductivity of Deep Sea Sediments by a Needle Probe Method." *J. Geophys. Res.* **64**, 1557 (1959).

25. J.A. Koski and D.F. McVey, "Application of Parameter Estimation Techniques to Thermal Conductivity Probe Data Reduction," *Thermal Conductivity*, **17**, (17th International Thermal Conductivity Conference), pp. 587–600 (1983).

7

B.S. 1902 Panel Test Method for the Measurement of the Thermal Conductivity of Refractory Materials

W.R. DAVIS

1. GENERAL INTRODUCTION

The B.S. 1902 panel test method[1] is basically similar to the ASTM C201-68 (1979) test,[2] the differences being mainly constructional. Both are equilibrium or "steady-state" methods involving the direct measurement of the amount of heat flowing linearly through a known area of the material under a known and steady temperature gradient. A water-flow calorimeter is used for the heat-flow measurement, the test piece being a plane parallel slab or panel of the material under test. Both methods have been used as the principal means of determining the thermal conductivity of refractory materials in the U.K. and U.S.A. over the past 30 years.

1.1. Range of Application

The method is commonly used over the temperature range of 400 to 1350 °C (hot-face temperatures) for thermal conductivity values in the range 0.05-15 W m^{-1} K^{-1}. Other limitations are mainly those of size. The test was originally designed to use the "standard square"—230 × 114 × 76 (or 64) mm, two or three such bricks being required. Where standard squares are not available the length and breadth of the samples should be sufficient to permit the cutting of two or more slabs 230 × 114 mm. The thickness should lie within the limits of 25 to 76 mm. Satisfactory measurements cannot be made with specimens of appreciable water content, the test pieces being oven dried at 110 °C before testing. Unfired materials, such as monolithics, are usually

W.R. DAVIS ● 28 Joseph Crescent, Alsager, Stoke-on-Trent, ST7 2RP, England.

prefired beforehand. The thermal conductivity of powders and nonrigid materials can be measured by the use of a refractory container approximating in size to that of the normal test panel (see Section 2.3.1).

The upper limit to the conductivity range is dependent on the maximum heat flux which can be handled by the calorimeter. This can be overcome to some extent by placing insulation between the calorimeter and the "cold face" of the test piece, but the thickness of such insulation is usually limited to about 10–15 mm. The lower limit to the conductivity range is set by the minimum heat flux measurable. This latter can be increased by the use of a thinner test piece (thickness > 25 mm).

Because the heat flow is linear, anisotropic materials can be tested (see Section 3.3 paragraph 3). In such cases the direction of heat flow through the test material should be specified.

1.2. Thermometry

As already stated, the basis of the method is the measurement of the heat flux through a known area of the test piece material under a known temperature gradient. It is essential that the heat flow should be linear through the section covered by the calorimeter. Any divergence will result in an erroneous value for the area used in the calculation of the thermal conductivity. The following precautions are taken to ensure that the heat flow is truly perpendicular to the plane of the test slab:

1. The area of the test piece is considerably greater than that of the calorimeter.
2. The calorimeter is surrounded by a water-cooled guard ring which ensures that the "cold" (unheated) face of the rest of the test panel is at the same temperature as the area lying under the calorimeter. This transfers any lateral heat flow to the edges of the test panel and, if the test panel is large enough, gives linear vertical heat flow in the central portion.
3. The "hot face" of the test piece is heated as uniformly as possible.

If the water calorimeter is placed directly onto the test panel then the temperature of the "cold face" of the latter cannot be much higher than that of the water in the calorimeter. With the higher conductivity materials (say above $1-2 \text{ W m}^{-1} \text{ K}^{-1}$) this would impose severe limitations on the hot-face temperature or require an impossibly high rate of water flow through the calorimeter. This restriction is overcome by interposing a thin slab of refractory insulating material between the cold face of the test piece and the calorimeter. This has the effect of raising the temperature of the "cold face" to a more reasonable value. Tests carried out on the same test piece, with and without this "backup insulation," have shown that the conductivity value obtained is the same in both cases.[10,12]

The hot- and cold-face temperatures reported are the mean of the three thermocouple readings in each face. With proper adjustment of heater power and when thermal equilibrium has been reached the maximum temperature difference between the couple readings in a given face is usually of the order of 5 °C (±2.5 °C). The couple thermo-emfs can be measured with either a potentiometer (±1 μV) or high-accuracy digital meter (±0.1 °C).

1.3. History (Table 1)

In 1927 F.H. Norton described[3] a guarded water calorimeter method in which the testpiece was in the form of a cylinder, 108 mm diameter and 228 mm long. The test piece rested on top of the calorimeter being surrounded by insulation and heated on the top face by a gas furnace. Thermocouples were placed in holes drilled at four vertically spaced intervals from the top. A form of peripheral heating was used to reduce lateral heat flow. Hot-face temperatures ranged from 200 to 1450 °C. A range of firebricks, fused alumina, and zirconia bricks were tested, having conductivities ranging from 0.3 to 6 W m^{-1} K^{-1}. Claimed accuracy was ±15% to ±25%. In 1933 Wilkes[4] used a setup very similar to modern equipment. A three-brick panel was heated from the top with silicon carbide rod heaters and a guarded water calorimeter measured the heat flow. Mean temperatures (average of the hot- and cold-face temperatures) ranged from 200 to 1050 °C. The tests were made on insulating and magnesite bricks with conductivities between 0.3 and 5 W m^{-1} K^{-1}. The same author[5] described further experiments in 1934 with the same apparatus on a range of firebricks where the mean temperatures lay between 200 and 1350 °C with conductivities 0.5 to 3 W m^{-1} K^{-1}. The results were compared with those obtained by other methods.

Austin and Pierce[6] in 1935 used a similar apparatus to test firebricks and various types of silica brick. Here the mean temperature range was 200–1000 °C with a conductivity range of 1–2 W m^{-1} K^{-1}. Reproducibility is given as ±5% and comparisons made with other methods. Reliability and reproducibility are discussed.

In a report for the ASTM Sub-Committee C-8 in 1936[7] Nicholls reviews various methods including the guarded water calorimeter, guarded hot plate, heat flowmeter, and radiation meter used in a series of cooperative tests by six laboratories on firebrick and silica brick samples over the mean temperature range 200–1350 °C. Possible sources of error in each method are discussed in detail, although the wide variation of 30% in the test figures permitted no firm conclusions to be made regarding absolute accuracy.

A similar series of test were carried out in the U.K. in 1937,[8] where four laboratories tested diatomite bricks. Results from three of the laboratories using guarded water calorimeter equipment had a scatter of 14%. Mean

TABLE 1
Summary of Guarded Water–Calorimeter Thermal Conductivity Tests

No.	Investigator	Ref.	Year	Material[a]	Mean temp. range[b] (K)	Conductivity range ($W\,m^{-1}\,K^{-1}$)	Reproducibility[c] (%)	No.
1	Norton, R.H.	3	1927	Fb, Al_2O_3 (f), ZrO_2	500–1700[d]	0.3–6	+15–±25	1
2	Wilkes	4	1933	I/Fb, MgO	500–1300	0.3–5		2
3	Wilkes	5	1934	Fb	500–1600	0.5–3		3
4	Austin et al.	6	1935	Fb, SiO_2	500–1300	1–2	±5	4
5	Nicholls	7	1936	Fb, SiO_2	500–1600	0.7–2		5
6	Oliver	8	1937	Diatomite Fb	500–800	0.09–0.14	5	6
7	Norton, C.L.	9	1942	Fb, I/Fb	400–1350	0.06–0.45	3–4	7
8	Patton et al.	10	1943	Fb, SiO_2, CC, I/Fb	500–1400	0.25–1.7	1–5	8
9	Clements et al.	11	1949	I/Fb, Sil.	700–1500	0.3–1.2	3	9
10	Watson et al.	12	1953	Al_2O_3 (f), I/Fb	400–1500	0.4–2	4.5	10

[a] Fb, aluminosilicate firebrick; I/Fb, insulating firebrick; MgO, magnesite brick; SiO_2, silica brick; Sil, siliceous brick; ZrO_2, Zirconia brick; Al_2O_3 (f), fused Al_2O_3 brick; CC, China clay brick.
[b] Arithmetic mean of hot and cold face temperature.
[c] Due to the high variability of test material accuracy figures cannot be quoted with any degree of confidence—these are therefore replaced by reproducibility values.
[d] Hot face temperatures (for this reference only).

temperature range was 100–540 °C. Reproducibility of better than 5% was claimed.

C.L. Norton[9] modified Wilkes's design in 1942 and the report gives considerable constructional detail and operational procedure. Test results are given for insulating bricks and firebricks over the mean temperature range of 100–1000 °C with a conductivity range of 0.06 to 0.45 $W m^{-1} K^{-1}$. Possible sources of error are discussed in detail and a reproducibility of 3%–4% is claimed.

Further tests in 1943[10] with this apparatus gave a reduction in the lateral heat flow and this arrangement formed the basis for the present ASTM C201-68 (1979) test. In these tests super-duty firebrick, silica brick, and China clay brick, having conductivities between 1.0 and 1.7 $W m^{-1} K^{-1}$, were measured over the range 200–1100 °C. Insulating bricks (~0.25 $W m^{-1} K^{-1}$) were also tested between 200 and 700 °C. Reproducibility of 1.5% is claimed.

Experiments carried out between 1949 and 1951 are reported by Clements and Vyse[11] and the equipment described was used in drafting the present B.S. 1902 test. Results are given for insulating bricks and siliceous bricks over the mean temperature range of 400 to 1250 °C, conductivity range 0.3 to 1.2 $W m^{-1} K^{-1}$. The total scatter of the results at a given temperature was better than 3%. Errors are discussed in some detail although a figure for absolute accuracy is not claimed due to the lack of reliable standards.

A cooperative test was made by Watson, Clements, and Vyse[12] in 1953 in which the ASTM C201-47 [now known as C201-68 (1979)] and the B.S. 1902 panel test equipments were compared. Tests were made on fused Al_2O_3 brick (2.0–1.5 $W m^{-1} K^{-1}$) over the range 150 to 1200 °C and high-temperature insulating brick (0.4 $W m^{-1} K^{-1}$) over the mean temperature range 300–1100 °C. Tests were made with and without "backup insulation" with a maximum discrepancy of 2%. The results from the two sets of tests agreed to within 4.5%. Errors and equipment differences are discussed.

The present standard is designated as B.S. 1902: Part 1A: Section 12, 1966 and is currently (1991) under revision. This revision is a minor one and mainly concerned with arrangement and control of heaters and thermocouple layout.

This revised version has been renamed and will be issued as B.S. 1902: Part 5: Section 5.5: 1991 "Determination of thermal conductivity (panel/calorimetric) method." The standard now covers most bulk refractory products including dense shapes and lightweight refractory insulating materials.

2. GENERAL OUTLINE OF THE METHOD

2.1. Description

A schematic outline of the equipment, roughly to scale, is given in Fig. 1. The test panel (TP) consists of a complete test brick [(1) in Fig. 1b] 230 × 114 × 76 (or 64) mm or approximating to these dimensions, with a half brick or slab

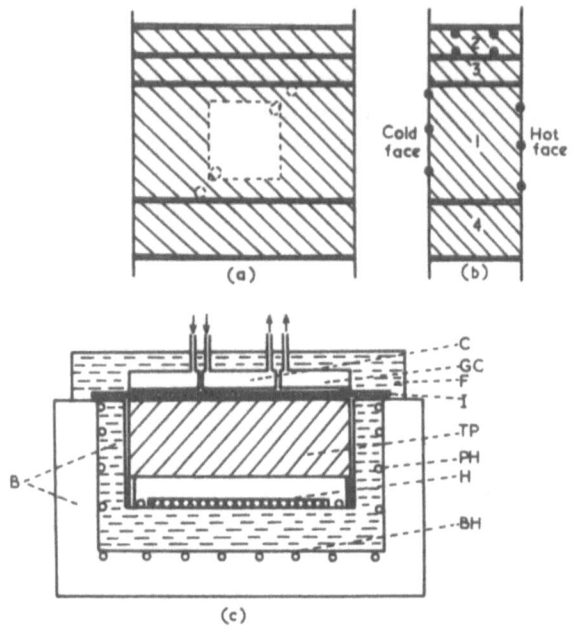

FIGURE 1. The B.S. 1902 test apparatus.

(4) $230 \times 55 \times 76$ (or 64) mm on one side and two slabs (2 and 3), $230 \times 30 \times 76$ (or 64) mm on the other side, making up a panel roughly 230×230 mm. This panel rests on a refractory box in the furnace chamber.

The top and bottom faces of the panel are dressed flat and parallel and six thermocouples inserted into grooves as in Fig. 1b with the couples running parallel to the 230-mm dimension. The grooves are of such a depth that the couples lie half in and half out, being cemented into place with a commercial high-grade alumina cement. Four other couples are placed in slabs 2 and 3 as shown. These are used to detect lateral heat flow.

The main heaters (H) are Pt–20% Rh wire thinly coated with alumina cement. A set of nickel-chrome alloy heaters (BH) are used to boost the main heaters and reduce heat flow through the bottom of the furnace case at temperatures above 500 °C. Peripheral heaters (PH) balance any lateral heat flow from the test panel, the thermocouples in slabs 2 and 3 being used for this purpose. The furnace casing has two-stage insulation, the inner lining being type 2800 insulating brick with a lightweight insulating brick layer between this and the casing.

The backup insulation is a 230×230 mm slab of insulating refractory (type 2600 or 2800) with a thickness of about 10 mm. This, however, can be altered to suit circumstances. This slab is bedded down onto the test panel with fine alumina powder to ensure good thermal contact. The calorimeter (C) and guard ring (GC) rest on thin strips (0.5×10 mm) of ceramic fiber placed on the insulating slab. This avoids the possibility of the calorimeter assembly touching at a few discrete points giving irregular temperature distribution.

To eliminate any ambient temperature effects the whole calorimeter assembly is covered with a 50 mm layer of ceramic insulating fiber. The design of the calorimeter and guard ring is shown in Fig. 2. Full constructional details of the apparatus are given elsewhere.[1]

2.2. Experimental Procedure

Before assembly the thickness (L) of the test brick is measured over the thermocouple junctions, using vernier callipers, to better than ± 0.2 mm. The panel is assembled and gaps between the edges and the furnace wall are filled in with alumina cement. The peripheral heaters are switched on to dry out the panel and then dry alumina powder (250 μm) is poured over the warm panel and the backup slab is bedded down to give a uniform powder layer about 1 mm thick. Ceramic fiber strips are placed on the insulating slab and the calorimeter–guard assembly positioned. The whole is covered with a 50-mm layer of ceramic insulating fiber.

The power input to the heaters is set to give the required hot-face temperature and the assembly is allowed to stabilize overnight (12–15 h). Readings are taken of the hot- and cold-face temperatures to ensure thermal equilibrium (less than 2 °C change in 30 min). The water flow through the calorimeter/guard is set to give a difference of about 3 °C between input and output water temperatures. Peripheral heat flow is checked with the lateral thermocouples

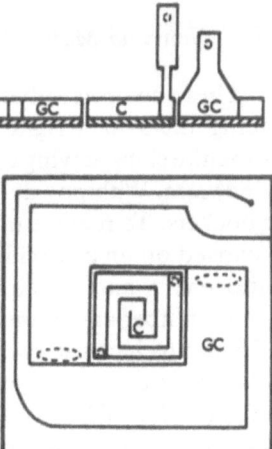

FIGURE 2. Design of calorimeter and guard ring.

and if necessary the peripheral heaters adjusted. If this is done then the measurements are delayed until thermal equilibrium is reestablished.

The heat flow rate is obtained by measuring the time taken for a known quantity of water to flow through the assembly, the input and output water temperatures being measured during this period. This gives the heat flow rate in calories per second.

The hot- and cold-face temperatures of the test pieces are measured before and after the heat flow measurement taking the mean of the three couples in each face to ±0.5 °C.

The conductivity (λ) can then be calculated from

$$\lambda = \frac{QL}{A(t_1 - t_2)} \times 418.67 \quad \text{watts per meter Kelvin } (\text{W m}^{-1}\,\text{K}^{-1})$$

where Q is the heat flow in cal s^{-1}, L is the mean distance between junctions of hot and cold face couples (cm), t_1 is the average temperature of hot face (°C or K), t_2 is the average temperature of cold face (°C or K), and A is the area of calorimeter plus one-half the area of the gap between calorimeter and the guard (cm^2). The conductivity is quoted as having been determined as a "mean" temperature of $(t_1 + t_2)/2$ (°C or K). Three determinations are made at this temperature at intervals of 1 h. These usually agree to ±2.5%. Measurements are normally made in ascending temperature steps usually agreed with the supplier although the "mean temperatures" will be dependent on the maximum permissible hot-face temperature.

2.3. Variations in the Standard Method

2.3.1. Nonrigid Materials

These comprise fiber materials, granules, and powders, and are tested by placing them in a rigid box of heat-resistant material, usually made from sillimanite slabs, having dimensions similar to the full test panel, 230 × 230 × 76 (or 64) mm. With powders and granular materials care must be taken to avoid air pockets. To remove these, light tamping or vibration is allowed but should be carried out in a specific manner, usually agreed with the supplier, to ensure homogeneity and with a known bulk density.

With fiber and granular materials the depth of the box is usually reduced to 25 mm to increase the heat flux. Because the thermal conductivity of these materials is dependent on bulk density, spacers are usually placed in the box to define the thickness and a known specified compressive load applied to produce the required bulk density.

2.3.2. Electrically Conducting Materials

In this case the thermocouples are insulated with close-fitting thin-wall pure alumina sheaths to avoid stray emfs or electrical short circuits between hot- and cold-face couples which would give erroneous readings.

2.3.3. Monolithics or Refractory Concretes

In service these materials are fired *in situ*, there being a loss of combined water during this process. With a lightweight insulating material ("castable") it is possible to have a temperature gradient such that the hot face is fully fired whereas the cold face is only in the "as-received–fully dried" condition. This can cause problems which are overcome by prefiring. This can be to a temperature slightly above the maximum hot-face temperature to be used in the conductivity test, the specimen being fired uniformly, or the complete panel can be gradient-fired in the conductivity apparatus to a temperature slightly above the hot-face temperature for any one set of measurements. This can be repeated at each temperature measurement. This technique gives conditions approximating to those encountered in service. The ASTM C201-68 test uses descending temperature steps so that the panel is usually uniformly prefired before testing.

2.3.4. Differences from the ASTM C201-68 (1979) Method

These are mainly constructional, the test panel being heated from above in the ASTM test. The B.S. 1902 test uses a smaller test panel with peripheral heaters to reduce lateral heat flow and there are also differences in calorimeter design. In the test procedure the ASTM test uses descending temperature steps as compared with ascending ones in the B.S. test. When compared, using the same temperature-stable test pieces, the two test methods usually agree to within ±5%.

2.3.5. Modifications

The main modification consists of replacing the calorimeter assembly with a heat-flow meter. This usually comprises a small slab of material of known conductivity with multiple thermocouples set in the upper and lower faces connected in a differential mode. The output emf is converted to heat flow either by a simple calibration constant or graphically. Errors can arise because of lateral heat flow, a guard ring not being normally incorporated.

2.4. Summary of Application Range

The method is used for refractory materials in block form, usually the "standard square"—230 × 230 × 76 (or 64) mm over the thermal conductivity limits of 0.05 to 15 W m^{-1} K^{-1}. Temperature range is restricted by the temperature gradient across the test piece since the hot-face temperature must not exceed the maximum working temperature of the test material. This may cause problems with lightweight insulating materials where the temperature gradient may be as high as 900 °C to give a "mean temperature" of 550 °C. This problem can be overcome to some extent by the use of thinner test pieces. With the present apparatus maximum hot-face temperature is around 1350 °C.

Nonrigid materials can be tested using a refractory container and electrically conducting refractories by the use of insulated thermocouples.

3. ERROR ANALYSIS

A more detailed review is given elsewhere.[11,12] The work quoted there was carried out around 1953 and since then there have been considerable improvements in temperature measurement and control. However, the error analysis section is still valid.

3.1. Measurement Accuracies

The various measurements made during a test run can be summarized as follows:

a. Heat path length (e.g., thickness of test piece). This is measured with vernier callipers to ±0.2 mm (roughly ±0.25%) and is dependent on the specimen top and bottom faces being flat and parallel.

b. Calorimeter area is usually measured to 0.1 cm^2—i.e., ±0.2%—and can be regarded as an equipment constant.

c. Measurement of temperature gradient—taken as the mean of three cold-face and three hot-face temperatures to ±0.5 °C or better.

d. The evaluation of the heat flow rate involves measurement of water volume, time, and temperature rise. The water is collected in a calibrated flask and with proper precautions the error should be negligible (<0.1%). Timing is usually to 0.2 s in 150 s, i.e., below 0.2% max. Temperature rise measurement is dependent on the type of detector used, e.g., mercury-in-glass restricted-range thermometers, multiple high-output thermocouples, platinum resistance thermometer. The overall temperature difference between the input and output water is normally set to about 3 °C with an expected measurement accuracy of ±0.01 °C. The timing period is usually between 90 and 240 s, and five to seven water temperature readings are made at 30-s intervals.

3.2. Possible Error Sources

Some of these have already been discussed in the preceding section.

3.2.1. Departure from Linear Heat Flow

Linearity of heat flow is essential since any lateral heat flow in the measurement section will give an error in the derived conductivity value. It is difficult to achieve zero lateral heat flow in practice but it can be reduced to a low value by use of the peripheral heaters. The lateral thermocouples can be used to estimate the actual value, and this has been quoted as being typically below 0.5% of the total heat flow.[12] The error from this source usually gives a low value for the conductivity. The linearity of the heat flow through the panel is primarily dependent on attaining an even temperature distribution across both hot and cold faces. There is a gap of about 40 mm between the main heaters and the test pieces hot-face, and in practice this gives good heat distribution shown by the fact that the readings on the hot face usually agree to ±2 °C.

The alumina powder layer between the backup insulation and the test panel ensures good thermal contact provided that there are no voids or hollows, that the layer is of uniform thickness and that the fiber strips ensure that there is no direct contact with the calorimeter to give a possible uneven temperature distribution.

With some insulating firebricks and silica bricks color changes occur on heating to high temperatures to give what could be termed "heat contour lines."[10] The presence of colored zones with perfectly straight boundaries indicates uniform temperature distribution and whether or not the heat flow is accurately perpendicular to the hot face.

3.2.2. Measurement of the Temperature Gradient

This has been covered to some extent in Section 3.1 (c) and involves the thickness and the surface temperatures. With steep temperature gradients a slight displacement of a thermocouple in or out of the grooves could give an error of 5 °C or more. The finite size of the junction may also have an effect. The use of three couples in each face will reduce this error. Nicholls[7] suggests that if a steep temperature gradient exists at a surface then a thermocouple lying in that surface may not register the true temperature. This would apply if the calorimeter were in direct contact with the "cold" face. The presence of the backup insulation has the effect of a marked reduction in the temperature gradient. With modern temperature measuring equipment the actual temperature measurement error should be well below 0.5 °C.

The effect of the backup insulation has been studied[10,12] and it was concluded that any errors introduced by this technique lie within the ±2% reproducibility region. It is difficult to estimate the maximum overall error that can arise in the temperature gradient measurement, but it has been given as about 1.5%.[11]

3.2.3. Measurement of Heat Flow [See Also 3.1 (d)]

The calorimeter system should not introduce any serious error. The design of the water channels in the guard ring and calorimeter results in "equivalent cooling areas" ensuring that the water flow rates in both are similar and can therefore be varied over a wide range without affecting the heat measurement. The thermal mass of the assembly gives a time lag of up to 60 s between a change in the temperature of the inlet water and the corresponding change in the outlet water temperature, but this is offset by taking readings every 30 s. The total water flow measurement time is usually between 90 and 240 s, 250 ml of water being measured in this time interval for a water temperature rise of about 3 °C. With low flow rates, as with insulating refractories, this volume can be reduced to 100 ml. The total calorimeter error, based on an analysis of the individual errors in measuring rate of flow and temperature, and taking into account the possible exchange of heat between the calorimeter and the surrounding air, has been given as less than 1%.[11,12]

A further possible source of error in the heat flow measurement is that due to direct radiation at high temperatures (say above 900 °C hot-face temperature), the test material being regarded as semitransparent to infrared radiation. Calculation of the heat flow due to radiation is extremely complex, being affected by such parameters as effective refractive index, absorption coefficient, thermal gradient in the specimen, and the nature of the surfaces of the test piece and calorimeter (smooth or rough). At the present time the magnitude of the radiation effect, for refractory materials under practical experimental conditions, has not been reliably established.

3.2.4. Attainment of Thermal Equilibrium

The panel is normally held at the operating temperature for 12–15 h unless the peripheral heaters are adjusted, in which case up to a further 24 h may elapse. Equilibrium is defined as not more than 2 °C change in the hot or cold face temperatures over a 30-min period.[1] These temperatures are also measured at the beginning and end of the heat flow measurements. With care any error arising from failure to reach thermal equilibrium should be very small.

3.2.5. Total Error

The individual errors listed above are of two kinds. The first are random errors which tend to cancel out when averaged, e.g., calorimeter and temperature measurement errors. The second type are constant or systematic errors arising from faulty setting up, e.g., positioning of backup insulation and/or calorimeter, lack of parallelism between hot and cold faces, faulty measuring equipment, contaminated thermocouples. In this respect the accuracy of a measurement is often confused with the precision with which it is made—equipment calibration should be checked at regular intervals—and reproducibility as absolute accuracy. The cooperative test in 1953[12] indicated that the ASTM and B.S. tests agreed to within 5%. However, no claims have been made for absolute accuracy although it is thought that the possible error in measuring thermal conductivity by either method is less than ±5%.

3.3. Comparison with Other Methods

When comparing the B.S. 1902 method with another method such as the split-column comparative method it is essential that the comparison is a valid one. It is preferable that the same test piece(s) should be used in both tests and that the temperature gradient should be of the same order. This is particularly important where, as in the case of some basic refractories, the conductivity–temperature curve is nonlinear. It is also important where physical changes occur with rising temperature such as with monolithics (refractory concretes).

The thermal conductivity value obtained from the B.S. 1902 test is given as being at a "mean temperature" which is the arithmetic mean of the hot- and cold-face temperatures. This thermal conductivity value is actually equal to the integral of the true conductivity over the range defined by the hot and cold face temperatures divided by the temperature difference. The difference between this derived value and the true conductivity at the "mean temperature" will depend on the shape of the conductivity–temperature curve over the temperature interval. The error will be negligible where the curve is linear or slightly curved but considerable errors can arise where the curve exhibits a minimum as in the case of some magnesite refractories or an "elbow" as with ceramic insulating fibers.

Heat flow should be linear and in the same direction in both tests. Problems can arise where the heat flow is radial (as in the hot-wire and radial-flow methods) and the material is anisotropic. The actual conductivity value can also be important. The B.S. 1902 method can be used with confidence over the range 0.05 to 15 $W\,m^{-1}\,K^{-1}$, and if comparison was made with the split-column comparative method on high-alumina firebricks with conductivity in the range 2–5 $W\,m^{-1}\,K^{-1}$ good agreement would be expected. However, if the comparison was made with carbon refractories ($>20\,W\,m^{-1}\,K^{-1}$) or insulating refractories ($\sim0.3\,W\,m^{-1}\,K^{-1}$), then in the first case the range would be outside that normally used with the B.S. 1902 test and in the second case well below that for the comparator test which gives best results when the conductivity value is of the same order as the steel standard (10–25 $W\,m^{-1}\,K^{-1}$).

Where the comparison is valid, i.e., with isotropic materials in the range 0.1–1.5 $W\,m^{-1}\,K^{-1}$ good agreement (±5%) is found with the hot-wire method.

NOTATION

λ   Thermal conductivity
Q Heat flow (Heat flux)
L Mean distance between hot and cold faces of the test piece
A Area through which the measured heat flow passes
t_1 Average temperature of the test piece hot face
t_2 Average temperature of the test piece cold face

ACKNOWLEDGMENTS. The author wishes to acknowledge the assistance given by his colleague A.M. Downs. Acknowledgment is also due to Dr. D.W.F. James. Chapter 7 is based on section 12 of B.S. 1902: Part 1A: 1966, with the permission of the British Standards Institution. Complete copies of the standard can be obtained through national standards bodies.

REFERENCES

1. British Standards Institution, "Methods of Testing Refractory Materials: Sampling and Physical Tests," B.S. 1902: Part 5: Section 5.5. 1991.
2. American Society for Testing and Materials, "Standard Test Method for Thermal Conductivity of Refractories," ANSI/ASTM C201: 68 (1979) supplemented by C182: 72 (1978), C202: 71 (1977), C417: 72 (1978), C767: 73 (1979).
3. F.H. Norton, "The Thermal Conductivity of Some Refractories," *J. Ceram. Soc.* **10**(1), 30 (1927).
4. G.B. Wilkes, "The Thermal Conductivity of Magnesite Brick," *J. Am. Ceram. Soc.* **16**(3), 125 (1933).
5. G.B. Wilkes, "The Thermal Conductivity of Refractories," *J. Am. Ceram. Soc.* **17**(5), 173 (1934).
6. J.B. Austin and R.H.H. Pierce, "The Reliability of Measurements of the Thermal Conductivity of Refractory Brick," *J. Am. Ceram. Soc.* **18**(2), 48 (1935).
7. P. Nicholls, "Determination of Thermal Conductivity of Refractories—Report for American Society for Testing and Materials, Sub-Committee C-8," *Bull Am. Ceram. Soc.* **15**(2), 37 (1936).
8. H. Oliver, "A Note of the Reliability of Thermal Conductivity Measurements for Insulating Materials," *Trans Br. Ceram. Soc.* **37**, 49 (1937/8).
9. C.L. Norton, "Apparatus for Measuring Thermal Conductivity of Refractories," *J. Am. Ceram. Soc.* **25**(15), 451 (1942).
10. T.C. Patton and C.L. Norton, "Measurement of the Thermal Conductivity of Fire-Clay Refractories," *J. Am. Ceram. Soc.* **26**(10), 350 (1943).
11. J.F. Clements and J. Vyse, "A New Thermal Conductivity Apparatus for Refractory Materials," *Trans. Br. Ceram. Soc.* **53**, 134 (1954).
12. A. F. Watson, J.F. Clements, and J. Vyse, "A Co-operative Test on Thermal Conductivity," *Trans. Br. Ceram. Soc.* **53**, 156 (1954).

8

The Variable-Gap Technique for Measuring Thermal Conductivity of Fluid Specimens

J.W. COOKE

1. INTRODUCTION

The variable-gap technique for measuring thermal conductivity of fluids using one directional heat flow is a significant improvement over other parallel wall methods in that it takes advantage of the fluidity of the specimen. By use of this technique the specimen thickness can be varied continuously during the operation with a minimum disturbance to the specimen composition or to the system temperature distribution. Also, by varying the specimen thickness, the undesirable effects of several factors, including the errors caused by specimen voids or inhomogeneities, natural convection, radiative heat transfer, corrosion, deposit formation, radial heat flow, thermocouple location, and thermocouple drift, can be greatly reduced. Since only the change in the specimen thickness and the change in the temperature across the specimen is measured, the potential errors of these measurements are smaller and the influence of convection, radiation, and heat losses can be detected and minimized. In addition, the apparatus can be used with little or no modification to measure the thermal conductivities of solids and gases as well as liquids. Considering the advantages of the method, it is surprising that only limited use has been made of the variable-gap technique.[1-4]

2. PRINCIPLE OF METHOD

The technique is shown schematically in Fig. 1. Heat from the main heater travels downward through the liquid sample region (labeled "variable gap"

J.W. COOKE ● Energy Programs Division, Oak Ridge Field Office, Department of Energy, Oak Ridge, Tennessee 37831, USA. Formerly with Union Carbide Nuclear Division at the Oak Ridge National Laboratory, Oak Ridge, Tennessee, USA.

in the figure) to a heat sink. Heat flow in the upward and radial directions is minimized by appropriately located guard heaters, and the heat flux into the sample is measured by the voltage and current of the DC power to the main heater. The temperature drop across the gap is determined by thermocouples located on the axial center line in the metal surfaces defining the sample region. The sample thickness is varied by moving the assembly containing the main heater and is measured by a precision dial indicator. The system temperature level is maintained by a surrounding zone-controlled furnace.

2.1. Idealized Model

The measured temperature difference can be resolved into the temperature drop across the sample gap; the temperature drops in the metal walls defining the test region; the temperature drops in any solid or gaseous films adhering to the metal surfaces; and errors associated with thermocouple calibration, lead-wire inhomogeneities in thermal gradient regions, and instrument malfunctions. Neglecting the error term, we can write

$$\Delta T = \Delta T_s + \Delta T_m + \Delta T_f \tag{1}$$

where subscripts are sample, metal, and surface film, respectively.
For the sample region, the temperature difference is

$$\Delta T_s = (Q/A)\Delta x_s / k_s \tag{2}$$

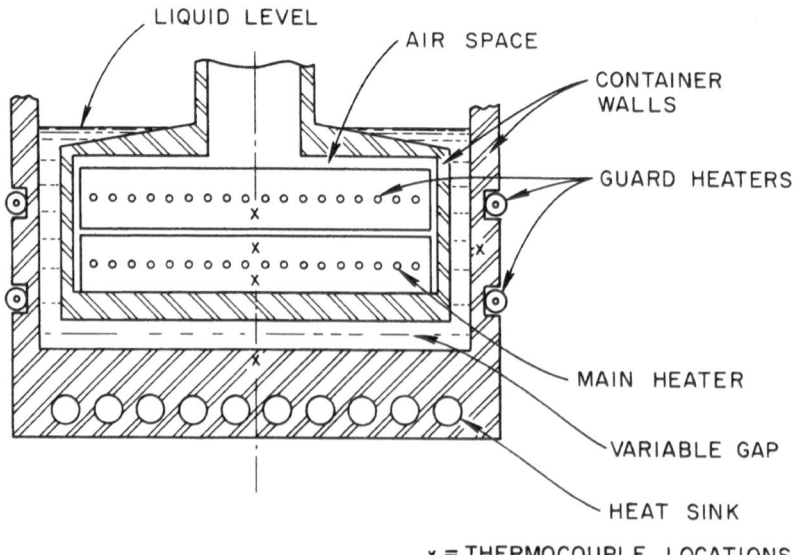

LIQUID LEVEL

AIR SPACE

CONTAINER WALLS

GUARD HEATERS

MAIN HEATER

VARIABLE GAP

HEAT SINK

x = THERMOCOUPLE LOCATIONS

FIGURE 1. Schematic drawing of a variable-gap thermal conductivity cell.

where Q/A is the heat flux, Δx_s is the gap thickness, and k_s is the thermal conductivity of the liquid sample. It is assumed that no natural convection exists in the sample region.

Similarly, the temperature drop in the confining horizontal metal walls can be written as

$$\Delta T_m = (Q/A)\Delta x_m / K_m \tag{3}$$

where Δx_m is the heat-flow path length in the metal walls and k_m is the thermal conductivity of the metal walls. The heat flux Q/A is the same as in equation (2), assuming no radial heat flow and no bypass heat flow through the side (vertical) walls of the sample cup. Since k_m is a function of temperature, equation (3) can be written separately for the upper and lower metal walls; however, for the purposes of this analysis, the two regions are combined.

The film temperature difference is of the same form as the ΔT quantities given in equations (2) and (3). If surface films are present but of constant and known thickness, Δx_f, during the experiment, there is no effect on the derived sample thermal conductivity or on the associated error. However, a film that grows or decays in an unknown way during the course of the measurement introduces an error in Δx_s.

On combining the above expressions, we obtain

$$\Delta T = \frac{Q}{A}\left(\frac{\Delta x_s}{k_s} + \frac{\Delta x_m}{k_m} + \frac{\Delta x_f}{k_f}\right)$$

or for thermal resistance

$$\frac{\Delta T}{Q/A} = \left(\frac{1}{k_s}\right)\Delta x_s + \left(\frac{\Delta x_m}{k_m} + \frac{\Delta x_f}{k_f}\right) \tag{4}$$

or, simplifying the notation,

$$\frac{\Delta T}{Q/A} = \left(\frac{1}{k}\right)\Delta x + \left(\frac{\Delta T}{Q/A}\right)_0 \tag{5}$$

where $[\Delta T/(Q/A)_0]$ combines all the fixed thermal resistances. This is of the form

$$y = ax + b \tag{6}$$

where a is the slope of this linear expression and is the reciprocal of the sample thermal conductivity. The intercept b combines all other resistances. In operating the apparatus, Q/A is kept constant and ΔT is recorded as Δx is varied. If other modes and paths of heat transfer exist within the specimen, the thermal resistance will not be a linear function of the specimen thickness. However, the effect of these other forms of heat transfer will be reduced as the specimen thickness is decreased. Thus, the thermal conductivity can be determined from the reciprocal slope evaluated at zero specimen thickness.

Another approach to the determination of the sample thermal conductivity can be obtained by rearranging equation (6) as

$$k = \frac{1}{a} = \frac{x}{y - b} \tag{7}$$

Again, if other modes and paths of heat transfer exist within the specimen, the value of thermal conductivity obtained from equation (7) will be the effective value, which will approach the true value as the specimen thickness approaches zero.

2.2. Effect of Radiation

Many investigators consider only the radiation emitted by the wall surfaces when evaluating the heat transfer through a medium separating the walls. This assumption may be correct when the medium is a gas whose mean absorption coefficient $\bar{\kappa}$ is small and whose mean refraction index $\bar{n}$ is near unity. Many media, however, absorb and emit significant amounts of radiation. This internal radiation can contribute more to the heat transfer from wall to wall than the radiation emitted by the wall surfaces. Indeed, even at room temperature, the heat transferred by radiation can approach 5% of that transferred by conduction in some organic fluids whose specimen thickness is as small as 0.1 cm.

If some simplifying assumptions are made, an expression for the radiant heat transfer can be derived. These assumptions are the existence of a constant temperature gradient within the medium and the use of mean values $\bar{n}$ and $\bar{\kappa}$ independent of wavelength. The following equation was derived by Poltz[5,6] for the radiant heat flux:

$$\frac{Q_r}{A} = \frac{16}{3} \frac{\bar{n}^2}{\bar{\kappa}} \sigma T^3 \left(\frac{\Delta T}{\Delta x}\right) Y \tag{8}$$

where

$$Y = 1 - \frac{3}{\tau}(2 - \varepsilon) \int_0^1 \frac{1 - \exp[-(\tau/v)]}{1 + (1 - \varepsilon)\exp[-(\tau/v)]} v^3 \, dv$$

τ is the optical thickness of the medium and equals $\bar{\kappa}\Delta x$, T the average medium temperature (K), ε the emittance of the wall surface, σ the Stefan–Boltzmann constant, 5.71×10^{-12} W cm^{-2} K^{-4}, and v is the dummy integration variable.

In Fig. 2, where Y is plotted as a function of τ for various values of ε, the curve for $\varepsilon = 0$ represents the hypothetical case in which the radiation heat transfer between the walls is accomplished solely by the inner radiation within the medium. The distance between the curve for $\varepsilon = 0$ and any one of the higher curves for the appropriate plate emittance represents the relative contribution of the radiation emitted by the wall surfaces to the total radiated heat flow from wall to wall.

Equation (8) can be combined with the previously derived equation (5) to obtain an expression for the total thermal resistance across a medium separated by two parallel walls when the heat is being transferred simultaneously by conduction and radiation. That is,

$$\frac{\Delta T}{Q/A} = \left(\frac{1}{k + \frac{16}{3}(\bar{n}^2/\bar{\kappa})\sigma T^3 Y} \right) \Delta x + \left(\frac{\Delta T}{Q/A} \right)_0 \tag{9}$$

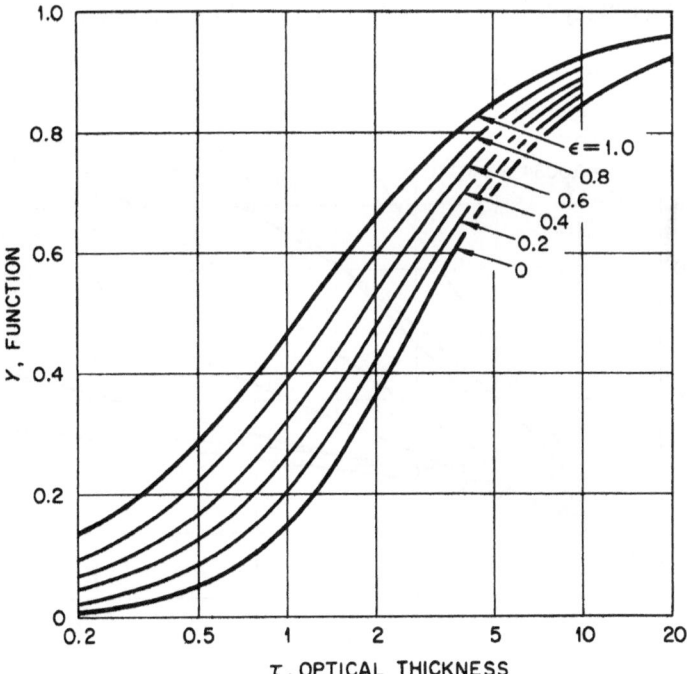

FIGURE 2. Radiative function (Y) vs plate emittance and optical thickness of the specimen.

In the limiting case where the optical thickness τ approaches zero (i.e., very small infrared-absorbing medium),

$$Y_{\tau \to 0} = \tfrac{3}{4}\varepsilon_r \tau \qquad \text{(after Poltz}^{(5)}\text{)}$$

where

$$\varepsilon_r \equiv \left(\frac{1}{\varepsilon_1} + \frac{1}{\varepsilon_2} - 1\right)^{-1} = \frac{\varepsilon}{2 - \varepsilon}$$

for $\varepsilon = \varepsilon_1 = \varepsilon_2$. Thus equation (9) simplifies to

$$\frac{\Delta T}{Q/A} = (k + 4\bar{n}^2\varepsilon_r\sigma T^3\Delta x)^{-1}\Delta x + \left(\frac{\Delta T}{Q/A}\right)_0 \qquad (10)$$

Figure 3 is a plot of the thermal resistance as a function of specimen thickness for various values of the absorption coefficient for a specimen assumed to have a $k = 0.0034 \text{ W cm}^{-1}\,^{\circ}\text{C}^{-1}$, $\bar{n} = 1.5$, $\varepsilon_{\text{wall}} = 0.5$, and $T = 1000\,^{\circ}\text{C}$. For these values of thermal conductivity and temperature, the percent of heat transferred by radiation is quite large. As the absorption coefficient $\bar{\kappa}$ decreases from $\bar{\kappa} = \infty$ (pure conduction) to $\bar{\kappa} = 0$, the percent of radiated heat increases to a maximum at about $\bar{\kappa} = 2$ and decreases until $\bar{\kappa} = 0$. Within the interval $\infty > \kappa > 0$ these curves have an inflection point producing what could be described as "lazy S" curves. Also shown in Fig. 3 is the resistance curve

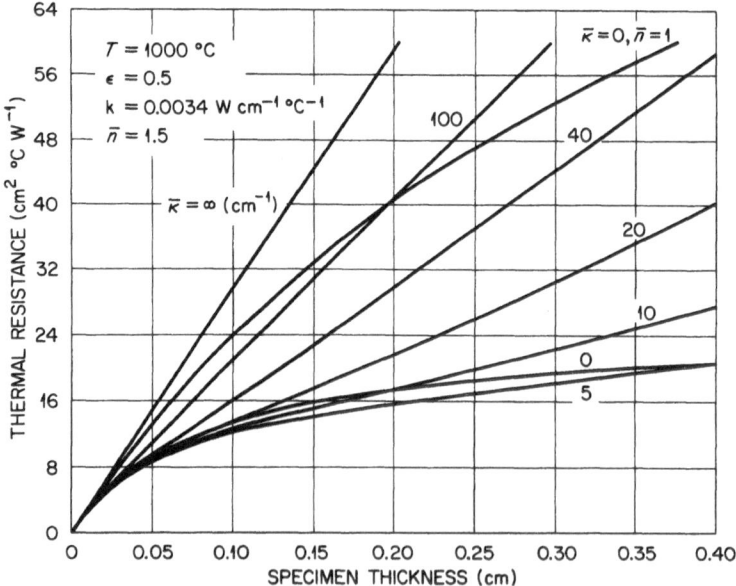

FIGURE 3. Thermal resistance of an infrared absorbing fluid having assumed properties at various values of absorptivity, $\bar{\kappa}$ (cm^{-1}), vs specimen thickness.

for a gas whose absorptivity is near zero and whose index of refraction is near unity. If the optical properties of a specimen are known, equation (9) can be fitted to the experimental data to obtain the slope (and thus the thermal conductivity) at a specimen thickness approaching zero; however, the mean optical properties must be used and the temperature gradient within the specimen must be nearly linear.

2.3. Effect of Natural Convection

Under ideal conditions, no natural convection would be expected in a fluid enclosed between two horizontal, parallel plates with one-dimensional downward heat flow. In the real situation, however, small departures from the ideal conditions can initiate and sustain convection currents within the fluid. If the plates are not horizontal or parallel, or if a temperature gradient exists in the horizontal direction, convection cells can occur. If, in addition, the vertical temperature distribution within the specimen is not linear but distorted by interfluid infrared absorption, the natural convection can be enhanced. Finally, vibrations, especially those in resonance with the natural frequency of the enclosed fluid, can induce and enhance natural convection.

In order to initiate and sustain buoyancy convection cells within enclosed spaces, certain instability criteria must be satisfied. Rayleigh was one of the first investigators to recognize that the instability criterion could be related to certain limiting values of the dimensionless moduli N_{Ra} known as the Rayleigh number, which is defined by

$$N_{Ra} = N_{Gr}N_{Pr} = \left(\frac{g\rho^2\beta\Delta T\Delta x^3}{\mu^2}\right)\left(\frac{C_p\mu}{k}\right) \tag{11}$$

where, in a self-consistent set of units, g is local acceleration due to gravity, ρ density of fluid, μ viscosity, Δx the gap distance, k thermal conductivity, C_p (specific) heat at constant pressure, and β is the coefficient of bulk expansion.

The Rayleigh number, in essence, is the ratio of the product of the buoyancy and inertial forces to the viscous forces. The limiting value of the Rayleigh number to initiate and sustain convection cells has been calculated to be 1700 when the fluid layer is bounded on both sides by solid parallel and horizontal walls and is *heated from below*.[7] Experimental studies by Norden and Usmanov[8] using an interferometer technique show, however, that the departure from a conductive to a convective mode of heat transfer can occur at $N_{Ra} < 1700$ for small specimen thicknesses heated from below.

Figure 4 shows a portion of the data taken from the above experimental studies in which the critical temperature difference, ΔT_c (the temperature difference above which convection occurs), is plotted as a function of the

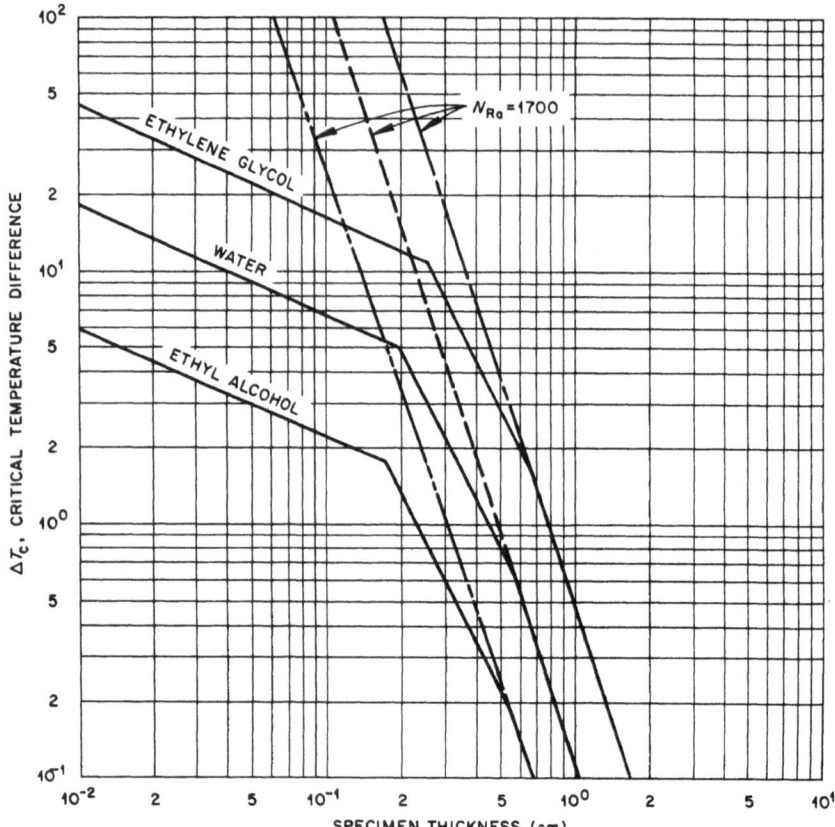

FIGURE 4. Critical temperature above which convection occurs as a function of specimen thickness for three liquids heated from below showing the departure from theoretical criteria, $N_{Ra} = 1700$.

specimen thickness for three liquids: ethylene glycol, water, and ethyl alcohol. Also plotted are the curves for the theoretical values of $N_{Ra} = 1700$. The region below each of the curves plotted in Fig. 4 is stable (i.e., conduction only) and above the curves is unstable (i.e., convection occurs). The experimental data is seen to have two linear slopes ($n = 0.45$ and 2.0) which merge with the theoretical curve ($n = 3$) where

$$\Delta T_c \Delta x^n = \text{constant}$$

The $N_{Ra} = 1700$ criteria could lead to a gross overestimate of ΔT_c for small specimen thicknesses.

Heat-transfer measurements made during Norden and Usmanov's studies showed the ratio of the effective thermal conductivity to the true thermal conductivity for water to be 1.10 at a specimen thickness of 0.15 cm and an N_{Ra} of only 260. Thus, considerable care must be exercised to prevent natural convection in fluids contained between parallel plates *heated from below* as well as inclined, cylindrical, and spherical annuli heated from either side.

A similar experimental study by Berkovsky and Fertman[9] was conducted in which the specimen *was heated from above* and had a nonuniform upper-plate temperature distribution and a uniform lower-plate temperature, T_0. This study showed that even when the nonuniformity of the upper-plate temperature distribution ($T_{max} - T_{min}$) equals ($T_{max} - T_0$), convective heat transfer is not significant at a Rayleigh number of less than 10^4.

No experimental results are reported for the amount of convection taking place when the specimen is heated from above and the plates are not exactly parallel or are slightly tilted. If we consider this case analogous to the Berkovsky and Fertman study, convection would be avoided at an N_{Ra} of less than 10^4 if the ΔT of the tilted layer above that of the horizontal layer did not exceed that of the average ΔT across the plates. Since $\Delta T \propto \Delta x$, the difference in the edge-to-edge separation distance between the plates with respect to each other, or with respect to the horizontal, should not exceed their average separation distance.

To minimize convection due to vibrations, the conduction cell should be well isolated from all sources of vibrations, particularly those within the resonance frequency of the cell.

2.4. Effect of Heat Shunting

Some shunting of heat around the specimen is unavoidable even for the most carefully designed thermal conductivity cells. The percent of shunted heat as compared to heat flow through the specimen can be minimized by careful use of insulating materials, by guard heating, by using large cell diameter-to-thickness ratios, and by using zoned heat sources and sinks. The shunting problems becomes most acute for low-thermal-conductivity specimens at elevated temperatures.

The apparatus described in the present study is designed to minimize the shunting error with specimens having estimated thermal conductivities in the range of 0.05 to 0.10 W cm^{-1} °C^{-1}. The cell wall thickness and the ratio of cell diameter to sample thickness are optimized to reduce the heat shunted to less than 1% of the total heat flow in the absence of heat guards. If the specimen thermal conductivity is significantly (an order of magnitude) lower than the range for which the apparatus was designed, the radial guard heating may not be adequate to prevent some heat shunting, and corrections will be required.

Figure 1 shows the complexity of the possible heat-transfer modes and paths within the thermal conductivity cell. Since neither the temperature distribution along the cell wall nor the heat-transfer coefficients are well known, the simplified model shown in Fig. 5 was selected as an appropriate model for calculating the amount of heat shunting in the system.

By assuming a uniform heat flux and a uniform sink and wall temperature equal to 0 °C, an easy solution for the center-line heat flux may be obtained from the generalized heat conduction equation [equation (12) below]. The radial heat-transfer coefficient U_2 may then be decreased to account for the guard heating.

The temperature distribution within the cylindrical solid can be described by the general heat conduction equation

$$\frac{\partial^2 T}{\partial r^2} + \frac{1}{r}\frac{\partial T}{\partial r} + \frac{\partial^2 T}{\partial z^2} = 0 \tag{12}$$

where T is the temperature (°C) and r and z are the radial and axial coordinates (cm) measured as shown in Fig. 5. Dividing the model into two regions, the boundary conditions for either region can be written in the form

$$k'\frac{\partial T(r, 0)}{\partial z} = -C \tag{13}$$

$$k'\frac{\partial T(r, L)}{\partial z} = -U_1[T(r, L)] \tag{14}$$

$$k'\frac{\partial T(R, z)}{\partial r} = -U_2[T(R, z)] \tag{15}$$

where k' is the thermal conductivity (W cm^{-1} °C^{-1}) of the cylindrical solid within the region, L is the thickness of the solid (cm), R is the radius (cm), C is a constant, while U_1 and U_2 are the axial and radial overall heat-transfer coefficients, respectively.

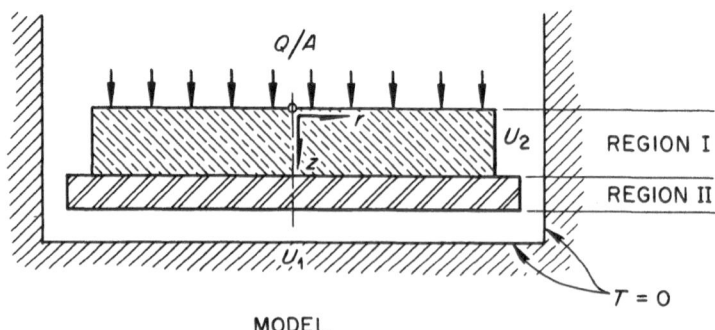

FIGURE 5. Model of the thermal conductivity cell for heat shunting calculation.

The solution of equation (12) for the ratio of the axial center-line heat fluxes entering and leaving region I, using the above boundary conditions, is

$$F_{\mathrm{I}} \equiv \frac{Q/A(0,0)}{Q/A(0,L)} = \sum_n \frac{2a_n J_1(a_n)}{(N_2 + a_n^2)J_0^2(a_n)} \left(\cosh \frac{a_n L}{R} - D_n \sinh \frac{a_n L}{R} \right) \quad (16)$$

where

$$D_n = \frac{N_1 \sinh(a_n L/R) + a_n \cosh(a_n L/R)}{N_1 \cosh(a_n L/R) + a_n \sinh(a_n L/R)} \quad (17)$$

and a_n are the roots of

$$a_n J_1(a_n) - N_2 J_0(a_n) = 0 \quad (18)$$

with

$$N_1 = RU_1/k' \quad \text{and} \quad N_2 = RU_2/k' \quad (19)$$

There is a similar equation for F_{II}.

The heat-transfer coefficients U_1 and U_2 were calculated assuming series and parallel paths of all three heat-transfer modes (convection, conduction, and radiation). The ratio F of the heat flux entering region I to the heat flux leaving region II was calculated as

$$F = F_{\mathrm{I}} F_{\mathrm{II}} \quad (20)$$

To account for various degrees of guard heating, a factor G was defined as

$$G = \frac{T_{\text{heater}} - T_{\text{wall}}}{T_{\text{heater}} - T_{\text{sink}}} \quad (21)$$

such that

$$N_2 = RGU_2/k' \quad (22)$$

Plots of the percent of heat shunted around the specimen, $1 - F$, vs the specimen thickness, Δx, for various specimen thermal conductivities as determined by a computer solution of equation (16) are shown in Figs. 6 and 7 for temperature levels of 300 and 900 °C, respectively. Both of these plots assume partial guard heating; $G = 0.5$ was employed. The dashed curves assume the specimen to be transparent to infrared radiation and the wall emittance to be

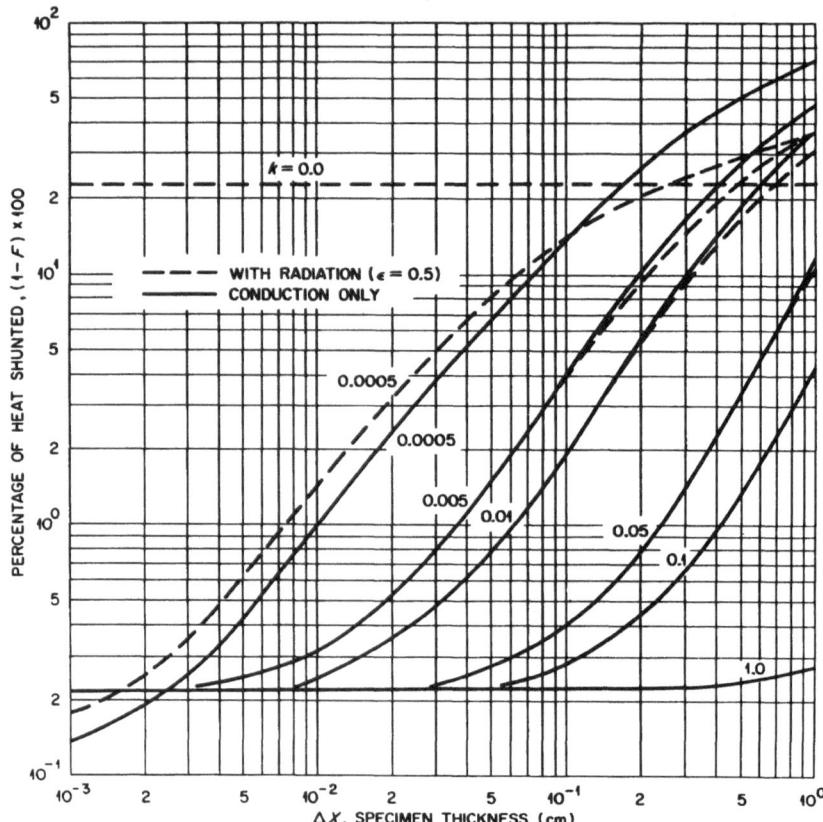

FIGURE 6. Percent of heat shunted around specimen vs specimen thickness for various specimen thermal conductivities with and without radiative heat transfer at 300 °C for $G = 0.5$.

0.5. From these plots, it can be seen that the amount of heat shunted around the specimen can approach 100% for large Δx and very small specimen thermal conductivities. In our thermal conductivity measurements, the guard heating factor G was near zero and a specimen thickness of <0.1 cm was used in determining the thermal conductivity, although larger thicknesses were used to test and analyze the apparatus.

2.5. Method of Calculation

In the variable-gap technique, the thermal conductivity coefficient can be determined (1) from the reciprocal slope of the total thermal resistance across the specimen (including metal walls, deposits, etc.) as a function of specimen thickness as it approaches zero, or (2) from the effective thermal conductivity as the specimen thickness approaches zero. Except for a few spot checks,

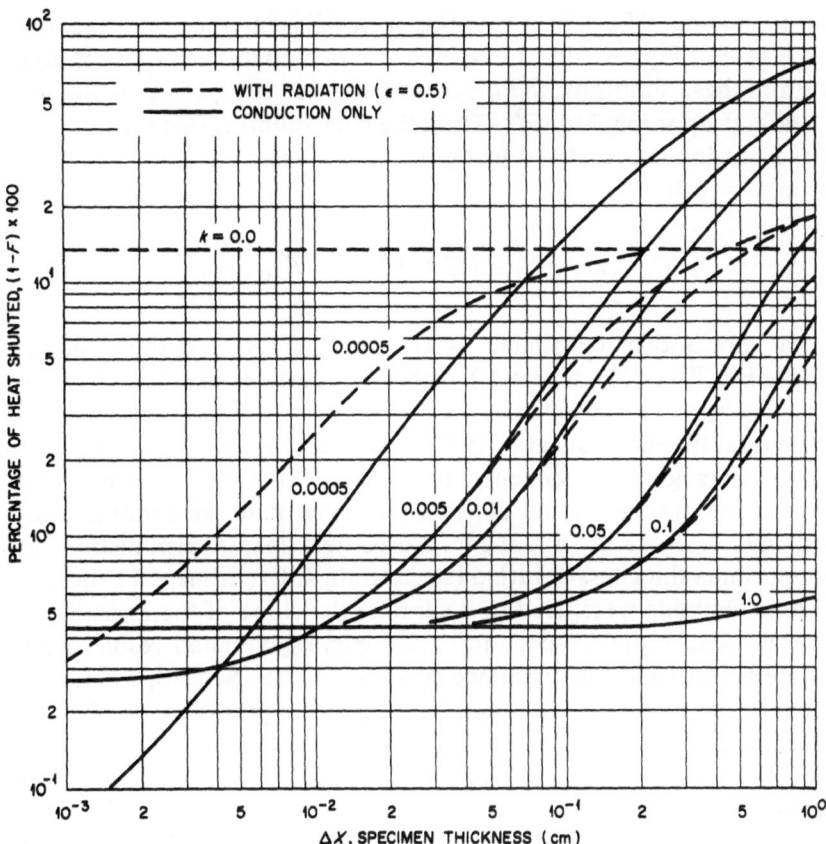

FIGURE 7. Percent of heat shunted around specimen vs specimen thickness for various specimen thermal conductivities with and without radiative heat transfer at 900 °C for $G = 0.5$.

method (1) was used to reduce the data in this study. The thermal resistance is calculated from the measured heat flux Q'/A and from the measured temperature difference, where

$$\frac{Q'}{A} = \frac{4EIV}{\pi D^2}$$

and I is the heater current (amps), V the heater voltage (V), D the diameter of the upper heater plate (cm), and E is the ratio of the effective to total heater wire length* (0.972).

The heat flux Q/A is obtained from the measured heat flux by correcting for the heat shunting. If the guard heating factor G [see equation (21)] is

*The total wire length between voltage taps includes two 2.22-cm-long lead wires.

greater than about 0.01, a heat shunting factor F is interpolated from the plots of $1 - F$ vs Δx (Figs. 6 and 7) obtained from computer solutions of equation (16). From these solutions, the percent of shunted heat, $1 - F$, is found to be very nearly proportional to $G^{0.8}$. Thus, the heat shunting factor for any degree of guard heating is calculated using the results from only one heat shunting factor at $G = 0.5$:

$$(1 - F)_G = (G/0.5)^{0.8}(1 - F)_{G=0.5} \tag{23}$$

The heat flux is then calculated as $Q/A = F(Q'/A)$ and the total thermal resistance is $\Delta T/(Q/A)$, where ΔT is the previously defined total temperature difference across the specimen.

The total thermal resistance is then plotted as a function of the specimen thickness for a given specimen temperature. The specimen temperature is defined as the average of the upper- and lower-plate temperature and both the specimen temperature and the measured heat flux are kept nearly constant as the specimen thickness is varied (see Section 4. Experimental Proceedures).

Three methods were used to determine the slope of the resistance curve at $\Delta x = 0$. First, visual inspection of the curve gave good results when the data were smooth and the resistance curve was linear. Second, when the curve was not linear, numerical finite-difference techniques were used to obtain the slope at $\Delta x = 0$. In the third method the data were fitted to equation (9) [or equation (10) if $\bar{\kappa} = 0$] if adequate information concerning the optical properties of the specimen were known.

Most of the data reported in this study were analyzed by visual inspection or the third method mentioned above. The data were fitted to equation (9) in the following way. The fixed resistance, $[\Delta T/(Q/A)]_0$, is found by extrapolating the thermal resistance vs Δx to $\Delta x = 0$; the plate emittances are determined by carrying out the experimental procedure with the conductivity cell evacuated; and the index of refraction is taken from the literature, an average $\bar{n}$ over the range of infrared wavelengths. From an estimate of the thermal conductivity by visual inspection of the data for $\Delta x < 0.1$, the absorptivity, $\bar{\kappa}$, can then be found from fitting equation (9) to the larger values of Δx. Using these values of the fixed resistance, the plate emittance, the average index of refraction, and the calculated absorptivity, the thermal conductivity is determined by the best fit of the data over the complete range of Δx.

3. EXPERIMENTAL APPARATUS

The apparatus and auxiliary equipment discussed here were designed and constructed for use with molten fluoride salts at temperatures up to 1000 °C.

The description of the thermal conductivity cell itself is sufficiently complete to permit duplication; however, only unusual auxiliary equipment is discussed in detail. To preserve dimensional tolerances and minimize rounding errors, the English units of inches used in the actual fabrication of the apparatus are given in this section.

The complete system can be considered to consist of three parts: the thermal conductivity cell, the furnace, and the electrical, instrument, and control system. The connections and relationships between the various components are shown schematically in Fig. 8.

3.1. Thermal Conductivity Cell

The thermal conductivity cell is shown assembled in Fig. 9 and in detail in Figs. 10 and 11. The cell is made up of two components: the cylindrical-shaped component, which consists of a sink and the radial heaters and contains the specimen, and the piston-shaped component, which is mobile in the vertical direction and contains the main heater and the guard heater assemblies. This cell is designed to minimize heat shunting around the specimen without sacrificing structural integrity and to facilitate the assembly/disassembly and the filling, draining, and cleaning of the apparatus.

The cylindrical component is machined from three pieces of stainless steel type 304 and welded together as shown in Fig. 10. The lower section, which contains the specimen, has holes drilled through the component for cooling air passages, and a sink heater which is a 1/8-in.-diameter sheath-type element pressed into grooves machined in the lower section. An air gap between the cylinder and the sink provides a uniform heat flux distribution to the sink. Similar sheath-type heater elements are placed around the cylinder at the level of the specimen for the radial guard heaters. Several dimensions of the cylinder component are maintained in close tolerances to insure that the specimen thickness is uniform across the diameter. The final machining of these surfaces was made after the entire cylinder had been dimensionally stabilized by heat treatment.

The mobile piston component (Fig. 11) contains the main heater assembly and is designed to minimize the upward flow of heat from the specimen area. The main heater is constructed with 10-mil Pt–10% Rh wire, wound, embedded, and buried in a high-density Al_2O_3 insulator. A duplicate heater, the top heater, sits above the main heater. Between these two heaters is a 1/16-in. gap containing two platinum-foil radiation shields. The temperature across this gap is monitored by two thermocouples and balanced with the top heater to prevent axial heat loss. A gold foil provides high contact conductance between the main heater and the bottom of the piston container. All the electrical wiring, the platinum wire, and the thermocouples extend through a hollow rod connecting the piston with the upper plate and support. A flexible bellows

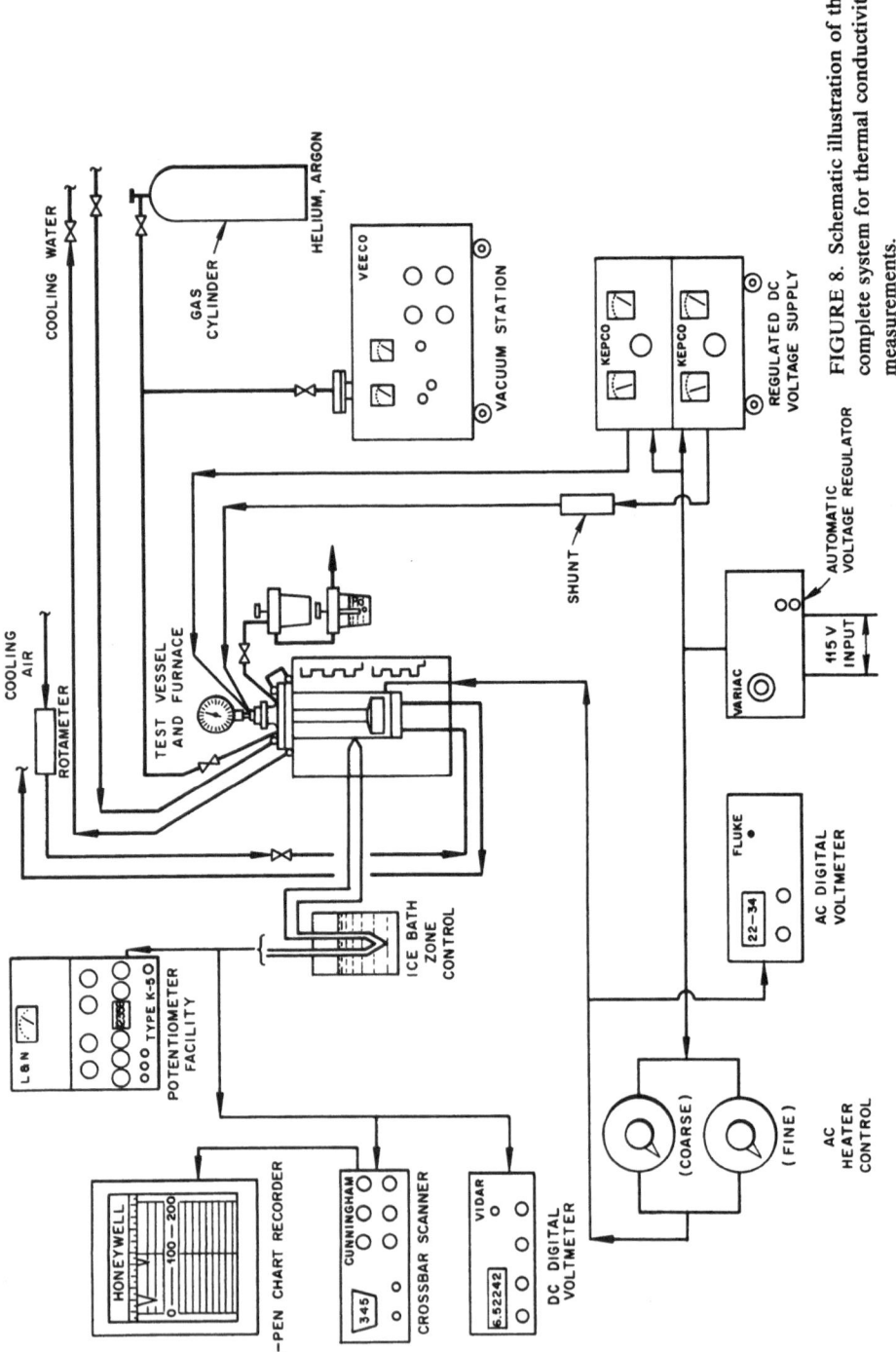

FIGURE 8. Schematic illustration of the complete system for thermal conductivity measurements.

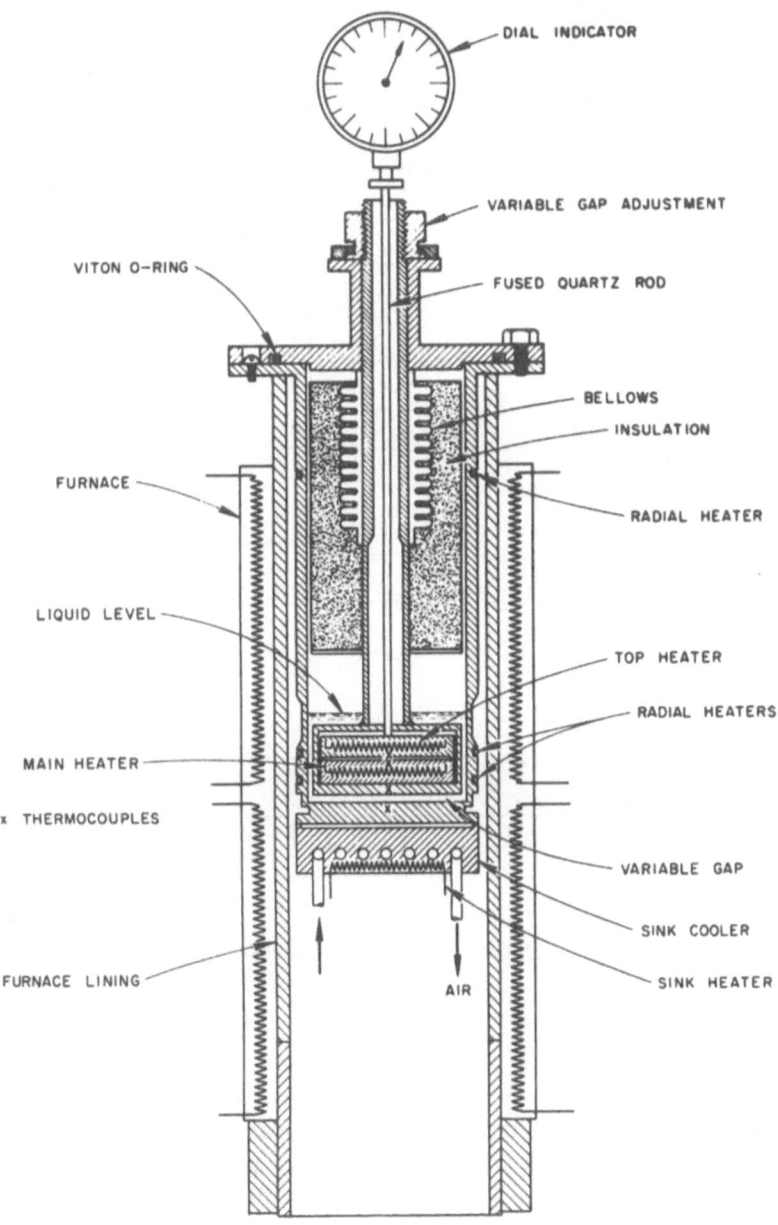

FIGURE 9. Schematic cross section of the thermal conductivity cell.

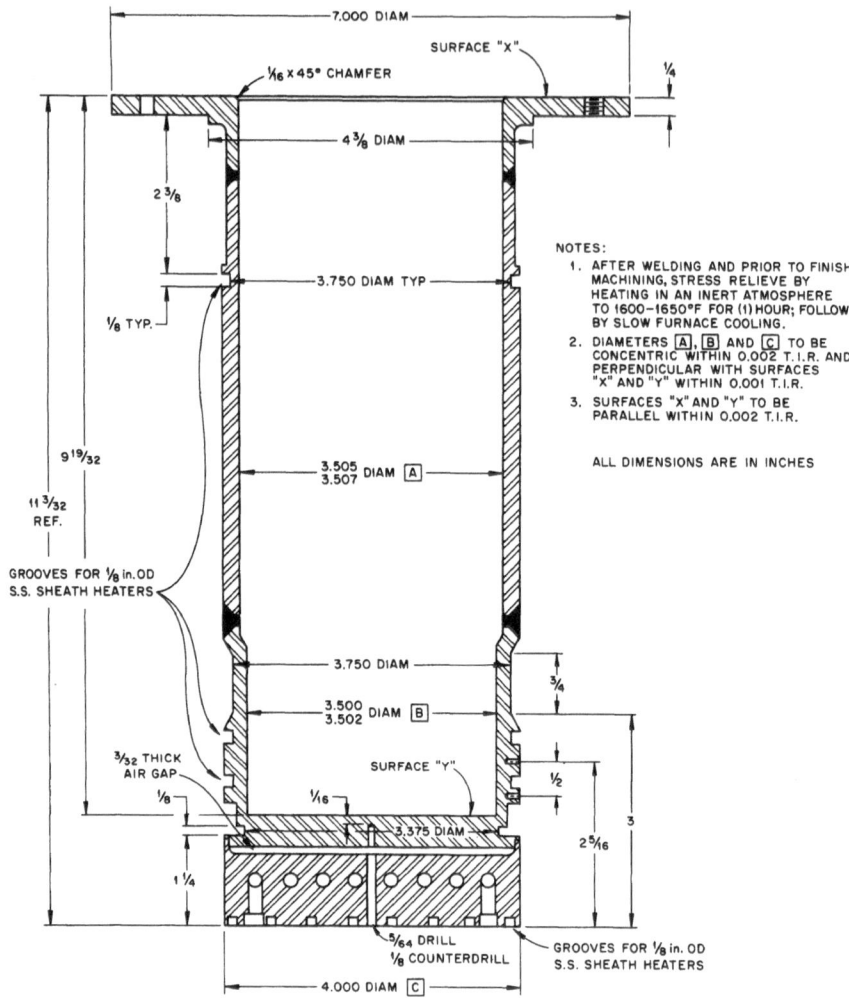

FIGURE 10. Detail of the thermal conductivity cell cylinder.

welded to the piston rod and the upper flange allows mobility of the piston while providing a vacuum tight seal. The vertical movement of the piston is adjusted by a threaded nut whose microthreads are precision machined. The vertical movement of the piston is measured with a dial indicator connected to the bottom of the piston with a fused quartz rod to minimize the effect of thermal expansion. Several dimensions of the mobile piston component are also maintained in close tolerances (see Fig. 11). The specifications of these

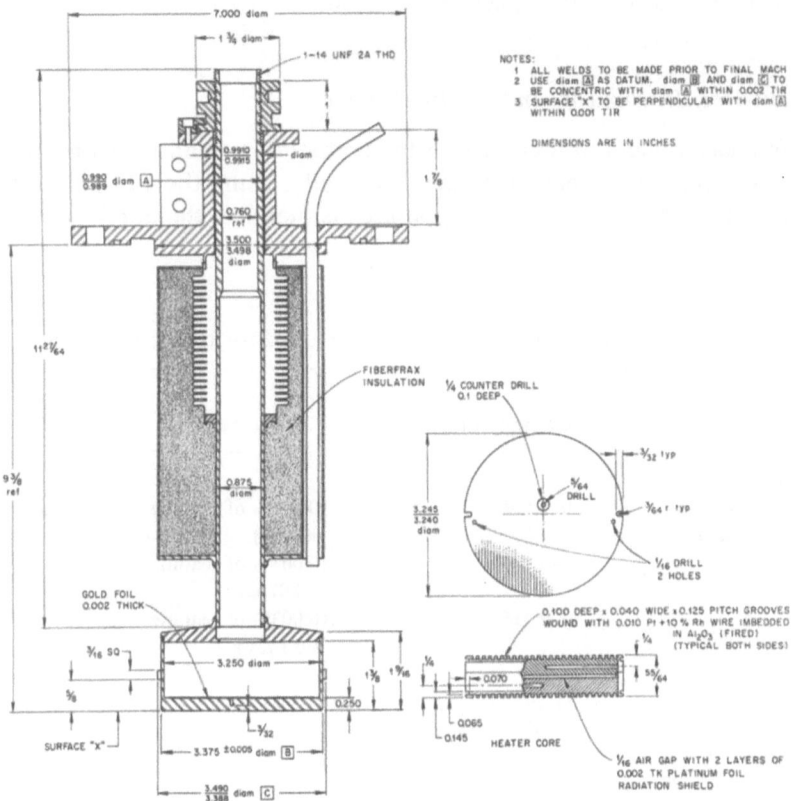

FIGURE 11. Detail of movable piston assembly of the thermal conductivity cell.

tolerances insure that the parallelism of the two plate surfaces does not vary by more than 0.005 in. This tolerance was checked *in situ* by placing a ball bearing in the gap of the specimen, rotating the apparatus in an inclined position, and measuring the variation in the gap thickness around the circumference of the cell with the dial indicator. The maximum edge-to-edge variation of the thickness for the present apparatus was 0.0017 in.

3.2. Furnace

The furnace consists of two 6-in.-ID × 8-in.-long individually controlled clamshell heaters of the embedded-wire type. The annular space between the heater and the 12.5-in.-OD water-cooled furnace shell was filled with high temperature ceramic fiber insulation. This furnace is capable of raising the ambient temperature of the specimen to 1000 °C.

3.3. Electrical, Instrument, and Control System

The measurement and control of the temperatures and guard heaters for the present apparatus were performed manually, using the equipment listed in Table 1. If extensive use of the apparatus is planned, the application of an

TABLE 1

List of Pertinent Experimental Equipment Used for the Present Study

Equipment	Capacity or range	Accuracy	Least count
Potentiometer			
L&N 7555 type K-5, 177362, Leeds &	0–1.6 V	±(0.001% of reading +2 μV)	2.0 μV
Northrup	0–0.16 V	±(0.003% of reading +0.02 μV)	0.2 μV
	0–0.016 V	±(0.003% of reading +0.1 μV)	0.02 μV
Voltage regulator			
Variac, automatic Model 1581-A	Output 115 V, adj. ±10%. 50 A	±0.25%	
Null detector			
Leeds & Northrup 9834-1	0.07 μV mm^{-1} for source resistances up to 2000 Ω		0.1 μV
DC digital voltmeter			
Vidar 521 integrating voltmeter	±10 μV to ±1000 V in six-decade stages	±0.01% of full scale	1.0 μV
Thermocouples			
Pt vs Pt–10% Rh		±0.2%	
Chromel–Alumel		±0.75%	
Precision resistor			
Leeds & Northrup Model 4360	0.1 Ω, 15 A	±0.04%	
DC voltage supply			
Kepco Model SM 75-8MX	Input 105–125 V, 60 cps, 9.6 A max; output 0–75 V, 0–8 A	Line: <0.01% voltage variation or 0.002 V, whichever greater after stabilizing	
Dial indicator			
Federal Model E3BS-R1	20 revolutions (0.4 in.)	one-half of one division	0.0001 in.

automated, computer operated, data acquisition and control system,[10-14] utilizing the updated components listed in Table 2, is recommended.

The heater electrical system is diagrammed in Fig. 12. The AC supply voltages to the sink, guard, and furnace heaters are regulated with a Variac automatic voltage regulator. The voltage through each of the heaters is adjusted by two Variacs, coarse and fine, and by a filament transformer. The fine adjustment extends the coarse Variac setting with the addition of 6 V of increased sensitivity (0.038 V). A Fluke AC digital voltmeter is used for setting and reading the voltages.

Two Kepco DC voltage supplies provide the electrical power to the main and top guard heaters. The supply voltage to the two Kepco units is also

TABLE 2

List of Components/Suppliers for Two Suggested Automated, Computer Operated, Data Acquisition and Control Systems

Components	Suppliers	
	System 1	System 2
Computer/Controller	Hewlett-Packard (9800 series)	IBM-PC/Compatible
EMF Measurement Digital Multimeter 1 to .01 microvolt resolution	Hewlett-Packard (34xx Series) Keithley Fluke Guildline	Same as System 1 with IEEE-488 interface adapter Or with Analog I/O Card: Data Translation Metrabyte Analog Devices Burr-Brown
Low-thermal Scanner <1 microvolt thermal emfs	Fluke Keithley	Same as System 1 with IEEE-488 interface adapter
Power Supplies with digital control interface (size to fit power requirements)	With IEEE interface: Hewlett-Packard (3495 Series) Kepco	With voltage programming or with D/A available or analog I/O cards or in IEEE-488 module: Hewlett-Packard Kepco

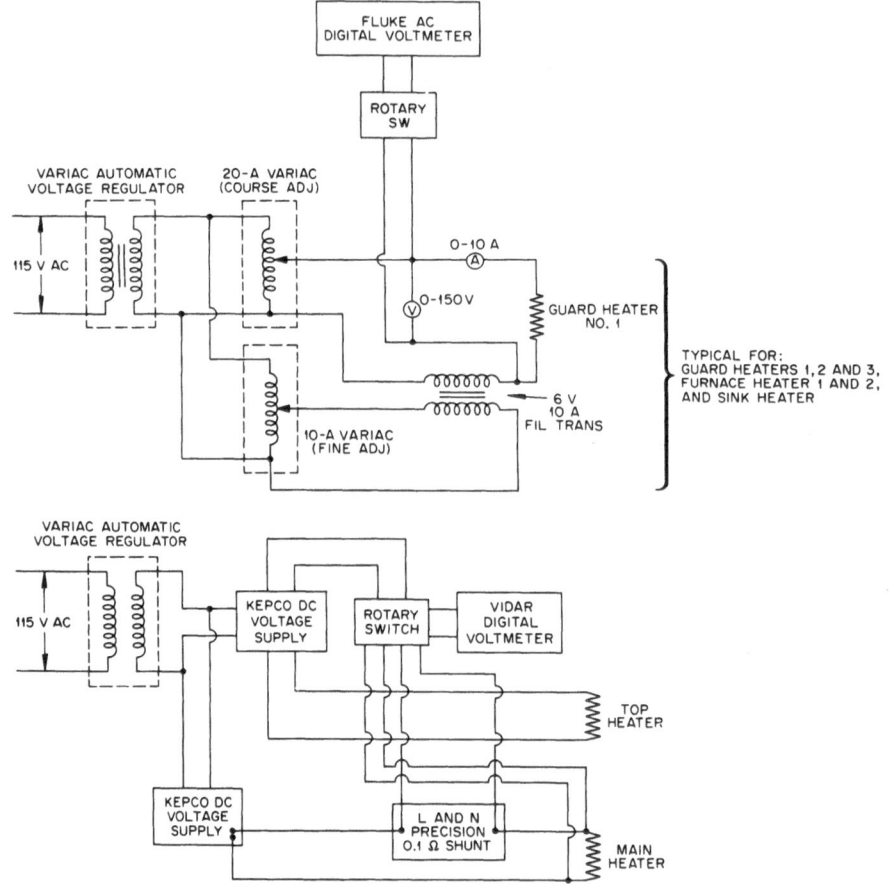

FIGURE 12. Diagram for electrical heater system.

regulated by the Variac automatic voltage regulator. The power to the main
heater is determined from measurements of the voltage and the current with
a Vidar digital microvoltmeter and a Leeds and Northrup precision resistor.

The thermocouple circuit diagram is shown in Fig. 13. The temperature
measurements are made with grounded 1/16-in.-OD chromel–alumel thermocouples or with ungrounded 1/16-in.-OD Pt vs Pt–10% Rh thermocouples;
both are sheathed in 304-type stainless steel. Lead wires from each thermocouple are joined by arc welding to high-purity copper circuitry wire. To
minimize and cancel extraneous thermal emfs, each of these junctions is
contained in a mineral-oil-filled glass tube inserted in a hole drilled into a
high-purity copper block, 3×2 in. diameter, and the whole assembly is

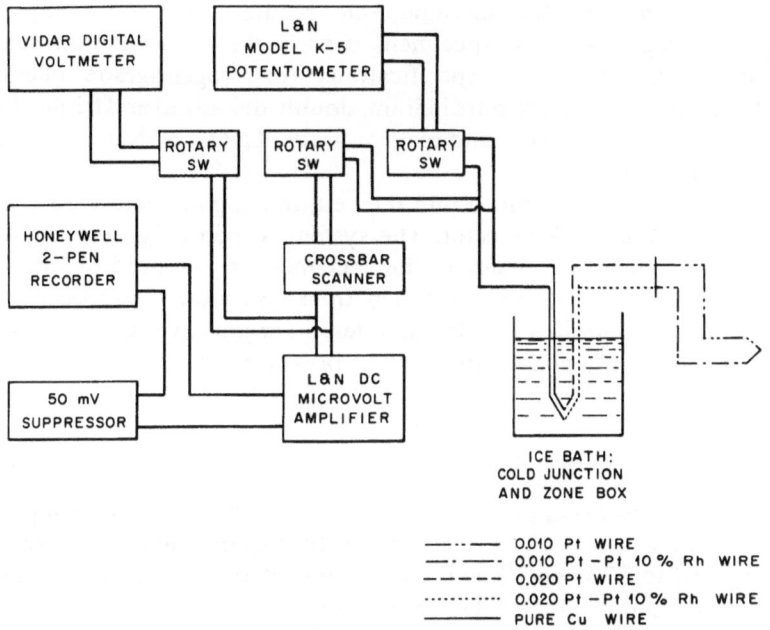

FIGURE 13. Thermocouple circuit diagram.

immersed in a distilled water ice bath. Low thermal emf solder is used in other circuitry junctions. Each thermocouple emf can be read by either a Vidar digital microvoltmeter, a Honeywell pen recorder, or a Leeds and Northrup K-5 potentiometer facility. All thermocouples, except the two imbedded in the main and top heaters, were arc-welded to the bottom of their respective thermal wells to insure their location remained fixed during the measurements.

4. EXPERIMENTAL PROCEDURES

4.1. Preliminary Procedures

The thermal conductivity cell surfaces are cleaned with detergent, rinsed with demineralized water, and dried with filtered air. After use, the corrosion products are removed by polishing with rotary stainless-steel brushes, 600-grit cloth, and crocus cloth to restore the surfaces. During assembly, the apparatus is checked for parallelism of the plates (<0.005 in.), as described previously, and for uniformity of the thermocouple readings at room temperature ($<0.2\,\mu V$). The purity of the specimens used in the determination meet the American Chemical Society specifications for reagent-grade chemicals: 99.997% pure argon, 99.995% pure helium, doubly deionized and triple distilled water, triple distilled mercury, and reagent-grade KNO_3, $NaNO_2$, and $NaNO_3$ salts dried in vacuum.

The specimen is introduced into the cell, the apparatus leveled and leak tested with a helium leak detector. The system is then outgassed by heating overnight at about 150 °C. Voids in the specimen are removed by alternately increasing and decreasing the specimen thickness after it has been melted under vacuum. For liquid and solid samples, an argon cover gas is introduced into the cell at slightly above atmospheric pressure.

4.2. Operating Procedure—Fluid Specimen

The furnace heat is applied to the system to reach the desired temperature level. The main heater is adjusted to obtain the required heat flux through the specimen (0.02 to 0.7 W cm^{-2}), and the guard heaters are adjusted to maintain a close temperature balance between the top and main heater thermocouples ($<\pm 0.5$ °C) and between the main and radial guard heater thermocouples ($<\pm 1$ °C)* to minimize heat losses (see Fig. 9). After steady-state conditions are reached, the measurements are recorded; the criterion for the steady-state conditions is a temperature drift of less than 1 °C per hour. Recordings are made of all thermocouple readings, air flow rate in the cooling sink, dial indicator reading, and voltages and currents from power panel meters and digital voltmeters at each gap distance. The gap is then changed, steady-state conditions are reestablished, and the recordings repeated. Four to 20 gap spacings varying from 0 to 0.4 cm are used, depending on the linearity of the data. The heat flux and average specimen temperature are kept constant for each gap spacing ($<\pm 0.005$ W cm^{-2}, $<\pm 1$ °C) by adjusting the cooling air flow rate and the electrical power to the sink.

The above procedure is repeated for each specimen temperature level.

4.3. Operating Procedure—Solid Specimen

If the measurements are made with the specimen in the solid state, the specimen is melted before each gap spacing is selected, and then the gap spacing is fixed by cooling the specimen to the desired temperature.

---

*In some cases the radial guard heating was inadequate to reduce the radial ΔT to $<\pm 1$ °C, and a correction for heat shunting was necessary.

5. EXPERIMENTAL RESULTS

To evaluate the accuracy, durability, and versatility of the variable-gap technique, the thermal conductivities of five materials were determined over a temperature range from 38 to 946 °C for a total of 31 series of measurements. Argon, helium, water, and mercury were selected to calibrate the apparatus because their thermal conductivities are well established and cover a wide range of values: 0.4×10^{-3} to 100×10^{-3} W cm^{-1} °C^{-1}. The heat-transfer salt, HTS,* was selected to evaluate the application of the technique to high-temperature molten salts. Nearly 350 measurements were made of the thermal resistance as a function of specimen thickness to obtain the 31 determinations of thermal conductivity. In addition, another 50 measurements of thermal resistance vs thickness were made with the system evacuated to evaluate the surface emittance.

5.1. Thermal Resistance Curves

Several representative plots of total thermal resistance as a function of specimen thickness are shown in Figs. 14–18. At the lower temperature the thermal resistance is a linear function, as shown in Figs. 14 and 15 for mercury at 60.8 °C and HTS at 197 °C. At the higher temperatures, the influence of

*HTS is KNO_3-$NaNO_2$-$NaNO_3$ (44–49–7 mol%).

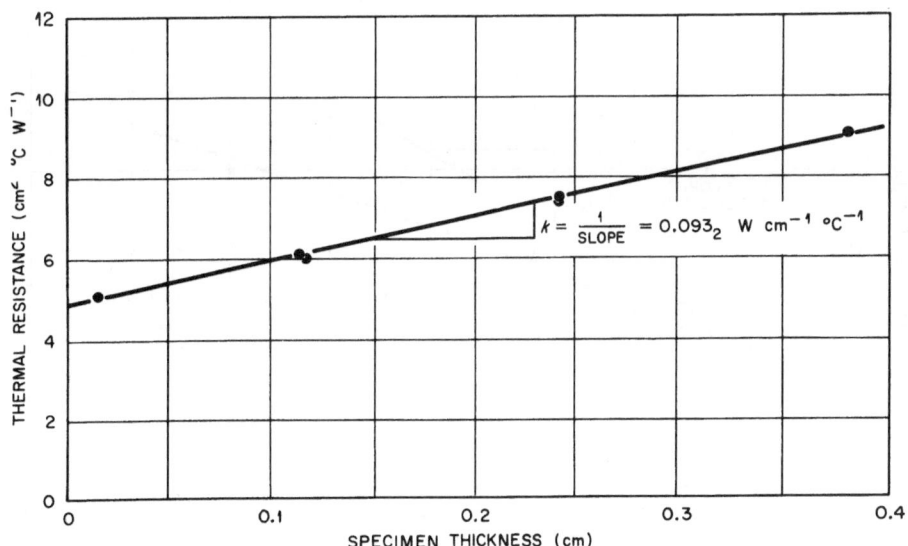

FIGURE 14. Total thermal resistance vs specimen thickness of mercury at 60.8 °C.

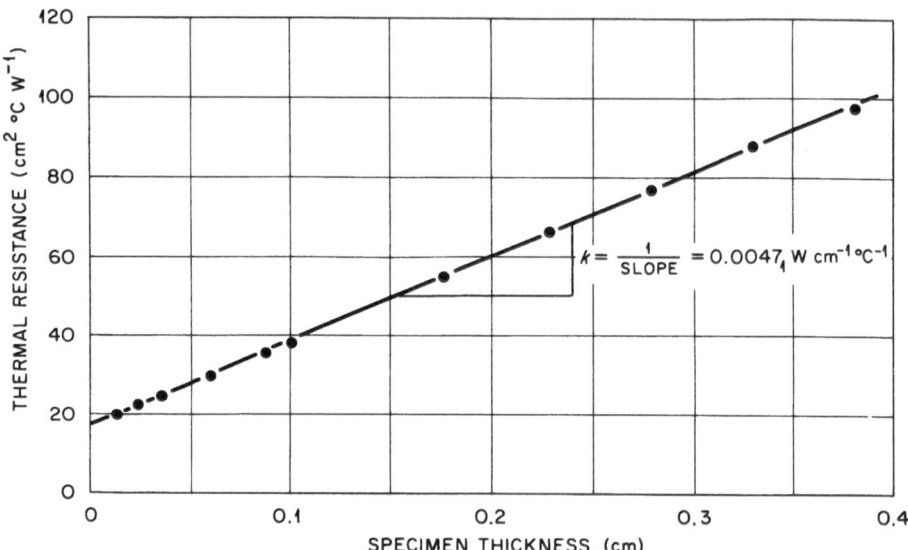

FIGURE 15. Total thermal resistance vs specimen thickness of HTS at 197 °C.

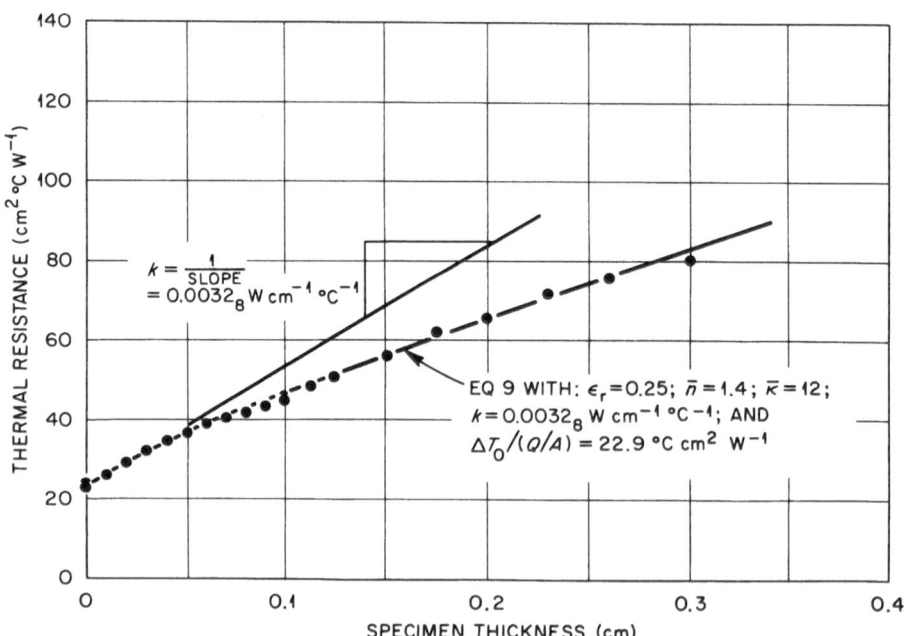

FIGURE 16. Total thermal resistance vs specimen thickness of HTS at 526 °C.

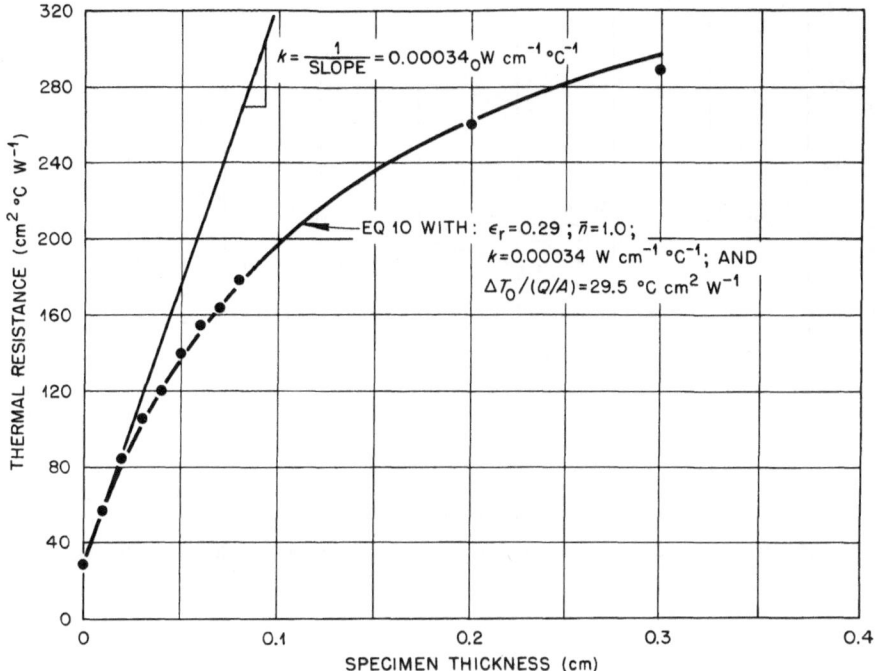

FIGURE 17. Total thermal resistance vs specimen thickness of argon at 503 °C.

infrared radiation on the heat transfer can be seen by the curvature of the resistance curve for HTS and argon at 526 °C and helium at 946 °C (Figs. 16–18).

The fixed thermal resistance of the thermal conductivity cell varies from $5\ \mathrm{cm^2\ {}^\circ C\ W^{-1}}$ for mercury to $30\ \mathrm{cm^2\ {}^\circ C\ W^{-1}}$ for argon. Thermocouple drift (particularly chromel–alumel) and interfacial corrosion and deposits account for most of the variation in the fixed thermal resistance. This variation has little affect on the thermal conductivity measurement using the variable-gap technique, while it could cause considerable uncertainty and error in the measurement using fixed-gap techniques.

Thermal resistance as a function of gap spacing with the cell evacuated is shown in Fig. 19 for two temperature levels (200 and 510 °C) and two values of heat flux (0.100 and 0.224 W cm^{-2}). The values of the specimen resistance (i.e., total minus fixed resistance) were used for the evacuated runs. The expected lack of variation in the resistance as a function of the gap spacing can be seen in the figure. A slight change in the emittance of the plate surfaces will account for the small difference in the resistance between the two heat flux values at 200 °C. The total hemispherical emittance ε, calculated from the thermal resistance using equation (10) (assuming equal emittance for the two plates), varied from 0.4 to 0.5.

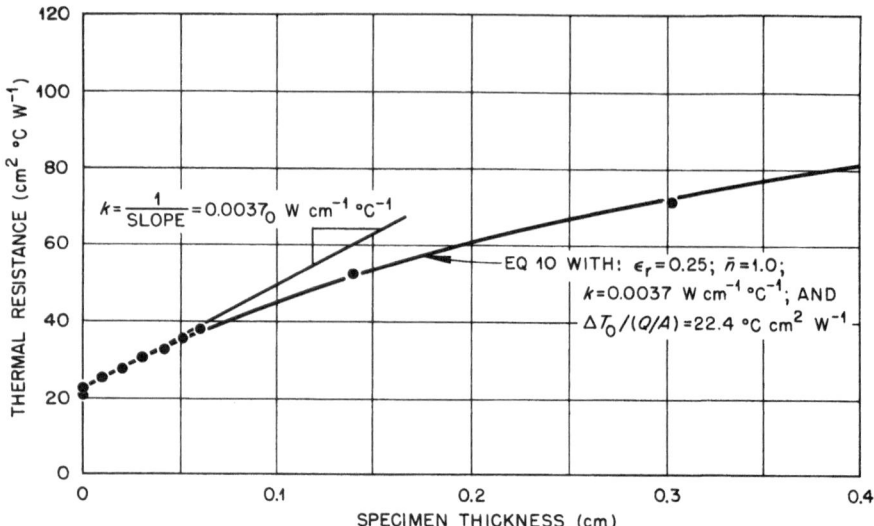

FIGURE 18. Total thermal resistance vs specimen thickness of helium at 946 °C.

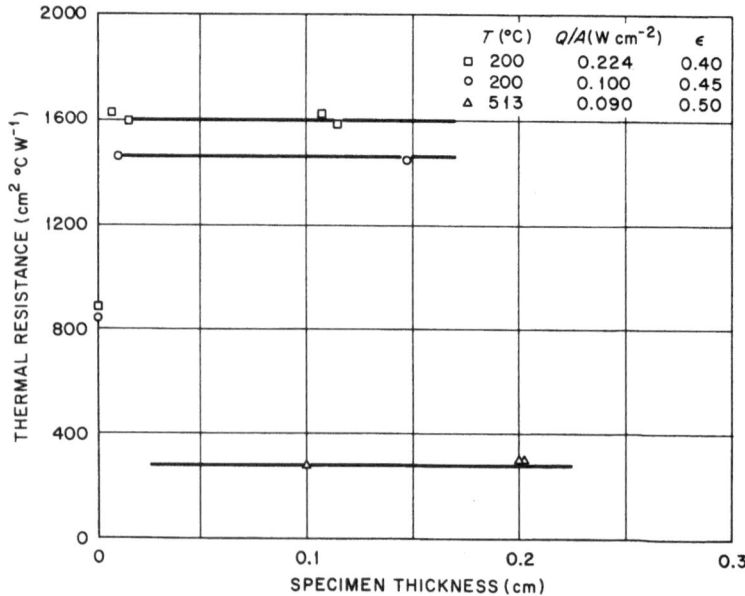

FIGURE 19. Thermal resistance vs gap spacing with the conductivity cell evacuated.

The inflection in the slope of the resistance curve for HTS at 526 °C (Fig. 16) is similar to those shown in Fig. 3 for small infrared absorbing materials. The solid curve drawn through the data was derived from equation (9), using the procedure outlined previously. A value of $\bar{n} = 1.4$ was derived from the *Molten Salts Handbook*[15] for the eutectic composition of $NaNO_3$-KNO_3 (47–53 mol%) at 525 °C and the previously measured value of $\varepsilon = 0.45$ was used for the plate emittance. Incorporating these values in

TABLE 3
Experimental Results

Specimen	Run No.	Temperature (°C)	Thermal conductivity ($W\ cm^{-1}\ {}^{\circ}C^{-1}$)		Difference (%)	Reference
			Exptl	Literature		
			$\times 10^3$	$\times 10^3$		
H_2O	1	45.2	6.7_1	6.38	5.2	19
	2	51.9	6.7_1	6.47	3.7	19
	3	38.0	6.7_8	6.30	7.6	19
Hg	1	60.2	$83._1$	93.2	−10.8	18
	2	60.8	$93._2$	93.2	0.0	18
HTS	1	307	4.2_4	4.30	−1.4	16
	2	308	4.1_5	4.30	−3.5	16
	3	546	3.4_5	3.24	6.5	16
	4	545	3.3_4	3.24	3.1	16
	5	197	4.7_8	4.80	−0.4	16
	6	277	4.4_3	4.38	1.1	16
	7	549	3.4_7	3.22	7.7	16
	8	285	4.6_4	4.40	5.5	16
	9	199	4.8_3	4.80	0.6	16
	10	198	4.7_1	4.80	1.9	16
	11	552	3.0_7	3.21	−4.4	16
	12	554	3.4_8	3.20	8.7	16
He	1	172	2.0_5	1.93	6.2	17
	2	509	3.0_4	3.02	0.7	17
He	1	519	2.7_0	3.04	−11.2	17
	2	946	3.7_0	4.21	−12.1	17
Ar	1	503	0.34_0	0.361	−5.8	17
	2	860	0.49_3	0.463	6.5	17
	3	859	0.45_2	0.463	−2.4	17
HTS	1	526	3.2_8	3.33	−1.5	16
He	1	525	3.2_0	3.07	4.2	17
	2	527	3.0_0	3.07	−2.3	17
	3	105	1.9_4	1.77	9.6	17
HTS	1	517	3.3_4	3.37	−0.9	16
	2	317	4.4_9	4.26	5.4	16
	3	120	6.2_7	5.5	14.0	16

equation (9), a best fit of the experimental data over the entire range of specimen thicknesses ($0 < \Delta x < 0.3$ cm) gave values of the specimen conductivity and mean absorptivity of 3.28×10^{-3} W cm^{-1} °C^{-1} and 12 cm^{-1}, respectively.

The thermal resistance curves shown in Figs. 17 and 18 resemble those shown in Fig. 3 for nonabsorbing gases ($\bar{n} = 1$, $\bar{\kappa} = 0$). Using the previously determined plate emittance, the values of the thermal conductivity were found from a best fit of the resistance data to equation (9) to be 0.34×10^{-3} and 3.3×10^{-3} W cm^{-1} °C^{-1} for argon and helium at 503 and 946 °C, respectively.

5.2. Thermal Conductivity

The experimental results for the thermal conductivity for the different calibration specimens are presented in Table 3. One value for the thermal conductivity of solid HTS at 120 °C is 14% larger than a published value.[16] The maximum deviation of the other results from published values are +9.6% and −12%. The average of the 18 positive deviations is 4.9%, and the average of the 12 negative deviations is −4.4%.

6. DISCUSSION OF THE RESULTS

The accuracy of the results, comparison with previously published values, and theoretical correlations are discussed here.

6.1. Comparison with Published Values

The values of the thermal conductivity of the substances used to calibrate the variable-gap apparatus were well established with the exception of HTS. Touloukian's recommended values[17] for argon and helium and Powell's values[18,19] for water and mercury are representative of accepted thermal conductivities for these materials. Of the two studies on the thermal conductivity of HTS, we selected the more recent results of Turnbull[21] over those of Vargaftik[20] because, as Turnbull points out, Vargaftik's calibration tests with "Dowtherm A" do not agree with other published values. However, Turnbull employed the transient-hot-wire technique with its attendant current-shunting concerns when used with electrically conducting specimens; consequently, the thermal conductivity of HTS cannot be regarded to be as well established as that of the other specimens measured.

The individual deviations between the experimental results and the published values were examined as a function of the specimen type, specimen thermal conductivity, and specimen temperature. No obvious correlation could be found. The average deviation of ±5% from the published values for the thermal conductivities is considerably less than expected considering the specialized design of the apparatus.

6.2. Comparison with Theory

The experimental results for HTS at 526 °C (Fig. 16) are an example of a test of the theory for infrared absorbing materials. Equation (9) in conjunction with the values of $\bar{\kappa}$, $\bar{n}$, ε, and k (selected in the manner previously described) agrees well with the experimental thermal resistance data over the range of specimen thicknesses examined. In addition, the inflection point of equation (9) and the apparent inflection point in the thermal resistance data both occur in the vicinity of $x = 0.1$ cm. Such agreement gives us confidence in the use of equation (9) to represent the thermal resistance data of mildly infrared absorbing specimens like HTS at 500 °C ($\bar{\kappa} = 12$ cm^{-1}). However, as Poltz[6] points out, the Kirchhoff theory ceases to be valid if the reciprocal of the absorption coefficient is of the same magnitude as the wavelength in some regions of the absorption spectrum. Fortunately, when this condition occurs, the thermal resistance is very nearly a linear function of the specimen thickness.

From a theoretical standpoint, the disruption of the lattice continuity by melting should lower the thermal conductivity significantly. If we extrapolate our one result for the thermal conductivity of solid HTS at 120 °C to the melting point at 142 °C by assuming that thermal conductivity is proportional to the inverse of the absolute temperature $k \propto 1/T(\text{K})$ (according to the Debye phonon-scattering theory), a ratio of the liquid-to-solid thermal conductivity at the melting point is found to be 0.85. Turnbull[21] reports the average ratio of liquid-to-solid thermal conductivities at the melting point for a large number of salts to be 0.86 ± 0.13. However, Turnbull's reported values for HTS[16] do not show a discontinuity at the melting point. Accordingly, our value of the thermal conductivity of solid HTS, 14% higher than Turnbull's, appears to be the more reasonable one.

6.3. Uncertainties in the Results

The general form of the equation used to determine the thermal conductivity in this investigation is equation (9), rearranged as

$$k = x \left\{ \frac{Q'/A}{[(\Delta T/F)_x - (\Delta T/F)_0]} - \tfrac{16}{3} \bar{n}^2 \sigma T^3 \left(\frac{Y_{\tau,\varepsilon}}{\tau} \right)_x \right\} \qquad (24)$$

where the measured heat flux, Q'/A, is assumed to be constant during the measurements and the specimen thickness is $x \equiv \Delta x$. Since the Rayleigh number for the present measurements was below 10^4, the effect of convective heat transfer is not included in equation (24).

By taking the total derivative of equation (24), neglecting the small change of k with temperature, and rearranging terms, the change in k due to a change in any of the quantities in equation (24) is

$$\frac{dk}{k} = \frac{dx}{x} + \frac{Q'/A}{JK}\left[\frac{d(Q'/A)}{Q'/A}\right] - \frac{Q'/A}{JK^2}\left(\frac{\Delta T}{F}\right)_x\left(\frac{d\Delta T_x}{\Delta T_x} - \frac{dF_x}{F_x}\right)$$

$$+ \frac{Q'/A}{JK^2}\left(\frac{\Delta T}{F}\right)_0\left(\frac{d\Delta T_0}{\Delta T_0} - \frac{dF_0}{F_0}\right) - \frac{L}{J}\left[\frac{2d\bar{n}}{n} + \frac{3dT}{T} + \frac{d(Y/\tau)}{Y/\tau}\right] \quad (25)$$

where J denotes the quantity enclosed by braces in equation (24), K the quantity enclosed by brackets in equation (24), and $L = (16/3)\bar{n}^2\sigma T^3(Y/\tau)_x$.

By using certain approximations, equation (25) can be simplified to express quantities more readily measured,

$$w \equiv \left[\frac{\Delta T}{F(Q'/A)}\right]_0$$

$$h_r = \frac{16}{3}\frac{\bar{n}^2}{\varepsilon}\sigma T^3(Y/\tau)_x \underset{\tau \to 0}{=} 4\bar{n}^2\sigma T^3$$

and

$$\Delta T_x = \Delta T_{sx} + \Delta T_0, \qquad (\Delta T_s/F)_x = \frac{Q'/A}{(k/x) + \varepsilon h_r}, \qquad \tau = \bar{\kappa}x$$

where ΔT_{sx} is the specimen ΔT as a function of the gap thickness, x.

In the present studies, the heat shunting was usually small so that F_x and F_0 can be assumed to be unity in the weighting functions. Furthermore, the error in $Y_{\tau,\varepsilon}/\tau$ can be evaluated as

$$\frac{d(Y/\tau)}{Y/\tau} = \left(\frac{\tau}{Y}\frac{dY}{d\tau} - 1\right)\left(\frac{d\bar{\kappa}}{\bar{\kappa}} + \frac{dx}{x}\right) + \frac{\varepsilon}{Y}\frac{dY}{d\varepsilon}\frac{d\varepsilon}{\varepsilon}$$

Finally, the total error in the conductivity would be approximately

$$\frac{dk}{k} = \frac{dx}{x} + [(1 + \varepsilon h_r x)/k]$$

$$\times \left\{\frac{d(Q'/A)}{Q'/A} - \frac{d\Delta T_{sx}}{\Delta T_{sx}} + \frac{dF_x}{F_x} + w[(k/x) + \varepsilon h_r]\left(\frac{dF_x}{F_x} - \frac{dF_0}{F_0}\right)\right.$$

$$-\left(\frac{\varepsilon h_r x}{k + \varepsilon h_r x}\right)\left[\left(\frac{2d\bar{n}}{\bar{n}} + \frac{3dT}{T}\right) + \left(\frac{\tau}{Y}\frac{dY}{d\tau} - 1\right)\left(\frac{d\bar{\kappa}}{\bar{\kappa}} + \frac{dx}{x}\right)\right.$$

$$\left.\left.+\left(\frac{\varepsilon}{Y}\frac{dY}{d\varepsilon}\frac{d\varepsilon}{\varepsilon}\right)\right]\right\} \tag{26}$$

Using equation (26), the maximum and standard error limits in the conductivity measurements were estimated for the accuracy limits listed in Table 1, the tolerances shown in Figs. 10 and 11, and the typical values of $Q/A = 0.5$ W cm^{-2} and $x = 0.05$ cm. These error limits are plotted as a function of the specimen thermal conductivity in Fig. 20. The effects of the specimen temperature range (300 to 900 °C) and of the degree of radiation effects (zero to maximum radiation at 900 °C) do not contribute greatly to the error and are shown by the areas included within the small bands in Fig. 20. It is apparent from the figure that the error is sensitive to the magnitude of the specimen thermal conductivity except at larger values of the thermal conductivity. Much of the increase in error at low values of thermal conductivity results from an increased influence of the uncertainty in the measurement of the change in the specimen thickness. This uncertainty, caused by the differential thermal expansion between the mobile piston and cylindrical components as the guard heating is adjusted, can be minimized by a dual quartz rod, dial indicator system. Such a system was not considered necessary for the present apparatus,

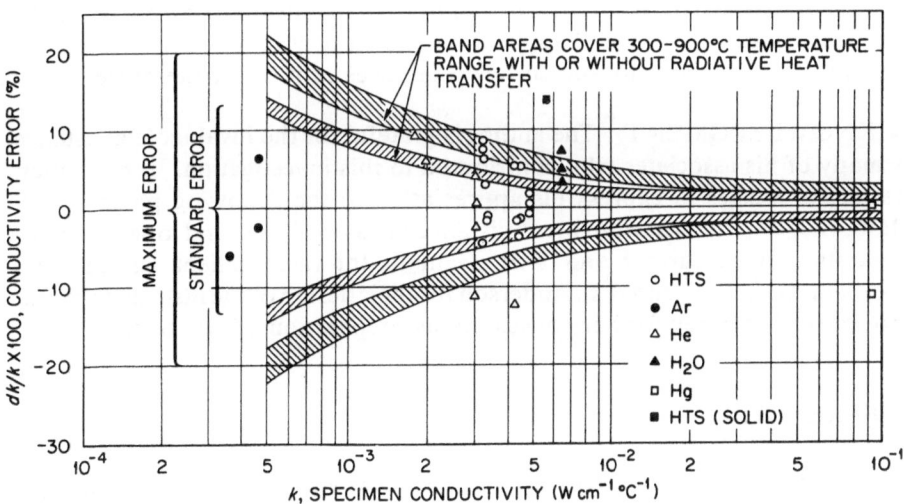

FIGURE 20. Estimated standard and maximum error limits in the thermal conductivity measurements vs specimen thermal conductivities with and without radiative heat transfer over the 300 to 900 °C range. Also shown are experimental deviations from published values.

since it was designed for use with molten fluoride salts whose conductivities are in the vicinity of $0.01 \ \mathrm{W \ cm^{-1} \ °C^{-1}}$.

Figure 20 shows also the deviations of each of the experimental results from the published values. Several points are outside the estimated error limits (the thermal conductivity of solid HTS was discussed in the previous section). Data for water and mercury were obtained using chromel–alumel thermocouples. Since the uncertainty of the temperature measured with the chromel–alumel thermocouples is four times that from Pt vs Pt–10% Rh thermocouples, the error limits are considerably larger, especially at the higher specimen thermal conductivities. Excluding these points, 93% of the data lies within the maximum error limits and 68% within the standard error limits.

7. CONCLUSIONS

The variable-gap apparatus designed for specialized thermal conductivity measurements with molten fluoride salt mixtures from 500 to 1000 °C has demonstrated remarkable accuracy, versatility, and durability. Experimental results agree with published results within an average deviation of ±5% for a wide variety of specimens (solids, liquids, and gases), specimen thermal conductivities (0.4×10^{-3} to $100 \times 10^{-3} \ \mathrm{W \ cm^{-1} \ °C^{-1}}$), and temperature levels (40 to 950 °C). The apparatus was still in operating condition after 3000 hours of operation with molten fluoride salts at temperatures from 500 to 960 °C. In addition, the variable-gap apparatus has demonstrated its ability to distinguish and evaluate the internal radiation within small infrared absorbing fluids. Considering the wide range of application and the accuracy of the variable-gap method, it is surprising to find so few references to it in the literature.

ACKNOWLEDGMENT. The author is grateful for the invaluable assistance of many of his associates who contributed to this investigation. In particular, the author wishes to express his appreciation to the following persons: S.J. Claiborne, Jr., for his valuable assistance in assembling and operating the apparatus; W.K. Sartory and W.R. Gambill for their patient advice; J.W. Krewson, W.A. Bird, and R.L. Anderson for their design of the instrumentation; and Patty Humphrey for her conscientious preparation of this text.

REFERENCES

1. W. Fritz and H. Poltz, "Absolutbestimmung Der Wärmeleitfähigkeit von Flüssigkeiten—I," *Int. J. Heat Mass Transfer* **5**, 307–316 (1962).
2. J. Matolich and H.W. Deem, "Thermal Conductivity Apparatus for Liquids; High Temperature, Variable-Gap Technique," pp. 39–51, *Proc. 6th Conf. on Thermal Conductivity, Oct. 19–21, 1966, Dayton, Ohio.*

3. J.W. Cooke, "Thermal Conductivity of Molten Salts," pp. 15-27, *Proc. 6th Conf. on Thermal Conductivity, Oct. 19-21, 1966, Dayton, Ohio.*

4. J.W. Cooke, *Development of the Variable-Gap Technique for Measuring the Thermal Conductivity of Fluoride Salt Mixtures,* ORNL 4831 (1973).

5. H. Poltz, "Die Wärmeleitfahigkeit von Flüssigkeiten II, Erschienen in der Zeitschrift," *Int. J. Heat Mass Transfer* **8**, 515-527 (1965).

6. H. Poltz and R. Jugel, "The Thermal Conductivity of Liquids—IV. Temperature Dependence of Thermal Conductivity," *Int. J. Heat Mass Transfer* **10**, 1075-1088 (1967).

7. H. Gröber and S. Erk, *Fundamentals of Heat Transfer,* 3rd ed. (rev., U. Grigull, Transl., J. R. Moszynski), p. 315, McGraw-Hill, New York (1961).

8. P.A. Norden and A.G. Usmanov, "The Inception of Convection in Horizontal Fluid Layers," *Heat Transfer Sov. Res.* **4**(2), 155-161 (1972).

9. B.M. Berkovsky and V.E. Fertman, "Advanced Problems of Free Convection in Cavities," *4th Int. Heat Transfer Conference, Paris, September 1970,* Vol. 4, Paper NC 2.1, E.L. Seiver Publishing Co., Amsterdam (1971).

10. B.G. Eads and B.C. Duggins, "Experience with Direct Digital Control of Several Temperature Processes," in: *Temperature, Its Measurement and Control in Science and Industry,* Vol. 4, Part 2, pp. 1435-1444, Instrument Society of America, Pittsburgh (1972).

11. R.K. Adams, "Application of Small Computers as Thermometry Research Tools," in: *Temperature, Its Measurement and Control in Science and Industry,* Vol. 4, Part 2, pp. 1445-1456, Instrument Society of America, Pittsburg (1972).

12. T.G. Kollie *et al.,* "Measurement Accuracy of a Computer-Operated-Data-Acquisition System," in: *Temperature, Its Measurement and Control in Science and Industry,* Vol. 4, Part 2, pp. 1457-1466, Instrument Society of America, Pittsburg (1972).

13. M.H. Cooper, R.L. Anderson, and C.A. Mossman," Automatic Temperature Measurements from −183 to 2300 °C," in: *Temperature, Its Measurement and Control in Science and Industry,* Vol. 5, Part 2, pp. 1287-1292, American Institute of Physics, New York (1982).

14. M.S. Conner, "Analog-I/O Boards and Software for IBM PCs," *EDN* **31**(12), 116-137 (1986).

15. G.J. Jantz, *Molten Salts Handbook,* p. 92, Academic Press, New York (1967).

16. A.G. Turnbull, "The Thermal Conductivity of Molten Salts," *Aust. J. Appl. Sci.* **12**(1), 30-41 (1961).

17. Y.S. Touloukian, P.E. Liley, and S.C. Saxena, *Thermal Conductivity; Nonmetallic Liquids and Gases* (Vol. 3 of *Thermophysical Properties of Matter,* TPRC Data Series), Plenum Press, New York (1970).

18. R.W. Powell and R.P. Tye, "The Thermal and Electrical Conductivity of Liquid Mercury," *Proc. Heat Transfer Conf., 1961-62; University of Colorado, 1961 [and] Westminster, England, 1961-62, Int. Developments in Heat Transfer,* Paper 103, pp. 856-862, ASME (1963).

19. A.R. Challoner and R.W. Powell, "Thermal Conductivities of Liquids: New Determinations for Seven Liquids and Appraisal of Existing Values," *Proc. R. Soc. London* **A238**(1212), 90-106 (1956).

20. N.B. Vargaftik, B.E. Neimark, and O.N. Oleshchuck, "Physical Properties of High Temperature Liquid Heat Transfer Medium," *Bull. All-Un. PWR Eng. Inst.* **21**, 1 (1952).

21. A.G. Turnbull, "The Thermal Conductivity of Molten Salts," *Aust. J. Appl. Sci.* **12**(3), 324-329 (1961).

II

ELECTRICAL RESISTIVITY

9

Methods for Electrical Resistivity Measurement Applicable to Medium and Good Electrical Conductors

B. CALÈS and P. ABÉLARD

1. INTRODUCTION

Among the various methods of investigating physical properties of solids, the measurement of electrical conductivity is one of the most currently used and a large variety of procedures have thus been developed over the last ten years. Hence the experimenter will often be faced by the choice of the most suitable technique owing to the nature of the conduction process, the range of electrical conductivity expected to be reached during experiments, the size and shape of the samples to be investigated, or also their chemical reactivity with respect to the sample holders.

Charge transport phenomena arise from diffusive processes of ionic (cation or anion) and/or electronic charge carriers, including electrons and holes, so that the total apparent conductivity σ_t of a given material can result from the contributions of both ionic (σ_i) and electronic (σ_e) partial conductivities. It is thus obvious that, in addition to total conductivity measurements, partial, ionic or electronic, conductivity investigations will also be performed when thorough analyses of transport properties are required.

Moreover, the partial conductivities are directly dependent on the concentration and mobility of the charge carriers according to the classical relationship

$$\sigma_k = n_k z_k q \mu_k$$

B. CALÈS • Céramiques Techniques Desmarquest, 27025 Evreux Cédex, France.
P. ABÉLARD • Ecole Nationale Supérieure de Céramiques Industrielles, 87065 Limoges Cédex, France.

where σ_k denotes the partial conductivity due to the charge species k of concentration n_k, q the electronic charge, z_k the valence of species k, and μ_k their mobility.

Therefore, the various environmental conditions that can affect the concentration or mobility of the charge carriers, such as the temperature or composition of the surrounding atmosphere, must also be taken into account when selecting a conductivity measurement procedure.

The aim of this chapter is to provide a comprehensive survey of the different techniques that may be developed for the measurement of either total or partial conductivities, and to point out the main sources of errors that can be encountered by the experimenter.

Both DC and AC procedures for total conductivity measurements, including two-probe and four-probe arrangements, are first described below. Special attention has been focused on the AC procedures, especially complex impedance spectroscopy. Indeed the latter, which has been more and more frequently used in the last few years, provides a very accurate investigation of electrical conductivity in solids and allows one to study not only the bulk properties, but also the grain boundaries or electrode polarization phenomena.

Special techniques for partial, ionic or electronic, conductivity measurements are reported below. They often involve relatively complex experimental apparatus and procedures, so that various specialized reviews are also indicated in this section.

2. DIRECT-CURRENT MEASUREMENTS OF TOTAL CONDUCTIVITY

Direct-current conductivity measurements have been used extensively for many years and are still widespread owing to the great ease of implementation and the relative simplicity of the different equipment required. Therefore, a large variety of experimental setups have been proposed including two-probe, three-probe, and four-probe arrangements, which are the most currently used. In some cases, the particular shape of the specimens implies that special techniques be developed, such as the Van der Pauw or point-probe methods.

2.1. The Two-Probe Arrangement

2.1.1. Experimental Procedure

The simplest technique for DC conductivity measurements is the two-probe method that consists of simultaneously measuring the current I flowing through the sample and the DC voltage drop V_S across it. The DC conductivity is then deduced from the following classical relationship:

$$\sigma = \frac{I}{V_S}\frac{l}{S} \tag{1}$$

where S is the cross-sectional area of the sample and l the distance between the voltage probes.

The voltage drop across the sample is read by either a DC voltmeter, or a recorder, or an electrometer having input impedance higher than the sample resistance by at least a factor of 10^3. The value of the DC current is deduced from the voltage drop V_R across a standard resistor or directly read by a DC amperometer (Fig. 1a, b). A convenient measurement circuit is obtained by combining a DC amperometer and a DC steady current source which may be fixed to known current values ranging from 10^{-6} to 1 A (Fig. 1c). Although various DC current sources are commercially available, such an apparatus is easily built with the aid of usual electronic components. As an example, a suitable circuit diagram is shown in Fig. 2.

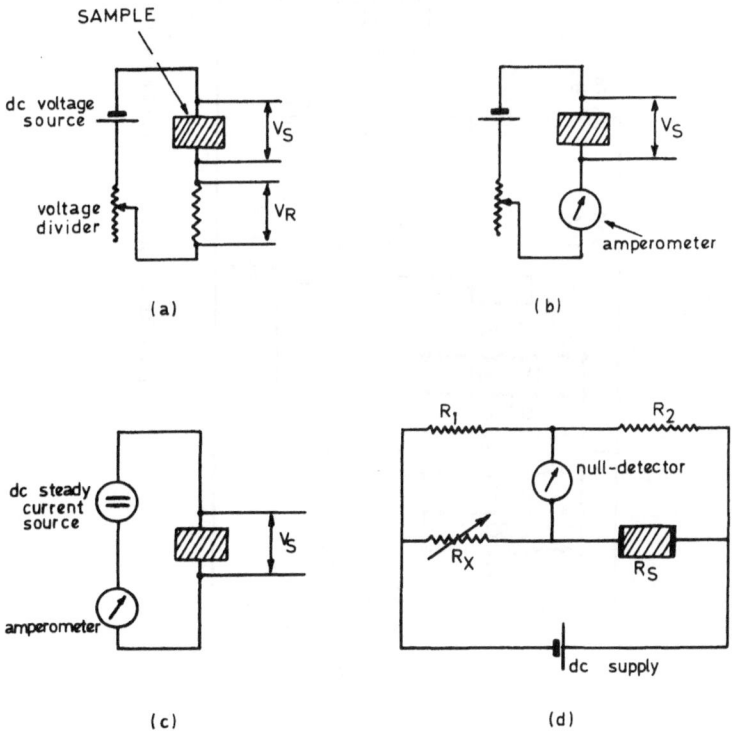

FIGURE 1. Measurement circuits with two-probe arrangement (see text).

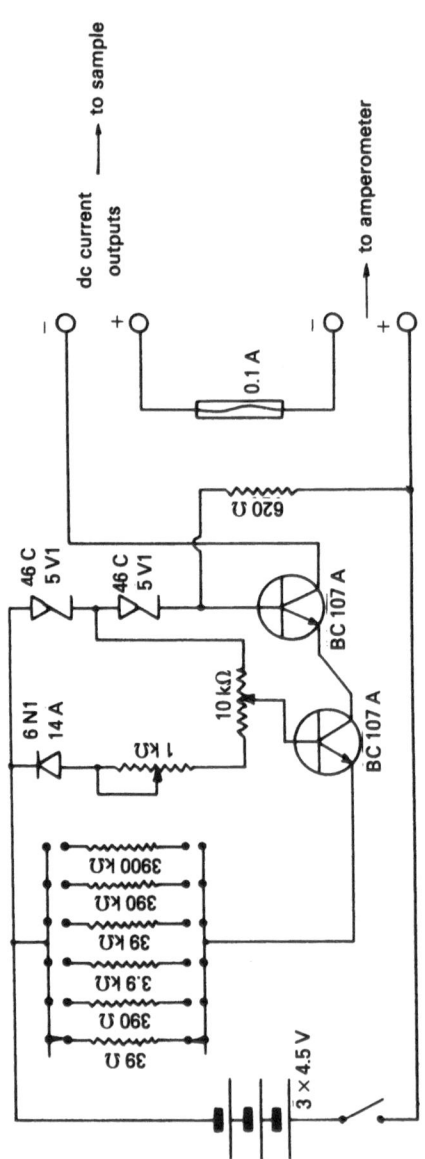

FIGURE 2. Circuit diagram of a DC steady current source.

Another experimental procedure is provided by the use of the Wheatstone bridge (Fig. 1d). Under bridge balance, which is achieved on adjusting the resistor R_X, the sample resistance R_S is given by

$$R_S = R_X \frac{R_1}{R_2} \qquad (2)$$

where R_1 are R_2 are calibrated resistors. However, the bridge technique is employed less and less while DC voltage–current procedures are preferred.

2.1.2. Sample Design

The geometrical characteristics of the specimens will be chosen with respect to the magnitude of the conductivity, so that the sample resistance ranges from about 10 to $10^9 \, \Omega$. Thus, in the case of materials having electrical conductivity lower than about $10^{-3} \, \Omega^{-1} \, m^{-1}$, disk-shaped samples are preferable. Typical dimensions are a few $10^{-3} \, m$ thick and about $10^{-2} \, m$ in diameter. They give rise to geometrical factors $1/S$ of the order of $10^{-1} \, m^{-1}$. Two electrodes are deposited on the two faces of the disk (Fig. 3a). The electrode material must remain chemically inert with respect to the specimen and the surrounding atmosphere and generally noble metals are used, such as silver, gold, or platinum. The electrodes may be either painted (on using commercial metallic pastes), or evaporated, or sputtered. The characteristics of the electrodes obtained by these various techniques are reported elsewhere.[1]

For good electrical conductors, the specimens are cut into rectangular bars or cylinders, the flat ends of which are covered by the electrodes (Fig. 3b, c). Typical dimensions of the samples are a few $10^{-3} \, m$ in diameter and $2-3 \times 10^{-2} \, m$ in length, so that the geometrical factor l/S amounts to about $1000 \, m^{-1}$. The electrodes are generally provided by thin metallic films (Ag, Au, Pt) obtained by painting, evaporating, or sputtering processes. Thin sheets of metal, directly contacted onto the flat ends of the sample by a mechanical device, may also be used.[2] The electrical leads may be tightly wrapped around the sample in small grooves (Fig. 3b) or directly pressed onto the electrodes (Fig. 3c). If small grooves are used, they are coated with metallic paste to improve the contacts, the same metal being used for the leads, the electrodes, and the metallic paste. The whole assembly is then annealed at high temperature in order to provide good adhesion.

In the case of powdered materials, special arrangements must be used and some examples are reported in the literature.[3,4]

Some attention will be devoted to the design of the sample holders in order to prevent any chemical contamination of the specimens, especially when high-temperature measurements are performed. Various sample holders are described in the following sections; they may be used for two-probe investigations after minor modification.

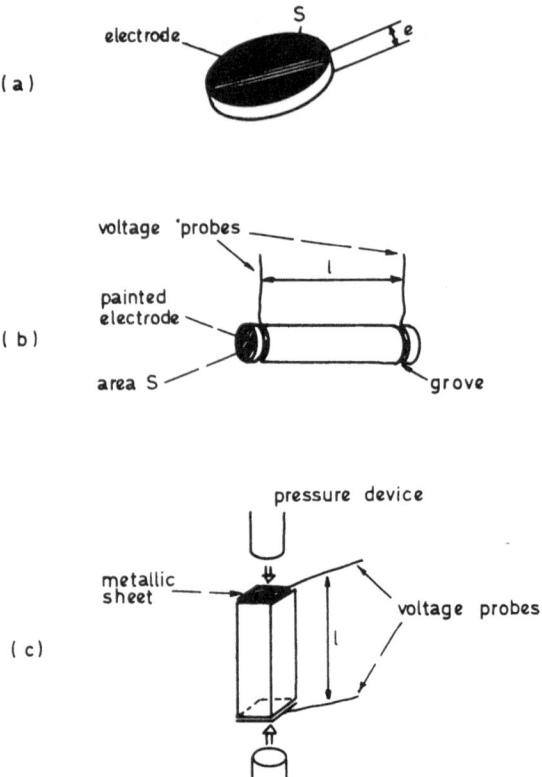

FIGURE 3. Schematics of two-probe mounted samples.

2.1.3. Main Errors Sources and Recourses

First, it should be noted that the conductivity measurement must not be confused systematically with a resistance measurement. Indeed, it consists of measuring a voltage drop V_S and current I, the ratio of which has the same dimension as a resistance, $R_S = V_S/I$. Moreover, the total sample resistance R_S is not necessarily homogeneous and can include, in addition to the material resistance itself, different contact resistances, interfacial resistances at the electrodes, or other contributions associated, for instance, with surface conductivity phenomena.

Thus, in spite of its simplicity, the two-probe arrangement can give rise to various experimental errors, which occur likewise in many other DC conductivity measurement methods and will be analyzed subsequently in detail.

The contact resistances are mainly attributable to the electrical leads. They do not exceed in the worst cases 1 to 5 Ω and can reasonably be neglected when the resistance of the specimen is higher than about 100 Ω. For very low sample resistances, a preliminary estimate of the lead resistances will probably be required.

Direct-current conductivity measurements may also be complicated by polarization phenomena taking place at the material–electrode interfaces.[5] These effects are generally removed by ensuring good contacts between the electrodes and the material and selecting the electrode material with respect to the conduction mechanism and the nature of the charge carriers. The electrodes may be either reversible or blocking the exchange of ionic carriers through the interface.[6,7] Reversible electrodes should be more suitable for predominant ionic conductors. Thus, in the case of M^{m+} cationic conductors, the metal M itself has sometimes been proposed as the electrode.[8,9] However, such electrodes concern only a few ionic mobile species and blocking electrodes, provided by thin films of noble metal (Ag, Au, Pt), are more frequently used.

In practice, the polarization phenomena are associated with a time dependence of the voltage drop V_S across the sample. Therefore, the experimenter must verify that the value of V_S is constant over long periods of time, i.e., a few minutes. If this is not the case, two recourses may be considered: (1) the direction of the current may be reversed periodically by using an automatic reversing switch; (2) the conductivity measurement may be performed with the aid of a four-probe arrangement.

Another source of errors lies in the surface leakage paths, which may arise when the conductivity of the material is higher in the surface layer than in the bulk. Such phenomena are encountered especially for low electrical conductors ($\sigma < 10^{-6}\,\Omega^{-1}\,m^{-1}$) and may be noted by the dependence of the voltage drop V_S on the geometrical factor l/S: a direct proportionality between V_S and l/S should be observed. Alternatively, a three-probe arrangement will be preferred.

Finally, nonlinear behavior must be carefully minimized. Thus, the experimenter makes sure of direct proportionality between the voltage drop V_S and the current I. Indeed, beyond a critical value of the voltage drop across the sample, generally a few volts, electrolysis phenomena can induce a nonlinear dependence of V_S on I. Thereby, the current flowing through the sample will be fixed at a sufficiently low value that V_S does not exceed 0.1 to 0.2 V.

2.2. The Three-Probe Arrangement

Direct-current conductivity measurements will be performed with the aid of a three-probe arrangement whenever surface conductivities are expected to predominate. The specimens are disk-shaped and an additional guard ring is deposited on one of the faces of the disk (Fig. 4a). The three electrodes can be conveniently prepared by evaporating or sputtering processes, using a mask to separate the guard ring and the central electrode.

The measurement circuit is then slightly modified and includes a variable resistor R_X to adjust the guard ring to the same potential as the central electrode (Fig. 4b). This condition is achieved with the aid of a null detector, such as a galvanometer or an electrometer, and prevents any leak current between the two faces of the sample.

After measuring the current I flowing through the sample and voltage drop V_S, the conductivity of the material is given by

$$\sigma = \frac{I}{V_s} \frac{4e}{\pi(D+h)^2} \tag{3}$$

where e is the thickness of the sample, D the diameter of the central electrode, and h the distance between the central electrode and the guard ring (Fig. 4a).

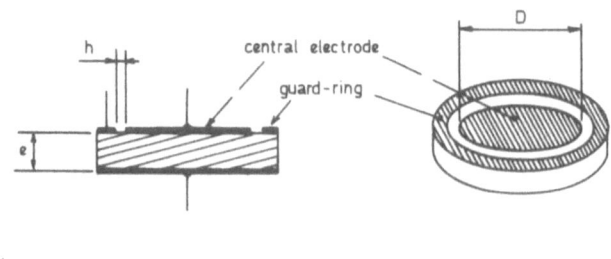

(a)

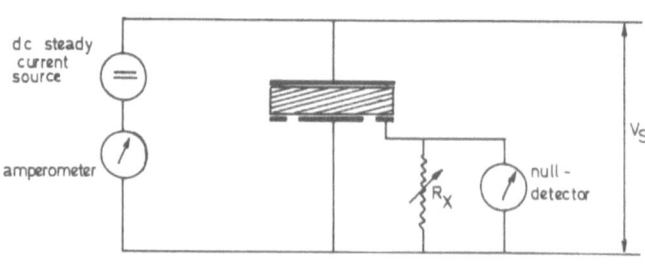

(b)

FIGURE 4. The three-probe arrangement with a disk-shaped sample (a) and corresponding measurement circuit (b).

2.3. The Four-Probe Arrangement

The most popular procedure for DC conductivity measurement is provided by the four-probe arrangement. In such an arrangement the potential drop V_S is measured with the aid of two extra probes located between the current leads. A very common design is obtained by cutting the sample into a cylinder (Fig. 5a).[10] Four small grooves are distributed at approximately equal distances (about $5-10 \times 10^{-3}$ m) along the cylinder and four metallic wires (Au, Pt) are tightly wrapped around the sample in the grooves and coated with metallic paste. The flat ends of the sample are also coated with metallic paste.

Another possibility, before sintering, consists of drilling two small holes through the largest face of a rectangular bar into which metallic wires are thread to provide voltage probes (Fig. 4b).[11] The whole assembly is then fired and the shrinkage of the material provides good contact and adhesion. The two other electrodes are then painted onto the flat ends of the sample.

Two measurement procedures may be considered: either the voltage-current or the bridge procedure. The most convenient is the voltage–current procedure, which is similar to that used with the two-probe arrangement. Nevertheless, in order to avoid nonlinear phenomena, the voltage drop across the total length of the sample should not exceed a few 10^{-1} V. The four probes being spaced at an approximately equal distance, the voltage drop V_S between the central probes should be lower than about 5×10^{-2} V. Anyhow, it is strongly recommended to preliminarily verify the linear dependence of the voltage drop on current. The DC conductivity is then derived from equation (1) after replacing the distance l between electrodes by the distance d between voltage probes (Fig. 5a). It may be advisable to include an automatic reversing switch in the measurement circuit in order to periodically reverse the direction of the current and to avoid experimental errors due to zero shift, especially when

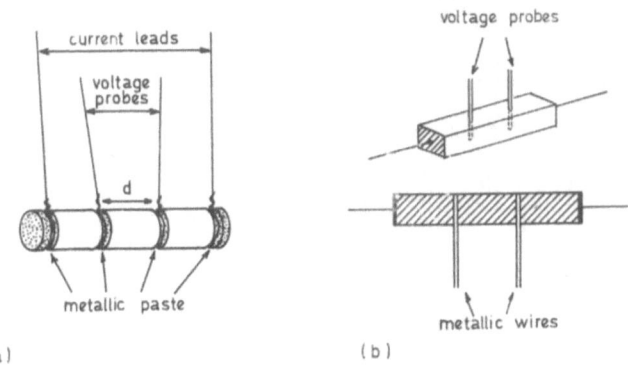

(a) (b)

FIGURE 5. Schematics of four-probe mounted samples.

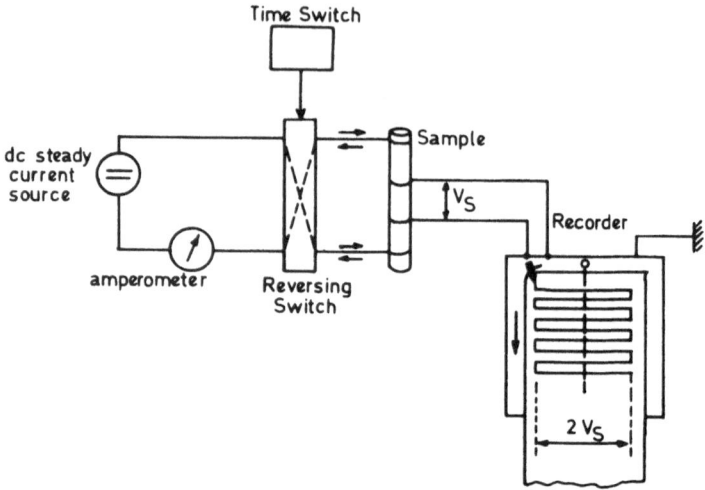

FIGURE 6. Four-probe experimental setup.

recorders are used for the voltage drop (V_S) measurement. Such a four-probe experimental setup is shown in Fig. 6.

Four-probe DC measurements may likewise be performed with the aid of double bridges, such as the Kelvin bridge. The main difference with respect to Wheatstone bridges lies in the bridge balance procedure. The latter is checked by a null detector located between two extra bridge arms including four standard resistors R_1, R_2, R_3, and R_4 (Fig. 7).[12] They are fixed to known values such as

$$\frac{R_1}{R_3} = \frac{R_2}{R_4} = \text{Cte} \qquad (4)$$

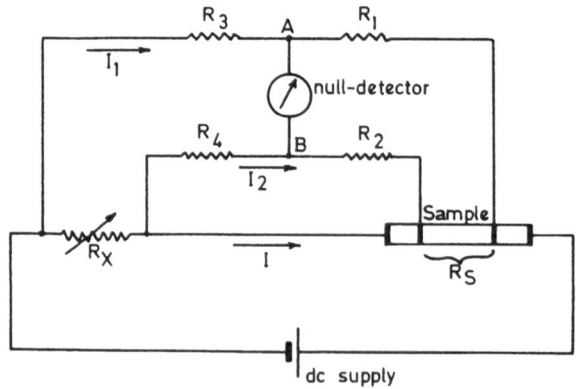

FIGURE 7. Electrical connection for the Kelvin double bridge.

Other multiple bridges have also been developed[12,13] in order to minimize errors due to contact resistances. However, owing to the delays associated with the bridge balance, voltage–current procedures are generally preferred.

The sample holder shown in Fig. 8 is convenient for simultaneously investigating two samples from low to high temperature, under various controlled atmospheres. It is located in an upright position at the center of a high-temperature furnace, which may be removed vertically in order to have easy access to the samples. The latter are situated in the isothermal zone of the furnace in order to avoid spurious effects due to temperature gradients, such as thermoelectric voltages. The thermal profile of the furnace is improved by two thermal screens made of porous alumina. The various leads are made of thin platinum wires, about 2×10^{-4} m in diameter, thread through small alumina tubes so that high-temperature measurements may be conveniently obtained.

2.4. Special Direct-Current Conductivity Measurements

In some cases, owing to the small dimensions or the particular shapes of samples, the previous four-probe arrangements cannot be developed with enough reliability and special devices have been proposed.

2.4.1. The Van der Pauw Method

The Van der Pauw method[14] is a very useful procedure in measuring flat samples with uniform thickness e. Although they can have arbitrary shapes, disk-shaped and very thin samples ($e \leqslant 10^{-3}$ m) must be preferred. Four point-probes are placed along the circumference of the specimens as shown in Fig. 9a. The sample mounting may be executed as follows:[15] Four small holes are drilled mechanically through the sample (Fig. 9b) and thin metallic rods and diamond paste used. Four platinum wires of about 2×10^{-4} m in diameter, melted at one end, are thread through these holes and coated with platinum paste. Good adhesion is obtained after annealing at high temperature (~ 1300 K). The sample is then hung in a furnace with the aid of a holder similar to that shown in Fig. 8.

The conductivity of the material is obtained by a voltage–current procedure. First, a DC current I_1 is passed through terminals 1 and 2 (Fig. 9a) and the voltage drop V_1 across terminals 3 and 4 is measured. The ratio V_1/I_1 defines a resistance R_1. Then, in the same manner, a DC current I_2 flows through terminals 2 and 3 and a resistance R_2 is derived from the voltage drop V_2 across terminals 1 and 4. The value of the conductivity is deduced from R_1 and R_2, according to the following relationship[14]:

$$\sigma = \frac{2 \ln 2}{\pi e} [(R_1 + R_2)f(R_1/R_2)]^{-1} \tag{6}$$

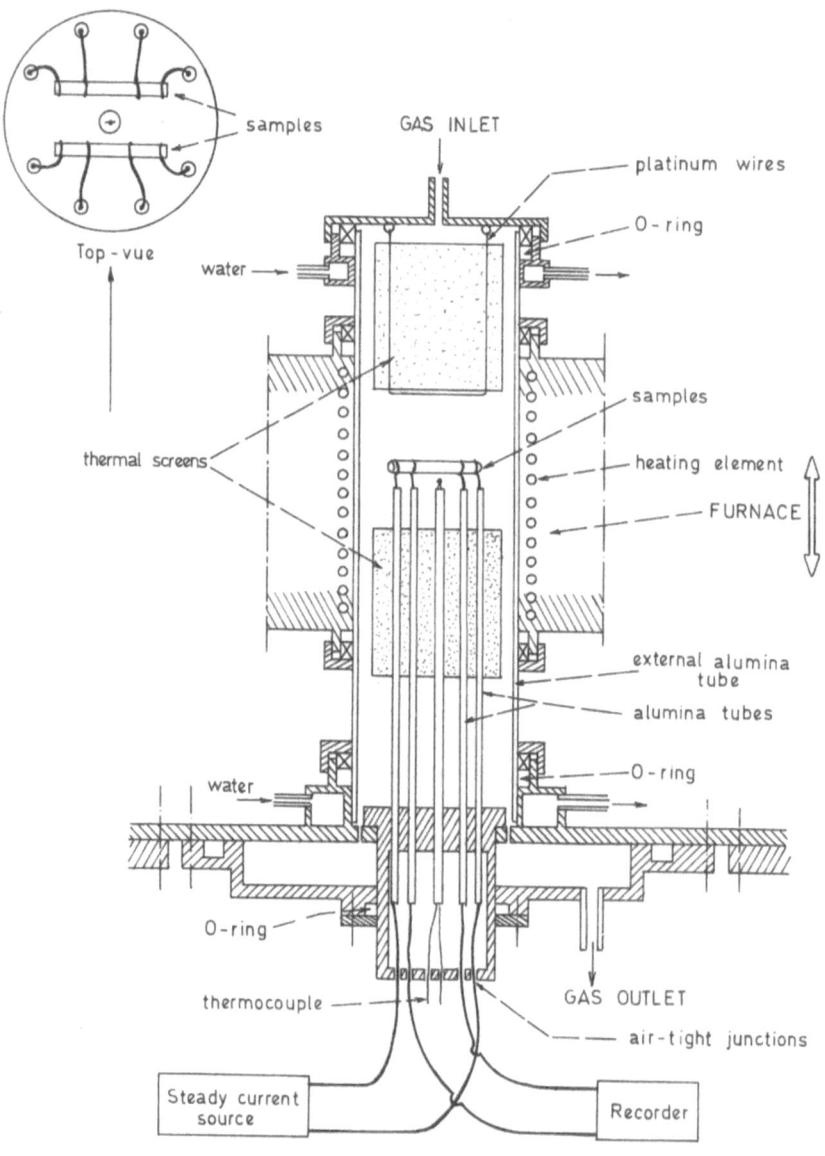

FIGURE 8. Schematic representation of a sample holder for high-temperature measurements by a DC four-probe method.

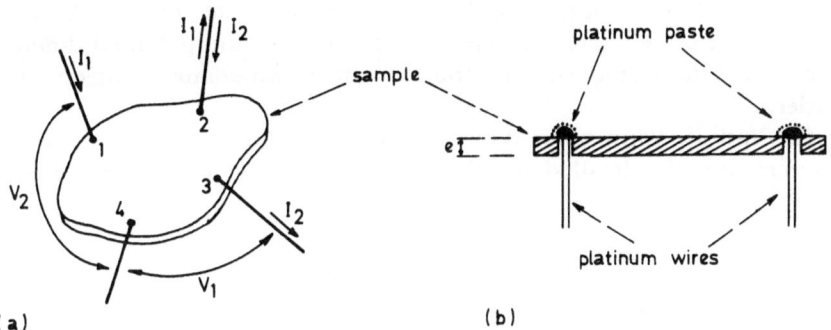

FIGURE 9. Principle of the DC Van der Pauw method: (a) electrode arrangement, (b) cross section of the sample.

where e is the sample thickness, while $f(R_1/R_2)$ is a function calculated and tabulated by Van der Pauw[14] and equal to unity when $R_1 = R_2$.

A convenient measurement circuit is shown schematically in Fig. 10. A DC current is delivered by a stabilized supply and the current direction is inverted periodically by a reversing switch. The current probes and the voltage

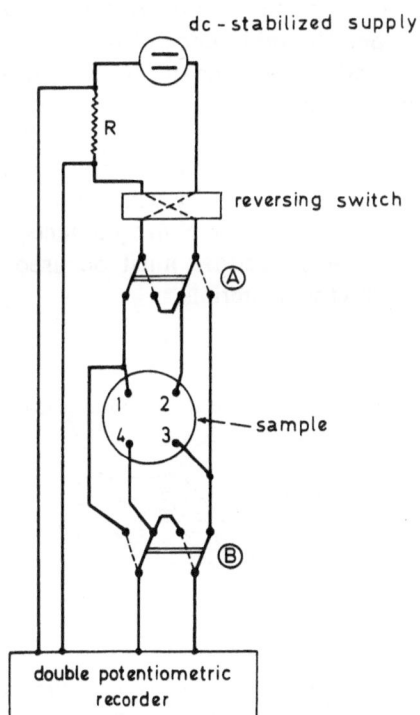

FIGURE 10. Measurement circuit for the DC Van der Pauw method.

probes are selected with the aid of two double commutators A and B. The voltage drop V_R across a standard resistor and the potential difference V_1 or V_2 are measured simultaneously with a two-channel potentiometric recorder.

The Van der Pauw method has also been developed for AC conductivity measurement and a detailed description of the experimental setup is given by Panis.[15]

2.4.2. The Point-Probe Method

This technique should be reserved for the investigation of very small samples, provided they offer at least one flat surface.[12]

Four probes, which must be point contacts, are placed on a line along a flat surface of the sample. The distance d between the different point probes must be constant and very small in comparison with the dimensions of the sample, including the thickness.[12]

In the case of small monocrystalline specimens, an original device is obtained by putting the sample onto an insulating plate[16] (Fig. 11a). Four fine-drawn metallic wires (Au, Pt) are attached to the sample with the aid of metallic paste. This may be achieved conveniently by means of a light microscope. The other ends of the metallic wires are coated on four snap contacts located at one edge of the insulating plate (Fig. 11a). The current I is passed through the sample and the voltage drop V_S is read through the agency of the snap contacts.

For larger specimens, four point probes may be pressed directly onto the sample (Fig. 11b) by using a mechanical device and sharp-pointed metallic rods. However, pressure contacts may develop high contact resistances, so that this procedure must be used carefully and only in the case of good conductive materials.

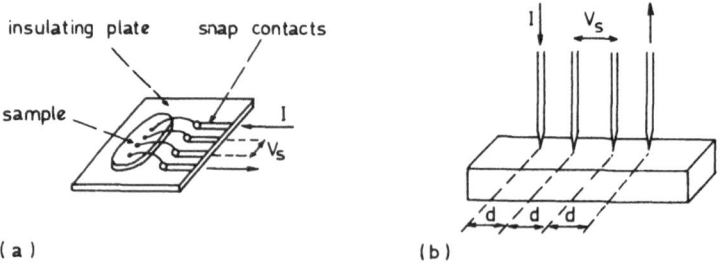

FIGURE 11. Schematics of the four point-probe arrangements: (a) small monocrystalline sample, (b) four sharp-pointed electrodes.

In both arrangements, provided the contact areas are small enough to be neglected and do not enter the conductivity calculation,[12] the conductivity of the sample is given by

$$\sigma = \frac{I}{V_S} \frac{1}{2\pi d} \tag{7}$$

One of the main experimental errors encountered when employing such devices arises from the lack of precision in the arrangement of the four point-probes, especially in the case of very small samples. Another source of errors lies in the principle of the measurement itself. Indeed, the contacts being located at the surface of the sample, the current density will be larger in the vicinity of the surface than in the bulk, making the measured conductivity only representative of the surface layer.

2.5. Experimental Requirements for High-Temperature Measurements

At high temperature, in addition to the increasing chemical reactivity of the materials, several factors can disturb the reliability of conductivity measurements and some experimental requirements are needed.

First, the high-temperature furnaces will be connected to DC-stabilized supply in preference to AC supply in order to minimize self-induction phenomena in the measurement circuit.

The temperature of the furnace must be carefully controlled and temperature fluctuations should be strictly avoided. This is of prime importance when the electrical conductivity is thermally activated with activation energies higher than 1 eV. Most of the PID regulated furnaces provide high-temperature stability (± 1 K at 1273 K). Measurements of temperatures higher than 800 K may be achieved with the use of Pt/Pt–Rh thermocouples. For lower temperatures, either chromel–alumel or iron–constantan thermocouples are more suitable.

The samples are introduced into the furnace via the sample holder, in the zone of constant temperature, in order to prevent spurious voltages due to thermoelectric power. In any case, the latter would be compensated by reversing periodically the direction of the current with the aid of an experimental setup like that shown in Fig. 6.

Finally, at very high temperature (higher than about 1800 K), thermal emission of electrons from the inner surface of the furnace can induce extraneous currents in the measurement circuit.[17] This effect can be removed by intercalating an electrical shield between the sample holder and the furnace. It may be provided by a refractory ceramic tube, covered by a thin metallic layer and grounded to earth.

3. ALTERNATING-CURRENT MEASUREMENTS

It has been emphasized in the preceding section that the measurement cell is a heterogeneous system comprising at least two electrodes and a sample. The sample itself may be polycrystalline, made up of grains and grain boundaries. However, through DC measurements only a total resistance is determined which may be very different from the bulk resistance of the material. Moreover, better knowledge of the electrical properties of, for instance, the grain boundaries or the interfaces between the sample and its electrodes could be the main focus of interest of the investigation. This is made possible with complex impedance spectroscopy, first used by Bauerle[18] in a study of the electrical properties of a calcia-stabilized zirconia solid electrolyte.

An alternating voltage, of such small amplitude that Ohm's law may be applicable, is applied to the cell using a two-, three-, or four-probe arrangement and the real and imaginary parts of the impedance are obtained for different values of the frequency. Then, the experimental data points are plotted in the complex impedance plane (Re Z-Im Z). In favorable cases, as depicted schematically in Fig. 12, the experimental curve may be regarded as the superposition of several arcs of circle and it may be shown that each one is associated with a definite transport process, through the bulk or across the grain boundaries, for instance.

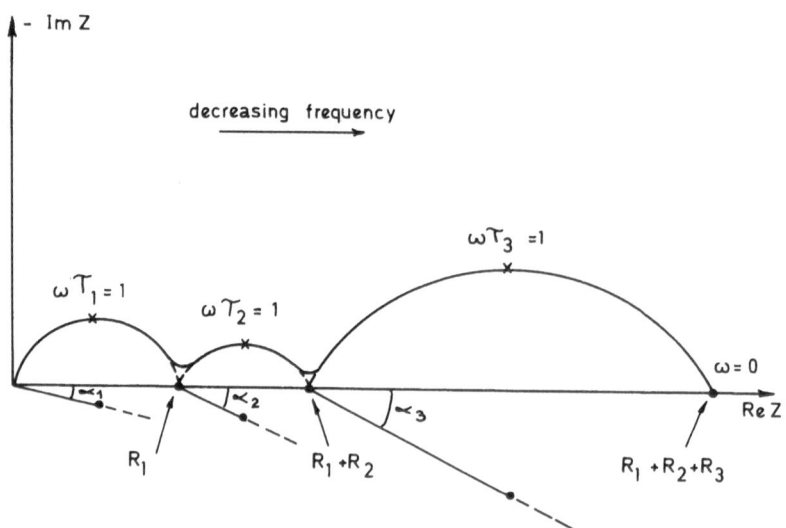

FIGURE 12. Schematics of an impedance spectra obtained with a polycrystalline sample. The first arc of circle, when going from left to right, is usually associated with bulk properties, the second with grain boundaries effects and the third with polarization phenomena at the electrodes. Definitions of the parameters α_i, R_i, and τ_i are given in the text.

3.1. Analysis of the Spectra

3.1.1. Phenomenological Approach

Let us recall that the impedance of an electrical circuit consisting of a pure resistor R_i in parallel with a pure capacitor C_i is given by

$$Z_i = R_i/(1 + j\omega\tau_i) \qquad (8)$$

where $\tau_i = R_iC_i$ is the time constant of the circuit. The corresponding figure in the complex impedance plane is a semicircle centered on the real axis. However, as depicted in Fig. 12, mostly the centers of the circles are not located on the real axis but shifted below it by some angle α_i. In order to simulate the experimental spectra, the basic formula, equation (8), must be replaced by the following one:

$$Z_i = R_i/[1 + (j\omega\tau_i)^{\eta_i}] \qquad (9)$$

where the exponent η_i is equal to $(1 - 2\alpha_i/\pi)$, if α_i is expressed in radians. Then, the resistors and capacitors can no longer be considered as pure, i.e., they are functions of the frequency. The general shape of the spectra suggests that the sample and its electrodes may be associated with an electric circuit made up of several such cells connected in series:

$$Z = \sum_i Z_i \qquad (10)$$

If the experimental spectra can be approximated by only one arc of circle, then the determination of the two parameters (R, τ) is straightforward: R is equal to the intersection of the circle with the real axis, while τ^{-1} is the value of the pulsation ω $(=2\pi f)$ at the top of the circle.

These simple rules have to be modified when several arcs of circle appear to be present. Under the condition that the overlap of two successive circles is small, which implies that the time constants τ_i are quite different, then τ_i^{-1} can still be taken as the value of the pulsation at the top of the ith arc of circle. One would expect that the resistances R_i could be calculated from the intersections of the fitted arcs of circle with the real axis. However, mathematical developments show that a better approximation is obtained if one considers the values of the real part of the impedance at the minima between two successive arcs of circle as depicted in Fig. 12. This is clearly exemplified in Fig. 13, where results are presented of a numerical simulation performed on a two-RC-cell electrical circuit.

A least-squares analysis of the experimental data using equations (9) and (10) permits a more precise determination of the parameters, although much computational effort is needed.[19] A special case must be mentioned, i.e., when

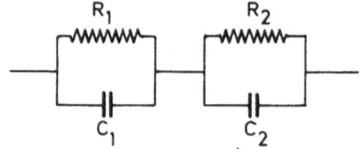

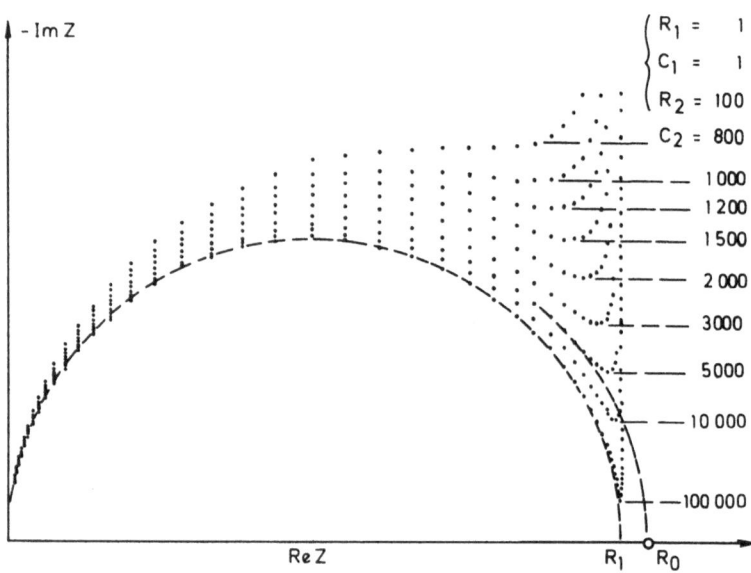

FIGURE 13. Numerical calculations performed on a two-RC-cell electrical circuit. They clearly show that the value of Re Z at the minimum between the two circles is a good evaluation of R_1 compared to R_0, the intersection of the fitted arc of circle with the real axis.

the arc of circle passes through the origin. Then, with a judicious choice of the ideal point M_{i0} associated with the experimental point M_i, as shown in Fig. 14, the least-squares analysis leads to a linear system of equations with two unknowns, x_c and y_c, which are respectively the abscissa and ordinate of the center of the circle[20]:

$$\left. \begin{array}{l} \sum_i x_i = 2x_c \sum_i (x_i^2/r_i^2) + 2y_c \sum_i (x_i y_i/r_i^2) \\ \sum_i y_i = 2x_c \sum_i (x_i y_i/r_i^2) + 2y_c \sum_i (y_i^2/r_i^2) \end{array} \right\} \tag{11}$$

where $r_i^2 = x_i^2 + y_i^2$.

Finally, one must keep in mind that, although the use of expressions (9) and (10) allows a good fit of the experimental data with a mean standard

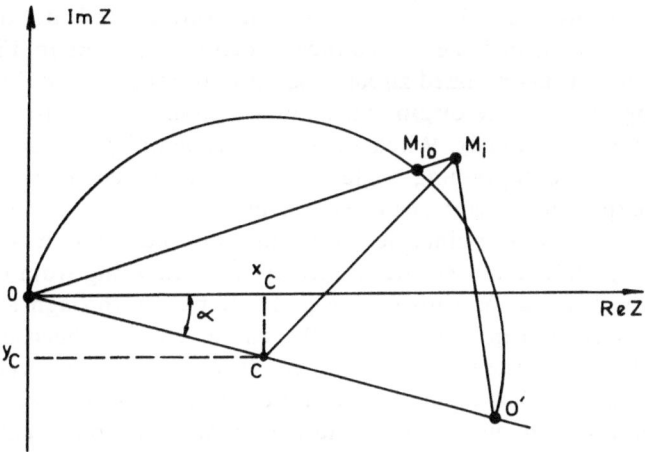

FIGURE 14. The experimental data (M_i) are fitted by an arc of circle (M_{i0}) passing through the origin and the center of which is depressed below the real axis by some angle α.

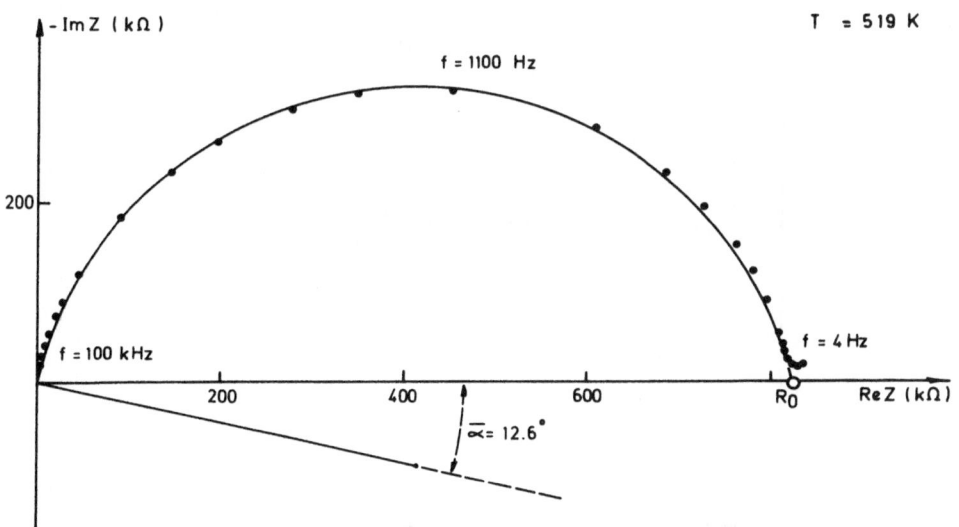

FIGURE 15. Impedance spectra obtained on a yttria-stabilized zirconia sample at 519 K: R_0 is the DC resistance of the sample while $\bar{\alpha}$ is the average depressing angle, calculated with the aid of least-squares analysis.

deviation of the order of 1% in many cases, this procedure is not unique. For instance, let us consider the impedance spectra presented in Fig. 15 and obtained on an yttria-stabilized zirconia sample. It consists of only one arc of circle passing through the origin, which was analyzed according to the procedure described previously. When all the experimental points are taken into account, an average depressing angle of 13° was obtained. However, a local study of the experimental curve can be performed by selecting a limited number of nearest-neighbor data points, seven in this example. Then, it is found that the depressing angle is not constant but varies when going from the right to the left of the spectra; as shown in Fig. 16, it passes through a maximum approximately at the top of the arc of the circle. This has been found to be quite general as far as the first circle, passing through the origin, is concerned. It seems in this case that the ellipse would be a better starting basis than the arc of circle, although susceptible of no electrical circuit representation.

3.1.2. Physical Interpretation

Interest in the complex impedance representation is focused on the fact that each arc of circle is associated with a definite transport process.

A first indication is given by the magnitude of the pseudocapacitances C_i deduced from a formal analysis of the impedance spectra. Very high capacitances, typically in the microfarad range, are usually characteristic of electrodes phenomena while low capacitances, in the picofarad range, must be associated with bulk properties, hence the interpretation given in the caption of Fig. 12.

However, before any detailed physical interpretation can be given, this point must be carefully substantiated. The easiest way is to vary the geometrical

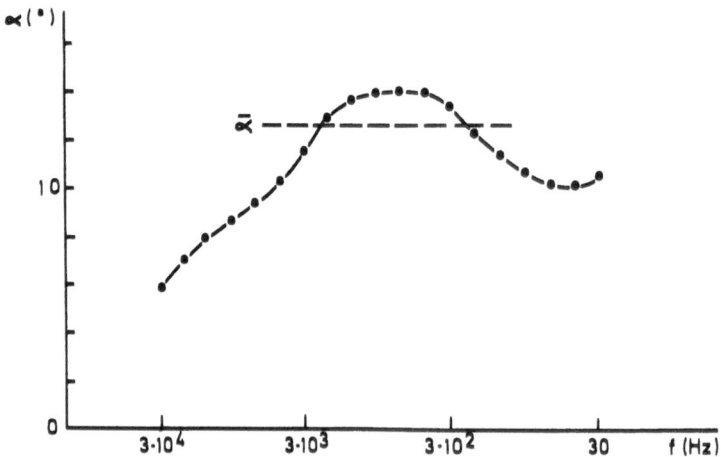

FIGURE 16. A local study of the curve presented in Fig. 15 is performed, i.e., only a group of seven neighboring points is considered when using the least-squares analysis. It may be seen that the local depressing angle α is not constant over the whole arc of circle.

dimensions of the sample, its width and diameter if it is disk-shaped; the bulk resistance is known to be proportional to the geometrical factor, equal to the ratio of area to width, while we expect that the width is of no importance if the main contribution to impedance arises from polarization effects at the electrodes. Other tests can be performed; for instance, different metals, evaporated aluminum or gold, silver or platinum paste, can be used in the preparation of the electrodes. In order to determine if grain boundaries play a significant role, one may compare data obtained on single crystals and polycrystalline materials or vary the mode of elaboration of the ceramic, such as the granulometry of the initial powder, the sintering time, or temperature.

We are now ready to employ more complete theories in order to interpret separately each contribution to the total impedance. However, the reader must keep in mind that some overlap of the arcs of circle is always present and can be a source of confusion and errors.

Most of these theories concern the bulk properties of homogeneous materials. According to the Maxwell's equations and in agreement with Ohm's law, the admittance $Y(\omega)$ of the sample is the sum of two contributions, a conductive, $\sigma(\omega)$, and dielectric, $\varepsilon(\omega)$, one:

$$Y(\omega) = L[\sigma(\omega) + j\omega\varepsilon_0\varepsilon(\omega)] \qquad (12)$$

where L is the geometrical factor of the sample and ε_0 the permittivity of the vacuum. The impedance $Z(\omega)$, equal to $Y(\omega)^{-1}$ and calculated from equation (12), is that of an RC electrical circuit and therefore the use of this analogy, equation (8) or (9), is justified on a theoretical basis. However, the proper dependence of $\sigma(\omega)$ and $\varepsilon(\omega)$ on frequency is a function of the type of involved charge carriers, ions or electrons, and of the detailed description of the mode of transport. For more details, the interested reader will consult Böttcher and Bordewijk[21] devoted to dielectric polarization, Abélard[20] and Wong and Brodwin[22] in case of ionic conductivity, and Schirmacher,[23] Srivastana and Chaturvedi,[24] and Chekunaev and Fleurov[25] if charge carriers are electrons moving through hopping in a disordered medium. Most often these theories lead to pseudoarcs of circle depressed below the real axis in the complex impedance plane and therefore this type of representation of the data is no longer adequate. A detailed analysis of the variations in the real and imaginary parts of the impedance, admittance, or any derived quantity against frequency is necessary, and a logarithmic system of coordinates appears to be most suitable.

Studies of the electrical properties of grain boundaries are actually in progress.[26] Most of the work is performed on semiconductors, such as silicon or zinc oxide. The existence of an interface between two crystals, a few angströms wide, is associated with a Schottky barrier. The application of an alternating field causes trapping and detrapping of electrons (or holes) from deep impurity levels in the gap and leads to a frequency-dependent conductivity.[27] In polycrystalline materials, the interpretation is obscured by the fact that all the grain boundaries constitute a complicated network and that only averaged properties are measurable. In many ceramics, grain boundaries are much thicker owing to the presence of a second phase, crystalline or amorphous. In this case, the sample must be regarded as a heterogeneous system and theories such as the Maxwell–Wagner model may be applicable.

What has been called "polarization effects at the electrodes" is in fact a complex process, the details of which depend on the nature of both the material and the electrodes. For instance, metal semiconductor interfaces may give rise to Schottky barriers while the case of ionic conductors appears to be more complicated.[28] Let us mention a study devoted to the oxygen electrode reactions on stabilized zirconia[29] where several arcs of circle could be distinguished, associated with different mechanisms. When diffusion of the charged species is the limiting step, the electrode characteristic is a well-defined Warburg impedance[30]:

$$Z_{Wb} \approx (1 + j)\omega^{-1/2} \tag{13}$$

The corresponding figure in the complex impedance plane is then a straight line making angle of 45° with the real axis.

3.2. Experimentation

3.2.1. Specifications

In order to obtain impedance spectra of good quality, one must care about the following requirements:

- A broad frequency range extending over several decades. Often, the available range is not wide enough to cover the whole spectra. However, the time constants which enter relations (8) and (9) are usually temperature dependent and obey an Arrhenius law:

$$\tau_i = A_i \exp (E_i/k_B T) \tag{14}$$

where E_i is an activation energy and A_i a preexponential factor. Therefore it is highly desirable to place the cell in a regulated furnace and record the spectra for several temperatures; different parts of the spectra will be made visible in different temperature domains as depicted in Fig. 17.
- A very high dynamic range, because the impedance of the measurement cell may change by several powers of ten when going from high to low frequencies.

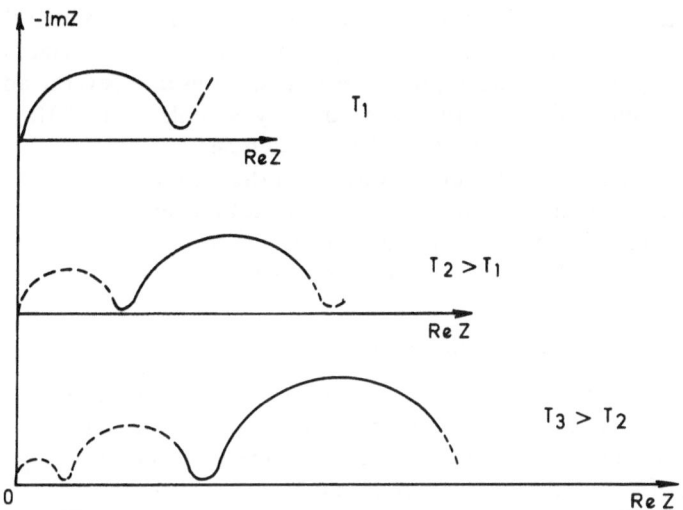

FIGURE 17. Due to an insufficiently available frequency domain, only a limited portion of the impedance spectra can be obtained at a given temperature. Different parts of the spectra are made visible at different temperatures (schematic).

- A broad phase range. The phase angle ν defined by the relation

$$\tan \nu = -\mathrm{Re}\,Z / \mathrm{Im}\,Z \qquad (15)$$

varies between $0°$ and $90°$ over the spectra. In particular, at the highest (respectively lowest) frequencies, ν is very close to $0°$ (respectively $90°$) and much sensitivity is necessary. Bridges are especially suited for the investigation of bulk properties because they allow the measurement of very small angles ($\nu \sim 0.001°$) over a broad frequency range. They are largely used in dielectric studies and the interested reader may find further details elsewhere.[31]

All these considerations and the necessary compromise they imply explain that the best results are obtained in the frequency range 1 Hz–1 MHz, although measurements can be performed in the 10^{-4} Hz–10^{9} Hz interval.

Cross-correlation, which ensures the measurement of small voltages with a low signal-to-noise ratio, is most popular and is the working basis of many commercially available analyzers. An alternating voltage of variable frequency is applied to the measurement cell using any of the different procedures described in Section 2. The resulting current is cross-correlated with two signals, one in phase, the other out of phase with the applied voltage. After averaging, both the real and imaginary components, or the modulus and the phase angle of the impedance or admittance, are displayed at the same time. Digital as well as analog or mixed electronic circuits are used to produce the output voltage to be applied to the cell and perform the cross-correlation. Digital equipment, which is being rapidly developed, presents several advantages, such as the synthesis of very low frequency signals ($\sim 10^{-4}$ Hz) and easy processing, but cannot be used at high frequencies.

Measurements can also be conducted in the time domain.[32] The impedance spectra are obtained after application of a Fourier transform to the data. However, a voltage of very high amplitude, of the order of 100 V, must be applied and Ohm's law might no longer be verified.

3.2.2. Description of a Versatile Set-Up

In the following, we describe a typical experimental setup devised in our laboratory which presents good specifications: a 1 Hz–100 kHz frequency range, a wide measurable range of impedance ($10\,\Omega$–$10^9\,\Omega$), a good phase sensitivity of 0.2° with an accuracy of ±0.2°.

A block diagram of the installation is given in Fig. 18. The AC voltage supply (ADRET 3100) provides two outputs. The first one, V, may be varied between 0 and 10 V and is applied to the experimental cell, while the other constitutes a reference voltage fixed at one volt. An adjustable standard resistor (VISHAY) R ($10\,\Omega$–$10^6\,\Omega$), characterized by a high precision and very small stray capacitances (~ 1 pF), is inserted in series with the cell and permits measurement of the current flowing into the circuit. The voltage drop v, developed across the resistor, constitutes the input of a double lock-in amplifier (PAR 5204) which performs cross-correlation and an averaging procedure. The real and imaginary components of the impedance are deduced from the following relations:

$$\left.\begin{array}{l} V = (R + Z)I \\ v = v_a + jv_b = RI \end{array}\right\} \tag{16}$$

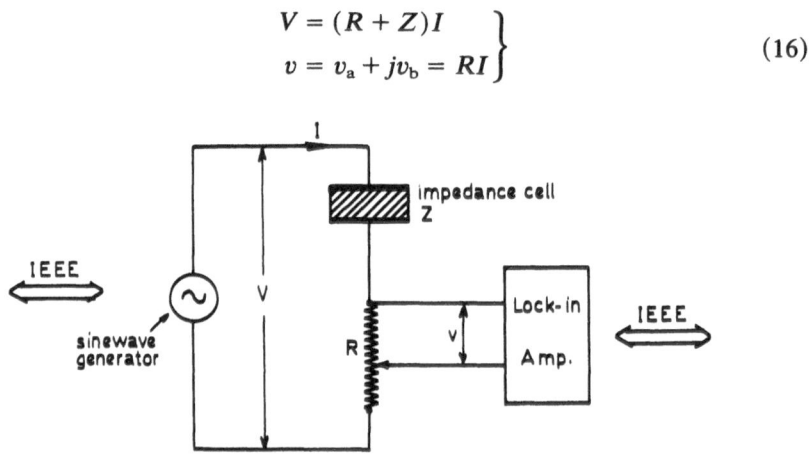

FIGURE 18. Block diagram of the experimental setup.

or

$$\left.\begin{aligned}
\operatorname{Re} Z &= R\left[\frac{Vv_a}{v_a^2 + v_b^2} - 1\right] \\
\operatorname{Im} Z &= R\left[\frac{Vv_b}{v_a^2 + v_b^2}\right]
\end{aligned}\right\}
\qquad (17)$$

Both the sine-wave generator and the lock-in amplifier are controlled by a microprocessor through an IEEE-488 parallel interface, making possible automatic measurements and easy treatment of the data.

The sample is a disk, the two faces of which, coated with a metal, serve as electrodes (see Fig. 3a). Dimensions are chosen in such a way that the bulk capacitance is not lower than 50 pF. By means of a spring, the specimen is firmly clamped between two insulating alumina blocks. The sample holder, depicted in Fig. 19, is placed inside a PID regulated furnace. If a detailed analysis of the spectra is sought, the temperature must be very stable during measurements. Indeed, according to equation (14), a fluctuation of 0.25 K at room temperature and for an activation energy of 1 eV corresponds to a variation of 3% in the impedance. Below 500 K, good performances are obtained on using a caloric or freezing fluid to establish the temperature within the cell.

Shielded BNC leads must be used if very small currents have to be measured. However, they have the disadvantage that their capacitance per

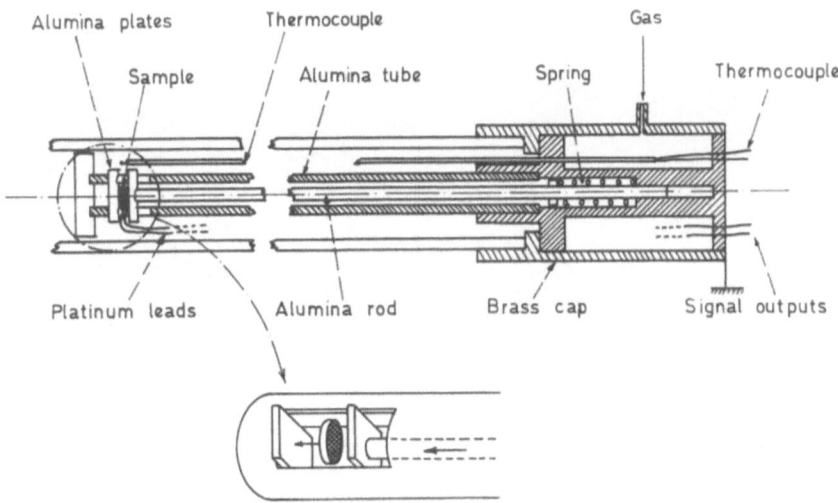

FIGURE 19. Description of the sample holder.

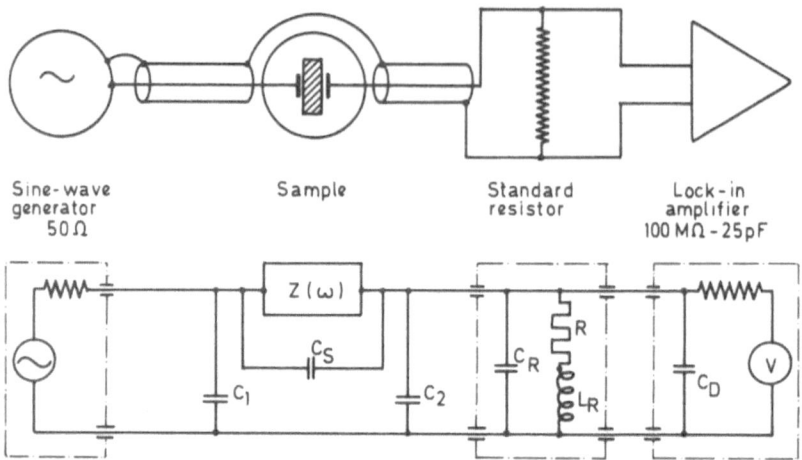

FIGURE 20. Analysis of the parasitic capacitances and inductances present in the experimental setup.

unit length is not negligible when compared to that of the bulk material. The configuration depicted in Fig. 20 is such that the parasitic capacitance in parallel with the sample is as small as possible. Using different specimens of the same material with different geometrical factors, it has been estimated to be of the order of 1 pF,[20] i.e., a few percent of the bulk capacitance.

The various ground contacts present in the measuring system constitute a low resistive electrical loop. Induced emfs having the same frequency as that of the applied voltage are not averaged to zero by the lock-in amplifier and may contribute significantly to the voltage drop measured across the standard

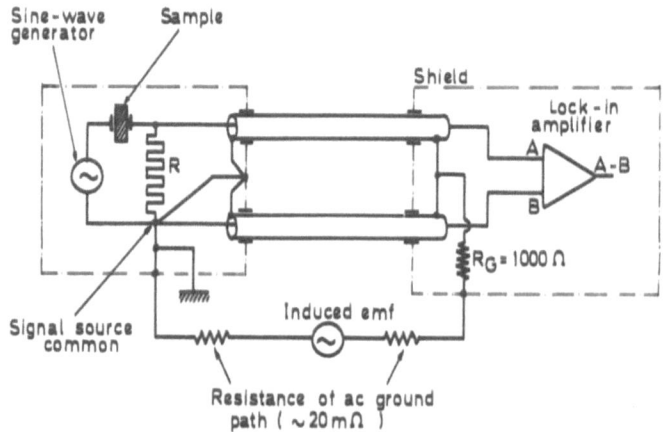

FIGURE 21. Description of the grounding procedure.

resistor. This parasitic effect is made negligible through the use of a unique ground point, a differential input to the lock-in amplifier, and the insertion of a 1000-Ω resistance as shown in Fig. 21.

4. MEASUREMENTS OF PARTIAL CONDUCTIVITIES

A detailed analysis of transport properties necessitates, in addition to the measurement of total conductivity σ_t, other complementary studies, especially when the material under investigation involves different types of charge carriers, i.e., ionic and electronic. The total electrical conductivity is then the sum of the partial conductivities associated with each type of charge carrier and may be written as

$$\sigma_t = \sum_i \sigma_i + \sum_e \sigma_e \tag{18}$$

where subscript i refers to ionic species, i.e., anions or cations, and subscript e to electronic species, i.e., electrons or holes.

The contribution of each partial conductivity may also be expressed in terms of transference number t_k, which is defined as the ratio of the partial conductivity σ_k due to species k to the total conductivity:

$$t_k = \sigma_k / \sigma_t \tag{19}$$

By definition, the sum of ionic and electronic transference numbers is equal to unity:

$$\sum_i t_i + \sum_e t_e = 1 \tag{20}$$

Depending on the nature of the mobile species, various procedures have been developed to evaluate the magnitude of the partial conductivities. Thus, ionic transference numbers may be derived from electrolysis experiments with reversible electrodes or on using galvanic cells, while electronic partial conductivities can be obtained with the aid of the polarization method or by performing permeation rate measurements.

4.1. The Electrolysis Method

This method is conceptually very simple and was orginally proposed by Tubandt[33,34] for cationic transference number measurements on halides. In such experiments, a DC current is passed through a cell stack made of three or more sample pellets pressed together and clamped between two electrodes, reversible with respect to the exchange of ionic carriers through the material-electrode interface (Fig. 22). In the case of cationic, as well as anionic, $M_n X_m$ conductors, the metal M itself may be used as electrode material. Different cell arrangements and cell holders are described elsewhere.[9,35] Each element of the cell, electrodes and sample pellets, is carefully weighed before being assembled.

Under the application of a DC voltage to the cell, the mobile species migrate through the material to either the cathode or the anode depending on the sign of their electric charge. Thus, the passage of a suitable amount of current induces some weight change of the electrodes and the adjacent sample pellets, the chemical composition of which may be modified. In the case of M^{m+} cationic conductors with identical reversible electrodes, a small quantity of metal M is deposited at the cathode while an equal quantity is dissolved at the anode. Nevertheless, various ionic mobile species may simultaneously contribute to the charge transport and an experimental procedure has been described elsewhere[9] to determine the nature of the charge carriers.

The transference numbers of anions or cations are then deduced from the ratio of the extent of charges ΔC_i transported via the ionic species during time Δt to the total amount of charges ΔC flowing during the same time through the measurement circuit:

$$t_i = \frac{\Delta C_i}{\Delta C} \tag{21}$$

The quantity ΔC is read by a coulometer or deduced from intensity-time measurements, while the extent of charges ΔC_i is derived from the weight change of either the cathode or the anode. In the case of M^{m+} cationic

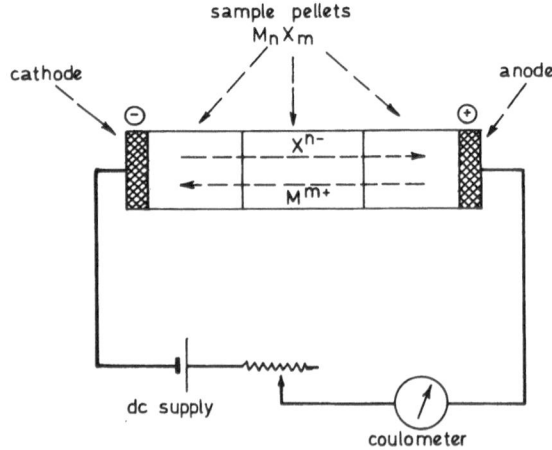

FIGURE 22. Measurement of ionic transference number by the electrolysis method. Electrical connecting and cell arrangement.

conductors, the amount of charges ΔC_i associated with a weight change Δp of the cathode is given by

$$\Delta C_i = \frac{\Delta p}{P} zF \tag{22}$$

where P denotes the molar weight of the metal M, $z = +m$, the valence of the cations, and F is the Faraday constant.

Another procedure consists in deriving the ionic transference number from dilatometric measurements of the dimensional changes in the various elements of the cell.[36,37] This technique and the cell arrangement are detailed by Morin.[35]

Suitable cell assemblies are generally achieved under mechanical pressure or by hot-pressing the different pellets together, after taking much care in machining and polishing the surface of the pellets. However, this problem must be treated cautiously in order to obtain acceptable ionically conducting interfaces.[38]

After current flow, the different sample pellets and the electrodes must be separated and analyzed by a chemical technique or a gravimetric method. A suitable process appears to be the quantitative chemical analysis that may be carried out after dissolution of the different pellets into appropriate solvents. However, the gravimetric technique is the most frequently used, though it may be sometimes complicated by the difficulties in separating the various elements of the cell.

Some applications of the electrolysis technique including halides or oxides are described elsewhere.[9,35,39,40]

4.2. The Polarization Method

The polarization method, also called the Wagner method, may be applied to the study of various mixed conductors. However, an advantage of this method lies in the fact that very low electronic conductivities, lower than $10^{-15} \, \Omega^{-1} \, m^{-1}$, are measurable in a reproducible manner.[41] In this technique, the ionic contribution to the charge transport is blocked with an appropriate electrode, reversible to electrons but blocking to ions. A very common device is obtained by clamping a sample pellet between one blocking electrode and one reversible electrode which fix the chemical potential of the ionic species at the electrode–material interface[42] (Fig. 23).

Upon application of a small DC voltage V, lower than the decomposition potential, both ionic and electronic charges migrate through the material. However, the ionic current is blocked at the nonreversible electrode so that, after some time, a steady state is reached.[43] The cell is then polarized and the current flowing through the sample only results from the motion of

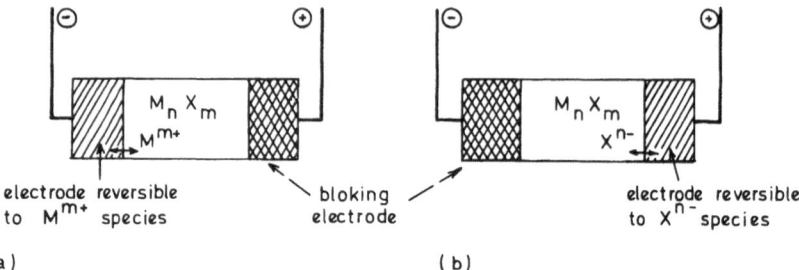

FIGURE 23. Schematics of the cell arrangement for the polarization method in the case of cationic (a) and anionic (b) conducting electrolytes.

electronic species. Then, it may be shown[44] that the steady-state current in cationic conductors is given by

$$I_C = \frac{kTS}{3ql}\left\{\sigma_n^0\left[1 - \exp\left(\frac{-qV}{kT}\right)\right] + \sigma_p^0\left[\exp\left(\frac{qV}{kT}\right) - 1\right]\right\} \qquad (23)$$

while for anionic conductors

$$I_A = \frac{kTS}{3ql}\left\{\sigma_p^0\left[1 - \exp\left(\frac{-qV}{kT}\right)\right] + \sigma_n^0\left[\exp\left(\frac{qV}{kT}\right) - 1\right]\right\} \qquad (24)$$

where k is the Boltzmann constant, S the surface of the electrode, and l the length of the sample; σ_n^0 and σ_p^0 denote the partial electronic conductivities due to electrons and holes, respectively, at the given chemical potential of ionic species fixed by the reversible electrode. Thus, they refer to the n-type and p-type electronic conductivities which would be observed in the cell M/MX/M with two identical reversible electrodes. These conductivities cannot be measured directly because of the predominant ionic current which occurs when a DC voltage is applied to such a symmetrical cell. Furthermore, as pointed out by Kennedy,[45] σ_n^0 represents the maximum value of electronic conductivity via electrons while σ_p^0 is the minimum electronic conductivity via holes.

Equations (23) and (24) permit the calculation of the partial electronic conductivities if one type of electronic carrier dominates, so that either σ_n^0 or σ_p^0 may be neglected. However, in many cases, it is not possible to make such an approximation and, as proposed by Patterson et al.,[46] the values of σ_n^0 and σ_p^0 may therefore be derived from the plots of $(zql/kTS)I[1 - \exp(-qV/kT)]^{-1}$ against $\exp(qV/kT)$, i.e., the intercept on the vertical axis and the slope, respectively, for cationic conductors and the reverse in the case of anionic conductors.

Several conditions required to perform consistent and reliable measurements are discussed by Wagner[41] and applications of the polarization method to various solid electrolytes and mixed conductors are given elsewhere.[7]

4.3. The Galvanic Cell Method

The galvanic cell method is quite different from the two previous techniques, since no current flows through the sample, and mostly applies to anionic conductors.

Let us consider an anionic conducting electrolyte M_nX_m sandwiched between two reversible electrodes which fix two different chemical potentials for the mobile species. In the absence of any current through an external circuit, the emf developed by such a cell is given by[47]

$$E = -\frac{1}{2zq} \int_{\mu X_2(I)}^{\mu X_2(II)} t_i \, d\mu X_2 \tag{25}$$

where $\mu X_2(I)$ and $\mu X_2(II)$ denote the chemical potential of the nonmetallic component X_2 at the two electrolyte-electrode interfaces, q is the electronic charge, and z the valence of the anions X^{n-}.

The ionic transference number t_i may then be derived from the value of the emf, E, after appropriate integration of equation (25).

4.3.1. Electromotive Force Measurement Procedure

The galvanic cell method has been used quite extensively in studies of transport properties of nonstoichiometric oxides.[47,48]

The emf of the cell is then expressed in terms of the oxygen chemical potentials at the two interfaces:

$$E = \frac{1}{4q} \int_{\mu O_2(I)}^{\mu O_2(II)} t_i \, d\mu O_2 \tag{26}$$

For small oxygen chemical potential differences, or when t_i is expected to depend slightly on μO_2, the latter is often considered to be a constant and replaced by a mean transference number $\bar{t}_i$, the value of which is obtained by integrating equation (26):

$$E = \frac{kT}{4q} \bar{t}_i \ln \left[pO_2(II)/pO_2(I) \right] \tag{27}$$

where $pO_2(I)$ and $pO_2(II)$ are the oxygen partial pressures at the interfaces ($\mu O_2 = \mu O_2^0 + kT \ln pO_2$[49]).

This approximation is sometimes inaccurate[48] and the variation of t_i versus oxygen partial pressure may then be checked out after differentiation of equation (26) at the constant $pO_2(I)$ value:

$$t_i = \frac{4q}{kT}\left[\frac{\partial E}{\partial \ln pO_2(II)}\right]_{pO_2(I)} \tag{28}$$

The value of the ionic transference number for a given oxygen partial pressure $pO_2(II)$ is thus derived from the slope of the E-ln $pO_2(II)$ plots at that $pO_2(II)$.

The emf measurements are performed with high input impedance apparatus (10^8 to $10^{12}\,\Omega$), such as DC voltmeters, electrometers, or potentiometric recorders, in order to prevent any electrical current in the external circuit. Indeed, the latter would be counterbalanced by an equivalent flow of ionic carriers through the electrolyte which could modify the oxygen activities at the two electrolyte-electrode interfaces and induce erroneous emf.[50]

Reversible electrodes may be provided by the use of metal–metal oxide mixtures (M'/M'_aO_b). The oxygen partial pressure at the electrolyte-electrode interface is hence fixed by the equilibrium between the metal and the oxide:

$$aM' + \frac{b}{2}O_2 \rightleftharpoons M'_aO_b \tag{29}$$

and is derived from the thermodynamic data relative to equilibrium (29).[51]

However, many metal–metal oxide mixtures are required when a large range of oxygen partial pressures must be covered, so that more popular arrangements are obtained by combining electronically conducting porous electrodes and gaseous phases with given oxygen chemical potentials. The use of inert gas–O_2, CO–CO_2, or H_2–H_2O mixtures allows one to vary the oxygen partial pressure over many orders of magnitude.[52]

A useful device has been perfected by Fabry et al.[53] using a thin electrolyte tablet, clamped between two ceramic tubes (generally alumina tubes) and two metallic rings made of gold or platinum (Fig. 24). The latter may be obtained by soldering the two extremities of gold or platinum wires of about 10^{-3} m in diameter[54] or directly cut in gold or platinum sheets of about 1 to 3×10^{-4} m thick.[10] Mechanical pressure is applied to the ceramic tubes and the metallic rings so that, after firing at high temperature, very good gas tightness is ensured.[55,56] A convenient mounting is described in detail elsewhere.[54] Two gaseous phases, with different oxygen partial pressures, flow into small alumina tubes up to the vicinity of the porous electrodes. The electrical leads made of gold or platinum wires are threaded into the small alumina tubes and pressed onto the electronically conducting porous electrodes obtained by painting the two flat ends of the sample with a metallic paste. An additional alumina tube is also necessary to minimize local short-circuit effects at the metallic rings and to prevent direct transfer of oxygen through the external gas phase.[50] Therefore, a very low oxygen partial pressure is maintained in the annular guard space on using pure inert gases or vacuum. The alumina guard tube is

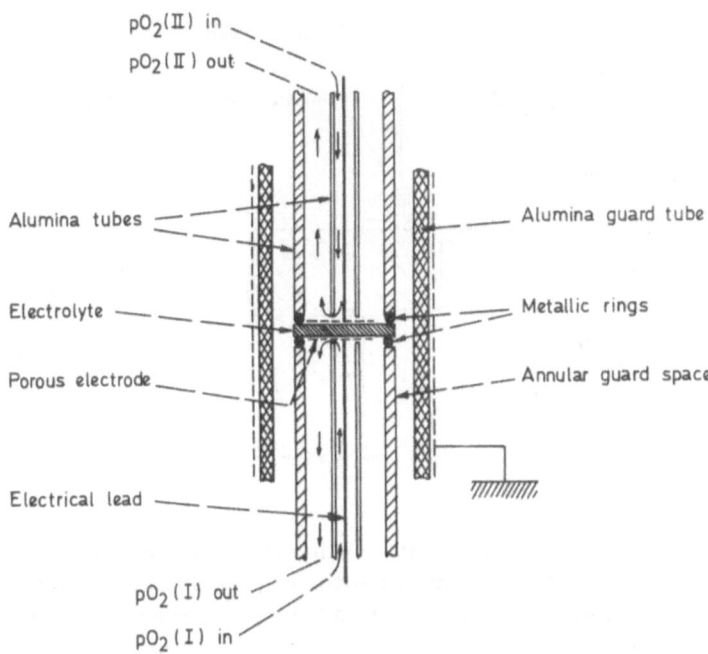

FIGURE 24. Schematics of a galvanic cell for ionic transference number measurement, after Fabry *et al.*[53]

also covered by a thin metallic film and grounded to earth, in order to prevent induction phenomena when the cell is heated in a furnace.

Such an arrangement allows one to perform quite reproducible determinations of ionic transference number, although several factors can affect the reliability of the cell emf.[57] One frequent source of errors lies in the electrochemical semipermeability of the electrolyte which induces an oxygen flow through the sample pellet and can modify the thermodynamical equilibria at the electrolyte-electrode interfaces.[57] The oxygen semipermeability arises from the partial electronic conductivity σ_e of the electrolyte, and the local charge flow of ionic species in the material is given by[47]

$$\vec{j_i} = \frac{1}{4q} \sigma_e t_i \overrightarrow{\text{grad } \mu O_2} \tag{30}$$

As proposed by Fouletier *et al.*,[50] the errors due to the oxygen semipermeability may be largely minimized by using a point electrode having the same composition as the electrolyte. Such an electrode is shown schematically in Fig. 25. The whole point electrode is supposed to be surrounded by the same

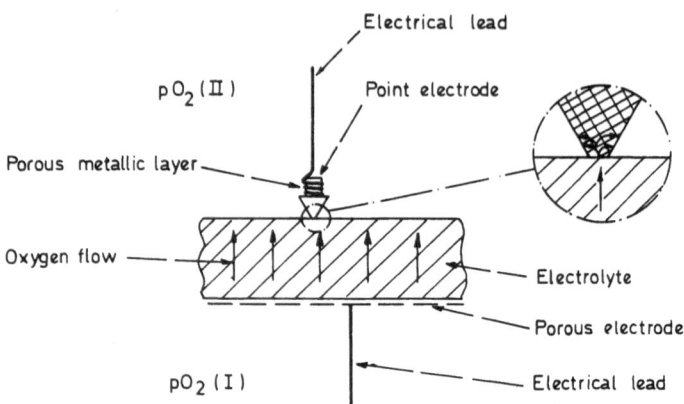

FIGURE 25. The electrolyte point-electrode and its principle of operation, after Fabry et al.[53]

oxygen partial pressure. Thus, let us assume that the electrolyte exhibits an oxygen semipermeability which modifies the oxygen chemical potential at the contact surface between the electrolyte and the point electrode. The resulting oxygen flow in the point electrode is directed toward its closest free surface[50] (Fig. 25), so that no oxygen flows up to the electronically conducting electrode.

In some cases, the electronic conductivity is high enough to make the material pervious to oxygen. Therefore, the previous technique is no longer accurate and the partial conductivities may then be obtained from measurement of oxygen permeation rates.

4.3.2. Measurements of Oxygen Permeation Rates

Let us consider an oxide electrolyte having almost pure ionic conductivity ($t_i \sim 1$), located in a gradient of oxygen partial pressure. The oxygen flow through the material is induced by its electronic conductivity and obtained by integration of equation (30):

$$J_{O_2} = \frac{kT}{16qFe} \int_{pO_2(I)}^{pO_2(II)} \sigma_e \, dpO_2 \tag{31}$$

where e is the sample thickness.

It is emphasized that the value of the electronic conductivity σ_e may be derived from measurement of the oxygen permeation rate J_{O_2}, once a convenient model has been chosen to describe the variations of σ_e versus pO_2.[48]

Various experimental setups have thus been developed. Fouletier et al.[50] have used a galvanic cell similar to that shown in Fig. 24. Air streams in the lower compartment while an inert gas is used in the upper one (Fig. 26). The gas circulating mainly includes two electrochemical pumps (EP1, EP2) and two oxygen gages (OG1, OG2). The principle of the apparatus is described elsewhere.[58] Two measurement procedures have been perfected. In the first,[59] the oxygen partial pressure pO_2^* in the inert gas supplied to the cell is fixed with the aid of the electrochemical pump EP1 and read by the oxygen gage OG1, while the oxygen gage OG2 reads the oxygen partial pressure pO_2^{**} at the gas outlet. The permeation rate of oxygen through the sample is then given by[59]

$$J_{O_2} = \frac{D}{22.4} \cdot \frac{pO_2^{**} - pO_2^*}{pO_2^{**}} \cdot \frac{1}{S} \tag{32}$$

where D is the flow rate of the inert gas and S the permeation area of the sample.

The procedure developed by Fernandez[60] consists of decreasing the oxygen partial pressure pO_2^{**} with the aid of the electrochemical pump EP2, until $pO_2^{**} = pO_2^*$. The oxygen permeation rate through the electrolyte is then equal to the quantity of oxygen extracted by the pump and is given by

$$J_{O_2} = I/4F \tag{33}$$

where I is the reduction current in the electrochemical pump EP2.

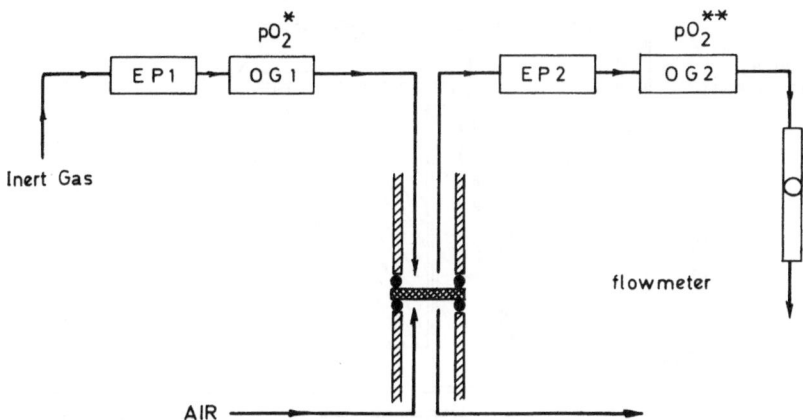

FIGURE 26. Experimental setup for oxygen permeation rate measurement developed by Kleitz and co-workers.[59]

Another experimental setup has been perfected by Iwase and Mori[61] using gas-tight electrolyte tubes, closed at one end and filled with pure oxygen under atmospheric pressure (Fig. 27). The oxygen partial pressure of the surrounding atmosphere is fixed by N_2-O_2 mixtures, with variable N_2/O_2 ratio. The oxygen permeating through the electrolyte induces a decrease of the total pressure inside the tube, which is read by a mercury manometer. However, the pressure of pure oxygen is kept at a constant atmospheric value by adjusting a mercury leveller. Hence, the oxygen permeation rate is given by

$$J_{O_2} = \frac{\pi r^2 P}{RT} \frac{1}{S} \frac{\Delta h}{\Delta t} \tag{34}$$

where r is the inner radius of the mercury manometer, P the atmospheric pressure, R the gas constant, S the effective permeation area of the tube, and Δh the change of mercury level during time Δt.

Finally, permeation rate measurements at high temperature have also been reported.[62] In this device, described in detail elsewhere,[10] a high-temperature furnace, the heating elements of which are made of partially stabilized zirconia,[63] allows an electrolyte tube to be heated up to 2100 K (Fig. 28). A water vapor–argon mixture is injected into the electrolyte tube and the oxygen partial pressure $pO_2(I)$ in the inner part of the tube is fixed by the water

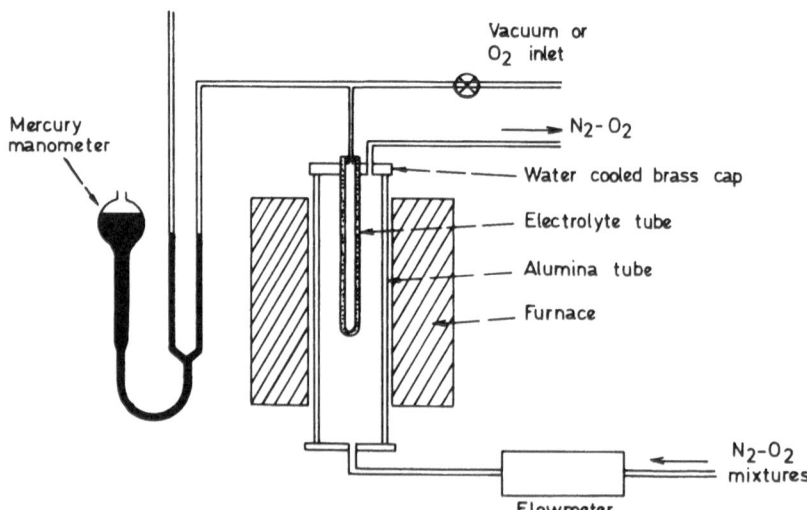

FIGURE 27. Apparatus for oxygen permeation rate measurement used by Iwase and Mori.[61]

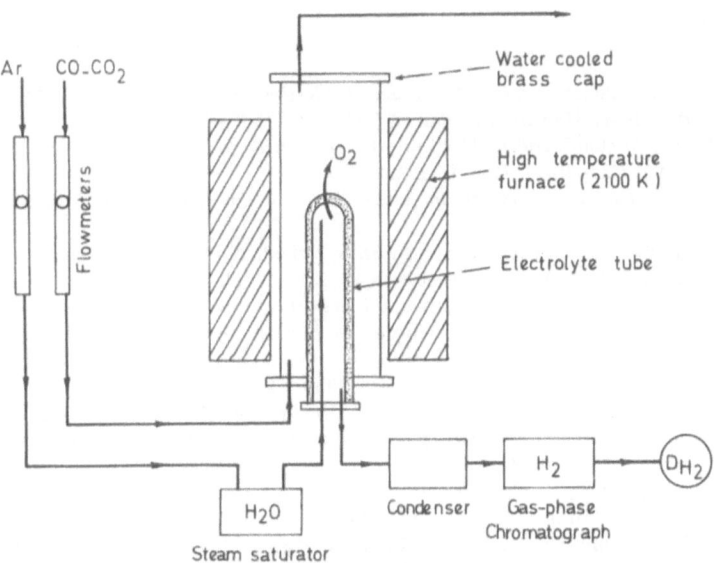

FIGURE 28. Gas circuit for oxygen permeation rate measurement on electrolyte tubes at high temperature.[10]

dissociation equilibrium occurring at high temperature. The outer compartment is fed with an equimolar CO–CO_2 mixture, so that the oxygen partial pressure $pO_2(II)$ in this compartment is much lower than $pO_2(I)$. The oxygen pressure gradient thus provided allows the oxygen arising from the steam decomposition to diffuse through the electrolyte. The permeation rate of oxygen is then derived from the amount of excess hydrogen evolved from the electrolyte tube and measured with the aid of a gas-phase chromatograph. Indeed, owing to stoichiometric considerations, under stationary conditions, the oxygen permeation rate is given by[62]

$$J_{O_2} = D_{H_2}/2S \tag{35}$$

where D_{H_2} is the flow of evolved hydrogen and S the effective permeation area of the electrolyte.

It should be noted that the last two procedures for oxygen permeation measurement are performed on electrolyte tubes so that the results are not relevant for isothermal samples, like the procedure developed by Fouletier *et al.*[50] This implies that the effective permeation area S of the tube must be defined accurately before experiments.[64]

REFERENCES

1. R.J. Brook, W.L. Pelzman, and F.A. Kröger, *J. Electrochem. Soc.* **118**, 185 (1971).
2. J. Maluenda, Thesis, University of Paris 6 (1979).
3. K.J. Euler, R. Kirchhof, and H. Metzendorf, *Mat. Chem.* **4**, 611 (1979).
4. K. Shahi and S. Chandra, *J. Phys. C* **8**, 2255 (1975).
5. S. Pizzini, in: *Fast Ion Transport in Solids* (W. Van Gool, ed.), p. 461, North-Holland, Amsterdam (1973).
6. F.A. Kröger, *The Chemistry of Imperfect Crystals*, 2nd ed., North-Holland, Amsterdam (1974).
7. R.G. Linford and S. Hackwood, *Chem. Rev.* **81**, 327 (1981).
8. B. Scrosati, G. Germano, and G. Pistoia, *J. Electrochem. Soc.* **118**, 86 (1971).
9. J.S. McKechnie, L.D.S. Turner, C.A. Vincent, M. Lazzari, and B. Scrosati, *J. Chem. Educ.* **55**, 418 (1978).
10. B. Calès, Ph.D. Thesis, University of Orléans, France (1983).
11. N.H. Chan, R.K. Sharma, and D.M. Smyth, *J. Electrochem. Soc.* **128**, 1762 (1981).
12. W. Crawford Dunlap, in: *Methods of Experimental Physics* (K. Lark Horovitz and V.A. Johnson, eds.), Vol. 6, Part B, p. 32, Academic Press, New York (1959).
13. I. Warshawsky, *Rev. Sci. Instrum.* **26**, 711 (1955).
14. L.J. Van der Pauw, Philips Res. Rep. **13**, 1 (1958).
15. D. Panis, Thesis, University of Orléans, France (1976).
16. A. Casalot, Ph.D. Thesis, University of Bordeaux, France (1968); J.P. Bonet, Ph.D. Thesis, University of Bordeaux, France (1980).
17. J.P. Loup and A.M. Anthony, *Rev. Int. Hautes Temp. Refract.* **1**, 15 (1964).
18. J.E. Bauerle, *J. Phys. Chem. Solids* **30**, 2657 (1969).
19. R. Fletcher and M.J.D. Powell, *Computer J.* **5**, 163 (1963).
20. P. Abélard, Ph.D. Thesis, University of Orléans (1983).
21. C.J.F. Böttcher and P. Bordewijk, *Theory of Electrical Polarization*, Vol. 3, Elsevier, Amsterdam (1978).
22. T. Wong and M. Brodwin, *Solid State Commun.* **36**, 503 (1980).
23. W. Schirmacher, *Solid State Commun.* **39**, 893 (1981).
24. V. Srivastana and M. Chaturvedi, *Z. Phys. B* **48**, 351 (1982).
25. M.I. Chekunaev and V.N. Fleurov, *J. Phys. C* **17**, 2917 (1984).
26. L.M. Levinson (ed.), *Grain Boundary Phenomena in Electronic Ceramics, Advances in Ceramics*, Vol. 1, The American Ceramic Society, Columbus, Ohio (1981).
27. D.V. Lan, *J. Appl. Phys.* **45**, 3023 (1974).
28. A.D. Franklin, *J. Am. Ceram. Soc.* **58**, 465 (1975).
29. E. Schouler, Ph.D. Thesis, University of Grenoble (1979).
30. E. Warburg, *Ann. Phys.* (*Leipzig*) **6**, 125 (1901).
31. A.H. Sharbough and S. Roberts, in Ref. 12, p. 1.
32. F. I. Mopsik, *Rev. Sci. Instrum.* **55**, 79 (1984).
33. C. Tubandt, *Z. Anorg. Allgem. Chem.* **115**, 105 (1920).
34. C. Tubandt and A.B. Lidiard, in: *Handbuch der Physik* (S. Flugge, ed.), Vol. 20, p. 246, Springer-Verlag, Berlin (1957).
35. F. Morin, *Solid State Ionics* **12**, 407 (1984).
36. M. Gauthier, M. Duclot, A. Hammou, and C. Déportes, *J. Solid State Chem.* **9**, 15 (1974).
37. M. Duclot and C. Déportes, *J. Solid State Chem.* **30**, 231 (1979).
38. F. Bénière, in: *Physics of Electrolytes* (J. Hladik, ed.), Academic Press, New York (1972).
39. L. Heyne, *J. Electrochim. Acta* **15**, 1251 (1970).
40. W.L. Worrell, *Top. Appl. Phys.* **21**, 143 (1977).
41. J.B. Wagner, in: *Electrode Processes in Solid State Ionics* (M. Kleitz and J. Dupuy, eds.), p. 185, D. Reidel, Dordrecht (1976).

42. C.Z. Wagner, *Elektrochem.* **4**, 60 (1956).
43. H. Rickert, *Angew. Chem., Int. Ed. Engl.* **17**, 37 (1978).
44. D.O. Raleigh, in: *Progress in Solid State Chemistry* (H. Reiss, ed.), p. 3, North-Holland, Amsterdam (1963).
45. J.H. Kennedy, *J. Electrochem. Soc.* **124**, 865 (1977).
46. J.W. Patterson, E.C. Bogren, and R.A. Rapp, *J. Electrochem. Soc.* **114**, 752 (1967).
47. L. Heyne, *Mass Transport in Oxides*, National Bureau of Standard, Special Publication No. 296, p. 149 (1968).
48. H.L. Tuller, in: *Nonstoichiometric Oxides* (O.T. Sørensen, ed.), p. 271, Academic Press, New York (1981).
49. P. Kofstad, *Nonstoichiometry Diffusion and Electrical Conductivity in Binary Metal Oxides*, Wiley Interscience, New York (1972).
50. J. Fouletier, P. Fabry, and M. Kleitz, *J. Electrochem. Soc.* **123**, 204 (1976).
51. O. Kubaschewski, E.LL. Evans, and C.B. Alcock, *Metallurgical Thermochemistry*, 4th ed., Pergamon Press, London (1967).
52. K. Schwerdtfeger and E.T. Turkdogan, in: *Physico-chemical Measurements in Metals Research* (R.A. Rapp, ed.), Vol. 4, Part 1, p. 321, Wiley, New York (1970).
53. P. Fabry, M. Kleitz, and C. Déportes, *J. Solid State Chem.* **5**, 1 (1972).
54. P. Fabry, Ph.D. Thesis, University of Grenoble, France (1976).
55. H.J. de Bruin, A.F. Moodie, and C.E. Warble, *Gold Bull.* **5**, 62 (1972).
56. H.J. de Bruin, A.F. Moodie, and C.E. Warble, *J. Mater. Sci.* **7**, 909 (1972).
57. J. Fouletier, Ph.D. Thesis, University of Grenoble, France (1976).
58. M. Gauthier, A. Belanger, Y. Meas, and M. Kleitz, in: *Solid Electrolytes* (P. Hagenmuller and W. van Gool, eds.), p. 497, Academic Press, New York (1978).
59. M. Kleitz, E. Fernandez, J.F. Fouletier, and P. Fabry, in: *Science and Technology of Zirconia* (A.H. Heuer and L.W. Hobbs, eds.), p. 349, The American Ceramic Society, Columbus, Ohio (1981).
60. E. Fernandez, Ph.D. Thesis, University of Grenoble, France (1980).
61. M. Iwase and T. Mori, *Metal. Trans.* **9B**, 365 (1978).
62. B. Calès and J.F. Baumard, *J. Mater. Sci.* **17**, 3243 (1982).
63. A.M. Anthony, in: Ref. 59, p. 437.
64. A. Ounalli, Thesis, University of Orléans, France (1981).

III

THERMAL DIFFUSIVITY

10

The Apparatus for Thermal Diffusivity Measurement by the Laser Pulse Method

K.D. MAGLIĆ and R.E. TAYLOR

1. INTRODUCTION

The laser pulse or laser flash method is a popular and productive method for measuring thermal diffusivity. It can be applied to a variety of solid materials in the 100 to 3300 K temperature range and 1×10^{-7} to 1×10^{-3} $m^2\,s^{-1}$ thermal diffusivity range. The method needs small specimens, 6 to 16 mm in diameter and thickness from a fraction to a few millimeters.

Although the method is primarily designed for characterization of homogeneous, isotropic, opaque materials, it has been successfully applied in the study of disperse nonhomogeneous materials, materials with two-dimensional anisotropy and multidirectional anisotropy, layered materials and translucent materials. The layered arrangement enabled the study of different temperature-sensitive materials, thin films, or other materials which would be very difficult to measure using other transport property measurement methods. During the past few decades, the method has been the major instrument in generating the thermal diffusivity data published in the scientific literature.

2. PHYSICAL MODEL AND MATHEMATICAL INTERPRETATION

The physical model is founded on the thermal behavior of an adiabatically insulated infinite slab initially at constant temperature whose one side has

K.D. MAGLIĆ • Boris Kidrič Institute of Nuclear Sciences, Institute of Thermal Engineering and Energy Research, 11001 Belgrade, Yugoslavia. R.E. TAYLOR • Thermophysical Properties Research Laboratory, Purdue University, West Lafayette, Indiana 47906, USA.

been subjected to a short pulse of energy. The model assumes:

1. One-dimensional heat flow.
2. No heat loss from the slab surfaces.
3. Uniform pulse absorption at the front surface.
4. Very short pulse duration compared with the thermal response of the slab.
5. Absorption of the pulse energy in a very thin layer.
6. Homogeneity and isotropy of the slab material.
7. Property invariance with temperature within experimental conditions.

Assumptions (6) and (7) reduce the general heat diffusion equation

$$\rho C_p \frac{\partial T}{\partial t} = \frac{\partial}{\partial x}\left(\lambda \frac{\partial T}{\partial x}\right) \tag{1}$$

to the simpler form

$$\frac{\partial T}{\partial t} = a \frac{\partial^2 T}{\partial x^2} \tag{2}$$

which contains the parameter

$$a = \lambda/\rho C_p \tag{3}$$

representing the thermal diffusivity of the slab material (in $m^2\,s^{-1}$). The other notation includes λ, the thermal conductivity (in $J\,m^{-1}\,K^{-1}\,s^{-1}$), ρ, the density (in $g\,m^{-3}$), and C_p, the specific heat capacity (in $J\,g^{-1}\,K^{-1}$). Quantities T and t denote the temperature and time, respectively.

The boundary conditions resulting from assumption (2) yield

$$\frac{\partial T(0, t)}{\partial x} = \frac{\partial T(L, t)}{\partial x} = 0, \qquad t > 0 \tag{4}$$

The general solution of equation (2) giving temperature T at time t in position x within a slab of thickness L takes the form[1]

$$T(x, t) = \frac{1}{L}\int_0^L f(x)\,dx + \frac{2}{L}\sum_{n=1}^{\infty} \exp\left(-n^2\pi^2 at/L^2\right)$$

$$\times \cos\frac{n\pi x}{L}\int_0^L f(x)\cos\frac{n\pi x}{L}\,dx \tag{5}$$

where function $f(x)$ represents the temperature distribution in the slab caused by a short pulse of laser light energy Q absorbed in a thin surface layer g. The initial conditions defining temperature distribution are:

$$f(x) = Q/\rho C_p g, \qquad 0 \leq x \leq g$$

$$f(x) = 0, \qquad g < x \leq L \tag{6}$$

For these initial conditions, expression (5) takes the form

$$T(x, t) = \frac{Q}{\rho C_p L} \left| 1 + 2 \sum_{n=1}^{\infty} \exp(-n^2 \pi^2 a t / L^2) \cos \frac{n\pi x}{L} \sin \frac{n\pi g}{L} \bigg/ \frac{n\pi g}{L} \right| \tag{7}$$

which, for a sufficiently small ratio g/L typical for opaque solids, allows the approximation $\sin(n\pi g/L) \simeq n\pi g/L$, leading to the following expression for the temperature at the slab rear face:

$$T(L, t) = \frac{Q}{\rho C_p L} \left| 1 + 2 \sum_{n=1}^{\infty} (-1)^n \exp(-n^2 \pi^2 a t / L^2) \right| \tag{8}$$

After infinite time the rear-face temperature will reach its maximum value:

$$T_{L\,max} = Q/\rho C_p L \tag{9}$$

For a slab initially at reference temperature T_r, the relative increase in its rear-face temperature $V(L, t)$ is then given by

$$V(L, t) = \frac{T(L, t) - T_r}{T_{L\,max} - T_r} = 1 + 2 \sum_{n=1}^{\infty} (-1)^n \exp(-n^2 \pi^2 a t / L^2) \tag{10}$$

establishing a relation between the percent rise in the rear-face temperature and the material thermal diffusivity.

This, in fact, is the analysis proposed originally by Parker et al. in 1961[1] when the method was established. Considering the level of measurement instrumentation and the recording/analysis facilities in the early sixties, as well as their measurements near room temperature only, they rightfully proposed the point corresponding to 50% of the temperature rise to its maximum value as the reference point for the calculation of thermal diffusivity. Then $V = 0.5$ and $\pi^2 a t_{0.5}/L^2 = 1.37$, leading to the expression

$$a = 0.1388 \frac{L^2}{t_{0.5}} \qquad (11)$$

where $t_{0.5}$ is the time from the initiation of the pulse until the rear-face temperature rise reaches one-half of its maximum value. Actually, one may use any percent rise:

$$a = K_x L^2 / t_x \qquad (12)$$

where K_x is a constant corresponding to an x percent rise and t_x is the elapsed time to an x percent rise.

Departures from the model assumptions encountered in a real experiment may distort the temperature response compared with the form given by equation (10). In the calculation of thermal diffusivity, it is inadvisable to rely on a single point from the whole curve, as is the case with expression (11). Expression (12) permits the use of information contained in the whole rear-face response curve, indicating deformations caused by departures from specific assumptions, or allowing the use of the portion of the response curve which has been least affected by interferences. The use of expression (12) is therefore strongly recommended, but this subject will be dealt with in the section devoted to the error analysis. Table 1 contains the values of K_x calculated for various x percent rises.

TABLE 1
Values of K_x in Equation (12)

$x(\%)$	K_x	$x(\%)$	K_x
10	0.066108	60	0.162236
20	0.084251	66 (2/3)	0.181067
25 (1/4)	0.092725	70	0.191874
30	0.101213	75 (3/4)	0.210493
33 (1/3)	0.106976	80	0.233200
40	0.118960	90	0.303520
50 (1/2)	0.138785		

3. MEASUREMENT SYSTEM

The pulse thermal diffusivity measurement system consists of a specimen, an experimental chamber, a furnace with power supply, energy pulse source, control and measuring circuits and instruments. Details of the system are described and discussed in the following subsections.

3.1. Specimen

The specimen usually has the shape of a disk. Other specimen shapes may be imposed by the nature of the material and the process of its manufacturing. The diameter of disk-shaped specimens is usually between 6 and 16 mm, depending on the material and the temperature range of measurement. Measurements at high temperature favor smaller specimen diameters, which are also more convenient for the pulse sources with poorer energy-beam uniformity. The less homogeneous materials might be better suited with larger-diameter specimens, which in turn require the energy pulse sources with good energy-beam spatial distribution.

The specimen thickness is defined by the nature of the material, its thermal diffusivity, and the fabrication process. It can vary between a fraction of a millimeter and 6 to 7 mm. In order to reduce the influence of the finite length and shape of a real energy pulse and be close to the model assumption (4), the specimen thickness should be such as to keep the pulse length close to 1% of $t_{0.5}$. Deviations will require corrections to expression (12), which will be discussed in other sections.

3.2. Experimental Chamber

The experimental chamber has to accommodate the specimen in a holder, provide the desired vacuum or other ambient conditions, ensure stable temperature boundary conditions, and provide access for energy pulses, contactless temperature measurement and control and other circuits.

The specimen holder is used to locate the specimen within the most homogeneous temperature zone of a furnace or cryostat, and to ensure the model assumption (1) mentioned above of one-dimensional heat flow in the specimen. The specimen holder material should be compatible with the specimen material to prevent possible chemical reaction between them and to account for thermal expansion over the whole temperature range of measurement.

In electroresistive heated furnaces, specimen holders made of machinable ceramics give good service for many specimen materials to 1300 K. Higher temperatures require a good grade alumina or some other refractory oxide ceramic material. When oxide ceramics translucent to the laser light are used, care should be taken to cover such specimen holders with a cap made of refractory metal foils, such as tantalum. Graphite or carbon-based composite specimens require graphite specimen holders which are adequate for the whole temperature range of work. In high-frequency induction furnaces, the specimen is frequently supported by the graphite susceptor body itself, unless a separating sleeve must be inserted to prevent possible chemical contamination of the specimen.

In operation below room temperature, the specimen holder made of copper is connected to a liquid-nitrogen-cooled heat sink, provided with a heater for realization of intermediate equilibrium temperatures.

Thermal contact between the specimen and its holder should be as poor as possible. In this sense, a holder made of refractory metal, where the specimen is held by three pointed screws of the same material,[2] is very suitable. Furnaces with a vertical axis, where the specimen rests on a narrow supporting ring,[3] are also very suitable in the same respect.

Orientation of the furnace axis is a compromise between the simplicity of the pulse supply and optical systems, and the ease and convenience of supporting the specimen. Furnaces with a horizontal axis are very convenient from the first point of view, having the laser, furnace, and transient temperature response detector located along the same horizontal axis, easily accessible for optical alignment. The specimen holder must then either be supported by refractory metal rods[2] or supported at one end by the supporting tube.[4] The horizontal-axis systems have to cope with adverse requirements of good specimen fixation within the system and the need for its poor thermal contact with the holder.

Vertical-axis systems have the advantage of the specimen resting on its seat by gravity, with hardly any radial thermal contact with the holder. The pulse and transient-temperature detection system are more intricate, as the laser pulse beam must be deflected by a prism or a laser mirror. Problems are similar in focusing the specimen rear-side thermal radiation onto the optical transient-temperature detector-sensitive area. These difficulties are significantly offset by much smaller problems due to thermal expansion of the supporting system at very high temperatures.

The experimental chambers are usually made of stainless steel, polished inside for easy evacuation and maintenance, and for lower absorption of thermal radiation from the furnace. For fabrication reasons, chambers are usually cylindrical, with optical and other ports at the flat bases, except the vacuum connections. All the walls are water-cooled to enable maintenance of a stable temperature at experimental boundaries. Figure 1 shows a sketch of the experimental chamber of the Boris Kidrič Institute thermal diffusivity apparatus, operating successfully between 250 and 2000 K.[4]

3.3. Furnace and Power Supply

It is desirable that the furnace establishes the desired specimen temperature within the shortest period of time, enabling a series of measurements

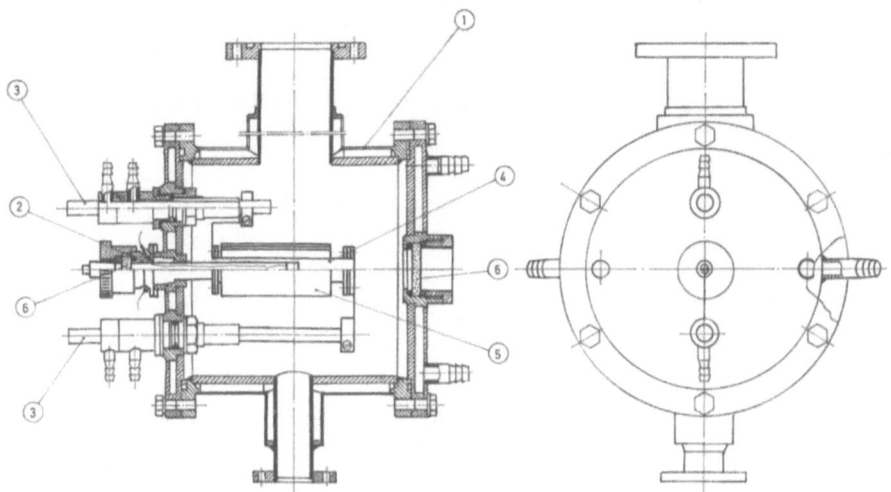

FIGURE 1. Experimental chamber with furnace and specimen holder: 1—experimental chamber, 2—specimen holder, 3—current lead-ins, 4—furnace, 5—radiation shields, 6—windows.

at different temperatures. As the measurement itself takes less than a second, the productivity of the method depends upon the thermal inertia of the furnace and specimen holder/specimen assembly. With a good, low-inertia furnace, the latter dominates the efficiency of the heating system at low temperatures, its role decreasing rapidly as the radiation heat transfer begins to prevail near 1000 K.

The competitive low-inertia furnace systems are the thin-walled electro-resistive furnaces and the high-frequency induction furnaces. At very high temperatures, direct specimen heating by a continuous CO_2 laser has proved successful.[5,6]

An exploded view of the left end of a low-inertia electroresistive-type furnace made of tantalum foil 0.05 mm thick is shown in Fig. 2. The furnace consists of a tantalum tube made by spot-welding a piece of tantalum foil along its length, stretched between two pairs of tapered rings fixed to re-entrant water-cooled heavy current lead-ins. The copper lead-ins supply direct current to the furnace and support it at the same time. The ends of the furnace tube are slit at both sides approximately 5 mm deep at 5 mm separations, these fins fitting between the tapered rings for electrical and mechanical contacts. The rings might be made of refractory metals, or stainless steel for operation to 1800 K. Tungsten or molybdenum strips connect the rings to the copper clamps on the lead-ins.

Three or more radiation shields made of refractory metal or stainless-steel foil are held around the furnace, supported by a rod mounted onto the same base plate.

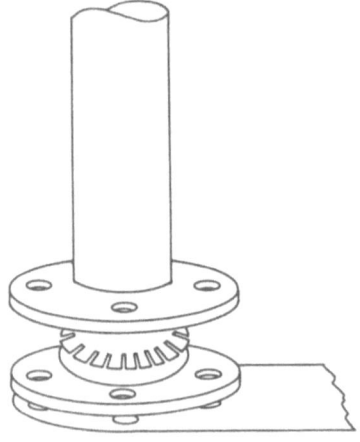

FIGURE 2. Fixation of the tantalum-foil furnace tube to current supply.

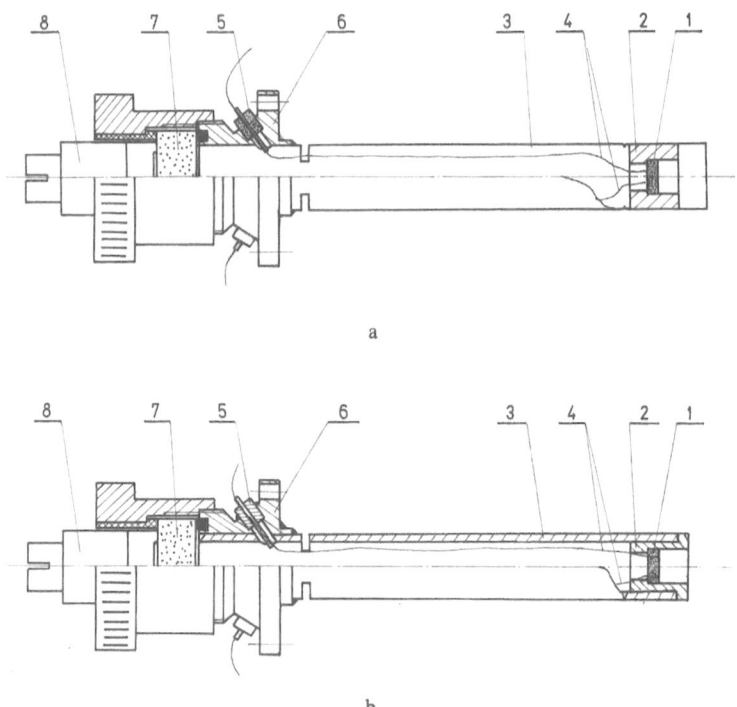

FIGURE 3. Specimen holder assemblies, (a) medium-temperature assembly, (b) high-temperature assembly: 1—specimen, 2—specimen holder, 3—supporting tube, 4—reference temperature thermocouple, 5—thermocouple lead-in, 6—brass flange, 7—window, 8—optical detector.

The specimen holders and their supporting tubes are shown in Fig. 3. For operation to 1300 K, a stainless-steel, thin-walled supporting tube with a machinable ceramic specimen holder (Fig. 3a) is adequate. For operation to 2000 K, a recrystallized high-grade alumina tube and a specimen holder of the same material (Fig. 3b) are necessary. Due to significant translucence of this material, it is necessary to sheath the specimen holder and the tube with a closely fitting tantalum foil sleeve. For the graphite base specimens and measurements at high temperatures, the supporting tube and the specimen holder must be made of graphite. Irrespective of the supporting tube variant, each of them must have a hole or a cut close to the cold end that will enable evacuation of the tube volume simultaneously with evacuation of the experimental chamber.

The supporting tubes are either brazed or cemented with a vacuum-tight cement into a brass flange, equipped with thermocouple wire lead-ins and an aperture for optical transient-temperature detection. The change of the specimen is effected by replacing one specimen support by another, i.e., by unscrewing and tightening four bolts which fix the flange to the baseplate.

Another apparatus with a horizontal electroresistive low-inertia furnace operating to 2800 K is described by Taylor.[2] Such an apparatus is now commercially available (Theta Industries, Port Washington, New York, USA), as is the new, highly automated apparatus with a low-inertia vertical furnace (Holometrix, Inc., Cambridge, Massachusetts, USA). Works describing productive high-temperature apparatuses operating to 3000 K, with a vertical-axis high-frequency induction furnace[3] or an electroresistive-type furnace[7] have been published. A furnace with a multiple specimen holder for twenty samples has been successfully used.[8]

3.4. Energy Pulse Source

The energy pulse supply has to provide a short pulse of energy and deliver it to a specimen inside a vacuum or a controlled-atmosphere enclosure. The specimen rear-face temperature increase due to the energy pulse should not exceed a few degrees, in order to maintain detector linearity and to comply with assumption (7) of the material property invariance. The energy falling on the specimen surface should also not cause excessive heating, which might damage the specimen material and its surface layer. The spatial distribution of the energy pulse should be reasonably uniform. The necessary degree of uniformity will depend on the technique of measuring transient-temperature response, and shall be discussed elsewhere.

As the energy pulse supply, ruby or Nd doped glass lasers are presently most frequently used. Any other source of energy capable of delivering the amount of energy sufficient to raise the specimen temperature 1 to 2 K, with a pulse not exceeding 1 ms, should also be adequate, provided the pulse shape is regular and reproducible. The term "regular" implies a form easily describable mathematically, such as square wave, sawtooth, etc.[9] Reproducibility of the pulse shape and its duration are significant in the measurements, where the pulse width and shape are determined occasionally for the whole set of thermal diffusivity measurements, and where the pulse width represents more than 1% of the characteristic time, $t_{0.5}$. In experiments where the pulse shape is analyzed in each experiment, many difficulties of this type have been overcome by allowing the pulse to be located on the time scale with a high degree of accuracy as described elsewhere.[10]

It is difficult to adjust the laser pulse energy for each specimen to suit the requirements of permissible specimen-temperature increase and to maintain the energy spatial distribution within acceptable limits. Thus it is customary to maintain the pulse energy at a level granting a fair uniformity of the beam profile and to employ techniques for its attenuation and even better homogenization. Attenuation has been achieved in different ways. Passage of the pulse through a cuvette with plane-parallel walls filled with a laser-light-absorbing fluid (such as copper sulfate solution) absorbs a portion of the pulse energy proportional to its concentration. Spreading of the most homogeneous portion of the pulse by means of a lens attenuates the pulse energy and may improve the beam homogeneity. The same effect has been obtained by introducing a fine stainless-steel wire mesh between the laser and the specimen, scattering the coherent laser light and mixing it at the same time.[11] There are many practical solutions. It is only important that their final effect provides an acceptable total specimen-temperature rise, and acceptable homogeneity of the laser-beam energy distribution.

3.5. Control and Measuring Circuits and Instruments

It follows from equation (12) that significant measured quantities include specimen thickness and the relative change in the specimen rear-side temperature as a function of time. The specimen temperature must also be known accurately, as well as the vacuum or inert gas-pressure conditions throughout the experiment. Measuring circuits and instruments have to provide information on the specimen temperature and its transient caused by the pulse, as well as the gas pressure inside the experimental chamber. The thickness of the specimen is measured before its installation in the holder at room temperature, and its magnitude at different temperatures is obtained from thermal-expansion measured or published data. The task of the control circuits is to attain and maintain the desired temperature during the experiment and to realize its change to the next temperature level in the most convenient way.

The specimen reference temperature is measured by a thermocouple or pyrometer. Thin thermocouple wires, 0.1 mm or less in diameter, are recommended for the least thermal disturbance of the specimen and true temperature readings. The short duration of the experiments prevents their severe contamination and change of calibration, which might be otherwise significant due to their dimensions. With metallic specimens, thermocouple wires should be directly welded to the specimen rear face, preferably with the specimen as an intermediate material between them ("intrinsic" or open thermocouple). With specimen materials which do not permit welding or which might contaminate the thermocouple, the hot junction is inserted in a hole in the specimen holder by means of a 1-mm-diameter twin-bore alumina tube. When the specimen holder material is incompatible with the thermocouple material, 1-mm-diameter sheathed thermocouples should be used. To 1300 K, the K- or E-type thermocouples should be used. The S or R types are recommended for temperatures to 1800 K. Bare wires need to be changed frequently. Sheathed thermocouples can last much longer subject to recalibration. Irrespective of the thermocouple type, for a given furnace, the reference thermocouple output should be calibrated *in situ* against a calibrated thermocouple positioned within the specimen as shown in Fig. 4, over the whole temperature range of measurement. This should be done with two specimens at least, having emissivities close to the opposite extremes. Sometimes, corrections of the measured reference temperatures will be necessary.

In contactless measurements of the specimen's absolute temperature, either a two- or multicolor pyrometer is used, or the specimen holder is provided with a blackbody-resembling cavity with a precisely known emissivity. In the latter case, this temperature should be calibrated against temperature readings on a specimen provided with a similar blackbody cavity.

A digital voltmeter or digital thermometer capable of measuring temperature with 0.1-K resolution is adequate for the measurement of reference temperature with thermocouple detectors. A pyrometer with 1-K resolution will be good for most measurements above 1500 K.

For the measurement of the small temperature rise, function requirements are very modest. It is only necessary to have a sensitive detector whose output is linear over temperature changes of a few degrees. In principle, all temperature detectors sufficiently sensitive over a given temperature range can be used. They include thermocouples, infrared photoresistors, silicon photodiodes, photomultipliers, etc., provided their response characteristics conform with

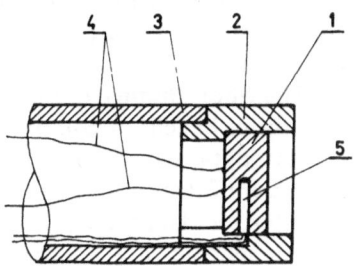

FIGURE 4. Schematics of calibrating an intrinsic reference temperature thermocouple: 1—dummy specimen, 2—specimen holder, 3—supporting tube, 4—intrinsic thermocouple wires, 5—calibrated thermocouple.

the needs of the particular measurement. Thermocouples introduce the highest degree of uncertainty into dynamic temperature measurement. In part it is due to the thermocouple bead inertia and heat conduction along thermocouple leads.[12-15] Dissimilar thermocouple pairs give emf problems.[14] In addition, thermocouples might be more sensitive to minor inhomogeneities in the energy pulse beam[11,16] because the measurements are made over a localized area of the sample. All these effects deform the rear-face response curve and directly influence the accuracy of thermal diffusivity measurement. Infrared photodetectors (lead sulfide, indium antimonide and arsenide, three-component alloys, etc.) are very convenient in respect of their dynamics and the range of application characteristics. Below 500 K, they require infrared transparent windows and/or lenses; above this temperature quartz glass windows are satisfactory. The variety of these detectors, including a silicon photodiode for the visible portion of spectra, provide means for appropriate transient temperature detection from low temperatures to 3000 K. The signal-to-noise ratio can be kept relatively low, requiring relatively little filtering and smoothing of the resulting signal. A significant convenience of these optical detectors is their ability to integrate a major portion of the rear-face temperature response, which can compensate for relatively large inhomogeneities in the energy pulse beam.[11,16-18] These detectors are therefore strongly recommended, except below room temperature and in special studies of nonhomogeneous media.[19] Commercial pyrometers can be also used, but they are often more complex than this purpose would require.

The measuring circuits of both thermocouple and photoresistor detectors are identical, except that the former must have an adjustable battery-operated DC source to offset the DC component of the thermocouple emf corresponding to the specimen reference temperature, while the latter has an adjustable resistance bridge scheme where the photoresistor represents one of the resistance branches. The rest of the measurement circuit consists of an amplifier and a recording and measuring instrument.

One data-processing procedure involves a digital oscilloscope with memory as the recording and measuring instrument. In this case, $t_{0.5}$ or some other percent characteristic time is read off the oscilloscope record, and thermal diffusivity values are calculated using equation (12) and the K_x values from Table 1. The necessary corrections are also calculated from recorded temperature transients. The overall measurement uncertainty in such an approach can be below 4%.

The whole thermal diffusivity experiment can be, and initially was, conducted without benefit of computers. However, with the availability of relatively low-cost data acquisition and computers, most researchers use them. The remainder of this section and all of Section 4 is devoted to describing a typical computerized data acquisition and processing system. Such a system is presented in a scheme (Fig. 5) showing the schematic of the Boris Kidrič Institute

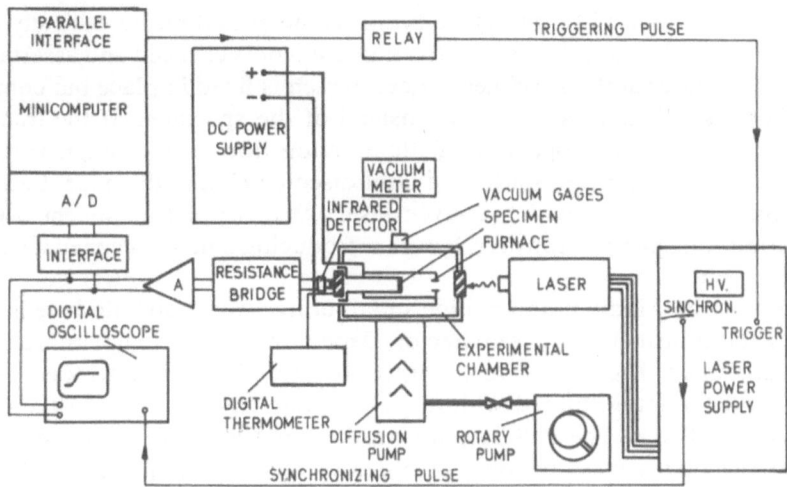

FIGURE 5. Schematics of the BKI laser-pulse apparatus.

(BKI) laser-pulse apparatus. The digital oscilloscope serves here as a parallel monitoring instrument. In such a scheme, all the operations are effected under computer control, except changing the measurement reference temperature. This operation can also be automated, but it should not lead to any significant improvement. It should, in a way, reduce the experimenter's freedom to make decisions on the spot, imposed by the measurement course.

Operations conducted under computer control include the whole measurement, calculation of thermal diffusivity, calculation of corrections, demonstration of their effects, and printing out the results. The correction procedures account for radiation heat losses, finite pulse duration, and pulse shape effect.

The computer-controlled approach insures higher reproducibility and reliability of results, reducing measurement uncertainty to below 2.5%.

4. MEASUREMENT PROCEDURE

The measurement procedure[20] described briefly in this section has been developed for thermal diffusivity measurements with the apparatus illustrated in Figs. 1–5.

A horizontal, low-inertia thin-walled tantalum furnace is heated directly by a bank of three highly stabilized 100-amp 10-volt power supplies connected in parallel. The specimen in its holder, equipped with a reference temperature measuring thermocouple, is introduced axially into the experimental chamber, which is then evacuated to 10^{-5} millibars, or better. The purpose of evacuation is to protect the specimen and furnace parts from oxidation and to eliminate convective heat exchange between the specimen and the ambient.

The ruby laser of 15-mm pulse-beam diameter is directed optically onto the front face of the specimen. Its proper alignment is checked and ascertained by firing the laser at the specimen holder, which is fixed in place but contains a disk of laser-light-sensitive paper instead of the specimen. If the trace on the paper shows inhomogeneity of illumination due to laser alignment, the position of the laser head is adjusted until discoloration of the paper becomes as uniform as possible with the given laser. At this point alignment usually holds for a long period; however, before introducing a new specimen it should always be checked.

Photovoltaic indium-antimonide and photoresistive lead-sulfide cells were used as the transient response detector, depending on the temperature range of measurement. The former, in conjunction with a CaF window, was used to 500 K, and the latter, paired with a quartz-glass window, covered the range from 500 to 2000 K. The PbS detector was held in the specimen holder flange next to the quartz-glass window in order to receive the maximum thermal radiation from the specimen's rear face. A CaF lens was used to concentrate the energy for the InSb detector. A glass filter, opaque to the laser-light wavelength, was positioned in series with the window to protect the photodetectors.

The detector constituted a branch of a resistance bridge. When the bridge was balanced, the DC level corresponding to a given reference temperature was nulled and the maximum amplification of the temperature transient signal was achieved. A differential amplifier was used to amplify the bridge unbalance signal. Signal amplification was necessary for data transmission to the computer's data acquisition system and for its adjustment to the A/D convertor input range of −5 to +5 volts. In order to keep the signal-to-noise ratio close to its optimum, amplification was usually between 50 and 200 times. The amplifier's frequency range was usually set two orders of magnitude above the effective frequency of the temperature rise curve, i.e., if the rise curve frequency corresponded to 10 Hz, the amplifier frequency was equal to or more than 1 kHz. The amplified signal was fed to the computer, and fed simultaneously to a digital oscilloscope for immediate insight into the quality of the obtained transient response. After processing, the data are stored in data files on the disk. The measurement was initiated on a graphics terminal when the specimen had attained the desired, stable temperature and the laser capacitor bank had been charged manually. Activation of the laser trigger is effected via a line from the 16-bit parallel interface, which operates a relay commanding the trigger. Activation of the laser starts data acquisition and processing in real time, including its presentation in the form of tables or diagrams at the terminal printer or plotter. At each reference temperature, 3–5 measurements are repeated. The whole operation takes about five minutes.

The program consists of three basic programs named LASER, FILTER, and DIFUS. The programs enable data acquisition and experiment control in real time, smoothing of collected data and processing, and calculation of thermal diffusivity. The latter is calculated at 17 points according to equation (12) for different values of percent rise of the normalized transient response curve.

4.1. The Program LASER

This program effects acquisition of experimental data and controls the experiment in real time. It has a major role in the realization of the measurement procedure. Figure 6 shows the data collected in one experiment. The initial flat part of the record is the signal base corresponding to the specimen reference temperature. The rising portion of the sharp peak denotes the laser flash. The peak is caused either by electrical induction due to the capacitor bank discharge or penetration of the laser light into the photodetector. After the transient portion of the signal, another flat part corresponding to the stabilized specimen temperature is observed. The number of data points collected during each of these three characteristic parts of the temperature transient is fixed. The sampling frequencies, F_X, vary depending upon the specimen material and its dimensions. During the first interval, t_1, 50 data points are collected. This number is adequate in order to reconstruct the signal base and its slope. Three-hundred data points are collected during the rising part of the transient, t_2. In the last portion, t_3, 150 data points are recorded.

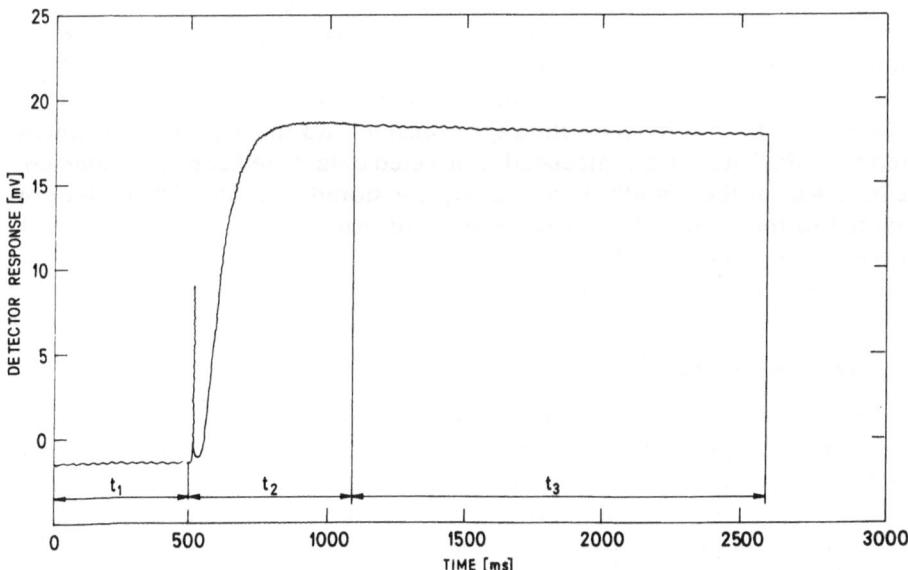

FIGURE 6. Experimental data collected in an experiment.

In the initial program dialogue the experiment identification and data storage file are given. The data are stored in a nonformated data file, with direct access. Its purpose is to retain all data which are unchanged in a series of experiments on a single specimen (date, specimen, thickness, channel, and amplification), adding in subsequent experiments only the experiment number and reference temperature. The sampling rate is selected by choosing one of six combinations of sampling frequencies, F_s, and the lengths of intervals t_2 and t_3, with F_{s1} and t_1 being fixed at 100 Hz and 500 microseconds, respectively. Selection of the combination depends upon thermal diffusivity and the dimensions of the specimen, as the transient must reach its maximum within interval t_2 and the rising portion of the curve should be described with the aid of as many points as possible.

The next program section performs acquisition of the initial set of data and establishes whether the offset of detector signal DC component has been accomplished successfully or will require further adjustment.

The following program block effects the start of the experiment, its control in real time, and the acquisition of data:

- Acquisition of 50 data points defining the signal base with frequency F_{s1}.
- Laser firing, initiated by a call to the sub-program via parallel interface and relay, followed by the acquisition of 300 data points with frequency F_{s2}.
- Acquisition of the last 150 data points with frequency F_{s3}.

In the next program step, collected signal data are converted into millivolts and the signal base slope determined. If the furnace temperature has not been stationary, the collected transient response data will be corrected for the slope observed in the signal base data. Graphical presentation of the primary collected data and their corrected values gives insight into the data set from which thermal diffusivity will be calculated. Corrected data, together with parameters defined within the initialization section, are stored in a file. Other files are devoted to the presentation of corrected and primary data on the plotter. If the sampling frequencies have not been chosen adequately, or for some other reason the experiment must be repeated, it is done at this stage.

4.2. The Program FILTER

The purpose of this program is to define the moment of laser discharge, to remove the peak or other disturbances on the record caused by laser discharge, and to effect smoothing of raw data delivered by the program LASER. Figure 7 shows the initial portion of the recorded transient signal, including the laser discharge disturbance. The end of interval t_1 marks the moment the command call for laser firing has been issued. Data acquisition starts immediately but, due to the delay in the control line, the laser discharge

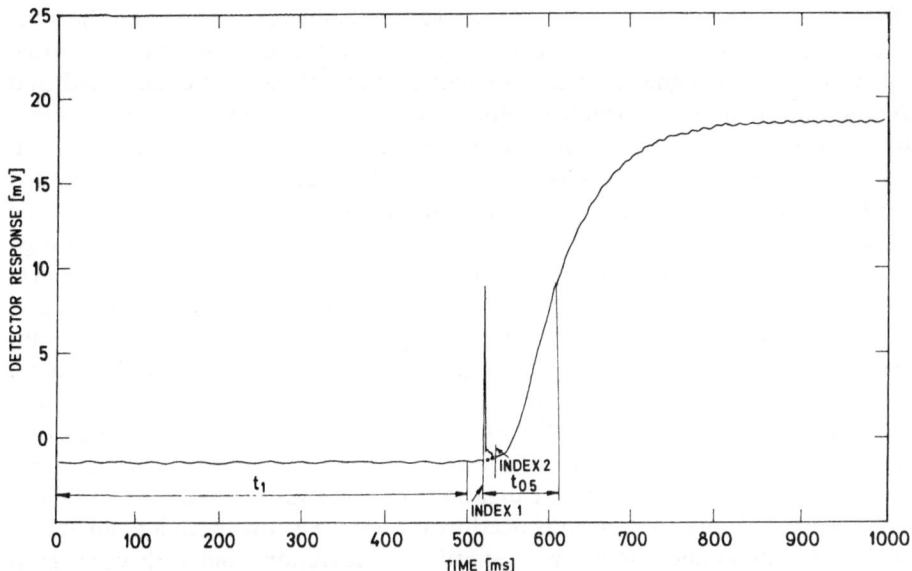

FIGURE 7. Initial portion of the recorded transient signal.

occurs some 20 microseconds later, marked in Fig. 7 by a sharp peak. In the program presentation, the experimenter denotes with a marker the moment of laser discharge (INDEX 1) as well as the first data point after disappearance of the effects of disturbance caused by the laser discharge (INDEX 2). Starting with the mean value of the signal base data (data collected during interval t_1) extrapolated to the moment of laser discharge (first rising point on the disturbance peak), linear extrapolation is made to the first point after the disturbance marked by the INDEX 2 marker, thus reconstructing the portion of the signal which was affected by the laser discharge disturbance. Subsequently, the signal is smoothed using the SMOOT routine. The number of smoothings is determined from the noise on the signal and the experimenter's insight into their effect. Processed data, starting with the point of the laser discharge defined as above and terminating with the end of interval t_3, is also stored.

4.3. The Program DIFUS

The program DIFUS calculates thermal diffusivity, effects graphical presentation of calculated values, and prints the table of results including all relevant data.

The program reads the processed data, the normalized data for the theoretical response curve, and the values of K_x corresponding to different percents of rise, which are stored in their respective files. For normalization of the measurement data, its maximum value is determined first and then the initial minimum value, equal to the signal base mean value. Next the time equivalent to the signal reaching its half-maximum value is determined and the measurement data normalized. In the following step, both the experimental and theoretical normalized curves are presented in the same graph. Deviation between these indicate departures in the real experiment from the assumed model conditions, the most frequent being those due to heat losses and the effect of finite pulse duration, or the nonhomogeneity of the laser-beam energy distribution. For accurate thermal diffusivity measurement, each of the effects must be recognized and accounted for by an adequate correction procedure.

Thermal diffusivity a is calculated for 17 values of percent temperature rise, between 10% and 90% of the maximum, and the data are stored. Graphical presentation of this data normalized at thermal diffusivity calculated for a 50% rise enables a check of the quality of thermal diffusivity values. In the case of no deviation from the model assumptions, all the values would be the same irrespective of the percent rise. Deviation from the ratio $a_x/a_{0.5} = 1$ indicates a necessity to apply correction procedures. Figure 8 shows such a comparison of data for two experiments: with significant deviations and with very small deviations from the ideal experimental conditions. The program has separate correction procedures for the heat loss and for the finite pulse time effect, to be discussed in the following section.

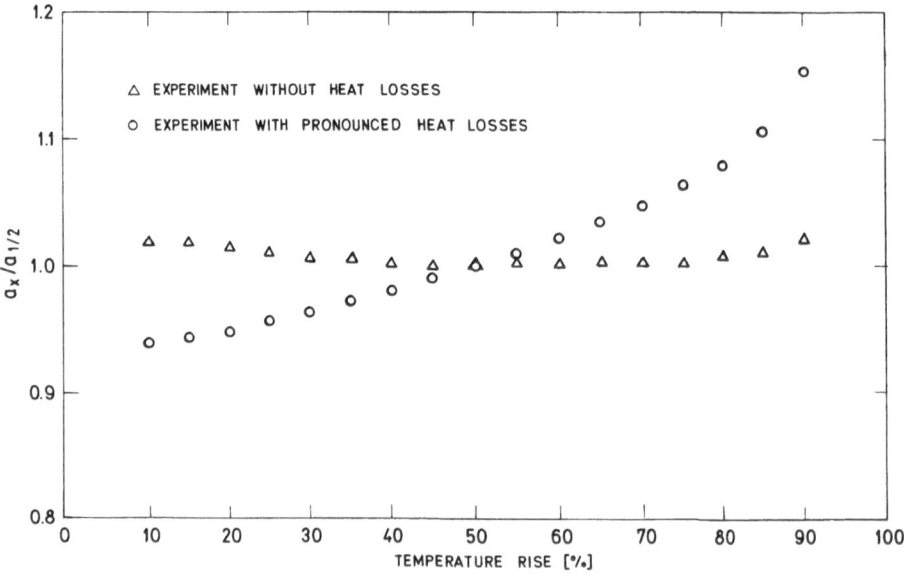

FIGURE 8. Thermal diffusivity calculated at different percents: effect of deviations from the model conditions.

After application of the necessary correction procedures, the table of final thermal diffusivity values is produced together with all pertinent information for a given measurement.

The described measurement procedure is being accomplished with a system consisting of a PDP11/34 minicomputer with 256KB memory and two 2.5MB disks, an AD11-KT fast acquisition system, DR11-K parallel interface, and VT-55 graphics terminal with access to printer and plotter facilities.

5. ESTIMATE OF ERRORS AND CORRECTIONS

In thermal diffusivity measurements using the laser pulse method, errors can be classified into two main groups. The first group, measurement or determinate errors, includes errors associated with determining the effective thickness of the specimen and errors associated with measuring the time during which the rear face attains a certain percent of the maximum rise. The second group, nonmeasurement or nondeterminate errors, includes errors due to departure from assumed model conditions.

5.1. Measurement Errors

For the ideal case, thermal diffusivity is given by equation (12) and the measurement error defined by

$$\frac{\Delta a}{a} = \frac{\Delta t_{0.5}}{t_{0.5}} + 2\frac{\Delta L_0}{L_0} \tag{13}$$

assuming $t_{0.5}$ to be a mean representative of times used in the calculation of thermal diffusivity.

The response time of the total circuit and its significance for measuring characteristic percent time was stressed earlier. Except when thermocouples are used, it may be assumed that the time constants of the measurement chain do not introduce any disturbance in the time domain as they are a few orders of magnitude less than the time constant of the transient signal itself. The main source of error here is the finite resolution of the data acquisition system, i.e., the sampling frequency of the analog signal. For typical characteristic half-times equal to 100 ms and for a sampling frequency of 1 kHz, the relative error in the half-time determination is 1% at most. For shorter characteristic times a system with higher sampling frequency must be used so that the error due to this source is kept below 1%. The uncertainty in defining the position of a point on the signal in the presence of electronic noise can be largely compensated by mathematical smoothing procedures. If the signal-to-noise ratio is low, errors from electronic noise can exceed the uncertainty due to limitations of the acquisition system.

Relative errors in determining the specimen thickness, L_0, measured on a specimen 1-mm thick with a micrometer of 0.005-mm resolution will be 1%. The typical thicknesses of 2.5 mm will give a much lower error, and *vice versa*. As it is difficult to machine a thin specimen of nonmagnetic solid material with plane-parallel faces, further error from this source could contribute to the total error in the determination of L_0. As regards considerations of the measurement error in thermal diffusivity determination employing computer-assisted measurement, one may assume that the total error from this source can be kept well below 2%, particularly if, at each temperature, measurements are repeated and mean values of a few experiments are taken. The use of thermocouples as transient temperature detectors may increase measurement uncertainty, so they should be used only whenever absolutely necessary, i.e., below 250 K, where infrared optical detectors become inefficient. In their use, precautions should be taken to avoid sources of error that are likely to be met in dynamic temperature measurements. For this purpose, an intrinsic thermocouple usually represents the best choice.

In the case of intrinsic thermocouples, the response time (time to reach 95% of the steady-state value) can be defined[13] as

$$t_{95} = \frac{25}{\pi} \frac{D_T^2}{a_s} \frac{\lambda_T}{\lambda_s} \tag{14}$$

where D is the thermocouple wire diameter (in m), a_s is thermal diffusivity of the specimen (in $m^2 s^{-1}$), while λ_T and λ_s are the thermal conductivities of the thermocouple and specimen materials (in $W m^{-1} K^{-1}$), respectively. Thus, a small-diameter thermocouple of low-conductivity material attached to a specimen of a high-conductivity and high-diffusivity material yields the fastest response time. Equation (14) is misleading in that it can postulate that the thermocouple response is a smooth rise. Actually, the response is a step change, followed by an exponential rise to the final value. This behavior is best represented by the equation

$$\frac{T_T - T_0}{T_\infty - T_0} = 1 - (1 - p)e^{p^2}t^* \text{ Erfc } (pt^*) \tag{15}$$

where T_0 and T_∞ are shown in Fig. 9, t^* is dimensionless time ($t^* = 4a_s t / D_T^2$), and p is approximated by $1/(1 + 0.667 \lambda_T/\lambda_s)$. In order to obtain the fastest response, a small-diameter thermocouple wire of an alloy having low thermal conductivity attached to a substrate of high thermal diffusivity should be used. For example, a 25-μm constantan wire on a copper substrate requires 3 μs to reach 95% of the steady-state value. However, for the converse of this example, i.e., 25-μm copper wire on a constantan substrate, it is found that 15 ms are required to reach 95% of the steady-state value. This is 5000-fold slower than

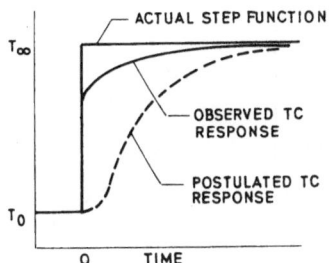

FIGURE 9. Dynamic response of an intrinsic thermocouple: comparison with model.

the first example. Thus the proper selection of materials, based upon their thermal properties and geometries, is essential for accurate measurement of transient responses using intrinsic thermocouples.[14]

Equations (14) and (15) relate to the minimum response time possible for a thermocouple. Proper attachment of the thermocouple is important since, if the thermocouple is attached poorly to the specimen, the effective response time can be much longer. The preferred method for electrical conducting materials is to spot-weld intrinsic thermocouples, i.e., nonbeaded couples where each leg is attached independently to the specimen about 1 mm apart. For electrical insulators, where spot welding is not feasible, it may be possible to spring-load the thermocouple against the back surface. For materials with low diffusivity values, it may be preferable to spot-weld thermocouples onto a thin, high thermal conductivity, metallic sheet and spring-load or paste this sheet onto the specimen. Metal-epoxy and graphite pastes have been used successfully to bond layers together. This eliminates the problem of using thermocouples of relatively high diffusivity to measure specimens of materials of low thermal diffusivity, which can lead to very large response times [equation (14)].

5.2. Nonmeasurement Errors

As defined earlier, the nonmeasurement errors include errors caused by deviation of real experimental, initial, and boundary conditions from those assumed by the model. The model assumes that:

1. The laser pulse has a triangular, delta-function shape and that its duration is much shorter than the characteristic transient response half-time.
2. Specimen boundaries are adiabatic and no heat exchange occurs after its exposure to the laser pulse.
3. The laser pulse is distributed homogeneously over the whole front surface of the specimen, resulting in one-dimensional heat flow through the specimen.

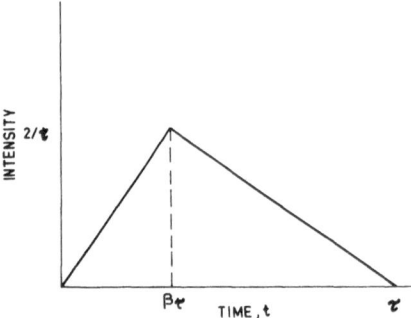

FIGURE 10. Postulated laser energy pulse shape.

The presence of errors is observed by a comparison of experimental and theoretical curves, i.e., by comparing thermal diffusivity values calculated from equation (12) at 25, 50, and 75% of the rear-face maximum temperature rise. If the values at 25, 50, and 75% of ΔT_{max} lie within $\pm 2\%$, the overall accuracy is probably within $\pm 5\%$ at the half-time. If the α values lie outside of this range, the response curve should be analyzed further to see if finite-pulse time, radiation heat loss, or nonuniform heating are present.

Finite-pulse time effects can usually be corrected by using the equation

$$a = K_1 L_0^2/(K_2 t_x - t_p) \tag{16}$$

For this to be valid, the energy pulse must be represented by a triangle of duration τ and time to maximum intensity of $\beta\tau$ as shown in Fig. 10. The pulse shape of the energy pulse for the laser should be determined using an optical detector which can detect the laser pulse as opposed to the flash lamp pulse. From this pulse shape β and t_p are obtained. Values of K_1 and K_2 for various values of β are given in Table 2 for correcting $\alpha_{0.5}$.

TABLE 2
Finite-Pulse Time Factors

β	K_1	K_2
0.15	0.34844	2.5106
0.28	0.31550	2.2730
0.29	0.31110	2.2454
0.30	0.30648	2.2375
0.50	0.27057	1.9496

Figure 11 shows a typical shape[20] of the laser pulse recorded with an optical detector directed at the specimen front surface. For this shape, numerical factors are $\beta = 0.50$, $K_1 = 0.27057$, and $K_2 = 1.9496$; and the duration $t_p = 1$ ms. Equation (16) is incorporated into the program DIFUS for the finite time effect correction. Procedures for correcting for finite pulse time effect are also elaborated in the literature.[9,10,21-25]

Thermal-radiation heat-loss effects are most readily determined from the temperature of the specimen and the rear-face temperature response after $4t_{1/2}$. The recommended procedure is to plot the experimental values of $\Delta T/\Delta T_{max}$ versus $t/t_{1/2}$ along with the values for the theoretical model. Some numbers for the theoretical model are given in Table 3.

A plot of the normalized experimental data and the theoretical model can be prepared readily on-line with a computer-based data acquisition system or by preparing graphs using the tabulated values of $\Delta T/\Delta T_{max}$ and $t/t_{1/2}$ and plotting the corresponding experimental data at several percent levels of the rise. All normalized experimental curves must pass through $\Delta T/\Delta T_{max} = 0.5$ at $t/t_{1/2} = 1.0$. Therefore, calculations including the 25 to 35% and 66.67 to 80% ranges are required to compare the experimental data with the theoretical curve.

Heat-loss corrections should be based on using both Clark and Taylor rise-curve data[26] and Cowan cooling-curve data.[27] Rise-curve data are especially affected by nonuniform heating effects. Cooling-curve corrections

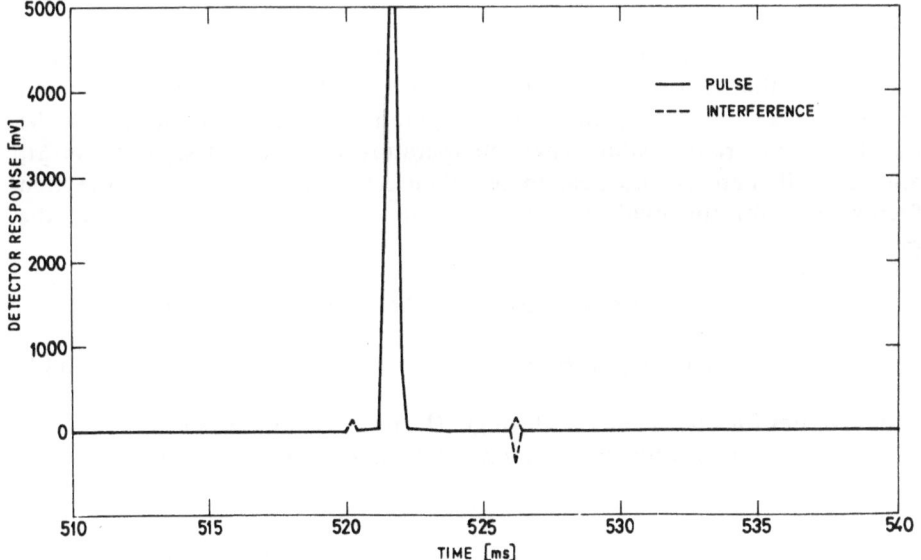

FIGURE 11. Recorded laser pulse shape.

TABLE 3
Values of Normalized Temperature Versus Time for Theoretical Model

$\Delta T/\Delta T_{max}$	$t/t_{1/2}$	$\Delta T/\Delta T_{max}$	$t/t_{1/2}$
0	0	0.7555	1.5331
0.0117	0.2920	0.7787	1.6061
0.1248	0.5110	0.7997	1.6791
0.1814	0.5840	0.8187	1.7521
0.2409	0.6570	0.8359	1.8251
0.3006	0.7300	0.8515	1.8981
0.3587	0.8030	0.8656	1.9711
0.4140	0.8760	0.8900	2.1171
0.4660	0.9490	0.9099	2.2631
0.5000	1.0000	0.9262	2.4091
0.5587	1.0951	0.9454	2.6281
0.5995	1.1681	0.9669	2.9931
0.6369	1.2411	0.9865	3.6502
0.6709	1.3141	0.9950	4.3802
0.7019	1.3871	0.9982	5.1102
0.7300	1.4601		

are affected by conduction losses to the holders in addition to radiation losses from the surfaces. Thus, the errors in the correction procedures are affected by different phenomena and comparison of diffusivity values corrected by the two procedures is useful in determining the presence or absence of these phenomena.

In order to use the Cowan cooling-curve corrections, one must determine the ratio of the net rise-time temperature values at times which are five and ten times the experimental half-time temperature value, to the net rise at the half-time temperature value. These temperature ratios are designated as Δt_5 and Δt_{10}. If there are no heat losses, then $\Delta t_5 = \Delta t_{10} = 2.0$. The correction factor (K_C) for the five and ten half-time cases are calculated from the polynomial fits:

$$K_C = A + B(\Delta t) + C(\Delta t)^2 + D(\Delta t)^3 + E(\Delta t)^4 + F(\Delta t)^5$$
$$+ G(\Delta t)^6 + H(\Delta t)^7 \tag{17}$$

where values for coefficients A through H are given in Table 4.

Corrected values for diffusivity are calculated from the relation

$$a_{corrected} = a_{0.5} K_C / 0.13885 \tag{18}$$

where $a_{0.5}$ is the uncorrected diffusivity value calculated using the experimental half-time.

TABLE 4
Coefficients for Cowan Corrections

	5 Half-times	10 Half-times
A	−0.1037162	0.054825246
B	1.239040	0.16697761
C	3.974433	−0.28603437
D	6.888738	0.28356337
E	−6.804883	−0.13403286
F	3.856663	0.024077586
G	−1.167799	0.0
H	0.1465332	0.0

Heat loss corrections based on the Clark and Taylor rise-curve data also employ the ratio technique.[26] For the $t_{0.75}/t_{0.25}$ ratio, i.e., the time to reach 75% of the maximum divided by the time to reach 25% of the maximum, the ideal value is 2.272. This ratio must be determined from the experimental data. Then the correction factor, K_R, can be calculated from the equation

$$K_R = -0.3461467 + 0.361578 \, (t_{0.75}/t_{0.25}) - 0.06520543 \, (t_{0.75}/t_{0.25})^2 \qquad (19)$$

The corrected value of thermal diffusivity at the half-time will be

$$a_{\text{corrected}} = a_{0.5} K_R / 0.13885 \qquad (20)$$

Corrections using many other ratios can be also used.

Examples of normalized plots for experiments which approximate the ideal case, in which there is a finite pulse time effect, and in which there are radiation heat losses, are shown in Figs. 12, 13, and 14, respectively. Various procedures for correcting for these effects are also given in the literature.[2,9,10,22-25]

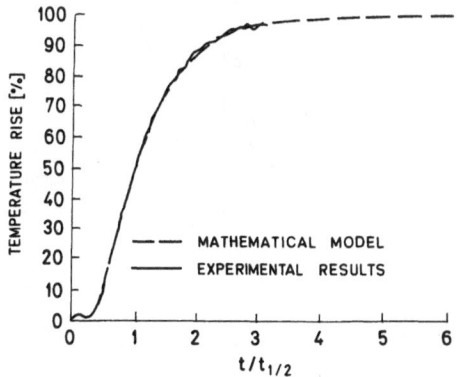

FIGURE 12. Comparison of the mathematical model with experimental results obtained under conditions approximating an ideal case.

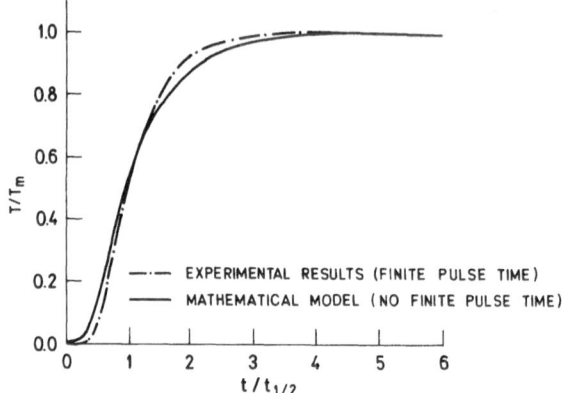

FIGURE 13. Comparison of the mathematical model with experimental results showing the finite pulse-time effect.

Nonuniform heating effects also cause deviations in the reduced experimental curve from the model because of two-dimensional heat flow. Since there are a variety of nonuniform heating cases, there are a variety of deviations. Hot-center cases approximate the radiation heat loss example. Cold-center cases result in the rear-face temperature continuing to rise significantly after $4t_{1/2}$. Nonuniform heating may arise from the nature of the energy pulse or by nonuniform absorption on the front surface of the specimen. The former case must be eliminated by altering the energy source, while the latter may be eliminated by adding an absorbing layer and using two-layer mathematics.[21,28] When the specimen thickness exceeds 1 mm, the effect of nonuniform heating can be largely compensated by integration of transient temperature response

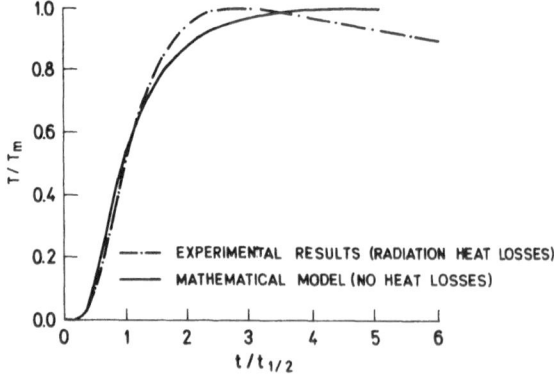

FIGURE 14. Comparison of the mathematical model with experimental results showing the effect of heat losses.

from the whole rear face[16-18]; see Section 3.5. With very thin specimens, nonhomogeneous heating causes significant deviations in the recorded signal, and every precaution should be taken to homogenize the pulse energy distribution.

The foregoing argument about nonmeasurement errors indicates that they can be largely compensated by adequate corrections and careful experimentation. In an ideal case, with sufficiently thick specimens, well-chosen sampling rates, favorable signal-to-noise ratio, and a qualified and experienced staff, the measurement error can be reduced to less than 1%. The nonmeasurement error can be also reduced below 1%, giving the overall uncertainty between 1 and 2%. Very thin specimens with surfaces not perfectly plane-parallel, or naturally rough surfaces, or in any way departing from the isotropic, opaque, and homogeneous material concept, will never permit such a high accuracy. For the least measurement uncertainty, efforts should be made to get acquainted with all possible sources of measurement and nonmeasurement errors and reduce their effect. Increasing the number of measurements and conducting measurements with more than one specimen thickness, etc., will help to attain increased reliability and reduced measurement uncertainty.

6. EXTENSION OF THE METHOD TO NONIDEAL MATERIALS

6.1. Anisotropic Media

The usual flash diffusivity experiment involves uniform exposure of the front surface and measuring the resulting temperature rise on the rear surface, and assumes one-dimensional heat flow. In the case of determining the diffusivity values of anisotropic materials, significant errors in the determination of the thermal diffusivity exist if the directions of the principal axis and of the imposed heat flux are different (He et al.[29]).

To measure the radial diffusivity a_r (parallel to the plane faces) of an anisotropic material, a radial-heat-flow pulsed method was developed by Chu, Taylor, and Donaldson.[30] The heat pulse is uniform over a circular area of radius r_p smaller than the sample radius R. The temperature is measured on the rear face at locations $r = 0$, T_1, and $r = r_m$, T_2. The radial diffusivity is deduced from the temperatures at these two locations, and the axial diffusivity a_z (perpendicular to the plane faces of the samples) is determined from the temperature at the center. The data reduction, based on a mathematical solution of Donaldson,[31] taking into account heat losses, was established by Chu, Taylor, and Donaldson.[32]

With the same configuration, Amazouz et al.[33] and Batsale and Degiovanni[34] employed a data-reduction procedure based on the use of order −1 and 0 temporal moments of the T_2/T_1 ratio, and applied it to relatively thin samples of two-dimensional carbon/epoxy composites. The method allows determination of the axial diffusivity and radial diffusivity a_r (parallel to the plane faces).

6.2. Layered Composites and Coatings

Layered composites are important for a number of cases: a second layer may be placed on a homogeneous sample to limit temperature rises or prevent radiation from the energy source penetrating the sample; one may wish to study coatings on a surface; one may study contact conductance between layers; one may form a cell to conduct measurements on liquids or powders; or one may add a high-conductivity layer to the rear surface in order to increase the temperature-response characteristics of thermocouples.

The application of the flash technique to layered samples was studied by Lee,[35] Lee and Taylor,[36] and Lee, Donaldson, and Taylor.[37] They developed data-reduction methods which calculate the diffusivity of one layer of a two- or three-layer composite from the half-time $t_{0.5}$ measured in the conventional manner by the flash technique. Programs were also written and tested to compute the contact conductance between two layers whose thermal properties are known. Lee et al.[37] also established criteria for distinguishing between a resistive and a capacitive layer. The method has been used extensively since then by Taylor[38] and Stark and Taylor.[39] Lee, Donaldson, and Taylor[37] established criteria to use a simple model based on the assumption of one layer being capacitive (no thermal gradient); it was applicable to metal deposited on poor conductor substrates. James[40] has solved equations for the case of unequal heat losses from the front and rear surfaces of materials with widely different properties. Heat-loss corrections are generally within 10% of those obtained using standard procedures, which ignore differences in emissivity, diffusivity, and specific heats of the two layers.[41] Since heat-loss corrections are generally less than 20% using the standard heat-loss correction method, the latter procedure usually results in errors of less than 2%. Thus a rather complex solution need only be used when heat losses are larger than 20%. The case of several layers with thermal contact resistance and heat losses was also analyzed by Degiovanni et al.[42] and Sweet.[16] The methods have been applied by a number of researchers to ceramic coatings.[43-45] However, as mentioned by Taylor,[38] the flash method is not well-suited in the case of thin films with highly conductive materials coated on relatively thick, poorly conductive substrates.

6.3. Multidimensional Composites

A class of composite materials is composed of fiber reinforcements imbedded in a more-or-less homogeneous matrix. Such composites are becoming increasingly important in new technological applications. Major subgroups of this class of composites include metal matrix composites, whisker-reinforced ceramics, and carbon/carbon materials. The latter consist of arrays of graphite fibers lined up in one direction and imbedded in a matrix of arrays of graphite fiber lined up in more-or-less perpendicular directions, the spaces being filled with graphite. A variety of geometries have been fabricated, including one-dimensional, two-dimensional, three-dimensional, and multidimensional arrays depending on the orientation of the fibers. The graphite fibers have high thermal conductivity and diffusivity values, while the matrix materials have relatively low values compared with those of the fibers. Thus the fibers oriented in the direction of heat flow act as preferred paths for heat transfer. When such composites are subjected to an instantaneous heat pulse on one surface, the temperature wave is not planar. The results of thermal diffusivity methods and the applicability of the concept of diffusivity for such materials can legitimately be questioned.

However, Taylor et al.[46] measured the thermal diffusivity, specific heat, and thermal conductivity of a fine-weave carbon/carbon composite. The normalized rear-face temperature-response curves measured in the diffusivity experiment followed the theoretical curve very closely; this demonstrates that the diffusivity experiment was valid. The diffusivity values calculated at different percent rise times for a typical experiment were all $0.838 \pm 0.009 \text{ cm}^2 \text{ s}^{-1}$. Furthermore, the diffusivity values calculated from the measured conductivities and specific heats were within $\pm 2\%$ of the measured diffusivity values. Thus the concept of effective diffusivity obviously applies to such fine-weave fiber-reinforced composites.

In the case of fiber-reinforced composites where the weave structure is much coarser (a factor of three or more), the situations may be considerably different. This is analyzed in various papers,[38,47-49] which show that such assumptions may lead to inadequate results like time- and sample-thickness-dependent diffusivities. Experimental approaches were made by Lafond-Huot and Bransier[50] and Taylor.[38] The latter[38] was able to measure in situ the properties of the fiber and of the matrix. A method for determining the equivalent homogeneous medium is presented by Balageas.[51] Pujolá and Balageas[52] proposed a method involving the volume content of the reinforcement and the data obtained on samples of different thickness. This information can be used to determine the diffusivities of the reinforcement (reinforcement parallel to the heat flux), of the equivalent matrix (real matrix plus transverse reinforcement), of the equivalent homogeneous medium and the contact thermal resistance of the reinforcement/matrix interface. In this paper,[52] examples are given concerning three-directional reinforced carbon/carbon composites, showing major discrepancies between such materials and homogeneous media. For sufficiently thick samples, however, the equivalent homogeneous medium assumption may be acceptable (see, for instance, Whittaker et al.[53] and Balageas[51]).

6.4. Dispersed Composites

Kerrisk[54,55] derived criteria for the heterogeneity of dispersed composites that can be tolerated by the flash method, based on geometrical considerations. H.J. Lee and Taylor[56] and T.Y.R. Lee and Taylor[57] demonstrated that the flash method was applicable to materials with heterogeneities more than fifty times greater than Kerrisk's criteria. They also investigated the effects of ratios of various property values in addition to volume ratios. Particle-to-matrix diffusivity ratios of 0.48–1137, specific heat ratios of 0.04–1.16, and volume ratios of 0–0.34 were studied. Even up to 35 volume percent, the agreement between measured diffusivity values was consistent with experimental conductivity values. As a result of these studies it is evident that the flash method is applicable to a very wide range of dispersed composites. The case of large grain samples, where single grains are of the order of the sample thickness, has also been studied,[58] and worthwhile data can be obtained even under these very adverse conditions.

6.5. Semitransparent Media

Semitransparent media may be coated on the front surface to prevent laser-beam penetration and on the rear surface to prevent the infrared detector from viewing the sample, or large opaque layers may be attached. The effects of the layers or coatings can be handled just like layered sample cases described in a separate section. In general, the attachment of a metal layer to a relatively poorer conducting ceramic presents little difficulty as far as the effects of interfacial resistance is concerned. For example, attaching a tantalum layer to a plasma-sprayed ceramic will usually cause no problems. However, if materials of relatively high conductivity are joined together, the attachment becomes critical and great care must be taken in order to obtain reliable results. Thus attaching a metal to a diamond film requires excellent bonding.

More recently, Tischler et al.[59] proposed a data-reduction procedure based on a multidimensional Newton–Raphson iterative algorithm and using the whole thermogram. This way the diffusivity can be determined at the same time as the absorption coefficient for the spectral length of the source for which the material is assumed to be partially transparent (exponential-like absorption). The material is opaque for all other spectral lengths and just the first internal reflection on the rear face is taken into account.

6.6. Liquids

Liquids are treated as three-layer cases when the first and third layers are opaque solid layers with known thermal properties. Then the usual analytical solutions apply, including the James[40] method for heat losses involving layers with different properties. Alternatively, Otter and Vandevelde[60] gave an analytical expression for the rear-face temperature. If the thermal properties of the external solid layers and the thicknesses of the three layers are known, this expression is dependent on a single parameter related to the liquid diffusivity. An iterative determination is proposed using the half-rise point. Starting from finite-difference numerical solutions, an iterative method is also proposed for the case of uniform heat losses. In this case times $t_{1/2}$ and $5t_{1/2}$ are used, as in the Cowan method,[27] and the iterations concern simultaneously the X parameter and a heat-loss parameter.

In the case of molten semiconductors, Taylor et al.[61] conducted successful measurements up to 1300 K and 100 atm using encapsulated samples. The quartz used to encapsulate the samples was transparent to the laser and infrared detector, while these particular semiconductors were not. Taylor has also measured molten superalloys encapsulated in sapphire cups.

Batsale and Degiovanni[34] present a flash method in cylindrical geometry suitable for liquids. The liquid is sandwiched between two coaxial metal cylinders. The heat pulse is generated by a flash lamp located on the axis of the system and the temperature is measured on the outer face. A three-layer model, based on the quadrupole method,[62] is used. It is shown that the metal wall may be considered as purely capacitive.

7. FRONT-FACE METHODS

The aim of this chapter is to discuss the techniques and apparatus used with the conventional flash diffusivity technique, in which the front surface is irradiated and the resulting rear-face temperature measured. However, the authors realize that important and exciting developments using front-face irradiation and temperature sensing are taking place. Lepoutre[63] surveyed the measurements of thermal properties using photothermal techniques, and Balageas[17] presents a thorough discussion of the present status of front-face techniques.

REFERENCES

1. W.J. Parker, R.J. Jenkins, C.P. Butler, and G.L. Abbott, "Thermal Diffusivity Measurement Using the Flash Technique," *J. Appl. Phys.* 32(9), 1679–1684 (1961).
2. R.E. Taylor, "Critical Evaluation of Flash Method for Measuring Thermal Diffusivity," *Report PRF-6764*, National Science Technical Information Service, Springfield, VA 22151 (1973).
3. R. Taylor, "Construction of Apparatus for Heat Pulse Thermal Diffusivity Measurements from 300–3000 K," *J. Phys. E* 13, 1193–1199 (1980).
4. K.D. Maglić, N.Lj. Perović, and Z.P. Životić, "Thermal Diffusivity Measurements on Standard Reference Materials," *High Temp. High Pressures* 12, 555–560 (1980).

5. G.I. Daniliants, A.V. Yevseev, A.V. Kirillin, and K.A. Khodakov, "Experimental Study of Radiation and Transport Properties of High-Temperature Materials," Proc. Second Asian Thermophysical Properties Conference, Sept. 20-22, 1989, Saporo, Japan.

6. A.V. Kirillin, A.V. Evseev, and K.A. Khodakov, "Measurement of Temperature Conduction of Graphite Materials at Temperatures of 2000-3900 K," (in Russian), *Teplofizika Vysokih Temperatur* (in press, 1989).

7. A. Cezairliyan, T. Buba, and R. Taylor, "A High Temperature Laser Pulse Thermal Diffusivity Apparatus," *Int. J. Thermophys*, **10**, (in press) (1989).

8. R.U. Acton and J.A. Kahn, "Thermal Diffusivity—A Multispecimen Automatic Data Reduction Facility," in: *Proc. 10th Thermal Conductivity Conference*, Boston (1970).

9. R.E. Taylor and L.M. Clark, III, "Finite Pulse Time Effects in Flash Diffusivity Method," *High Temp. High Pressures* **6**, 65-71 (1974).

10. T. Azumi and Y. Takahashi, "Novel Finite Pulse-Width Correction in Flash Thermal Diffusivity Measurement," *Rev. Sci. Instrum.* **52**(9), 1411-1413 (1981).

11. A.A. Zolotuhin, Private communication (1988).

12. K.D. Maglić and B.S. Maršićanin, "Factors Affecting the Accuracy of Transient Response of Intrinsic Thermocouples in Thermal Diffusivity Measurements," *High Temp. High Pressures* **5**, 105-110 (1973).

13. C.D. Henning and R. Parker, "Transient Response of an Intrinsic Thermocouple," *J. Heat Transfer* **39**, 146 (1967).

14. R.C. Heckman, "Intrinsic Thermocouples in Thermal Diffusivity Experiments," in: *Proc. Seventh Symposium on Thermophysical Properties (ASME)* (A. Cezairliyan, ed.), pp. 155-159, ASME, New York (1977).

15. N.R. Ketner and J.V. Beck, "Surface Temperature Measurement Error," *J. Heat Transfer* **105**, 312-318 (1983).

16. J.N. Sweet, "Effect of Experimental Variables on Flash Thermal Diffusivity Data Analysis," in: *Thermal Conductivity 20* (D.P.H. Hasselman and J.R. Thomas, Jr., eds.), Plenum Press, New York, pp. 287-304 (1989).

17. D.L. Balageas, "Thermal Diffusivity Measurement by Pulsed Methods," *High Temp. High Pressures* **21**(1), 85-96 (1989).

18. R.E. Taylor and K.D. Maglić, "Pulse Method for Thermal Diffusivity Measurement," in: *Compendium of Thermophysical Property Measurement Methods 1* (K.D. Maglić, A. Cezairliyan, and V.E. Peletsky, eds.), pp. 305-336, Plenum Press, New York (1984).

19. R.E. Taylor, J. Jortner, and H. Groot, "Thermal Diffusivity of Fiber-Reinforced Composites Using the Laser Flash Technique," *Carbon* **23**(2), 215-222 (1985).

20. N.Lj. Perović, A.S. Dobrosavljević, and K.D. Maglić, "Laser Pulse Method for Thermal Diffusivity Measurement," *Boris Kidrič Institute Int. Report IBK-ITE 566*, Belgrade (1986).

21. R.E. Taylor, "Heat Pulse Diffusivity Measurements," *High Temp. High Pressures* **11**, 43 (1979).

22. J.A. Cape and G.W. Lehman, "Temperatures and Finite Pulse-Time Effects in the Flash Method for Measuring Thermal Diffusivity," *J. Appl. Phys.* **34**, 1909 (1963).

23. K.B. Larson and K. Koyama, "Correction for Finite-Pulse-Time Effects in Very Thin Samples Using the Flash Method of Measuring Thermal Diffusivity," *J. Appl. Phys.*, **38**, 465 (1967).

24. R.C. Heckman, "Error Analysis of the Flash Thermal Diffusivity Technique," in: *Thermal Conductivity 14* (P.G. Klemens and T.K. Chu, eds.), pp. 491-498, Plenum Press, New York (1974).

25. R.C. Heckman, "Finite Pulse-Time and Heat-Loss Effects in Pulse Thermal Diffusivity Measurement," *J. Appl. Phys.* **44**, 1455 (1973).

26. L.M. Clark, III and R.E. Taylor, "Radiation Loss in the Flash Method for Thermal Diffusivity," *J. Appl. Phys.* **46**, 714 (1975).

27. R.D. Cowan, "Pulse Method of Measuring Thermal Diffusivity at High Temperatures," *J. Appl. Phys.* **34**, 926 (1963).

28. K.B. Larson and K. Koyama, "Measurement of Thermal Diffusivity, Heat Capacity and Thermal Conductivity in Two-Layer Composite Samples by the Flash Method," in: *Proc. 5th Thermal Conductivity Conference*, pp. 1-B-1 to 1-B-24, University of Denver, Denver, CO (1965).

29. G.H. He, X.Z. Zhang, Z. Wei, S.Q. Dong, Z.Q. Di, and B.L. Zhou, "Suggestions Regarding Thermal Diffusivity Measurements on Pyrolitic Graphite and Pyrolitic Boron Nitride by the Laser Pulse Method," *Int. J. Thermophys.* **7**(4), 789-802 (1986).

30. F.I. Chu, R.E. Taylor, and A.B. Donaldson, "Thermal Diffusivity Measurements at High Temperatures by the Radial Flash Method," *J. Appl. Phys.* **51**, 336-341 (1980).

31. A.B. Donaldson, "Radial Conduction Effects in the Pulse Method of Measuring Thermal Diffusivity," *J. Appl. Phys.* **43**, 4226-4228 (1972).

32. F.I. Chu, R.E. Taylor, and A.B. Donaldson, "Flash Diffusivity Measurements at High Temperatures by the Axial Heat Flow Method," in: *Proc. Seventh Symposium on Thermophysical Properties, ASME 1977* (A. Cezairliyan, ed.), pp. 148-154, ASME, New York (1977).

33. M. Amazouz, C. Moyne, and A. Degiovanni, "Measurement of the Thermal Diffusivity of Anisotropic Materials," *High Temp. High Pressures* **19**, 37-41 (1987).

34. J.C. Batsale and A. Degiovanni, "Extension de la méthode 'flash' à deux cas particuliers: Les matériaux anisotropes et les liquides," *Proc. Rencontre Société Française des Thermiciens 88*, Limoges, CPM-14-1 (1988).

35. H.J. Lee, "Thermal Diffusivity of Layered and Dispersed Composites," Ph.D. Thesis, School of Mechanical Engineering, Purdue University, West Lafayette, Indiana (1975).

36. H.J. Lee and R.E. Taylor, "Determination of Thermophysical Properties of Layered Composites," in: *Thermal Conductivity 14* (P.G. Klemens and T.K. Chu, eds.), pp. 423-434, Plenum Press, New York (1976).

37. T.Y.R. Lee, A.B. Donaldson, and R.E. Taylor, Thermal Diffusivity of Layered Composites," in: *Thermal Conductivity 15* (V.V. Mirkovitch, ed.), pp. 135-148, Plenum Press, New York (1978).

38. R.E. Taylor, "Thermal Diffusivity of Composites," *High Temp. High Pressures* **15**, 299-309 (1983).

39. J.A. Stark and R.E. Taylor, "Determination of Thermal Transport Properties of Ammonium Perchlorate," *J. Propulsion* **1**(5), 409-410 (1985).

40. H. James, "Theory of Pulse Measurement of Thermal Diffusivity on Two-Layer Slabs," *High Temp. High Pressures* **17**, 481-496 (1985).

41. J.A. Stark, "Measurement of Thermal Transport Properties of Selected Solid Rocket Propellant Components," Master's Thesis, School of Mechanical Engineering, Purdue University, West Lafayette, Indiana (1984).

42. A. Degiovanni, A. Gery, M. Laurent, and G. Sinicki, "Attaque impulsionnelle appliquée à la mesure des résistances de contact et de la diffusivité thermique," *Entropie* **64**, 35-43 (1975).

43. L. Pawlowski, D. Lombard, A. Mahkia, C. Martin, and P. Fauchais, "Thermal Diffusivity of Arc Plasma Sprayed Zirconia Coatings," *High Temp. High Pressures* **16**, 347-359 (1986).

44. P. Morrel and R. Taylor, "Thermal Diffusivity of Thermal Barrier Coatings of ZrO_2 Stabilized with Y_2O_3," *High Temp. High Pressures* **17**, 79-88 (1985).

45. K. Inoue and K. Ohmura, "Measurement by Laser Flash Method of Thermal Diffusivity of Ceramics Coating," *Proc. 8th Jpn. Symp. on Thermophysical Properties*, pp. 145-148, Japan Society of Thermophysical Properties (1987).

46. R.E. Taylor, H. Groot, and R.L. Shoemaker, "Thermophysical Properties of Fine-Weave Carbon/Carbon Composites," *Spacecraft Radiative Transfer and Temperature* **83**, 96-108 (1982).

47. A.M. Luc and D.L. Balageas, "Comportement thermique des composites à renforcements orientés soumis à des flux impulsionnels," *High Temp. High Pressures* **16**, 209-219 (1984).

48. R.L. Shoemaker, "Limitations of the Pulse Diffusivity Method as Applied to Composite Materials," *High Temp. High Pressures* **18**, 645-654 (1986).

49. D.L. Balageas and A.M. Luc, "Transient Thermal Behaviour of Directional Reinforced Composites: Applicability Limits of Homogeneous Property Model," *AIAA J.* **24**(1), 109-114 (1986).

50. M. Lafond-Huot and J. Bransier, "Caractérisation thermocinétique de materiaux composites fibreux soumis à un flux de rayonnement impulsionnel," *Lett. Heat Mass Transfer* **9**, 49-58 (1982).

51. D.L. Balageas, "Détermination par la méthode flash des propriétés thermiques des constituants d'un composite à renforcement orienté," *High Temp. High Pressures* **16**, 199-208 (1984).

52. R.M. Pujolà and D.L. Balageas, "Derniers développements de la méthode flash adaptée aux matériaux composites à renforcement orienté," *High Temp. High Pressures* **17**, 623-632 (1985).

53. A Whittaker, R. Taylor, and H. Tawil, "Thermal Diffusivity of Some Fine-Weave Carbon/Carbon-Fibre Composites," *High Temp. High Pressures* **17**, 225-231 (1985).

54. J.F. Kerrisk, "Thermal Diffusivity of Heterogeneous Materials," *J. Appl. Phys.* **42**(1), 267 (1971).

55. J.F. Kerrisk, "Thermal Diffusivity of Heterogeneous Materials, II. Limits of the Steady-State Approximation," *J. Appl. Phys.* **43**(1), 112 (1972).

56. H.J. Lee, and R.E. Taylor, "Thermal Diffusivity of Dispersed Composites," *J. Appl. Phys.* **47**(1), 148 (1976).

57. T.Y.R. Lee and R.E. Taylor, "Thermal Diffusivity of Dispersed Materials," *J. Heat Transfer* **100**, 720-724 (1978).

58. R.E. Taylor, "Thermal Diffusivity Measurements of Composites," in: *Proc. 1st Asian Thermophysical Properties Conf. Beijing, China*, pp. 24-30, Academic Publishers (1986).

59. M. Tischler, J.J. Kohanoff, G.A. Ranguni, and G. Ondracek, "Pulse Method of Measuring Thermal Diffusivity and Optical Absorption Depth for Partially Transparent Materials," *J. Appl. Phys.* **63**, 1259-1264 (1988).

60. C. Otter and J. Vandevelde, "Contribution à l'étude du problème thermocinétique lié à la mesure de la diffusivité thermique des materiaux liquides, à hautes températures par la méthode du 'flash laser'," *Rev. Int. Hautes Temp. Refract.* **19**, 41-53 (1982).

61. R.E. Taylor, L.R. Holland, and R.K. Crouch, "Thermal Diffusivity Measurements on Some Molten Semiconductors," *High Temp. High Pressures* **17**, 47-52 (1985).

62. A. Degiovanni, "Conduction dans un 'mur' multicouche avec sources: extension de la notion de quadripole," *Int. J. Heat Mass Transfer* **31**, 553-557 (1988).

63. F. Lepoutre, "Mesures thermiques par les méthodes photothermiques," *Rev. Gen. Therm*, **301**, 8-14 (1987).

11

Modulated Electron Beam Thermal Diffusivity Equipment

R. DE CONINCK

1. INTRODUCTION

Among the various possible ways of measuring thermal diffusivity, treated in Vol. 1 of this Compendium,[1] the sinusoidally modulated electron beam method with a disk-type specimen is the only one considered in this chapter. The apparatus described here allows measurement of the thermal diffusivity of most solid materials over a broad temperature range. High temperatures are intended in particular. Also spectral emissivity values are obtained as a byproduct.

As far as the author is aware of, today no equipment of this kind is commercially available as yet. Only around 1970 was M.J. Wheeler, from the Hirst Research Centre in Wembley, U.K., willing to sell some prototypes to interested people, including the author.

For many years now the modulated electron beam thermal diffusivity apparatus has reached a mature state of development. Nevertheless, it must be emphasized that this kind of equipment still belongs to the specialized laboratory domain only and presumably this will be the situation for many more years. Moreover, it should be realized that such apparatus, as long as it remains in the hands of laboratory people, will continually be subject to various, new improvements. Meanwhile several alterations have indeed been carried out with the aim of enlarging the measurement range, enhancing reliability and accuracy, and simultaneously simplifying operational handling.

2. PRINCIPLE OF THE METHOD

The specimen, which has the shape of a thin plane-parallel disk, is brought to the desired measuring temperature by bombarding one of its surfaces with

---

R. DE CONINCK • Materials Science Department, SCK/CEN, B-2400 Mol, Belgium.

an electron beam whose intensity can be adjusted to a given value. Superimposed is a sinusoidally modulating signal with an amplitude which is small compared to the magnitude of the mean steady-state signal. The temperature of the bombarded surface varies according to this modulation but is shifted in phase with respect to the beam modulation. The temperature of the other surface is modulated accordingly as a result of the energy transported through the sample, but this modulation is smaller in amplitude and its phase again lags with respect to the modulation at the bombarded surface.

The temperature phase shift created between either side of the specimen is unequivocally connected with the thermal diffusivity value of the specimen—see Section 5 below. Consequently, one need only measure that phase lag and the mean temperature to obtain diffusivity values as a function of temperature.

When C_p, the specific heat, and ρ, the density of the specimen, are known, thermal conductivity can be calculated.

3. MEASUREMENT RANGE CAPACITIES

In the past the equipment has been used to determine the thermal diffusivity of various materials. Values lying between approximately 10^{-7} and $5 \times 10^{-5} \, \mathrm{m^2 \, s^{-1}}$ have been quoted. The conductivity range covered was between some 0.05 and 200 $\mathrm{W \, m^{-1} \, K^{-1}}$. Most probably lower and higher values can be treated.

Generally, reproducibilities of the order of 1 to 1.5% were obtained. Uncertainty is assumed to be 2 to 5%, possibly attaining 10%, depending on the material measured.

The temperature range covered is presently between some 200–250 °C and 2500 °C. Here also a broader range is possible in principle.

All kinds of solid materials have been measured, namely metals, refractory metals, insulators, ceramics, semiconductors, porous materials, and so on. Highly porous materials and very poor electrically conducting materials can produce problems but these can generally be solved (see Section 6.2).

When a brightness pyrometer (one small wavelength band) is available as well as a two-color pyrometer (two wavelength bands), the apparatus can provide simultaneously spectral emissivity values. Spectral emissivities at 0.65 μm and 2.3 μm have been determined, but only above approximately 1000 °C. The reproducibility is generally better than 5% but uncertainty can easily rise up to 10 or 15%.

4. DESCRIPTION AND EXPERIMENTAL DESIGN FEATURES

4.1. Outline of the Equipment

A schematic drawing of the practical construction and composition of the modulated electron beam thermal diffusivity equipment using disk-type

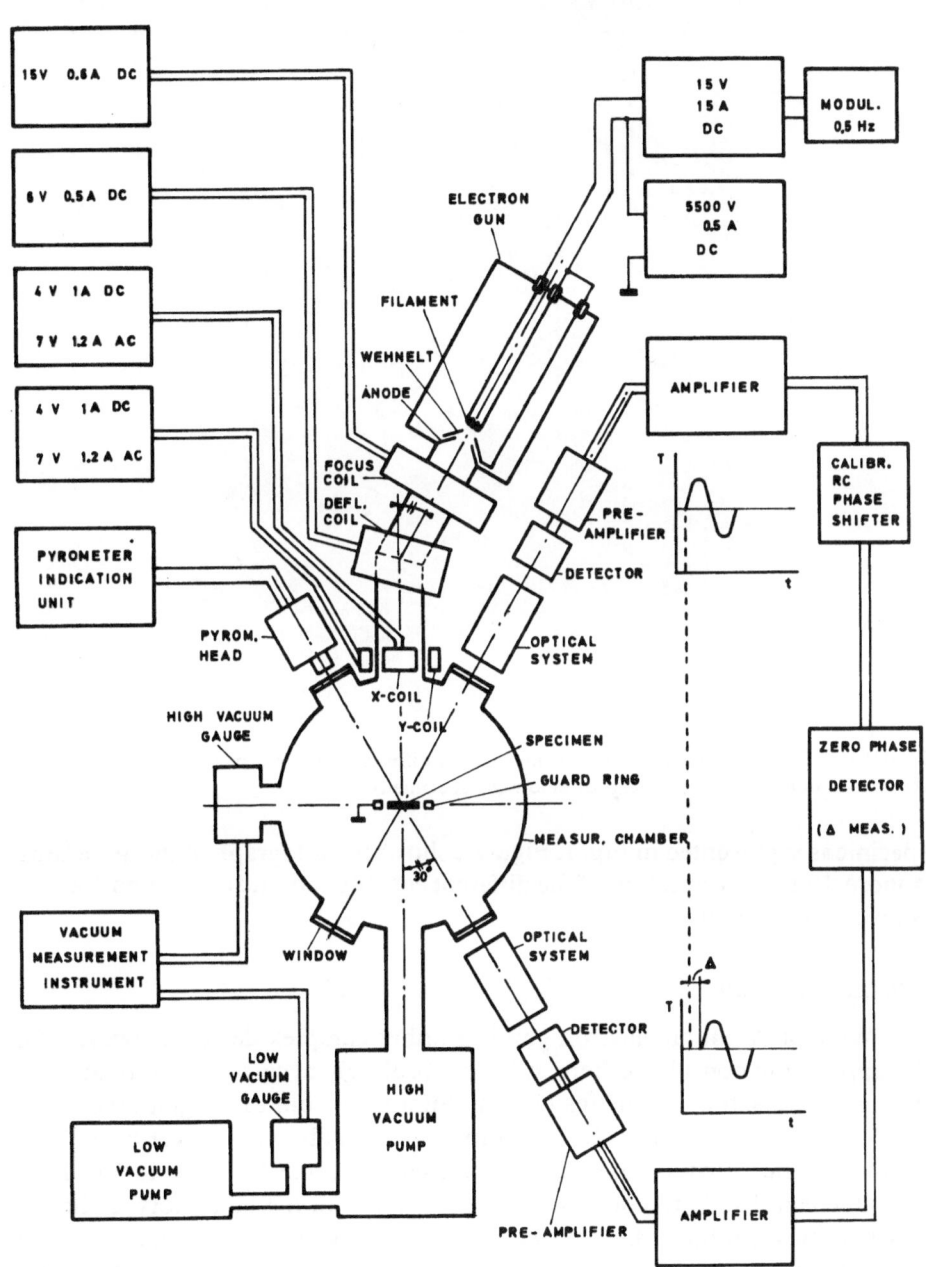

FIGURE 1. Schematic outline of the modulated electron beam thermal diffusivity equipment.

FIGURE 2. The modulated electron beam thermal diffusivity equipment (the three pyrometers are shown mounted above instead of below the specimen).

specimens is presented in Fig. 1. Figure 2 shows a photograph of the apparatus. A more detailed description of the different constituents and associated features is given subsequently.

4.2. Electron Gun

Without doubt an electron gun is a rather complex device. However, for the application considered here, it is not absolutely necessary to have at one's disposal a versatile, sophisticated gun. Simple custom-made gun assemblies are commercially available. Several suppliers exist all over the world. Without power supplies such a gun can be purchased for some US$5000 to 10,000.

One can decide to construct an electron gun in the laboratory, although this will most probably be considered by many scientists, not specialized in this domain, to be an almost unattainable task. It is possible to carry out a computer simulation program which calculates the optimum configuration of a cathode-Wehnelt-anode electrode system with the specific needs for this application in mind. See, e.g., the EGUN program from Stanford.[2]

A relatively simple gun that serves the required purpose sufficiently well is shown in Fig. 3. This configuration has been used for many years. The figure is not a fully detailed construction drawing but the most important parts can be easily identified. Vital dimensions are given to allow proper construction. Figure 4 is a photograph of the gun assembly with its outer vacuum housing removed.

Basically this gun is compoased of three parts; the filament cathode, the cathode shield or forming electrode, and the anode or accelerating electrode. The accelerating power-supply voltage is minus 5500 V. The maximum current is 0.5 A DC, although in most cases a current not larger than some 0.2 A will

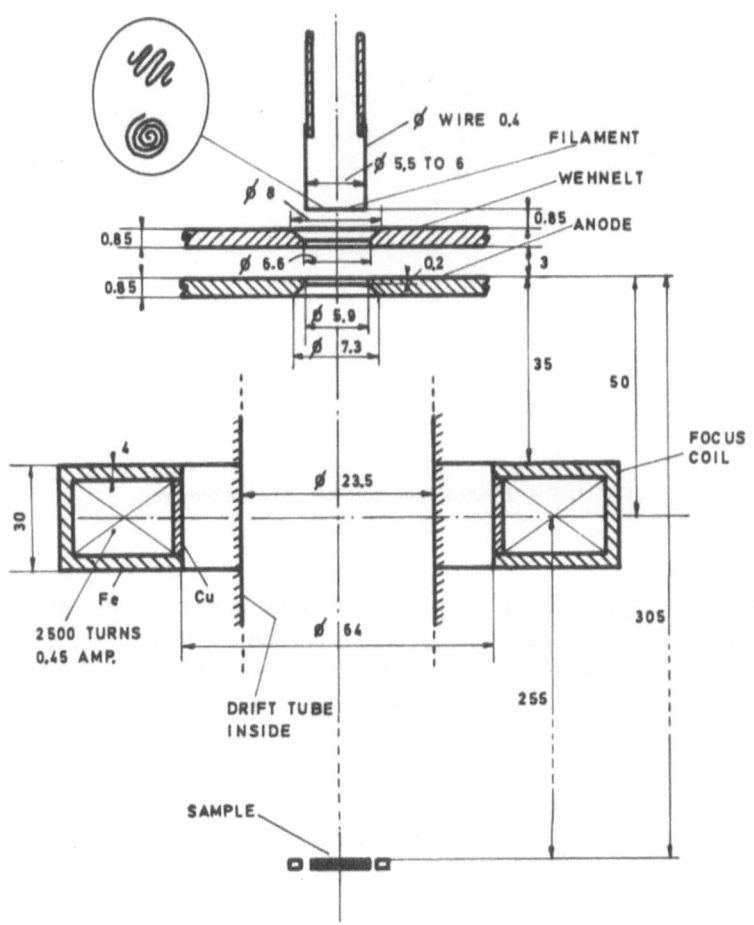

FIGURE 3. Schematic configuration of an electron gun with the major dimensions indicated (in mm).

FIGURE 4. Electron gun with outer cover removed.

suffice for normal operation. A negative voltage has been chosen for safety reasons. Indeed, in that case the specimen can be earthed.

Most parts are preferably made of polished stainless steel, because of its excellent properties with respect to adsorption–desorption and degassing, its mechanical strength, and thermal stability, and also it is nonmagnetic. The anode and the lower part of the forming electrode are preferably made of, e.g., tantalum as these parts may reach rather high temperatures.

The filament is either a tantalum or preferably a tungsten wire, 0.4 mm in diameter and flat-folded or spiral-wound (see inset in Fig. 3). A power supply of 15 V, 15 A DC is needed.

It is self-evident that such a cathode does not exhibit a homogeneous temperature distribution and that full symmetry with respect to the electrode openings cannot be expected. Moreover, the cathode will neither show equipotentiality nor will its surface show a continuous uniform thermionic emission profile. Thus homogeneous electron emission cannot be achieved and additionally, in the neighborhood of the filament, the equipotential planes of the accelerating electric field will be distorted. Therefore, for a more advanced construction, it is better to replace the simple filament either by a directly or by an indirectly heated small circular platelet, or by a small indirectly heated rod-shaped cathode.

A suitable directly heated cathode can be constructed as shown in Fig. 5. Electrical contacts are made by spot-welding.

Indirect heating is adequately performed also by electron heating and using an auxiliary primary tungsten filament cathode. For that purpose a power supply of 500 to 600 V rated at 1 A is certainly sufficient to accelerate the primary electrons toward the platelet or to the rod cathode. Such an indirectly heated cathode is shown in Fig. 6. More sophisticated constructions are possible but not compulsory.

The insulators supporting the shield electrode–filament assembly are made of Al_2O_3 bars. They can easily be cut to the desired length by a diamond wheel.

The relative positions of the cathode and shield, and their position with respect to the anode, determine the efficiency of the gun. Once the optimum positions have been established, possibly by trial and error, everything should be left undisturbed.

A new filament should be fitted from time to time. Therefore it is interesting to provide for an identical spare filament assembly. It is noteworthy that a new filament will almost always move slightly when heated for the first time. Consequently it is absolutely compulsory to heat a new filament up to the

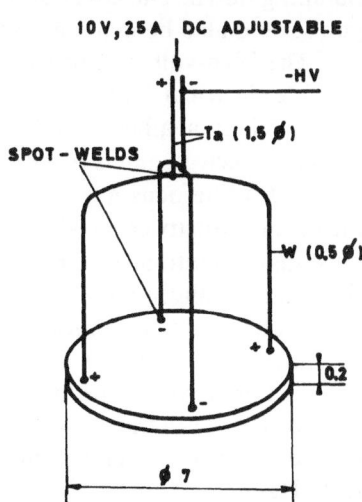

FIGURE 5. Example of a suitable directly heated cathode (measurements are in mm).

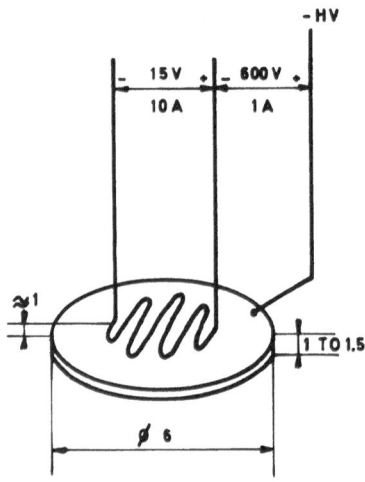

FIGURE 6. Example of an indirectly heated cathode.

maximum temperature that it will experience during later operation, either in a separate vacuum enclosure or mounted in the gun (without applying the accelerating voltage) for at least 15 minutes. Before its actual use it should be replaced to satisfy its optimum alignment position again. Flatness, interspacing, inclination, and centering have to be checked carefully.

Access to the filament assembly is made possible by sliding upward the cover of the gun, leaving all other components undisturbed. For better thermal stability, but also for safety, it is useful to cool the cover. Therefore it is double-walled, allowing water to flow between the walls. This also allows for hot-water heating, which considerably improves the degassing efficiency during pumping down. The cover and all other tubular and vacuum-tight parts of the gun should also be made of seamless, annealed, and smoothed stainless steel.

The high-voltage supply and filament current enter the vacuum through suitable electrical feedthroughs; almost all vacuum equipment firms have them commercially available.

The electron gun is in a certain sense a combination of a diode-type Pierce gun and a telefocus gun. It works in the saturation region. This means that the beam intensity can easily be controlled in accordance with the Richardson-Dushman equation, i.e., by varying the cathode temperature—see Vol. 1 of this Compendium, page 378.[1] This allows tuning to the desired measuring temperature and, by modulating the filament current, a modulation of the specimen temperature is obtained.

In this construction a restriction is set by the frequency response. The relatively large thermal inertia of the filament (or of the directly or indirectly heated cathode) generally limits the frequency to an upper value of some 5 to 10 Hz.

Here a fixed frequency of 0.5 Hz is used. A very simple construction allows one to obtain the desired settings. The shaft of a potentiometer, which controls the output of the filament supply, is driven back and forth by an electric motor with gear case and crank. The speed of the motor depends on the 50-Hz mains frequency.

The modulation depth can be adjusted. Its value should be chosen so that the specimen experiences temperature fluctuations which are as small as possible in practice (sufficiently noticeable signal). Normally, 1 to 2% would be sufficient and even 0.5% or less at higher target temperatures. When carrying out measurements at lower temperatures and for poorly conducting specimens, a depth of 5% is sometimes necessary.

4.3. Drift Tube and Beam Manipulation

The gun assembly is connected to the vacuum measuring chamber via the equally double-walled stainless-steel drift tube (see Fig. 7).

Immediately below the anode a magnetic lens coil serves as a beam focusing device. Since the beam is diverging when passing through the anode, it is beneficial but also necessary to focus the beam in order to obtain a beam section, at the surface of the specimen, that is not too large compared to the dimensions of that specimen.

Such a coil can be made very simply. A reel, made of mild steel or of pure iron, or possibly of a specific transformer core material, and on which the requisite windings are wound, is fitted with a closing ring at the periphery—see Fig. 3. Approximately 2500 turns of 0.4-mm-diameter copper wire is used. An adjustable power supply of 15 V, 0.6 A allows trimming to ideal focusing.

Further toward the measurement chamber the drift tube forms an elbow at an angle of 30°. A larger angle would even be advisable. The electron beam is deflected over this angle by a magnetic field. The purpose of this elbow is to prevent direct light from the filament reaching the chamber and thus the specimen. This light would give rise to unwanted signals on the detectors, especially at low specimen temperatures (see Section 4.6). Moreover, spectral emissivity determinations are now possible since emission from the specimen only is captured. Grooves in the inner surface of the tube will further reduce the amount of light still getting through.

An additional advantage of the elbow is that only a small proportion of the positive ions, formed by the bombardment and by collisions with the residual gas atoms, will hit the cathode.

The deflection field is formed by a coil of 2500 ampère turns and energized by an adjustable power supply of 6 V, 0.5 A. This field is created between two rectangular pole pieces. A photograph of the set-up is shown in Fig. 8. The special design of these pole pieces serves the purpose of achieving the necessary deflection, yet with a minimum of aberration. Indeed the field strength all over

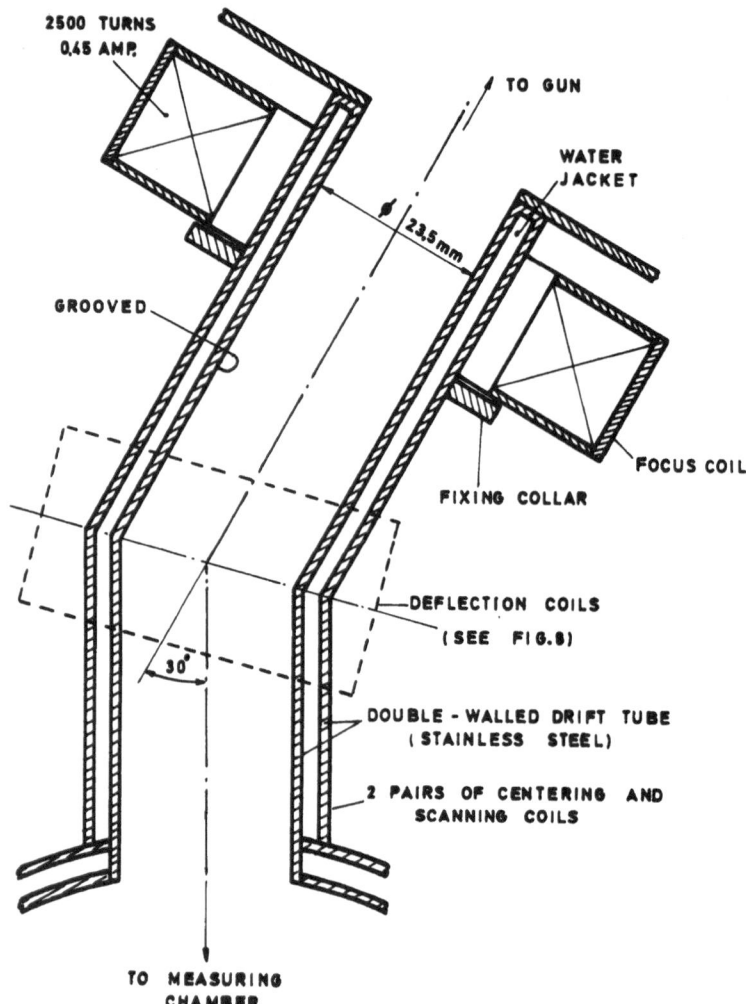

FIGURE 7. Double-walled drift tube connecting the gun with the vacuum measuring chamber.

the beam section is constant, and the distances covered when crossing this field are nearly identical for each individual electron trajectory.

Immediately before the connection with the chamber two pairs of coils are mounted around the tube at an angle of 90° with respect to each other. These coils allow exact centering of the beam onto the specimen surface and serve simultaneously as an X-Y beam-scanning means. They consist of simple air coils, without pole pieces. Each pair can be put either in series or in parallel. Two DC power supplies (4 V, 1 A) deliver to each pair of coils a current, which can be adjusted in the positive as well as negative sense. The number of necessary ampère turns depends on the distance between the coils and the specimen. Two power sine generators (7 V, 1.2 A) send also two adjustable AC currents through each pair. Their frequencies should be at least 100 times higher than the modulation frequency (thus at least 50 Hz) in order to avoid any appreciable interferences. Moreover, they should differ from each other by at least a factor of 50 to 100, in order to avoid relatively large distances between the nodal points of the Lissajous figure written over the specimen surface. Deciding about the practical construction of the coils, and choosing the frequencies and voltage and current, will in general be possible only after a few experimental tests.

A more sophisticated electronic beam scanning system has been described by Mayer and Neuer.[3] The beam is driven over the sample's surface following

FIGURE 8. Deflection coils. A few viewing ports with shutters are visible. The gun is removed from the drift tube.

an adjustable spiral. In this way the outer rim can be heated somewhat more than the central part in order to compensate for inevitable losses from the sides.

4.4. Measuring Chamber—Vacuum Equipment

The vacuum chamber is also made of a seamless, annealed, and smoothed stainless-steel tube. When measurements at very high temperatures (above 2000 K) are regularly envisaged, it might be better to select a polished pure copper tube, since stainless steel is a rather poor thermal conductor and might heat up too much by electrons passing alongside the specimen.

A large door in the front of the chamber allows the operator to mount a specimen on the holder. The specimen holder is positioned at the center of the chamber. The chamber and door are both double-walled to allow water cooling (or heating when pumping down).

Three viewing ports above and three viewing ports below the specimen are provided, equidistant from the center of the specimen. A fourth port, vertically above the specimen, is used for mounting the drift tube and electron gun. In this way the electron beam hits the upper surface of the sample perpendicularly and so will not heat up its sides. The upper and lower ports at the front of the chamber are used for mounting the upper and lower optical assemblies. The two remaining ports at the top of the chamber permit visual observations, or optical pyrometry, or the installation of any other sensing device the operator may wish to utilize. Generally one port is used for a two-color pyrometer and the other simultaneously for a brightness pyrometer (see also Sections 4.8 and 4.9).

The ports underneath the chamber can be used for pyrometry as well. Also, they can be used to serve various purposes, such as the introduction of an inert gas or additional water cooling, for inserting electrical feedthroughs for thermocouples or other supply or measuring wires, for fitting leak-detection equipment or additional vacuum gages, etc. All ports not in use are fitted with opaque stainless-steel or water-cooled copper stopping blanks.

The windows in the ports are made of quartz or calcium fluoride (CaF_2), which transmit visible and infrared wavelengths up to, respectively, 2.5 and 9.5 μm. They are removable for cleaning purposes.

All ports, except the vertical upper and lower ones, are positioned at an angle of 30° to the vertical. This corresponds to a loss of energy collection of only some 13% (1 − cos 30°). They are equipped with metal shutters, which mask the windows against unwanted vapor deposition between measurements.

The chamber, the drift tube, and the gun should be kept under vacuum when not in use. This prevents absorption of gases and vapors which would hinder sequential fast pumping down. It is also advantageous to fill the vacuum space with dry nitrogen gas and possibly warm up the chamber before exposing it to air. Absorption and adsorption of air molecules, but especially of moisture, will largely be reduced in consequence.

In order to allow one to partly close down the chamber when changing the specimen, and in order to avoid oil from the pumps climbing up to the chamber, it is necessary to put an isolating baffle valve with a low pumping resistance between the chamber and the vacuum pumps.

As electron-beam generation is used, a vacuum of at least 10^{-4} mbar is compulsory. The pumping system consists of a turbomolecular pump of some 500 liter per second, backed by a two-stage rotary vane pump of 30 m^3 per hour. A few valves are provided, allowing the various pumping sequences and other manipulations to be carried out. The necessary vacuum measuring gages are provided, together with their appropriate indicating instruments.

4.5. Specimen Dimensions—Specimen Holder

A disk-shaped sample has been chosen, as it is simple to make and exhibits the lowest ratio of side surface to total surface for a given thickness. Radiation losses from the sides will thus be reduced to a minimum, which is an advantage considering the necessity of linear heat transfer through the disk.

A specimen of diameter 8 mm is chosen when possible. The minimum diameter is about 6 mm, unless a decreased accuracy is tolerated or unless a very small thickness (such as 0.1 mm) can be justified.

Thickness is governed by the minimum mechanical strength requirement and by the modulation frequency or phase angle. Thickness should be constant over the whole surface and may be chosen around 1 to 1.5 mm for a good conductor and between, e.g., 0.2 and 0.6 mm for an insulator. For some porous materials larger thicknesses might be necessary. These figures are intended as a guide only and deviations are permissible, but generally at the expense of accuracy. When decreasing the thickness the phase-angle measurement is less accurate, since its absolute value decreases. Increasing the thickness lowers the modulation amplitude at the nonbombarded surface and increases the risk of side-radiation losses.

The specimen holder is very simple but fully adequate. As shown in Fig. 9, it consists primarily of two parallel, horizontal tungsten rods (20 mm apart and 3 mm diameter) resting upon notches in a supporting bar at the inner surface of the chamber. A tungsten wire, of 0.4-mm diameter, is spot-welded to one of these rods and two more 0.4-mm-diameter wires are spot-welded to the other rod. All three are positioned at 120° with respect to each other. Upon these three wires rests a guard ring (inner diameter 10 m, outer diameter 13 mm, and thickness 1.5 mm) made of tungsten or tungsten–10% tantalum. Underneath this ring three other tungsten (0.25-mm-diameter) wires are spot-welded, positioned also at 120° with respect to each other and at 60° from the ring supporting wires.

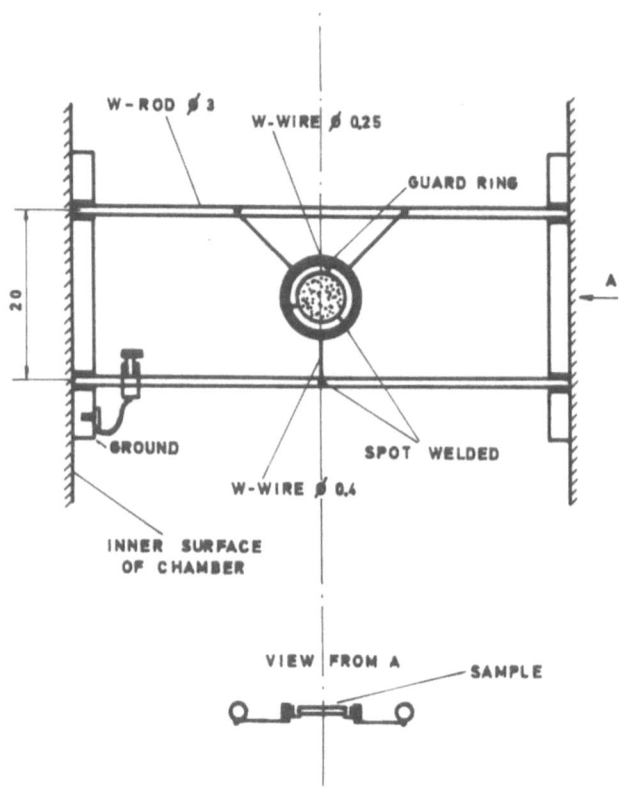

FIGURE 9. Sample holder with sample (measurements are in mm).

The specimen rests horizontally under its own weight on these wires, entirely within the ring. It might be necessary to choose some other material than tungsten when dealing with specimens which exhibit a certain incompatibility with tungsten.

As a whole the holder is constructed so that the center of the sample is in the center of the chamber. To facilitate centering, minute shifts in the x, y, and z directions are possible. The correct position is achieved when the specimen is perceived at the center of the field of view when looking through one of the vacant observation ports.

The ring is heated together with the specimen and its temperature will normally be only slightly lower than that of the specimen, thus reducing to a minimum radiation loss by the sides. Ideally, the ring can be made of the same material as the specimen so that its emissivity is identical. However, in that case often also a specimen with a diameter of 13 mm or more can be chosen instead. This will give results which are equally correct, except perhaps for very highly conducting materials.

Heat loss by conduction through the holder is negligible, taking into consideration the small diameter of the supporting wires and their lengths and the poor thermal contacts.

Electrical contact is generally sufficiently good although some electrical insulating materials may give a problem. So far, it has always been possible to solve this problem by the deposition of a conducting layer over the entire specimen's surface (see Section 6.2). The holder itself is automatically earthed. Nevertheless, for certainty and security (possible high-voltage buildup) a firm grounding wire is fastened to one of the tungsten rods.

In some cases (e.g., measurement of C_p, investigation of the form or efficiency of the beam, etc.) it may be interesting to measure separately the magnitude of the electron beam current passing through the specimen and/or holder. This is a difficult task, especially at very high temperatures where adequate insulation becomes a problem. Moreover, the thermal equilibrium may be disturbed, and especially secondary electrons will strongly falsify the data. Sometimes, one also wishes an individual extra voltage to be applied to the specimen and/or holder.

The holder and specimen as a whole can be isolated electrically from the chamber by simply slipping a small insulating tube, such as Al_2O_3, over the two tungsten rod extremities before placing them onto the supporting bars in the chamber. At that location the temperature remains low and thus insulation is guaranteed. It is self-evident that some compensation of the height difference caused by the insulation must be provided so that the specimen remains in the center of the chamber. Insulation of the specimen only can be done in an analogous way, e.g., by long, thin insulating rods (e.g., a minimum of 2–3 cm Al_2O_3) which are not bombarded over their full length. Of course an electrical path to earth must be provided to close the electron beam circuit. A wire attached to the holder (or to the specimen) and taken out via a special feedthrough allows one to measure the current or the voltage over a small series resistance, or to apply an extra voltage.

4.6. Optoelectronic Assembly

The upper and lower viewing ports at the front of the measuring chamber are utilized to detect the modulated temperature fluctuations created on either side of the specimen. The energy emitted by the specimen consists of infrared radiation at low temperatures and of infrared as well as visible light at high temperatures.

For high temperatures a drawing of the setup is shown in Fig. 10. Radiation passing through the window is focused by a commercial photography lens with a focal distance of 50 mm. Thus, an image is produced of the specimen surface some 50 mm behind the lens. This image plane is tilted over approximately 30° with respect to the axis of the lens, since this axis forms also an

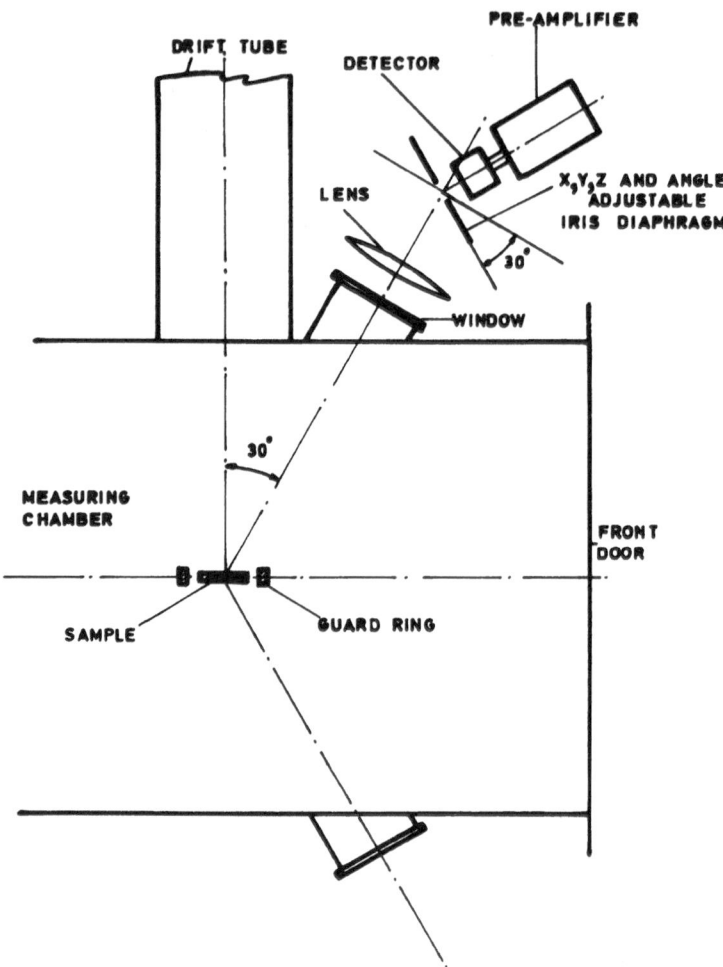

FIGURE 10. Principle of the optoelectronic detection system.

angle of 30° with respect to the specimen. The lens has an aperture of 1.8, adjustable down to 16. This is convenient to reduce the light intensity to reasonable values when measuring at very high temperatures.

A manually adjustable iris diaphragm is placed in the focus plane. This diaphragm is mounted on an x, y, z and angle-adjustable table in order to center and align exactly the image onto it. It cuts off the outer ring of the image, allowing only radiation from the central zone (e.g., 3-mm diameter) to pass toward the detector mounted behind. Maximum energy collection is obtained when diaphragm and detector are parallel to each other and when the distance between them is minimum. Alignment may be carried out visually using either the light emitted by the specimen at incandescence, or the light reflected from, e.g., a halogen lamp concentrated onto the specimen.

The main advantage of this assembly lies in the fact that only the central part of the specimen is perceived by the detector. Since the temperatures at the periphery and at the center of the specimen will be slightly different, the recorded modulation temperature fluctuations will not be representative when the detector covers the whole surface. Elimination of the guard ring and of the outer part of the specimen is necessary. Moreover, possible incidental stray light pick-up will be reduced to a negligible minimum.

The detector is a silicon PIN photodetector diode operating in the photoconductive mode. A United Detector Technology type PIN 10 D is used. It has a circular sensitive area of 1 cm^2 and a N.E.P. (10^3 Hz, 1 Hz, 0.85 μm) of 10^{-12} W Hz$^{-1/2}$ and a response time of 25 ns. However, other (and smaller) types are readily available and will do as well. In order to stabilize its temperature it is recommended that it be fastened to a solid copper block.

The spectral response of a silicon detector ranges from 0.35 to 1.15 μm, thus covering a part of the near-infrared. An apparatus fitted with Si detectors is not capable of delivering results below some 550 to 600 °C. Lower temperatures can be detected by PbS or InSb, InGaAs, HgCdTe, or similar detectors. It is self-evident that adequate optical systems and lower-temperature pyrometers must be provided.

There are two possibilities for the optical system. One can use either infrared lenses (e.g., CaF$_2$, Ge, KBr, sapphire, etc.) or concave–convex mirror collecting and imaging systems. Such a mirror system is more expensive and is larger and more inconvenient for an equal light-collecting efficiency, but it has the advantage of showing theoretically no chromatic aberration. Consequently, a mirror system is suitable for measurements at high as well as low specimen temperatures since all rays are focused to the same spot.

With such a mirror system, and using normal PbS, PbSe, or InAs detectors (possibly cooled by Peltier elements), temperatures down to some 150 to 200 °C are easily measured. For still lower temperatures one needs pyroelectric detectors or cryogenic cooled detectors (InSb, HgCdTe, GeAu, etc.), which are unfortunately rather expensive.

4.7. Phase-Shift Determination

Each detector will be hit by a quantity of AC energy superimposed on a DC level. The magnitude of the resulting AC and DC signals delivered by the detectors depends on the quantity and kind of energy emitted by each surface (dimensions, emissivity, mean temperature, and modulation depth), on the detector response curve, and on the optical conditions of the assembly (lens or mirror system characteristics, distance, F.O.V., diaphragm, etc.) A semiconductor detector is a very sensitive device which, moreover, has the favorable property of showing a linear response over five to seven decades. Since the AC output of the detector is small (a few μV up to some mV) compared to the irrelevant bias voltage and/or the DC output (e.g., 10 to 50 V), AC amplification is necessary. Therefore use is made of a variable high-gain, low-noise source-follower AC coupled with an operational pre-amplifier backed by an AC-coupled and DC-compensating amplifier. Particular attention must be paid to the detector bias voltage supply and to the shielding of the first stage, in order to reduce noise and ripple to a minimum. The output from each amplifier is indicated by an analog meter. This is an appropriate means of showing quickly a possible overload of the amplifiers at a particular gain setting or at a certain DC-compensation level.

Phase meters for a frequency as low as 0.5 Hz have been available for a few years. Several data acquisition systems and some lock-in amplifiers are suitable too, but often the accuracy is rather mediocre over that frequency range. Phase meters are widely available for frequencies above a few hertz.

A simple and cheap solution is still in use today, because it is fully adequate and its inaccuracy, which is of the order of 0.5 to 1% for small phase shifts (e.g., 10°), is becoming rapidly negligible for larger phase shifts. It consists of a precision multiple resistance decade (100 to 1.111×10^6 Ω) with a precision capacitor (1 μF) in parallel. This RC network, as integrated in the amplifiers, is shown in Fig. 11. It provides a means of introducing a phase lag into the 0.5-Hz wave from the upper detector until it is in phase with the lower detector wave. Therefore a null indicator is needed. In this case a Lissajous figure, using an X-Y oscilloscope, enables one to adjust the resistance of the RC network until a straight line, inclined at approximately 45° (depending on the relative amplitudes), is attained. A screen with a long persistence is an advantage.

The phase difference Δ is calculated using the expression

$$\Delta = a\tan\left(\frac{\omega CR^2}{R + R'[1 + (\omega CR)^2]}\right) \qquad (1)$$

where R and C are respectively the accurately known values of the adjustable resistance and of the fixed capacitance of the RC network, and R' of the fixed resistor (here $68 + 33$ kΩ). The maximal achievable phase shift with this network amounts to approximately 56°. Should the specimen considered produce a higher value of Δ, e.g., in the case of a very poor conductor, a fixed and known RC circuit can be added ahead of the amplifier-phase shifter.

An interesting feature to further reduce the residual ripple in the Lissajous figure is offered when an oscilloscope is chosen which has at the same time an AC and a DC input for both X and Y channels, since then the residual DC signal can easily be subtracted from the total signal.

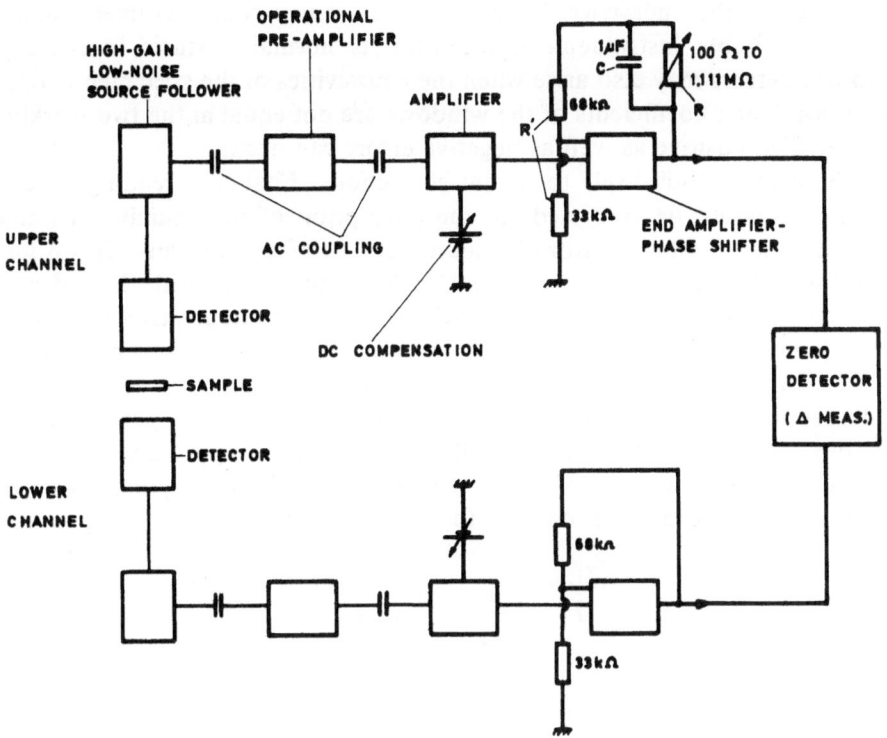

FIGURE 11. Upper and lower detector-amplifier channels and zero detection, with phase shifter in upper end-amplifier.

Sufficient attention should be paid to possible adverse interferences between the input and output impedances of the various parts. Also, no occasional erroneous phase shifts may occur anywhere else in the circuits.

4.8. Temperature Measurement

It is clear that knowledge of the specimen temperature is a necessity. A thermal diffusivity value should be linked to its corresponding temperature and both temperature and diffusivity should be known with comparable accuracy.

Radiation pyrometry is obviously indicated for this application, considering the numerous significant advantages of pyrometers versus thermocouples or other contact methods.

A two-color pyrometer is currently utilized, since a brightness pyrometer suffers from an inevitable uncertainty as a result of the frequently doubtful knowledge of the emissivity. Regretfully, also a two-color pyrometer is not always a reliable instrument. Apart from its normal intrinsic inaccuracy, significant errors may also arise when the emissivities of the specimen and/or the transmission coefficients of the windows are not equal at the two working wavelengths. Positive as well as negative errors can occur.

Pyrometry lends itself to a few corrections. First, the reading of any pyrometer should be corrected for the absorption of the chamber window through which the temperature is measured. Therefore the transmission loss ε_T at the wavelength concerned must be determined, e.g., with the aid of a spectrophotometer, before and after each series of measurements. When working at temperatures close to the melting point, the windows will get blurred quickly. In that case a correct judgement about the value of the transmission to be adopted for the various preceding measurements can be problematic.

For a two-color pyrometer as well as for a brightness pyrometer, window correction is carried out using the so-called "brightness pyrometer error equation," approximated according to Wein's radiation law:

$$T_T = \left(\frac{1}{T_B} + \frac{\lambda_B}{c_2} \ln \varepsilon_T \right)^{-1} \tag{2}$$

where T_T is the true temperature after correction (K), T_B the measured temperature (K) (by a two-color or brightness pyrometer), λ_B the wavelength at which the measurements are conducted (m), c_2 the second radiation constant ($1.43878^6 \times 10^{-2}$ m K), and ε_T the transmission loss coefficient (dimensionless). A second correction stems from the emissivity of the specimen.

When using a brightness pyrometer the same correction equation as equation (2) can again be used, except that ε_T is now replaced by ε_s, the spectral emissivity of the specimen:

$$T_T = \left(\frac{1}{T_B} + \frac{\lambda_B}{c_2} \ln \varepsilon_s \right)^{-1} \tag{3}$$

For the two-color pyrometer a correction equation analogous to equation (2) is appropriate:

$$T_T = \left[\frac{1}{T_R} + \frac{\lambda_2 \lambda_1}{(\lambda_2 - \lambda_1)} \times \frac{1}{c_2} \times \ln \frac{\varepsilon_{s1}}{\varepsilon_{s2}} \right]^{-1} \tag{4}$$

where λ_1 and λ_2 are the first and second selected wavelength, respectively, ε_{s1} and ε_{s2} are the spectral emissivities at λ_1 and λ_2, respectively, and T_R is the measured ratio temperature.

The two spectral emissivities ε_{s1} and ε_{s2} will not be known with better accuracy than the single spectral emissivity needed for the brightness pyrometer. However, only their ratio is important and that can be determined with much better accuracy by measuring the reflectivity ratio using a mirror adaptor mounted in a spectrophotometer.[4]

Measurement of the mean temperature of the specimen by a two-color pyrometer has undoubtedly an important advantage. Unfortunately most two-color pyrometers are incapable of measurements below some 1000 °C. Ircon Inc. (U.S.A.) put on the market a two-color pyrometer (R-Bicolor) which measures down to 600 °C. The "Red Eye" of Capintec International Inc. (U.S.A.) is claimed to go as low as 150 °C but unfortunately its field of view seems to be too large for this application. Some time ago Galai (Israel) announced his model IR 107, which should measure down to 200 °C.

At present use is made of a disappearing-spot pyrometer (Optix, Germany), working at 0.65 μm, and an electronic comparative pyrometer (Ircon 300, U.S.A.), working at a mean wavelength of 2.3 μm, together with a two-color pyrometer (Milletron-Capintec International, Thermo-O-Scope, U.S.A.), working at 0.53 and 0.62 μm.

The nonbombarded face of the specimen is recommended for pyrometry, since this simplifies the computer program. Nevertheless, e.g., mounting difficulties of the pyrometer can be a good reason to measure the temperature of the bombarded surface instead. This is also advisable when dealing with low-conductivity specimens. Indeed, for such specimens the bombarded surface can be much higher in temperature than the nonbombarded one. When measuring directly the temperature of the bombarded surface, accidental local melting can more easily be avoided.

Each pyrometer should be positioned so that the measured area corresponds to the center of the specimen. A prism or mirror can be placed at the observation window to facilitate correct mounting of the pyrometer. Additional corrections must then be made for inevitable supplementary losses.

4.9. Spectral Emissivity Determination—Assessment

In the range covered by the two-color pyrometer, temperature is measured simultaneously with a brightness pyrometer. Consequently, the spectral emissivity ε_s at the wavelength of this brightness pyrometer can be calculated from equation (3), with T_T and T_B respectively the two-color and brightness pyrometer indications, both after the necessary corrections, at least when assuming the temperature T_T to be exact (after correction).

Below the two-color pyrometer temperature limit, measurements rely on the brightness pyrometer only. This causes a problem since spectral emissivity is mostly not known with sufficient accuracy. There, another procedure is used. The plot showing the calculated spectral emissivity as a function of the accepted true temperature (two-color pyrometer T_T) may be extrapolated beyond the minimum true temperature. It is evident that this procedure cannot yield a high accuracy. Some 5 to 10% would represent a fair estimate which is, after all, in many cases a much better alternative to a pure guess.

4.10. Safety Provisions and Some Recommendations

Since the electron gun uses several high voltage–high current supplies, even up to 5500 V at 0.5 A, a warning of danger to life is certainly not out of place. All power circuits must be switched off prior to any maintenance work being carried out, except when necessary for tests and/or adjustments which require circuits to be "live." It should be emphasized that the high-voltage supply is also in contact with the filament and with its power supply and modulation unit. When the high-voltage supply is switched off, a relay is released discharging the voltage to earth via bleed resistors.

A mains isolation transformer feeds the equipment, and additional thin mica-film gaps separate the mains supply from earth to prevent it from being raised to 5500 V if a fault should develop in the circuitry. All removable panels around the high-voltage supply as well as the removable gun covers are provided with microswitches acting as secondary safety devices. They cut off the high-voltage supply when a panel or cover is opened.

The sample and its holder are both connected to earth, either directly via the chamber or indirectly over a small resistor. This forms an elementary but efficient safety provision on its own.

The chamber and its door, the drift tube, and the gun enclosure are all double-walled to allow water-cooling, not only in order to obtain better vacuum conditions but also for safety reasons. A pressure switch at the outlet of the water circuit cuts off the high-voltage supply in case of water failure.

Also, the circuit of the high-vacuum pump is protected against failure by a pressure switch. Switches are mounted on the separation valves, which cut off the high voltage in case of incorrect operation of the valves. Moreover the filament current cannot be switched on when the inlet valve is open.

When working at very high specimen temperatures, it is advisable to shut the window protectors as often as possible in order to reduce vapor deposition and to lower the risk of window damage as a result of excessive overheating.

4.11. Maintenance

This equipment involves many different and specialized techniques to be activated concurrently. One might therefore think that extravagant or arduous skills are required to maintain it in running order. Fortunately this is not the case. The vacuum pumps need normal prescribed maintenance, such as an oil change from time to time. For cleaning purposes all windows must be dismounted regularly and, in particular, the windows used for pyrometry certainly need maintenance before and after each new series of measurements. The detector windows do not need cleaning as frequently, since the absolute value of the signal delivered by the detector is not of prime importance. Cleaning the windows is simple, although some precautions must be observed since CaF_2 is a relatively soft material which scratches easily. Deposited coatings should be removed by polishing with, e.g., 1-μm diamond paste or any other equally fine abrasive.

The filament of the electron gun will probably demand closer attention. One might reasonably expect that the filament requires replacement after three to four months if the equipment is in daily use. A more frequent exchange may be necessary if specimens are heated regularly to temperatures whereby their vapor pressure rises above some 10^{-5} mbar. Renewal is necessary when the beam develops "hot spots" at the specimen surface. These "hot spots" are caused by shrinkage of the diameter of a certain part of the filament, by vaporization, and particularly by bombardment with positive ions when working at pressures above 10^{-4} mbar.

Sometimes the filament has to be replaced because it has moved from its initial central position due to internal stresses. In that case the beam can no longer be aligned correctly and will be distorted strongly. Section 4.2 indicates how to avoid this as much as possible.

5. MATHEMATICAL TREATMENT OF THE DATA

5.1. Mathematical Basis

The mathematical treatment is based on the original study of Cowan.[5,6] In Vol. 1 of this Compendium, on pp. 380–388,[1] a survey of his work together with some later improvements from others has been reported. This survey is rather comprehensive and, moreover, comprises too many items which are irrelevant for the particular equipment described here. The more practically oriented contributions of Wheeler[7,8] and of Penninckx[9] are of especial interest. Meanwhile, a more simplified and more advanced mathematical method to analyze the measured data has also been worked out by the author and Maene.[10,11] This method is based on the direct application of complex numbers, and has the merit of avoiding the use of the rather cumbersome earlier formulas and also extending the range of applicability.

If Cowan's theoretical treatment and both Wheeler's and Penninckx's contributions are combined with some supplementary improvements of the author and the new complex numbers method, then the subsequent mathematical outlines can be derived.

5.2. Practical Equations

Figure 12 illustrates the model used. An electron beam heats uniformly the surface $x = d$ of an infinite plate with thickness d. The intensity of this beam is modulated sinusoidally and the modulation depth is assumed to be small.

In a first approximation the temperatures of both surfaces are modulated sinusoidally. Phase shifts arise between the incident beam and, respectively, the heated surface (δ_d) and the nonheated surface (δ_0). The phase difference Δ between both surfaces obeys the relation

$$\Delta = \delta_0 - \delta_d \tag{5}$$

The heat applied to the plate is partly transported through it in a one-dimensional manner and is released by radiation through the surfaces. A temperature difference is created over the specimen with T_d the mean temperature of the bombarded surface and T_0 the mean temperature of the other surface.

The sample is characterized primarily by its thermal diffusivity a, thermal conductivity λ, density ρ, and specific heat C_p. These quantities are related to each other by

$$\lambda = a\rho C_p \tag{6}$$

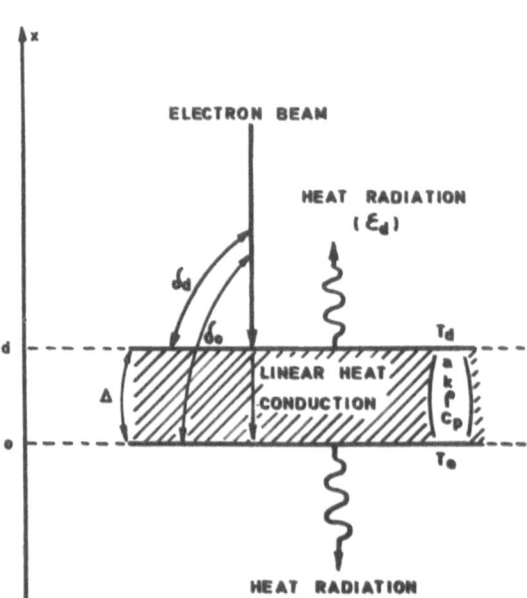

FIGURE 12. Schematic characterization of the sample as used in the theoretical treatment.

Prior to any specific diffusivity-conductivity calculations a subroutine giving the estimated spectral emissivity is worked out, at least when reliable values for the sample emissivity cannot be found in the literature. Therefore, both the temperatures of the two-color pyrometer and of the brightness pyrometer are needed and equation (3) is solved for ε_s. Values for the latter are possibly extrapolated to lower temperatures.

Cowan demonstrated that it is advisable to calculate diffusivity-conductivity values starting from measured values of Δ instead of from δ_0 or δ_d. However, the development of equation (5) is a most cumbersome and tedious task, since a multitude of often complicated parameters are involved, unless the mathematics of complex numbers is introduced. In this case the following equation, expressing the phase shift Δ directly as a function of only the two parameters c_0 and β (and of thickness d), is obtained:

$$\Delta = \text{phase} \left[(\beta + c_0) \, e^{\beta d} + (\beta - c_0) \, e^{-\beta d} \right] - \pi/4 \tag{7}$$

Parameter c_0 is defined by

$$c_0 = \frac{1}{\lambda} 4 \varepsilon \sigma T_0^3 \tag{8}$$

where T_0 is the mean temperature of the lower surface while quantities ε and σ are, respectively, the total thermal emissivity and the Stefan-Boltzmann constant. Further, it is assumed that ε and λ remain constant with time and space within the temperature-modulation range. The quantity β is defined by

$$\beta = \left(\frac{i\omega}{a} \right)^{1/2} = (1 + i) \left(\frac{\omega}{2a} \right)^{1/2} \tag{9}$$

A relatively simple iteration loop between equations (7) and (9) with the aid of equation (6) now enables one to compute values for a and λ. Indeed, when Δ is known after determination by the Lissajous figure method, a provisional value for β can be calculated from equation (7), while taking both equations (8) and (6) into account and selecting an appropriate provisional value for a and quantities ρ, C_p, and ε (either from the literature or from separate measurements). Subsequently, equation (9) allows calculation of a second provisional value for a and thus also for λ when again taking equation (6) into account. In this way the iteration loop is closed and may be repeated until convergence of the values for β, a and λ.

5.3. Computer Procedure

The essential elements and the various operations used are enumerated below.

5.3.1. Read Instructions

- Title, date, identification data, number of data, etc.
- Universal constants:
    $\pi = 3.14159265$
    $\sigma = 5.66961 \times 10^{-8}$ (W m^{-2} K^{-4}), Stefan–Boltzmann constant.
    $c_2 = 1.438833 \times 10^{-2}$ (m K), second radiation constant.
- Fixed equipment constants:
    $f = 0.5$ (Hz), modulation frequency.
    $C = 1 \times 10^{-6}$ (F), capacitance of RC phase shifting circuit.
    λ_B = wavelength at which temperature is measured with brightness pyrometer (m).
- Variable equipment constant (by prior determination):
    ε_T = transmission coefficient of window(s) at wavelength(s) applicable (dimensionless).
- Specimen constants (either measured or taken from literature):
    ε_s = spectral emissivity at wavelength(s) applicable (dimensionless).
    $(\varepsilon_{s1}/\varepsilon_{s2})$ = ratio of spectral emissivity at first and second wavelengths of a two-color pyrometer (dimensionless).
    ε = total hemispherical emissivity as a function of temperature (dimensionless).
    d_{20} = thickness at 20 °C (m).
    ρ_{20} = density at 20 °C (kg m^{-3}).
    C_p = specific heat (J kg^{-1} K^{-1}).
- Data measured:
    R = value of resistance of RC circuit when Lissajous figure is reduced to a straight line (Ω).
    t_B = mean brightness temperature of lower surface (°C).
    t_R = mean temperature of lower surface, measured with a two-color ratio pyrometer (°C).

5.3.2. Specific Calculation Sequences

$$\left.\begin{array}{l} T_B = t_B + 273.15 \\ T_T = t_R + 273.15 \end{array}\right\} \quad \text{conversion to degrees K.}$$

$$T_{BW} = \left(\frac{1}{T_B} + \frac{\lambda_B}{c_2} \ln \varepsilon_T \right)^{-1} \quad \text{correction for window transmission.}$$

$$T_{TW} = \left[\frac{1}{T_T} + \frac{\lambda_2 \lambda_1}{(\lambda_2 - \lambda_1)} \frac{1}{c_2} \ln \frac{\varepsilon_{T1}}{\varepsilon_{T2}} \frac{\varepsilon_{s1}}{\varepsilon_{s2}} \right]^{-1} \quad \begin{array}{l}\text{correction for window trans-} \\ \text{mission and ratio of emissivity} \\ \text{at } \lambda_1 \text{ and } \lambda_2.\end{array}$$

$\varepsilon_{\rm s} = \exp\left[(T_{\rm TW}^{-1} - T_{\rm BW}^{-1})c_2\lambda_{\rm B}^{-1}\right]$ calculation of spectral emissivity.

$T_{\rm TW}$ is the real, fully corrected temperature (in degrees K) of the lower surface. It is used for further calculations and subsequently called T_0.

$C_p = C_0 + C_1 T_0 + C_2 T_0^2 + C_3 T_0^3 + C_4 T_0^4$ specific heat as a function of temperature using the parameters C_0-C_4.

$d = d_{20}(1 + D_0 + D_1 T_0 + D_2 T_0^2 + D_3 T_0^3 + D_4 T_0^4)$ thickness corrected for temperature using thermal expansion data characterized by the parameters D_0-D_4.

$\rho = \rho_{20} \dfrac{(d_{20})^3}{(d_{T_0})^3}$ correction of the density as a function of temperature.

$\Delta = a\tan\left(\dfrac{\omega CR^2}{R + R'[1 + (\omega CR)^2]}\right)$

calculation of the phase difference employing the measured value R—see equation (1).

An iteration loop is now carried out. Such a loop is necessary because equation (7) is not an explicit function of the required quantities λ and a, since this equation contains the parameters β and c_0 which themselves depend on these quantities λ and a. Even d is not a constant parameter since it is a function of T_0, which in turn is a function of λ and a.

Iteration in this manner proceeds until the measured and calculated values of Δ converge. Generally it takes two or three iterations until consecutive values stabilize to within a residual difference of 0.01%.

An additional perfection of the calculations is advisable. It concerns the mean temperature. Indeed, when the mean temperature $T_{\rm m} = (T_0 + T_{\rm d})/2$ can be used instead of T_0, all intervening parameters can be recalculated for this $T_{\rm m}$ to yield results that are in better agreement with the real values. Knowing λ, the value of $T_{\rm d}$ can be computed directly when assuming one-dimensional heat flow through the sample. Since this flow will equal the heat lost by radiation at the nonbombarded surface, and assuming the temperature of the surrounding measuring chamber walls to be 16 °C, one can write

$$\frac{\lambda}{d}(T_{\rm d} - T_0) = \varepsilon\sigma[T_0^4 - (16 + 273.15)^4]$$

or

$$T_{\rm d} = T_0 + \frac{\varepsilon\sigma d}{\lambda}(T_0^4 - 7 \times 10^9)$$

If, for some particular reason, one wishes to measure $T_{\rm d}$ instead of T_0, a simple iteration loop can calculate T_0 from $T_{\rm d}$, using this latter equation. Some five iterations generally lead to a precision of 10^{-3} K.

Finally, a printout and plot subroutine concludes the program.

6. PRACTICAL OPERATION TECHNIQUES

6.1. Operational Procedure

The transmission of the windows should first be measured and then, after preparation of the specimen (diameter 8 mm, plane-parallel, estimated favorable thickness, measurement of $\varepsilon_{s1}/\varepsilon_{s2}$), it is mounted in the sample holder and placed at the center of the chamber.

During the pump-down sequence the detectors, diaphragms, and pyrometers are aligned for full focus and centering. This is generally carried out with the aid of an intense light source illuminating the specimen through one of the windows.

After switching on the filament supply and the accelerating voltage, the specimen is slowly warmed up to the desired temperature. Temperature uniformity is checked by shifting the pyrometer's viewing spot over the specimen surface and by adjusting the beam until all pyrometer readings are identical.

The adequate depth of modulation and the best settings of the amplifiers are selected, after which it is necessary to wait a few minutes until full stabilization is established. The Lissajous figure is adjusted by the resistance R of the RC circuit until a straight line is obtained, and then both R and T_0 (or T_d) are measured.

The next measurement is prepared by adjusting the temperature to a new mean value. The results are accumulated point by point. After the measurements, the transmission of the windows and possibly the ratio $\varepsilon_{s1}/\varepsilon_{s2}$ should be determined again. Thereafter, the data are processed.

6.2. Checks for Proper Operation—Error Assessment

Erroneous results are mostly due to poor knowledge of Δ. This, in turn, arises in the first instance from an insufficiently homogeneous temperature distribution on the sample surface. Exact adjustment of the electron beam and scanning settings prior to each series of measurements, and possibly also reexamination from time to time during such a series, are both absolutely mandatory. Specimens that have an inhomogeneous structure, that are very poor electrical conductors, or that are highly porous, will often cause trouble. It happens that certain specimens undergo a transformation of their emissivity, resulting in an inhomogeneity of the temperature distribution or temperature instability.

A recommendable test to detect a possible error in the determination of Δ consists in simply carrying out a few more measurements on specimens of the same material, but which differ in thickness or diameter.

It is self-evident that before measurements are started with a newly built apparatus, it is necessary to check for possible phase errors introduced by the detectors and amplifiers. Therefore it is convenient to test first the electronic circuits without the detectors. It is sufficient to inject a 0.5-Hz signal from a sine generator to the input of both the preamplifiers simultaneously and check on a possible phase shift between both the outputs. Thereafter, the complete setup should be checked with the detectors at their normal position. It is recommended that one looks at a small low-voltage incandescent lamp or LED, or observes both surfaces of a very thin (e.g., 0.025 mm or thinner) metal strip energized by a current modulated at the same 0.5-Hz frequency. It is even more advisable to carry out measurements with both detectors simultaneously mounted either above or below the same surface of a specimen.

During the measurements, attention must be paid to spurious variable magnetic fields in the vicinity of the electron trajectories. Troubles arising from poor electrical conduction or variations in emissivity and temperature can usually be avoided by the deposition of a very thin metal layer. Attention must be paid to vapor deposition on the windows, especially in connection with the temperature determination.

Comparative examinations are highly recommended. Results obtained with equipment measuring directly conductivity, or measurement of a reference material, or data culled from other sources should be compared.

In Vol. 1 of this Compendium (on pp. 411–418)[1] a rather comprehensive error evaluation of the modulated heat input method is presented. The many measurements carried out previously on various materials reveal that, in general, a reproducibility of the order of 1 to 2% is obtained. Moreover, comparative measurements and measurements from other sources enable one to make allowance for a probable overall uncertainty. An uncertainty between 2 and some 5% is most probably reached, possibly attaining some 10% when porous or inhomogeneous specimens are encountered.

7. CLOSING REMARKS—FURTHER IMPROVEMENTS

The equipment described here was put into service in 1970 and undoubtedly has substantial merits. It belongs to a specific generation of nonstationary or quasi-stationary equipment which covers particularly the high-temperature range. It has the advantage, in contrast to steady-state methods, that possible errors can be detected and also corrected much more easily. Furthermore, it is not too difficult to construct such equipment for less than some US$30,000 exclusive of labor.

The only well-known competitor of this method is the laser flash technique apparatus, which is moreover commercially available (such as Theta Industries, Inc., Port Washington, NY, USA; Conductronic IV; about US$110,000 without data acquisition; maximum temperature 2100 °C), while actually a modulated electron beam apparatus is not commercially available.

Occasionally, the fact that measurements are restricted to vacuum conditions can sometimes represent a disadvantage. It is, of course, feasible to convert the measuring chamber to two compartments and in this way isolate the sample from the beam and bring the sample compartment to the desired controlled atmosphere. However, as the sample will now be heated by radiation from the separation plate, the measurement of the upper specimen surface by an optical detector is nearly impossible. Moreover, the maximum attainable temperature will be some 1500 to perhaps 1700 °C only.

These last few years have been marked by a succession of new developments in electronics and optoelectronics, so each apparatus may appear doomed to obsolescence shortly after it has been put into service and inevitably a few improvements should possibly be taken into consideration when conceiving similar apparatus in the future.

1. A beam scanning system is advised equivalent to that designed by Mayer and Neuer.[3] The beam follows a spiral path over the sample surface whereby its speed is adjusted such that a homogeneous temperature distribution is reached.

2. It is much more convenient to modulate the beam by modulating the voltage of a Wehnelt cylinder instead of the filament temperature. This facilitates modulation at various frequencies and may be helpful when measuring specimens which are, "thermally" speaking, either too thick or too thin. Of course, a new electron gun will have to be designed that delivers a smaller beam diameter (1 mm or less) on target, and whose Wehnelt enables perfect beam control.

3. The phase-shift measurement setup can be made more handy and possibly more accurate by pure electronic means. A method analogous to the method mentioned by Brandt and Neuer[12] could be of interest here. They use zero-point triggers which deliver needle-shaped pulses at the output. Each time the sinusoidal signal passes through zero, the time difference is measured between the pulses by a digital counter. This time difference is converted to phase shift.

4. Acquisition of a high-speed data-processing system and coupling the reading of both the phase difference and temperature allows direct analysis of the transient data and possibly also smoothing over several cycles. This will improve the resolution and accuracy of the results.

5. Especially when carrying out measurements at lower temperatures, it is advisable to add an electronic stabilizer of the mean sample temperature in order to shorten the time needed for temperature stabilization. This can be done by picking up a signal from the pyrometer or from one of the detectors and feeding it back to the temperature adjusting unit, i.e., the filament supply or Wehnelt supply. Actually, without a stabilizer and at low specimen temperatures, often five or even ten minutes are needed to reach a more or less stable temperature.

6. Finally, it is also interesting to find out how to conduct trustworthy specific heat measurements simultaneously.

NOTATION

Symbol	Name	Unit of value
a	Thermal diffusivity	$m^2\,s^{-1}$
c_2	Second radiation constant	1.438786×10^{-2} m K
C	Capacitance of RC network	F
c_0	Boundary condition parameter	m^{-1}
C_p	Specific heat at constant pressure as a function of temperature	$J\,kg^{-1}\,K^{-1}$
C_0-C_4	Parameters used to calculate $C_p = f(T)$	Dimensionless
d	Thickness of specimen as a function of temperature	m
d_{20}	Thickness of specimen at 20 °C	m
D_0-D_4	Parameters used to calculate the thickness $d = f(T)$	Dimensionless
f	Frequency of modulation	0.5 Hz
i	Imaginary unit	Dimensionless
N.E.P.	Noise equivalent power of radiation detector	$W\,Hz^{-1/2}$
R	Electrical resistance of RC network when adjusted for zero phase difference	Ω
R'	Fixed resistor in phase shifting network	101 kΩ
t_B	Mean brightness temperature of lower specimen surface	°C
t_R	Mean two-color ratio pyrometer temperature of lower surface	°C
T	Absolute temperature—several indices are used. They are explained in the text	K
x	Distance	m
β	Complex number connected with diffusivity	m^{-1}
δ_d	Phase difference between bombarded face and applied modulated beam	Radian
δ_0	Phase difference between non-bombarded face and applied modulated beam	Radian
Δ	Phase difference between both faces ($= \delta_0 - \delta_d$)	Radian
ε	Total hemispherical emissivity	Dimensionless
ε_s	Spectral emissivity at wavelength(s) applicable	Dimensionless
$\varepsilon_{s1}/\varepsilon_{s2}$	Ratio of spectral emissivity at wavelengths λ_1 and λ_2 of two-color pyrometer	Dimensionless
ε_T	Transmission coefficient of window(s) at wavelength(s) applicable	Dimensionless
λ	Thermal conductivity	$W\,m^{-1}\,K^{-1}$
λ_B	Wavelength at which temperature is measured with brightness pyrometer	m
λ_1, λ_2	First and second wavelength at which the two-color pyrometer operates	m
ρ	Density as a function of temperature	$kg\,m^{-3}$
ρ_{20}	Density at 20 °C	$kg\,m^{-3}$
σ	Stefan–Boltzmann constant	$5.66961 \times 10^{-8}\,W\,m^{-2}\,K^{-4}$
ω	Angular frequency	s^{-1}

ACKNOWLEDGMENTS. Sincere thanks are due to Mrs. M.J. Webers who, besides her various other duties, found time to type the text. Mr. A. Gijs is given full credit for his invaluable and continual technical assistance and inventiveness during the construction and gradual improvement of the equipment. Thanks are also extended to Dr. N. Maene and Dr. J. Nihoul for their many interesting and encouraging discussions.

REFERENCES

1. R. De Coninck and V.E. Peletsky, in: *Compendium of Thermophysical Property Measurement Methods*, Vol. 1 (K.D. Maglić, A. Cezairliyan, and V.E. Peletsky, eds.), pp. 367–428, Plenum Press, New York (1984).
2. W.B. Herrmannsfeldt, "Electron Trajectory Program—EGUN," Stanford Linear Accelerator Center, Report SLAC-226 (1979).
3. R. Mayer and G. Neuer, "Isothermal Sample Heating Using a Modulated Electron Beam," *Rev. Int. Hautes Temp. Refract.* **12**, 191–196 (1975).
4. R. De Coninck, R. De Batist, and A. Gijs, "Thermal Diffusivity, Thermal Conductivity and Spectral Emissivity of Uranium Dicarbide at High Temperatures," *High Temp. High Pressures* **8**, 167–176 (1976).
5. R.D. Cowan, "Proposed Method of Measuring Thermal Diffusivity at High Temperatures," Los Alamos Scientific Laboratory Report LA-2460 (1960).
6. R.D. Cowan, "Proposed Method of Measuring Thermal Diffusivity at High Temperatures," *J. Appl. Phys.* **32**, 1363–1370 (1961).
7. M.J. Wheeler, "Thermal Diffusivity at Incandescent Temperatures by a Modulated Electron Beam Technique," *Br. J. Appl. Phys.* **16**, 365–376 (1965).
8. M.J. Wheeler, "Rapid Measurement of Thermal Diffusivity Using a Modulated Electron Beam," *J. Sci. Technol.* **38**, 102–107 (1971).
9. R. Penninckx, "Calculation of Thermal Diffusivity from Measurements with the Sine Wave Modulation Method: Modification of Cowan's Equations," *Appl. Phys. Lett.* **21**(2), 47–48 (1972).
10. R. De Coninck and N. Maene, "On the Analysis of the Diffusivity Measurement Method with Modulated Heat Input, in: *Thermophysical Properties*, Proceedings of the 1st Asian Thermophysical Properties Conference, April 21–24, 1986, Beijing, China (Wang Buxuan *et al.*, eds.), p. 391 (abstract), China Academic Publishers (1986).
11. R. De Coninck, "On the Analysis of the Diffusivity Measurement Method with Modulated Heat Input," *Intern. J. of Thermophysics* **11**, No. 5, 923 (1990).
12. R. Brandt and G. Neuer, "Thermal Diffusivity of Solids—Analysis of a Modulated Heating-beam Technique," *High Temp. High Pressures* **11**, 59–68 (1979).

12

Instruments for Measuring Thermal Conductivity, Thermal Diffusivity, and Specific Heat under Monotonic Heating

E.S. PLATUNOV

1. INTRODUCTION

This chapter deals with instruments that allow one to study solid materials of thermal conductivities 0.1–100 W m^{-1} K^{-1} at temperatures ranging from 150 to 700 K. Test specimens shaped as disks or short rods 15–20 mm in diameter and 1–40 mm in height are used for measurements performed under monotonic near-linear heating. The instruments are equipped with conventional devices for measuring temperature and heat flow, as well as heating and cooling systems.

The chapter examines, in much detail, structures of thermal cells, features of calibration tests and experiments, sources of errors, and requirements of the accessory equipment. Moreover, in view of the fact that Volume 1 of the Compendium contains no theory pertaining to the instruments in question, the initial sections of the present chapter are intended to fill that gap.

2. THEORY OF A MONOTONIC REGIME

A monotonic heating regime is assumed to be a regime of heating or cooling bodies when the rate of temperature changes throughout a body is maintained nearly uniform.[1,4] The theory of such a regime is the generalization of that of well-known quasi-stationary conditions.[2,3,5] The investigators' bias toward thermophysical measurements under monotonic conditions may be

E.S. PLATUNOV • Physics Department, Leningrad Technological Institute of the Refrigerating Industry, Leningrad 191002, USSR.

explained by their wish to simplify the techniques applied to quasi-steady-state conditions so that they can be employed for a proximate analysis of thermo-physical properties of materials over a wide range of temperatures without assuming that the equation of thermal conduction is linear.

Bodies possessing a simple geometry (plate, cylinder, sphere, thin film) are used to study thermophysical properties under monotonic conditions. A one-dimensional temperature field $t(r, \tau)$ of sufficiently low temperature drop $\theta(r, \tau)$ is developed inside the body relative to the base section ($r = 0$) at temperature $t_0(\tau)$. The nonlinear equation of thermal conduction

$$\text{div} (\lambda \text{ grad } t) = cp(\partial t/\partial \tau)$$

can be reduced to the equation[1]

$$\frac{d^2\theta}{dr^2} + \frac{f-1}{r}\frac{d\theta}{dr} = \frac{b_0}{a_0} + \left[(k_b - 2k_a)\frac{b_0}{a_0}\theta - k_\lambda\left(\frac{d\theta}{dr}\right)^2\right] \tag{1}$$

where f is the shape factor (for a plate $f = 1$, for a cylinder $f = 2$, and for a sphere $f = 3$), $a_0 = a(t_0)$, and $b_0 = dt_0/d\tau$; $k_i = (1/i_0)(di_0/dt_0)$ are the temperature coefficients, with $i = \lambda$, a, and b.

Equation (1) may be solved by the iteration method, because if the restrictions $|k_i\theta| < 0.1$ are satisfied during the monotonic stage of the experiment, then its nonlinear part, the terms within the square brackets, turns out to be a first-order infinitesimal correction.

Under the above conditions for symmetrical monotonic heating (cooling) of a plate, a cylinder, and a sphere, the following temperature distribution will be valid with an error less than 1%:

$$\theta(r, \tau) = [b_0r^2/(2fa_0)](1 + \sigma) \tag{2}$$

where

$$\sigma = \tfrac{1}{2}[f/(f+2)]\left(k_b - 2k_a - \frac{2}{f}k_\lambda\right)[b_0r^2/(2fa_0)]$$

is a relative correction for nonlinearity, with the zero point r taken at the body center.

In a thin plate asymmetrically heated monotonically subject to the above restrictions, the temperature distribution is given by

$$\theta(x, \tau) = [g_0x + b_0x^2/(2a_0)] + \Delta\theta(x, \tau) \tag{3}$$

where $g_0 = \partial\theta(0, \tau)/\partial r$ and $\Delta\theta(x, \tau)$ is a correction for nonlinearity:

$$\Delta\theta(x, \tau) = \tfrac{1}{3}(k_b - 2k_a - 2k_\lambda)\left(g_0x + \frac{b_0x^2}{4a_0}\right)\frac{b_0x^2}{2a_0} - \tfrac{1}{2}k_\lambda g_0^2x^2$$

The coordinate $x = 0$ is aligned with the central section. The error of the solution does not exceed 1%.

Equations (2) and (3) show that with the restrictions $|k_i\theta| < 0.1$ the relationships governing a monotonic regime approximate those of a quasi-steady-state regime, while the nonlinear factors slightly affect the temperature distribution and can be taken into account by corrections. Hence, all known methods of dealing with a quasi-steady-state regime are suitable, in principle, for thermophysical measurements under monotonic conditions. However, here only those commonly used for thermophysical measurements in the temperature range of 150–700 K will be scrutinized.

3. THEORY OF MEASUREMENT TECHNIQUES

The theory underlying the measuring techniques pertaining to $\lambda(t)$, $a(t)$, and $c(t)$ for materials under a monotonic regime can readily be demonstrated by the behavior of an asymmetrically heated (cooled) plate, for example. To simplify the mathematical treatment, the correction for nonlinearity in equation (3) may be assumed negligible, in which case the temperature field in the plate will obey the relationship

$$\theta(x, \tau) = g_0x + b_0x^2/(2a_0) \tag{4}$$

According to Fourier's law, we have from equation (4)

$$q(x, \tau) = -\lambda_0(g_0 + b_0x/a_0) \tag{5}$$

Hence the following expressions apply for $\theta_l(\tau) = t_l(\tau) - t_0(\tau)$, $\theta_{-l}(\tau) = t_{-l}(\tau) - t_0(\tau)$, $q_l(\tau)$, and $q_{-l}(\tau)$:

$$\theta_l = g_0l + b_0l^2/(2a_0), \qquad \theta_{-l} = -g_0l + b_0l^2/(2a_0) \tag{6}$$

$$q_l = -\lambda_0g_0 - c_0\rho_0b_0l, \qquad q_{-l} = -\lambda_0g_0 + c_0\rho_0b_0l \tag{7}$$

These relations allow one to obtain explicit analytical expressions for the thermophysical coefficients in which we are interested, namely,

$$\lambda_0 = \frac{q_{-l} + q_l}{\theta_{-l} - \theta_l}l, \qquad c_0 = \frac{q_{-l} - q_l}{2\rho_0b_0l}, \qquad a_0 = \frac{b_0l^2}{\theta_{-l} + \theta_l} \tag{8}$$

Thus the generalized approach to heat flow and temperature measurements makes it possible to simultaneously and independently determine all three thermophysical coefficients for a plate as functions of temperature, i.e., $\lambda(t)$, $c(t)$, and $a(t)$. The universal relationship $\lambda = ac\rho$ may be used to check the mutual consistency of the coefficients determined from equations (8) and to detect and evaluate the systematic errors originating from the instrument.

In addition, it follows from equation (8) that in order to measure $\lambda(t)$, $c(t)$, and $a(t)$ with a predetermined accuracy, it is necessary to observe the whole complex of interconnected restrictions as the parameters $\theta_l(\tau)$, $\theta_{-l}(\tau)$, $b_0(\tau)$, $q_l(\tau)$, and $q_{-l}(\tau)$ are inevitably measured with errors during the experiment. These restrictions are mostly contradictory, so that when a particular measurement technique is being developed and a relevant instrument constructed, it is vital that the admissible value of the errors be optimized for every thermophysical coefficient measured.

In this connection, of particular interest are methods that allow one to study experimentally one or two thermophysical coefficients independently. A very promising group of measuring techniques comprises those where a thermal measuring cell yields symmetrical heating of a specimen ($\theta_l = \theta_{-l}$, $q_l = -q_{-l}$). These methods make it possible to measure independently

$$c_0 = [q_l/(\rho_0 b_0 l)](1 + \sigma_c) \qquad \text{and} \qquad a_0 = [b_0 l^2/(2\theta_l)](1 + \sigma_{a\theta}) \qquad (9)$$

where σ_c and $\sigma_{a\theta}$ are relative corrections for nonlinearity calculated from equation (2) and given by

$$\sigma_c = -\tfrac{1}{3}(k_b - 2k_a + k_\lambda)\theta_l \qquad \text{and} \qquad \sigma_{a\theta} = \tfrac{1}{6}(k_b - 2k_a - 2k_\lambda)\theta_l$$

The specific features of any method depend mainly on the ways of setting-up and measuring heat flows $q_l(\tau)$ and $q_{-l}(\tau)$. The simplest approach is that in which a thermal measuring cell provides for the preset symmetrical heating (cooling) of a specimen only, but not for measuring heat flows. Such a thermal cell allows one to determine the thermal diffusivity of materials, $a(t)$. Two variants of measurement are feasible. In one, the temperature at the center, $t_0(\tau)$, and temperature drop, $\theta_l(\tau)$, in the specimen are determined during the experiment [see equation (9)]; in the other, the lag $\tau_l(t)$ of the temperature $t_0(\tau)$ at the center with respect to that of the outside specimen surface, $t_l(\tau)$, is measured directly. Subject to the correction for nonlinearity, the design formula for the latter variant will adopt the following form:

$$a(t) = \frac{l^2}{2\tau_l(t)}(1 + \sigma_{a\tau}) \qquad (10)$$

where

$$\sigma_{a\tau} = -\tfrac{1}{3}(k_b - 2k_a + k_\lambda)b_0\tau_l$$

This measurement technique is attractive because it eliminates complicated calculation of the heating rate $b_0(\tau)$.

Another group of measuring techniques comprises methods in which a thermal measuring cell gives the extremely asymmetrical monotonic heating (cooling) of a plate when the heat flow stored in the plate turns out to be small in comparison with the cross-sectional heat flow passing through it while the temperature field remains close to linear and in a steady state ($q_{-l} \approx q_l$, $\theta_{-l} \approx -\theta_l$). The methods applicable to this group are very practical, since they allow the thermal conductivity of materials $\lambda(t)$ to be readily studied over a wide range of temperatures by proximate analysis. In order to determine $\lambda(t)$ by such methods, according to equation (2), it suffices to measure the temperature drop $\theta_{-l,l}(\tau)$ in the specimen and one of the specific surface heat flows, $q_l(\tau)$ or $q_{-l}(\tau)$:

$$\lambda_0 = (2lq_l/\theta_{-l,l})(1 + \sigma_c) = (2lq_{-l}/\theta_{-l,l})(1 - \sigma_c) \tag{11}$$

where $\sigma_c \approx (c_0\rho_0 lb_l/q_l) \approx (c_0\rho_0 lb_{-l}/q_{-l})$ is a correction for the heat capacity of a specimen.

Equation (11) does not require allowance for nonlinear factors if the restrictions $|k_b| < 3 \times 10^{-3}\,\text{K}^{-1}$, $\sigma_c < 0.1$, and $|\theta_{-l,l}| < 50\,\text{K}$ are sufficiently lax in the experiment. This is an important advantage of this group of methods.

The asymmetric regime required ($\sigma_c < 0.1$) may be readily attained when the system incorporating a specimen (a plate) and a massive metal body (the heat sink) is heated by a heat flow penetrating the system only through the front face of the specimen. If the heat sink has a known heat capacity $C_{hs}(t)$ with the temperature distribution maintained uniform, and there is no direct heat transfer between the surroundings and the heat sink, which has good thermal contact with the "cold" face of the specimen, then it can be used as an enthalpy heat flowmeter to measure the specific heat flow $q_l = C_{hs}b_{hs}/S$. In this situation, equation (11) will assume the form

$$\lambda_0 = (2l/S)(C_{hs}b_{hs}/\theta_{-l,l})(1 + \sigma_c) \tag{12}$$

where $\sigma_c = c_0\rho_0 lS/C_{hs}$.

In this equation the ratio $\theta_{-l,l}/b_{hs}$ may be replaced by the lag $\tau_{l,-l}(t)$ of the temperature $t_l(\tau)$ relative to that of $t_{-l}(\tau)$, which somewhat simplifies the experimental procedure. However, this results in increasing the corrections for nonlinearity and requires another rigid restriction on the allowable temperature drop ($\theta_{-l,l} < 10\,\text{K}$).

A thermal system consisting of a thin plate and a metal heat sink has found application when measuring not only thermal conductivity but specific heat as well. Actually, if a plate of known heat conductivity $K_{hm} = \lambda_{hm}S/h_{hm}$

is a constant element of the system, there appears to be a chance of measuring the thermal capacity of a heat sink $C_{hs}(t)$ heated through the plate. This does not mean that only metals can be studied in such a manner. A sufficiently uniform temperature distribution in a specimen is readily attained by application of a metal container.

If bare portions of the surface of a specimen and a container are completely insulated from the surroundings, the heat capacity of a specimen may be calculated from

$$C = K_{hm}\theta_{hm}/b_c - C_c \qquad (13)$$

where $\theta_{hm}(\tau)$ is the temperature drop in the plate heat flowmeter, while $c_c(t)$ and $b_c(\tau)$ are the effective heat capacity of the container (the correction for heat capacity of the heat flowmeter being allowed for) and the rate of its heating (which is to coincide with the rate of heating the specimen), respectively.

Variants of the aforementioned methods are employed extensively for thermophysical measurements in practice. A number of home-made thermophysical instruments for investigations over the 150–700 K temperature range are based on these methods.[6,7] Below we describe cells of such instruments which can be manufactured in any thermophysical laboratory.

We introduce a convention whereby the instruments will be designated as a-, λ-, or c-calorimeters, the symbols a, λ, and c indicating the thermophysical coefficients measured.

4. STRUCTURAL FEATURES OF CALORIMETERS

4.1. a-Calorimeters

a-Calorimeters of flat and cylindrical geometry are employed very extensively. Flat a-calorimeters are mostly used to investigate solid homogeneous and finely dispersed materials with $0.1 \text{ W m}^{-1} \text{ K}^{-1}$ thermal diffusivity. Cylindrical a-calorimeters prove more convenient for studying highly effective thermal insulating powder and fibrous materials and liquids.

The main elements of a cylindrical a-calorimeter (Fig. 1) are a relatively massive metal cup (1) with a cover (2) and a heat shield enclosure (3). A Ni-Cr alloy spiral (4) is placed on the side wall of the cup to provide a uniform heat distribution on its inner surface during heating. For heating, with b almost constant, it is convenient to employ the automatic temperature control system (ATCS) and the programmed temperature regulator (PTR), which change the

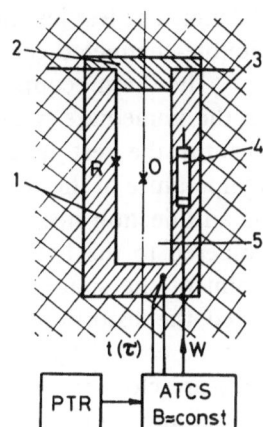

FIGURE 1. Schematic outline of a cylindrical a-calorimeter: 1—metal cup, 2—cover, 3—heat shield enclosure, 4—spiral heater, 5—test specimen.

power $W(\tau)$ of the heater (4) following the signal $t(\tau)$ of a thermocouple incorporated into the cup.

The tested material (5) is placed in the cup space. Thermocouples with junctions at points R and 0 are used to measure temperatures $t(R, \tau)$ and $t(0, \tau)$. The thermocouple wires, made of 0.15–0.20-mm-diameter wire, are insulated electrically and exit from the cup along the isothermal surfaces.

If the height H and diameter $2R$ of the cup space satisfy the condition $H/(2R) > 5$, the temperature distribution in the central part of the space remains practically one-dimensional cylindrical during the experiment and $a(t)$ may be calculated on the basis of the expression derived from equation (2) with $f = 2$:

$$a_0 = \frac{b_0 R^2}{4\theta_{R0}} (1 + \sigma_a) \tag{14}$$

where $\sigma_a = \frac{1}{4}(k_b - 2k_a - k_\lambda)\theta_{R0}$ is a correction for nonlinearity, which can readily be neglected for $\theta_{R0} < 20$ K and $|k_b| < 3 \times 10^{-3}$ K^{-1}.

The cup space radius R should be chosen so that the temperature drops θ_{R0} for the whole group of materials tested with an a-calorimeter fall in the interval 5–20 K, the heating rate $b = 0.01$ to 1 K s^{-1} being considered reasonable.

When studying materials with $\lambda < 1$ W m^{-1} K^{-1}, a cell with hollow space radius $R = 10$ mm and height $H = 100$ mm is found to be suitable. The cup of the cell may be made of copper, aluminum, or their alloys. The thickness of the cup walls and cover should not be less than 10 mm to ensure a uniform temperature of the inner surface of the space.

Temperatures are recorded adequately by copper–constantan, chromel–alumel, or nichrome–constantan thermocouples, the electrode diameters being

0.2 mm. The test junctions of the thermocouples are sheathed serviceably with double-channeled ceramic tubes of 1.2–1.6-mm diameter.

The thermocouple should be fitted into the axial hole drilled in the cup in the immediate proximity of the cup space. It is advisable to additionally sheathe the test junction of the thermocouple 0 (which is to record the temperature of the central zone of the specimen) with a thin-walled tube, such as an injection needle of suitable diameter. The base of the needle is silver-soldered to the bottom of the cup so that its upper tip reaches the central point of the space. If the thermocouples are manufactured identically and fitted, the distortions of the temperature fields of the cell and specimen (which may be caused by thermocouples 0 and R) are slight and similar, therefore they are close to the actual value and do not affect the measured values of $\theta_{R0}(\tau)$ and $b_0(\tau)$.

The basic elements of an a-calorimeter (Fig. 2) intended for investigating materials having $\lambda = 0.1$ to $10 \, \mathrm{W \, m^{-1} \, K^{-1}}$ are two symmetrical metal blocks $(1, 2)$ within which Ni–Cr spirals $(3, 9)$ are incorporated. The blocks are supplied with movable guard rings $(5, 7)$ and are protected from the surroundings by heat shields $(4, 8)$. When in working order the flat front faces of the blocks are situated snugly via a fluid lubricant against the end faces of specimen (6) having the shape of a disk or short rod. The air-filled gap between the specimen and guard rings $(5, 7)$ is an additional heat shield.

The most suitable specimens for measurements are those of diameter ca 20 mm. It is recommended that the specimen thickness be selected in line with the equation $h^2 \cong 10a/b$. Thus, if $b \approx 0.1 \, \mathrm{K \, s^{-1}}$ and $a < 10^{-5} \, \mathrm{m^2 \, s^{-1}}$, the optimal thickness of specimens is $h = 5$ to 30 mm. Each of the blocks $(1, 2)$ can be 40–60 mm in diameter and 15–40 mm in height. The thickness of the guard rings should be 1–3 mm. If the a-calorimeter is designed to study a group of materials having comparable properties, the thickness of the specimens measured can be kept the same. Hence, it is rational to make the blocks and guard rings as one piece. It is more workable to fabricate guard rings of

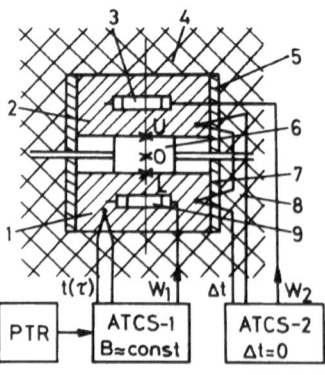

FIGURE 2. Schematic diagram of a flat a-calorimeter: 1,2—metal blocks, 3,9—heaters, 4,8—heat shields, 5,7—movable guard rings, 6—specimen.

various diameters so that they could overlap each other, thus ensuring a tight contact between the blocks and the flat faces of the specimen. As in the earlier cell (Fig. 1) both the blocks and guard rings may be made of copper, aluminum, or their alloys.

The automatic temperature control systems ATCS-1 and ATCS-2 afford symmetrical monotonic heating of the specimen through the end faces. The former regulates the power $W_1(\tau)$ of heater (9) via the programmed temperature regulator (PTR), thus providing a near-linear temperature change $t(\tau)$ in block (1). The latter regulates the power $W(\tau)$ of heater (3), maintaining equal temperatures in both the blocks (1, 2).

Thermocouples U and L are fixed in the immediate vicinity of the front faces of the blocks, in order to measure the temperature of the end faces of the specimen being measured. A flexible thermocouple is used to measure the temperature of the central section of the specimen ($x = 0$), the thermocouple junction being inserted into the specimen through a radial hole down to the point 0. It is important that the "cold" junctions of thermocouples U, L, and 0 be thermostatted, e.g., placed into a Dewar or into the hollow space of a massive metal block at room temperature. The permissible temperature drops and rates of temperature change in "cold" junctions should be negligibly small in comparison with the operating temperature drop and heating rate of the specimen.

The wires of the test junctions of thermocouples L, 0, and U are passed conventionally through double-channeled ceramic tubes 1.1–1.6 mm in diameter. The material of which the thermocouples should be made has already been referred to when the previous cell scheme was described (see Fig. 1). When mounting thermocouples L and U one should take care that there is good heat contact between the thermocouple wires, the tube, and the block. It is advisable to lubricate the tip of thermocouple 0 with silicon organic oil when it is fitted into the hole in the specimen. The flexible wires of thermocouple 0 should be as long as possible and placed loosely in the air-filled gap, without touching either the specimen or the walls of the gap.

Actually, the automatic temperature control systems of the blocks are likely to allow the temperature field to be somewhat asymmetrical throughout the specimen. Therefore, the design equations differ to some extent from equations (9) and (10) given earlier:

$$a(t_0) = \frac{b_0 h^2}{\theta_{L0} + \theta_{U0} - \Delta\theta_{cr}} (1 + \sigma_{a\theta} - \sigma_\alpha) \qquad \text{and}$$

$$a(t) = \frac{h^2}{\tau_{OL} + \tau_{OU} - \tau_{cr}} (1 + \sigma_{a\tau} - \sigma_\alpha) \qquad (15)$$

where $\Delta\theta_{cr} = cphb_0 P_{cr}$ and $\Delta\tau_{cr} = cphP_{cr}$ are corrections allowing for the contact thermal resistance $P_{cr}(t)$ between the face ends of the specimen and the front faces of the blocks, $h = 2l$ is the thickness and R the radius of the disk specimen, and $\sigma_\alpha = 0.57[h/(2R)]^2$. Quantity $Bi/[Bi + h/(2R)]$ is a dimensionless correction for the lateral heat transfer from the specimen through the air-filled gap,[1] where $Bi = \alpha R/\lambda$; $\sigma_{a\theta}$ and $\sigma_{a\tau}$ are dimensionless corrections for nonlinearity [see equations (9) and (10)].

The specific contact thermal resistance $P_r(t)$ was found to depend on the quality of grinding and on the materials comprising the contacting faces, on the value of the compressive load and the properties of the intervening media (air, powder, fluid film). If experiments are carried out under reproducible conditions, then quantity $P_r(t)$ may simply be regarded as a "constant" of the calorimeter. For example, investigations showed that the resistance of surface ground test specimens of diameter $2R = 10$ to 20 mm is usually $P_r \approx (0-1.0) \times 10^{-4}$ m^2 K W^{-1}, if the contacting surfaces are wetted with an oil film (such as silicon organic oil) and are compressed with a force of $(1-5) \times 10^5$ N m^{-2}.

Figure 3 presents another variant of a flat a-calorimeter which, unlike the earlier one, is intended for studying materials with $\lambda > 5$ W m^{-1} K^{-1}, including metals. Its basic element is a split massive metal block, its base (1) and cover (2) having thermal shields (5) and (8). As in the calorimeter variants discussed above, the ATCS-1 and PTR provide for the preset heating regime of the block with the help of a heater (7) incorporated into the base (1).

In the experiment a test specimen (6) having the shape of a short rod fits close to the surface of the base and is heated from the latter. A metal enclosure (4) eliminates heat transfer through the upper face of the specimen and reduces heat losses through the lateral surface of the specimen, due to the ATCS-2 and heater (3). The specimen temperature is measured at points 0 and L by

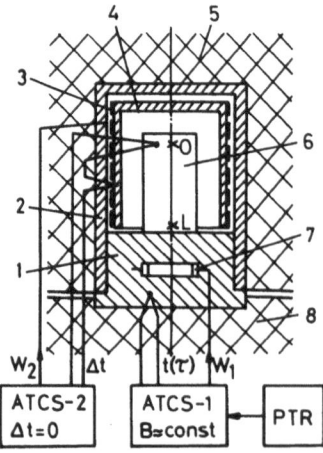

FIGURE 3. Schematic diagram of an a-calorimeter for metals: 1—base, 2—cover, 3,4—adiabatic enclosure with the heater, 5,8—heat shields, 6—specimen, 7—heater in the base.

thermocouples. Their junctions are electrically insulated and slide-fitted into the corresponding radial holes of the specimen.

The manner in which the thermocouples were fitted in the holes of the specimen was chosen with the aim of eliminating the effect of contact thermal resistances. The contribution of the latter is intolerably high when materials of high thermal conductivities are studied.

According to equations (9) and (10) thermal diffusivity is calculated by one of the equations

$$a(t_0) = \frac{b_0(x_L^2 - x_0^2)}{2\theta_{L0}}(1 + \sigma_\alpha) \qquad \text{and} \qquad a(t) = \frac{x_L^2 - x_0^2}{2\tau_{0L}}(1 + \sigma_\alpha) \quad (16)$$

where x_L and x_0 are coordinates of the thermocouple junctions L and 0 counted off from the upper specimen face; $\sigma_\alpha = [x_L^2/(6R^2)](\alpha R/\lambda)$ is a relative correction for heat leak through the lateral surface of the specimen.

Coefficient α may be evaluated in the air-filled gap between the specimen and the adiabatic enclosure (4) by the approximation relationship[1]

$$\alpha = \left(1 + \frac{R_e - R}{h}\right)\left(\frac{R_e}{R}\right)^{0.33} \frac{\lambda_a}{R_e - R} + 4\varepsilon_{red}\sigma_0 T_0^3$$

where R_e is the inner radius of the enclosure, h is the specimen height, λ_a is the thermal conductivity of the air, and ε_{red} is the reduced emissivity factor of the corresponding surfaces of the specimen and enclosure.

Recommended specimen sizes are 20-mm diameter and 20–50-mm height. Heating rate is $b \geqslant 10^{-1}$ K s^{-1}. The adiabatic enclosure (4) can be made of aluminum or copper with a wall thickness of about 1.0 mm. It is reasonable to choose the dimensions of the enclosure so that the air gap between it and the specimen occupies approximately 10 mm. The base (1) and cover (2) are made of copper, aluminum, or their alloys. The diameter and height of the base are about 50 mm. The cover wall thickness is approximately 3 mm, the cover being slide-fitted onto the base. The clearance between the cover (2) and enclosure (4) can be 2–3 mm. The heater (3) of enclosure (4) can be made of a uniformly wound nichrome spiral, electrically insulated with quartz fabric. It is promising to form a one-piece removable assembly consisting of the enclosure and cover.

The test junctions of thermocouples L and O are sheathed with thin (1.0–1.6-mm diameter) double-channeled ceramic tubes. The stationary parts of the thermocouple wires are inserted into the massive (3–5-mm diameter) ceramic tubes, which are fixed in the vertical holes in the base.

The flexible parts of the thermocouple wires should form a loop around the specimen so as to facilitate slide-fitting of the test junctions into the radial holes. Silicon organic oil or graphite lubricant can be used to promote thermal contact of the junctions with the specimen.

4.2. λ-Calorimeters

Today, very many designs of λ-calorimeters have been developed. The simplest are presented in Figs. 4-6. The first two designs are mainly intended for studying solid materials with $\lambda < 10\ \mathrm{W\,m^{-1}\,K^{-1}}$. However, they may be modified for investigating metals if need be. The third version (Fig. 7) is most promising for studying nonmetallic fluids and finely dispersed, thermally insulating materials.

The calorimeter shown in Fig. 4 consists of a split metal block comprising a base (1) and cover (2) enclosed by a heat shield (6, 9), a heat sink (4), and an adiabatic enclosure (5) with a heater (7) spread uniformly over the surface. A disk-shaped specimen (3) is placed between the heat sink and the base. In order to improve the heat contact, the flat faces of the specimen are wetted with silicon organic or graphite lubricant, if experimental conditions are favorable. Temperature drop in the specimen is measured by thermocouples L and S, incorporated in the base and heat sink in the immediate vicinity of the front surfaces.

The ATCS-1 and PTR system with heater (8) provide for the preset heating regime of the block. The ATCS-2 and heater (7) maintain the temperature of the enclosure (5) and heat sink (4) equal during the experiment, thus allowing one to heat the heat sink by the flow penetrating through the specimen from the base only. The heat capacity of the heat sink, $C_{\mathrm{hs}}(t)$, is

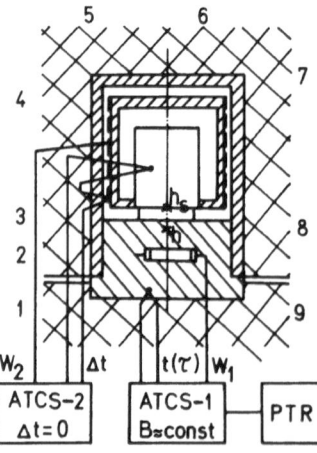

FIGURE 4. Schematic diagram of a simple λ-calorimeter: 1—base, 2—cover, 3—specimen, 4—heat sink, 5,7—adiabatic enclosure with the heater, 6,9—heat shields, 8—base heater.

known and chosen so as to fulfil the condition $C/C_{hs} < 0.2$ for all the specimens tested.

In accordance with equation (12), the following equation may be used to calculate the thermal conductivity of the specimen:

$$\lambda(t_0) = \frac{h}{S} \frac{C_{hs}b_{hs}}{\theta_{Lhs} - \Delta\theta_r} (1 + \sigma_c) \tag{17}$$

where $\sigma_c = cSh/(2C_{hs})$ is the relative correction for the heat capacity of the specimen, S is the surface area of the specimen, and $\Delta\theta_r = 2P_rC_{hs}b_{hs}/S$ is the correction for the contact resistance $2P_r(t)$ arising at the contact boundaries of the specimen with the base and heat sink.

If experiments are carried out under reproducible conditions, i.e., the flat faces of the specimen are ground and lubricated in the same say, their contact resistance $2P_r(t)$, as the first approximation, remains "the constant of the instrument" and is readily measured with the help of calibration tests when a thin metal (copper) disk is used as specimen, its thermal resistance satisfying the condition $P < 0.2P_r$. The values of $\theta^0_{Lhs}(t_0)$ obtained from such a test coincide with $\Delta\theta_r(t_0)$, so that it is possible to directly replace $\Delta\theta_r(t)$ by $\theta^0_{Lhs}(t_0)$ in equation (17). In addition, this allows one to account for all the characteristics of the mounting of the lower (L) and heat sink (hs) thermocouples and occasional heterogeneity of the thermocouple electrodes.

The requirements for the block thermostatting the "cold" junctions of the thermocouples L and hs are as usual: the nonuniformity of the temperature distribution and the permissible rate of its temperature change should be kept negligibly low in comparison with $\theta_{Lhs}(\tau)$ and $b_{hs}(\tau)$. The ATCS-2 should be adjusted so that occasional heat leaks through the air-filled gap between the enclosure (5) and heat sink (4) are negligibly small relative to the heat $C_{hs}b_{hs}$ infiltrating into the heat sink through the specimen.

In order to provide for a reliable, reproducible thermal contact of the specimen with the heat sink and the base, one is advised to force-press the former to the latter at a pressure of $p = (1-5) \times 10^5$ N m^{-2}. A thin needle with a certain load may be used for the purpose.

Disk-shaped specimens 20 mm in diameter are claimed to be most suitable. The specimen thickness should be chosen so that $h = (1-10) \times 10^{-3}\lambda$ (in m), where λ is the expected thermal conductivity value (in W m^{-1} K^{-1}). In practice, materials having $\lambda = 0.1$ to 5 W m^{-1} K^{-1} can be studied, specimen thicknesses varying correspondingly from 0.5 to 5 mm.

It is recommended that the heat sink be made from copper or any other metal or alloy possessing high thermal conductivity ($\lambda > 50$ W m^{-1} K^{-1}) and well-known heat capacity. It is important that the latter remain a smooth monotonic function throughout the operating temperature range. The recommended heat-sink dimensions are diameter 21 mm and height 20–25 mm.

The base diameter is 50 mm and its height is 50 mm too. The cover (2) and base (1) form, when ready for operation, the hollow space of size 50 × 50 mm. The air gap between the heat sink (4) and adiabatic enclosure (5) should be about 10 mm. The requirements as to the design and characteristic fitting of the base, the cover, and the adiabatic enclosure remain the same for all the cells.

The test junctions of the stationarily embedded thermocouples are placed in the holes in the base, the heat sink and the enclosure and are fixed with dental cement. The junctions of thermocouples L and hs should be placed in the immediate vicinity of the working faces of the base and heat sink.

The cover (2), enclosure (5), and heat sink (4), being a one-piece removable assembly, it is important for the latter two to have adequate play near the cover, while their connection is to be provided by thin (1.0–1.6 mm-diameter) tubes made of high-temperature steel or nickel, which can give minimum heat conduction together with adequate mechanical strength of the attachment.

The calorimeter shown in Fig. 5 differs from the previous one mainly due to the contact heat flowmeter (2) incorporated into base (1). The base of block (1) is heated in the experiment as usual with the help of the ATCS-1 and PTR and heater (8). A heat sink (4) is heated by a thermal flow passing through the heat flowmeter (2) and a specimen (3). A cover (5) functions as an adiabatic enclosure. The ATCS-2 with the aid of a heater (7) and a differential thermocouple maintains "zero" temperature drop between the heat sink (4) and the cover (5). An air-filled gap, as in the previous design, acts as a good heat shield for the heat sink.

Various designs of heat flowmeters are possible. However, in all of them it is practicable to place a thin thermally insulating layer of the heat flowmeter between two metal disks where the junctions of the thermocouples and thermopiles are located. The thermally insulating layers may be film-forming adhesives, air-filled spaces with metal tubes containing thermocouple elec-

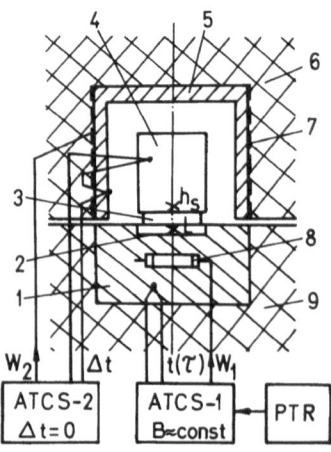

FIGURE 5. Schematic diagram of a λ-calorimeter with a contact heat flowmeter: 1—base, 2—contact heat flowmeter, 3—specimen, 4—heat sink, 5,7—adiabatic enclosure and a heater, 6,9—heat shields, 8—base heater.

trodes, or tubes connecting the plates of the heat flowmeter so offering additional thermal conductivity.[1,6] It is important that the thermal conductivity of the heat flowmeter, $K_{hm}(t)$, exceed that of the specimen S/h several times and remain stable in the temperature range of interest, with the heat capacity of the layer at least an order less than that of the specimen.

The computational equation has the form

$$\lambda(t_0) = \frac{h}{S} \frac{K_{hm}\theta_{hm}}{\theta_{Lhs} - \Delta\theta_r} (1 - \sigma_c) \tag{18}$$

where $\sigma_c = (0.5c\rho Sh + C_p)/C_{hs}$ is the relative correction for the heat capacity of the specimen $(c\rho Sh)$ and the upper contact plate of the heat flowmeter (C_p), $K_{hm}(t)$ and $\theta_{hm}(\tau)$ are the thermal conduction (W K^{-1}) and temperature drop in the layer of the heat flowmeter, while $\Delta\theta_r = 2P_rK_{hm}\theta_{hm}/S = \theta_{Lhs}^0$ is the correction for the contact thermal resistance $2P_r(t)$. In order to obtain the latter correction, one must conduct calibration tests with a metal plate (cf. the previous design).

The fact that the λ-calorimeter design discussed here allows much rougher adiabatization of the open surface of the heat sink, and does not require measuring the heating rate $b_{hs}(\tau)$ of the heat sink, may be regarded as its advantage. However, unlike the previous design it demands calibration of the heat flowmeter, which may be accomplished in two ways. In particular, to define $K_{hm}(t)$ of the heat flowmeter one can utilize a test with a standard specimen thermal conductivity $\lambda(t)$, which is well known, and make use of equation (18) in order to compute $K_{hm}(t)$. If there is no such specimen available, one can conduct an experiment without the specimen, recording $\theta_{hm}(\tau)$ and $t_{hs}(\tau)$ to calculate $K_{hm}(t)$ by the equation

$$K_{hm}(t) = \frac{C_{hs}b_{hs}}{\theta_{hm}} (1 + \sigma_c)$$

where $\sigma_c = C_p/C_{hs}$ is the correction for the heat capacity of the plate of the heat flowmeter.

The dimensions of the specimens, the base, the cover, and the heat sink are the same as in the earlier variant of the cell (see Fig. 4).

Suitable heat flowmeters were found to be those the main element of which consisted of a copper plate of 21-mm diameter and 1-mm thickness, with six 1.0–1.2-mm-ID thin-walled tubes welded onto it. The top ends of the tubes are pressed into blind axial holes evenly spaced throughout the square area of the plate. The fixing points are caulked by silver soldering from outside. The tube tips with the plate are pressed into the corresponding axial through holes in the base (or in the ancillary massive disk of 21-mm diameter). The air gap between the heat-flowmeter plate and the base should be 2–3 mm. The wires of a differential thermocouple are placed in the five tube holes, the junctions being situated at two levels: in the zone of the plate and lower; in the zone of the base, 5-7 mm away from the plate. The sixth hole is used for placing the test junction of the thermocouple L, which measures the plate temperature during the experiment.

It seems worthwhile to use the type of heat flow meter which has the tubes pressed into the radial holes of the plate. The open tips of the tubes are pressed into the corresponding radial grooves cut in the surface of the base.

It is expedient either to caulk the places where the tubes fit, or to seal them with dental cement. The bare sections of the tubes between the plate and the base should be as long as 3-5 mm, similar to the first variant. The conditions of mounting the differential thermocouple and the thermocouple L are similar to those mentioned earlier. An air gap 1-2 mm wide is necessary between the plate and the base. In order to prevent the plate sinking, one can use needle-like vertical supporting elements.

The demerit of such a type of heat flowmeter is that their air gaps must be protected from occasional contamination to avoid irreversible change in the heat-flowmeter sensitivity. In particular, this can be done by sealing the annular section of the air gap between the plate and the base with dental cement, or glass or quartz wool periodically removed.

If the upper temperature limit of the calorimeter does not exceed 100 °C, epoxy resin is likely to be the best choice for adhesion of the plate and the base, its layer being 0.5-1.0-mm thick. The thermal electrodes are best embedded into the radial holes or grooves with the aid of the same resin.

The characteristics of wiring the thermocouple hs and the differential thermocouple of the adiabatic enclosure remain as before (cf. Fig. 4).

The λ-calorimeter shown in Fig. 6 consists of a massive metal cup (1) with a cover (5) enclosed in a heat shield (4, 6). A metal heat sink (2) is placed

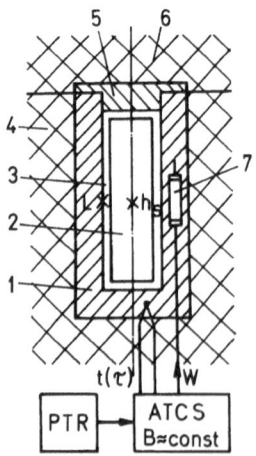

FIGURE 6. Schematic diagram of a cylindrical λ-calorimeter: 1,5—metal cup with the cover, 2—heat sink, 3—gap for a specimen, 4,6—heat shields, 7—cup heater.

concentrically in the cup. The material to be tested fills a gap between the core (heat sink) and the cup. The width of the gap (3) h is chosen so that the requirement $h < 0.1 R_{hs}$ is met, where R_{hs} is the radius of the heat sink. This allows the tested specimen to be regarded as a thin plate the surface area $\bar{F}$ of which is determined in terms of the central section, the end faces included. The ATCS, the PTR, and a heater (7) provide the preset regime of heating the calorimeter. The design equation has the form

$$\lambda(t) = \frac{h}{\bar{F}} \frac{C_{hs} b_{hs}}{\theta_{Lhs}} (1 + \sigma_c) \qquad (19)$$

where $\sigma_c = C/(2C_{hs})$.

Equation (19) does not include corrections for the curvature of the layer tested and for the nonlinearity of the equation of thermal conduction because, in general, they are successfully reduced to negligibly small values.[1]

The cell design can be made very similar to that of a cylindrical a-calorimeter (cf. Fig. 1). The metal heat sink (2) is put on the hs thermocouple needle, the latter passing through the central axial hole. The heat sink may be aligned with the aid of metal needles of diameter 1.00 mm, pressed into the blind radial holes on its lateral surface.

If the cup space diameter is chosen to be 20 mm, the heat-sink diameter may be 18 mm. It is advisable to make the heat sink of copper. A calorimeter of such a type allows one to test liquids, gases, finely dispersed powders, and fibrous materials. Liquids of low viscosity are likely to be studied in cells having a 10-mm-diameter heat sink and a 0.5-mm space gap. This will enable convection to be avoided in the material being tested. Optimum rates of heating are 0.01–0.1 K s^{-1} for cells of this type.

4.3. c-Calorimeters

When measuring specific heat, attempts are usually made to create calorimetric devices that furnish as uniform a distribution of temperature as possible throughout the thermal cell and the specimen. Therefore, c-calorimeters do not require limitations on the temperature regime in the experiment that are as strict as those required for λ- and a-calorimeters. Hence, one can reasonably assume that well-known scanning differential and adiabatic c-calorimeters are operating under a monotonic heating regime.

The scope of the present chapter confines us to a brief description of the simplest designs of such calorimeters. Figure 7 presents the schematic diagram of an adiabatic c-calorimeter. A metal container (3) is equipped with a heating

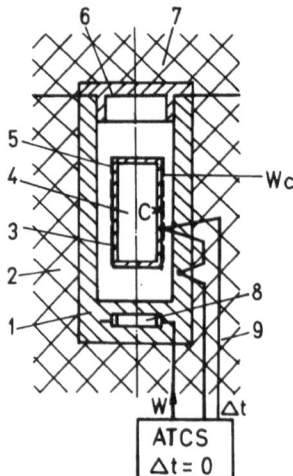

FIGURE 7. Schematic diagram of an adiabatic c-calorimeter: 1,6—metal cup with the cover, 2,7—heat shields, 3,5—metal container with a heater, 4—specimen, 8—cup heater, 9—differential thermocouple.

coil (5) spread uniformly over the surface and fixed by thin needle capillaries inside the metal cup space (1) with a cover (6). The cup is protected thermally from the container by an air-filled (or sometimes evacuated) gap. The space of the container is filled with the material to be tested (6). A thin metal foil may be placed radially inside the container to create a uniform temperature field in the specimen if the need arises. Both the cup and cover are protected by a thermally insulated enclosure (2, 7).

During the experiment the container and specimen are heated by a heater (5), its power W_c being maintained as constant as possible. Temperature changes of the container, $t_c(\tau)$, are recorded and the resulting temperature curve allows one to compute the variable heating rate of the container, $b_c(\tau)$. The ATCS, with the help of a differential thermocouple (9) (or a thermopile) and a heater (8), provides temperature equilbrium between the cup and the container.

The specific heat of the material to be tested is calculated from the equation

$$c(t) = \frac{1}{m}\left(W_c/b_c - C_c\right) \tag{20}$$

where m is the specimen mass and C_c is the heat capacity of the container.

The heat capacity of the container, $C_c(t)$, is, in practice, determined experimentally from the calibration test with an empty container (without a specimen). Such a test enables the effect of accessories and the systematic error of adiabatization to be taken into account.

The cell design may be similar to those presented in Figs. 1 and 6. The container is preferably made of copper, its optimal sizes being: diameter 10 mm, height 30–40 mm, wall thickness 0.5–1.0 mm. It is expedient to attach the heater and the thermocouple to the side surface of the container. The method of attachment is conventional and depends on the operating temperature ranges. Of importance is that the thickness of the container wall together with the heater and thermocouple be less than 1.0–1.5 mm, which will allow the required isothermal conditions ($\Delta t < 0.1$ K) to be preserved. In particular, it is suitable to place the heater and thermocouple wires in the axial and cylindrical grooves cut on the outside wall of the container. Both epoxy resin for $t < 120\,°C$ and dental cement for $t < 400\,°C$ may be used for fastening and electrical insulation of the wires.

The container can be attached to the cover (6). Thin-walled metal tubes with electrodes passing through them can be used for the purpose. However, it is sometimes possible to fabricate the container as an independent assembly and, when mounting it in the cell, to put it on the thermocouple needle and pin needle-like electrodes, which are incorporated rigidly in the bottom of the cell in the form of vertical supporting elements. With the needle diameters being 1.5 mm, the length of the bare sections of the supports, i.e., the gap between the container bottom and that of the cell, will not be less than 20 mm. This allows the thermal coupling of the container with the cell's cup to be kept negligible, and to effect the adiabatic heating of the container by means of the built-in heater.

The schemetic diagram of a differential c-calorimeter is shown in Fig. 8. The basis of the calorimeter is a massive metal block (1) with two identical cylindrical spaces where two strictly identical metal containers (3, 8) are fixed concentrically, the gap formed being filled with air. The block has a cover (5) and is protected by a thermally insulating enclosure (2, 6). One of the containers is filled with a reference material of well-known specific heat $c_{ref}(t)$, while the

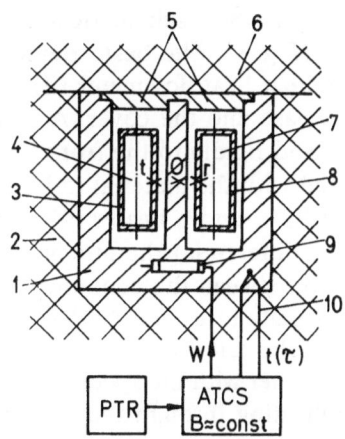

FIGURE 8. Schematic diagram of a differential c-calorimeter: 1,5—metal base with covers, 2,6—heat shields, 3,8—metal containers, 4—specimen, 7—reference specimen, 9—block heater, 10—thermocouple.

second container is filled with a material the specific heat of which, $c(t)$, is to be measured. If necessary, a thin metal foil is placed radially within the container.

The ATCS and PTR with heater (9) and thermocouple (10) furnish the preset, near-linear regime of heating the calorimeter. During the experiment it is necessary to measure the temperature of the block, $t_0(\tau)$, and the lag of the container temperatures relative to those of the block, $\theta_{0\,ref}(\tau)$ and $\theta_{0\,t}(\tau)$.

The specific heat of the material to be tested is calculated from the equation

$$c(t)m = (C_{ref} + C_{c\,ref})\theta_{0\,t}/\theta_{0\,ref} - C_{c\,t} \qquad (21)$$

The heat capacities of the containers, $C_{c\,ref}$ and $C_{c\,t}$, may be determined experimentally from a set of calibration tests with empty containers and containers filled with reference material.

It is desirable to make the block (1) of copper, its height and diameter being ~60 mm with the diameter of the space ~15 mm. The diameter and height of the containers are ~10 and 30–40 mm, respectively, the wall thickness being ~1 mm. One acceptable method of its mounting is as follows. A vertical thin-walled tube is soldered centrally onto the bottom of the container. The thermocouples t and r (see Fig. 8) are sheathed with thin-walled metal tubes and let through the block bottom along the space centers. The diameters and length of the tubes are chosen so that the tube of the container can be slide-fitted on the thermocouple needle for a preset depth. In such a way the concentricity of installing the containers inside the spaces can be attained, as well as their minimum thermal stray coupling with the block via the accessories.

It is very important that the sizes of both containers and spaces as well as their surfaces be strictly identical, because the design equations presuppose an identical effective thermal conduction of the air gaps.

One more schematic diagram, of a dynamic c-calorimeter, is worth discussing (Fig. 9). The basic elements of the calorimeter are a metal base (1) with gradient heat flowmeter (2) and metal cover (4). Heaters (7, 9) are built into the base and the cover. A metal container (3) with a cover and a heater (5) is fixed rigidly to the front face of the heat flowmeter. The container space is filled with the material to be tested. The heating regime is set by the ATCS-1, the PTR, and a heater (9) and thermocouple (10). The ATCS-2 with the aid of a differential thermocouple (8) and heater (7) maintains the temperature drop between the container and cover equal to zero. The base and cover are thermally insulated from the surroundings by a joint enclosure (6, 11).

The calorimeter may have two operating regimes. Under one of them the container with the specimen are heated during the experiment by the heat flow infiltrating through the heat flowmeter, the heater (5) being switched off. In

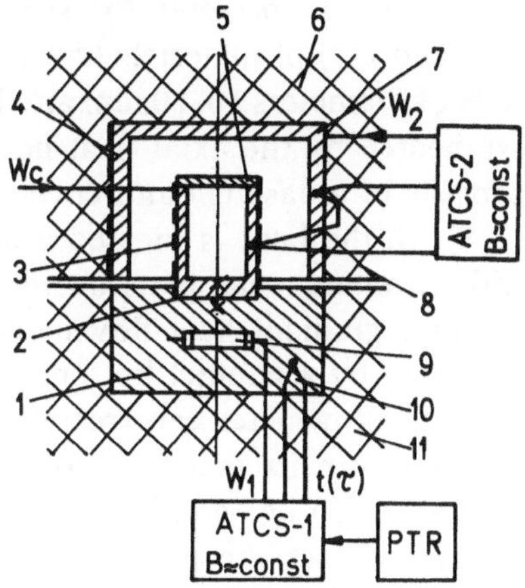

FIGURE 9. Schematic diagram of a c-calorimeter with a contact heat flowmeter: 1—metal base, 2—gradient heat flowmeter, 3,5—metal container with cover and heater, 4—metal cover, 6,11—heat shield, 7,9—heaters, 8—differential thermocouple, 10—thermocouple.

this case, equation (13) applies for calculating the specific heat of the material tested.

Another regime implies that the heater (5) is switched on, its power W_c being selected so that the heat-flowmeter readings are close to zero. This regime possesses the advantage of being less dependent on the errors caused by the heat flowmeter. Specific heat is calculated from the equation

$$c(t)m = (W_c + K_{hm}\theta_{hm})/b_c - C_c \qquad (22)$$

The calorimeter may be calibrated in terms of heater (5) without using the reference specimen of known specific heat. For this it suffices to carry out two experiments with the empty container, one with the heater switched on and the other with the heater switched off. The results of both experiments allow the thermal conduction $K_{hm}(t)$ of the heat flowmeter and effective heat capacity $C_c(t)$ of the container to be determined.

The cell design coincides with that of one of the λ-calorimeters (cf. Fig. 5) in all the parameters. The requirements pertaining to the heater (5) for the container remain the same as in Fig. 7. The container space (or specimen) can have diameter and height about 15 mm. The container should be made of copper with the thickness of the walls about 1.0 mm.

4.4. Multipurpose Calorimeters

This group of calorimeters allows two thermal physical characteristics of a specimen to be measured in one and the same experiment. Let us discuss two such calorimeters.

Figure 10 presents the schematic outline of a calorimeter for studying $\lambda(t)$ and $c(t)$ of materials having $\lambda = 0.1$–$10\ \mathrm{W\,m^{-1}\,K^{-1}}$. Specimens are short rods of diameters 10–20 mm and heights $h = 5$–30 mm. In the experiment they are heated by the axial heat flow. A metal heating block of the calorimeter consists of a base (1) and cover (3) protected by a joint insulating enclosure (2, 8). In the base is located a heater (9) and a gradient heat flowmeter (4) with a contact metal plate on which the bottom face of the specimen (5) rests. Another gradient heat flowmeter (6) is pressed against the top face of the specimen by a contact bottom plate. A heater (7) is located in its top plate. The heat flowmeter can freely travel vertically in the space of the cover (3) together with its thermocouples, depending on the height of the specimen. The cover (3) has good thermal contact with the base (1) of the block. Due to this contact the air-filled space surrounding the lateral surface of the specimen adiabatizes it reliably throughout the experiment.

The design equations can be simplified by employing heat flowmeters as identical as possible. The heat capacities of their contact plates ($C_{\mathrm{hm}_1} = C_{\mathrm{hm}_2} = C_{\mathrm{hm}}$) and the thermal conductions of their layers ($K_{\mathrm{hm}_1} = K_{\mathrm{hm}_2} = K_{\mathrm{hm}}$) should be equal. Subject to these restrictions, the following relationships can be stated between the flows $K_{\mathrm{hm}}\theta_{\mathrm{hmL}}$ and $K_{\mathrm{hm}}\theta_{\mathrm{hmU}}$ measured by the lower and upper heat flowmeters and the specific heat flows q_{-l} and q_l infiltrating through the corresponding faces of the specimen:

$$K_{\mathrm{hm}}\theta_{\mathrm{hmL}} = q_{-l}S + C_{\mathrm{hm}}b, \qquad K_{\mathrm{hm}}\theta_{\mathrm{hmU}} = q_l S - C_{\mathrm{hm}}b$$

where θ_{hmU} and θ_{hmL} are the signals of the heat flowmeters, and S is the cross-sectional surface area of the specimen.

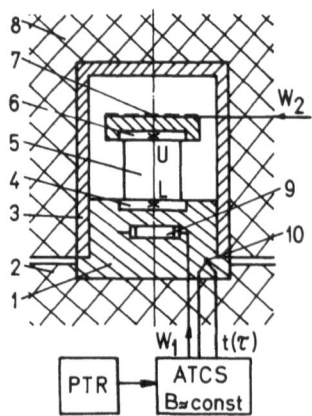

FIGURE 10. Schematic diagram of a c-calorimeter: 1—base, 2,8—joint insulating enclosure, 3—cover, 4,6—gradient heat flowmeters, 5—specimen, 7—plate with heater, 9—heater, 10—thermocouple.

For these relationships, design equations (8) for λ and c acquire the form

$$\lambda(t_0) = \frac{hK_{\text{hm}}}{2S}\frac{\theta_{\text{hmL}} + \theta_{\text{hmU}}}{\theta_{\text{LU}}}, \qquad c(t_0) = \frac{1}{m}\left(K_{\text{hm}}\frac{\theta_{\text{hmL}} - \theta_{\text{hmU}}}{b} - 2C_{\text{hm}}\right) \quad (23)$$

where $\theta_{\text{LU}}(\tau)$ is the temperature drop between the corresponding contact plates of the lower and upper heat flowmeters. The readings of the heat flowmeters are assumed positive when heat flows upward.

If the specimen thickness h is selected in terms of $P = h/\lambda > 10^2 \times 2P_{\text{r}}$, there is no need to introduce a correction for the contact thermal resistance. If it is difficult to meet this requirement, the procedure of allowing for the correction remains the same as that described for the λ-calorimeters.

Many assemblies of the calorimeter in question coincide with the corresponding ones of λ-calorimeters (see Figs. 4 and 5). Contact plates and layers of the heat flowmeters are to be made identical, because it will allow one (as already mentioned) to use equations (23) (which are quite simple) and, what is particularly important, to obtain the optimum temperature drop θ_{LU} (about 5–20 K) along the specimen height with the aid of heater (7) during the experiment. Unfortunately, the experience shows that it is difficult to provide for equal thermal conductivities K_{hm} of the heat flowmeters when they are fabricated individually. If this is the case, one can recommend constrained experimental conditions so as to maintain a zero temperature drop in the upper heat flowmeter ($\theta_{\text{hmU}} = 0$) due to heater (7). The upper heat flowmeter turns out to be the null indicator, its thermal conductivity being excluded from the design equations (23).

The calorimeter shown in Fig. 11 is intended for multipurpose investigation of materials having $\lambda > 10\ \text{W m}^{-1}\text{K}^{-1}$. It is based on an a-calorimeter for metals (Fig. 3) supplemented with a gradient heat flowmeter built into the

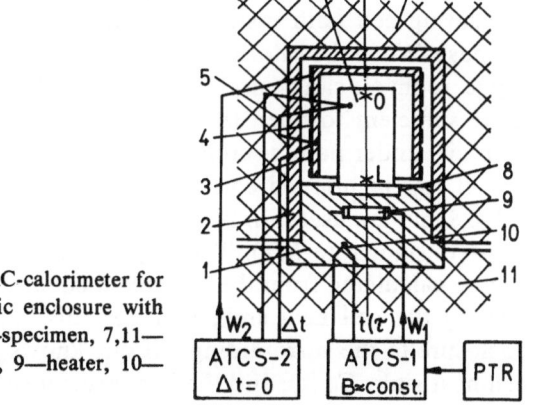

FIGURE 11. Schematic diagram of an AC-calorimeter for metals: 1—base, 2—cover, 3,4—adiabatic enclosure with heater, 5—differential thermocouple, 6—specimen, 7,11—heat shield, 8—gradient heat flowmeter, 9—heater, 10—thermocouple.

base of the block. A specimen is placed on the contact plate and heated by heat flowing through the heat flowmeter. Thus, the heat-flowmeter readings make it possible to measure the specific heat of the specimen.

The value of $a(t)$ can be computed from equation (17), while for calculating $c(t)$ the following expression is used:

$$c(t) = m = \frac{K_{hm}\theta_{hm}}{b}(1 - \sigma_\alpha) - C_{hm} \tag{24}$$

where $\sigma_\alpha = 2\pi Rh\alpha/(3K_{hm})\theta_{LU}/\theta_{hm}$ is a relative correction for heat dissipated by the lateral surface of the specimen.

The correction σ_α is necessitated due to the fact that the lateral surface of the specimen is not isothermal. Therefore, under the conditions discussed one can find the specific heat with an error of 3–5%. The multipurpose calorimeters are of little use for more accurate measurements of $c(t)$.

5. ACCESSORIES

All the calorimeters discussed are claimed to operate under the same temperature conditions. The thermal cell is precooled down to a certain lower level of the operating temperatures and then, during the experiment proper, it is smoothly heated up to the highest temperature level.

Special systems of forced cooling are necessary in two cases: (1) when the investigations are carried out at temperatures below those of the environment; (2) when there is a need to reduce the time between experiments. Cooling with liquid nitrogen vapors may be regarded as a universal technique. Liquid nitrogen vapors, passing through a pipe from the Dewar, directly cool the measuring cell both inside and outside via a set of holes provided for this purpose. When the calorimeter is expected to work in a reasonable temperature range, room air can be used to cool the cell after an experiment. It is sometimes quite sufficient to open the cell so that all the elements can be streamlined with air, under natural convection.

It was mentioned earlier that nichrome helical heaters of wire diameter 0.2–0.8 mm are useful in heating the thermal cell of the calorimeter at a preset near-constant rate $b(\tau)$. The required electrical power of the heater, $W(\tau)$, can be calculated from the simplified equation which allows for heat spent on heating the cell heat sink, and heat dissipation through the heat shielding enclosure to the surroundings. If one knows the effective heat capacity of the cell heat sink, $C_{hs}(t)$, and the heat resistance of the calorimeter heat shielding enclosure, $P_{en}(t)$, then the variation in heater power can be computed from the following equation:

$$W(\tau) = C_{hs}(t) \cdot b + \frac{1}{P_{en}(t)}[(t(\tau) - t_s]$$

where t_s is the temperature of the surroundings.

The heater power control may be effected either by altering the electric voltage or by suitable periodic energizing–deenergizing of the heater, the cycle being no longer than 10 s.

The well-known Ohm's law and data on the allowable current density through the nichrome helix ($<5\,A\,mm^{-2}$) are used to calculate the heater. Ceramic tubes may be applied as electric insulation. The nichrome wire of the heater can be faced with a quartz sleeve or any other flexible electrical insulation.

The automated regime of the experiment may be implemented in two ways: (1) by a micro- or mini-computer-assisted program; (2) by a set of specialized electronic devices. If the former is applied, the analog data input and output channels of the computer are supplemented with input precision amplifiers to normalize the signals from thermocouples and heat flowmeters, and the output power amplifiers.

The precision input amplifiers should possess 10^2–10^5 DC current gain. Time and temperature zero drift on the amplifier input should not exceed 1-2 V. The required performance can be obtained from amplifiers having M-DM channel. The signal-to-noise ratio is increased by placing the amplifiers close to the heat measuring cell in the thermostatted zone (<0.5 K).

The output amplifiers are to ensure increase of power of the heater in the 5-500-W range. The efficiency of the control channels is improved by running the output cascades of the power amplifiers in line with the setup of the pulse-width modulators, transistors being used as the key element. Their thermal load is the heater of the heat measuring cell.

The controlling computer provides for the operation of the adiabatic enclosures (ATCS-2), as the allowable inaccuracy of the temperature-drop control is 0.1 K, which corresponds to an electric signal of 4 V. The latter approach, i.e., controlling the preset conditions of the experiment by electronic devices, presupposes using special precision temperature controls, as well as arrangements for setting the linear heating regime and digital microvoltmeters, their sensitivity being as stipulated before. The experimental results can be processed with the help of any microcomputer.

More information on the theory of the calorimeters treated in this chapter, their applications, and their manufacture can be found elsewhere.[1,6,7]

NOTATION

Symbol	*Name*	*Unit of value*
a	Thermal diffusivity	$m^2\,s^{-1}$
b	Rate of heating	$K\,s^{-1}$
Bi	Biot number	Dimensionless
c	Specific heat	$J\,kg^{-1}\,K^{-1}$
f	Shape factor	Dimensionless
g	Temperature gradient	$K\,m^{-1}$
h	Specimen thickness, height	m
H	Height of the cell block	m
k	Relative temperature coefficient	K^{-1}
K_{hm}	Thermal conduction of heat flowmeter	$W\,K^{-1}$
l	Specimen size	m
m	Specimen weight	g
P	Specimen heat resistance	$m^2\,K\,W^{-1}$
P_{cr}	Contact thermal resistance	$m^2\,K\,W^{-1}$
q	Specific heat flow	$W\,m^2$
r	Coordinate	m
R	Radius of specimen (block)	m
S	Square area of specimen cross section	m^2
W	Power of thermal energy	W
x	Coordinate	m
α	Heat transfer coefficient	$W\,m^{-2}\,K^{-1}$
θ	Temperature drop	K
λ	Thermal conductivity	$W\,m^{-1}\,K^{-1}$
ρ	Density	$kg\,m^{-3}$
σ	Relative correction	Dimensionless
τ	Time	s

Subscripts

a	Air
c	Container
cr	Contact resistance
e	Enclosure
hm	Heat flowmeter
hs	Heat sink
L	Lower
r	Specific contact resistance
red	Reduced
ref	Reference
s	Surroundings
U	Upper

REFERENCES

1. E.S. Platunov, *Thermophysical Measurements Under a Monotonic Regime* (in Russian), Energy, Leningrad (1973).
2. A.V. Lykov, *Theory of Thermal Conduction* (in Russian), Moscow Vysshaya Shkola, Moscow.
3. V.P. Isatchenko, V.A. Osipova, and A.S. Sukomel, *Heat Transfer* (in Russian), Energy, Moscow (1969).
4. O.A. Krayev, "A Method of Determining Temperature Diffusivity vs. Temperature Relationship in One Experiment" (in Russian), *J. Tyeploenergetika*, No. 4, pp. 15-18 (1956).
5. A.A. Shashkov, G.M. Volokhov, T.N. Abramenko, and V.P. Kozlov, in: *Methods of Determining Thermal Conduction and Thermal Diffusivity* (in Russian) (A.V. Lykov, ed.) Energy, Moscow (1973).
6. S.E. Buravoy, V.V. Kurepin, *et al.*, "The Universal Set of Instruments for Thermophysical Measurements," *Journal of Engineering Physics* (*Minsk*) **38**(3), 420-428 (1980).
7. E.S. Platunov, S.E. Buravoy, V.V. Kurepin, and G.S. Petrov, in: *Thermal Physical Measurements and Devices* (E.S. Platunov, ed.), Mashinostroyenie, Leningrad (1986).

13

Apparatus for Measuring Thermophysical Properties of Liquids by AC Hot-Wire Techniques

L.P. PHYLIPPOV, S.N. KRAVCHUN, and A.S. TLEUBAEV

1. INTRODUCTION

The proposed apparatus is designed to determine simultaneously a series of thermal properties of liquids and gases, namely, thermal conductivity λ, heat capacity per unit volume $C_p\rho$ (C_p is the specific heat at constant pressure and ρ is the density), and consequently thermal diffusivity $a = \lambda/(C_p\rho)$ and the thermal activity coefficient $b = \lambda/\sqrt{a} = \sqrt{\lambda C_p\rho}$. The apparatus employs the principle of a probe, thin wire, or foil, which is heated to create a thermal disturbance in a substance, after which the thermal reaction of the substance to such a disturbance is determined. The thermal response carries information on the thermal properties.

The suggested technique takes into account all major trends in the development of thermophysical instrument engineering.[1] It carries maximum information on a series of properties being measured and enables it to be obtained at high speed. Small quantities of substances are needed for measurements over wide ranges of temperature and pressure. The measurements may be taken automatically. It will become clear from the text below that the technique possesses other merits as well, such as high noise stability and, consequently, sensitivity, the possibility of conducting measurements in fluid flows and under conditions of relatively rapid processes occurring in a substance, and negligible radiation heat transfer in liquids and gases.

L.P. PHYLIPPOV, S.N. KRAVCHUN, and A.S. TLEUBAEV • Physics Department, Moscow State University, Moscow 119899, USSR.

2. THE THEORY OF THE METHOD

When a low-inertial probe (thin wires or strips of metal foil) is heated with an alternating current (frequency ω), electric power oscillations W_0 cause probe temperature oscillations $\tilde{T}$ (at frequency 2ω). The amplitude and phase of the temperature oscillations are dependent on the thermophysical properties of the liquid into which the probe is immersed. In the case of a foil probe (liquid is probed with plane temperature waves), the amplitude and phase of temperature oscillations are governed by the thermal activity $b = \sqrt{\lambda C_P \rho}$. When a wire probe is used (liquid is probed with cylindrical temperature waves), both the amplitude and phase are dependent on two factors—thermal conductivity λ and heat capacity per unit volume $C_P \rho$. The relation of the thermal properties of a substance and the amplitude and phase of temperature oscillations is described elsewhere in the theory of the method.[2]

Let us first consider a plane probe. Probe temperature oscillations $\tilde{T} = \theta_0 \exp(2i\omega t)$ are defined by the heat balance equation

$$W_0 = 2\omega C'_p m' i\theta_0 - 2\lambda \bar{S} \frac{\partial \theta}{\partial x}\bigg|_{x=0} \tag{1}$$

where W_0 is the alternating component of the electrical heating power, $\bar{S}$ is the square of foil surface (on one side), C'_p is the specific heat of the foil metal, m' is the foil mass, and $(\partial\theta/\partial x)|_{x=0}$ is the alternating temperature gradient in a layer adjacent to the foil (assuming that the foil thickness is negligible compared with its length and width, and temperature gradients do not exist in the foil).

In order to find $(\partial\theta/\partial x)|_{x=0}$ in equation (1), one must solve the equation of heat transfer in a substance:

$$\frac{d^2\theta}{dx^2} = \frac{2i\omega}{a}\theta \tag{2}$$

The solution of the problem for a temperature wave propagating in a semi-infinite substance (actually, not reaching the walls) is given in the literature.[2] The amplitude of the probe temperature oscillation is given by

$$|\theta_0| = \frac{W_0}{2\bar{S}\sqrt{\omega}\sqrt{2b^2 + 2bd + d^2}} \tag{3}$$

where $d = \sqrt{\omega}\, C'_p \rho' h$, ρ' being the density of the foil material and h its thickness.

The equation governing the temperature oscillation phase is

$$\tan\varphi = -\left(1 + \frac{d}{b}\right) \tag{4}$$

It follows from equations (3) and (4) that the amplitude and phase of the probe temperature oscillations are determined by the thermal activity. Thus, the plane probe method is one for measuring the thermal activity b.

In the general case, the amplitude and phase of temperature oscillations are determined both by the thermal-inertial properties of a probe (magnitude d) and the thermal response of a medium (magnitude b). The condition $d \ll b$ (low-inertial probe) is favorable for the determination of thermal activity b by measuring the temperature-oscillation amplitude, as in this case

$$|\theta_0| \approx \frac{W_0}{2\sqrt{2\omega}\,\bar{S}b} \tag{5}$$

The foil should be made as thin as possible in order to provide for high sensitivity of the amplitude to b, i.e., maximum proximity of general expression (3) to the asymptotic equation (5).

The condition $d \gg b$ results in the temperature-oscillation amplitude depending only on the probe properties $|\theta_0| \approx W_0/(2C_p'm'\omega)$ and may thus be used for measuring the heat capacity of metals.

The condition $d \sim b$ is favorable for determining thermal activity by relationship (4) or (3).

The theory governing the method of alternating heating of a cylindrical probe is based on the solution of a problem related to the temperature oscillations of an infinitely long cylindrical conductor of radius r_0, placed in an infinite dielectric medium and heated by an alternating current of frequency ω. The probe heat balance equation is similar to that in equation (1):

$$W_0 = C_p'm'2i\omega\theta_0 - \lambda\bar{S}\frac{d\theta}{dr}\bigg|_{r=r_0} \tag{6}$$

where $(d\theta/dr)|_{r=r_0}$ is the temperature gradient in an adjacent layer.

The temperature field $\theta(r)$ in a substance is determined by the equation of heat transfer in cylindrical variables:

$$\frac{d^2\theta}{dr^2} + \frac{1}{r}\frac{d\theta}{dr} = \frac{2i\omega}{a}\theta \tag{7}$$

As in the case of a plane probe, equation (6) plays the role of boundary conditions for equation (7).

The solution to the proble enables one to establish a relation between the amplitude and phase of probe temperature oscillations and the thermal properties of a medium. For an amplitude the relation is as follows:

$$|\theta_0| = \frac{W_0}{2C_p'm'\omega}S(\kappa, \eta) \tag{8}$$

where $\kappa = r_0\sqrt{(2\omega/a)}$, $\eta = C_p'\rho'/2C_p\rho$, and $a = \lambda/(C_p\rho)$ is the thermal diffusivity of a medium, while C_p', m', and ρ' are the specific heat, mass, and density of the wire probe.

The phase difference between the oscillations of the probe temperature and probe power released in it is defined by the relation

$$\tan \varphi = F(\kappa, \eta) \tag{9}$$

Functions $S(\kappa, \eta)$ and $F(\kappa, \eta)$ take the forms

$$S(\kappa, \eta) = \sqrt{\frac{\text{kei}^2 \kappa + \text{ker}^2 \kappa}{(-\text{ker} \kappa + (1/\kappa\eta)\,\text{kei}'\,\kappa)^2 + (\text{kei}\,\kappa + (1/\kappa\eta)\,\text{ker}'\,x)^2}} \tag{10}$$

and

$$F(\kappa, \eta) = \frac{\kappa\eta(\text{kei}^2 \kappa + \text{ker}^2 \kappa) + \text{kei}\,\kappa\,\text{ker}'\,\kappa - \text{ker}\,\kappa\,\text{kei}'\,\kappa}{\text{kei}\,\kappa\,\text{kei}'\,\kappa + \text{ker}\,\kappa\,\text{ker}'\,\kappa} \tag{11}$$

where ker κ, kei κ, ker$'$ κ, and kei$'$ κ are the Kelvin functions and their derivatives.[3]

It follows from equations (8) and (9) that the amplitude and phase of temperature oscillations of a wire probe are dependent on two dimensionless parameters, κ and η, which include the properties of the substance (a and $C_p\rho$, with $\lambda = aC_p\rho$).

The parameters κ and η, which contain the required thermal properties of the substance, may be found by different ways: first, from the system of two equations (8), which correspond to measurements of temperature-oscillation amplitude $|\theta_0|$ at two different frequencies; second, from the system of two equations (9), which correspond to measurements of tan φ at two different frequencies; third, by solving system of equations (8) and (9), which correspond to measurements of $|\theta_0|$ and tan φ at one frequency.

The apparatus described below enables one to conduct measurements by any of these methods. Although the errors of thermal properties measured by all three approaches are comparable, the error of the third variant is somewhat smaller. Moreover, it is advantageous from the methodical viewpoint as measurements are taken at one frequency, an important factor for automation of the apparatus.

All the experimental data are obtained by using the amplitude-phase variant of the method, therefore only this variant will be examined below.

The measurement errors of the periodic heating method are largely determined by a value of parameter $\kappa = r_0\sqrt{2\omega/a}$ (parameter η is dependent on the heat capacity of the probe material per unit volume and the medium being measured, and its magnitude is about unity for a large variety of liquids over wide temperature and pressure ranges).

Figure 1 shows the relation of the sensitivities of thermal conductivity λ and heat capacity $C_p\rho$ to experimentally recorded values of S and F, for various values of η ($F = \tan\varphi$, $S = |\theta_0|2C'_p m'\omega/W_0$ is the reduced amplitude of temperature oscillations).

The values of the sensitivity coefficients K_λ and $K_{C_p\rho}$ show by what factor the relative error of the thermal parameters λ and $C_p\rho$ being measured exceeds that of the magnitude of S or F which caused the former.

When $\kappa \gg 1$, which corresponds to larger wire diameters or relatively high frequencies of periodic heating, both the amplitude and phase of temperature

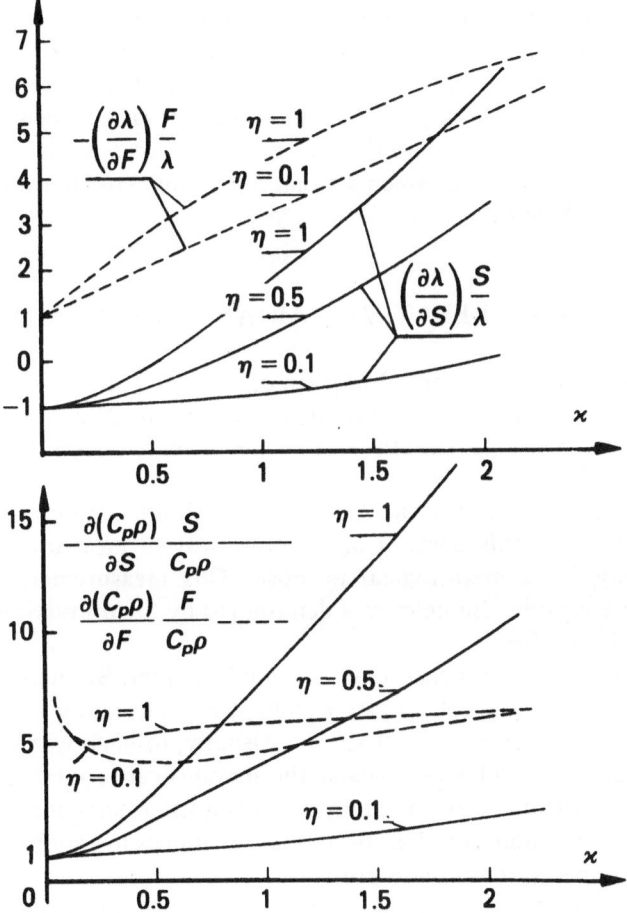

FIGURE 1. Dependence of sensitivity coefficients

$$K_{\lambda,S} = \frac{\partial\lambda}{\partial S}\frac{S}{\lambda}, \quad K_{\lambda,F} = \frac{\partial\lambda}{\partial F}\frac{F}{\lambda} \quad \text{and} \quad K_{C_p\rho,S} = \frac{\partial(C_p\rho)}{\partial S}\frac{S}{C_p\rho}, \quad K_{C_p\rho,F} = \frac{\partial(C_p\rho)}{\partial F}\frac{F}{C_p\rho}$$

on κ for different values of η.

oscillations of a wire probe are determined mainly by its properties, and to a lesser degree by the properties of the surrounding medium. This results in a growth of sensitivity coefficients due to increase of κ. In this case, even small error in measurements of tan φ and $S \sim |\theta_0|$ lead to large errors in λ and $C_p\rho$.

When $\kappa \to 0$, the sensitivity coefficients for thermal conductivity tend to a finite value while one of the coefficients for heat capacity $C_p\rho$ tends to infinity, owing to the exponential function $C_p\rho \sim \exp(1/\tan\varphi)$, which is true for $\kappa \ll 1$ and $\kappa\eta \ll 1$. It is evident from Fig. 1 that the most suitable range for thermal conductivity measurement from the viewpoint of errors is $0 < \kappa < 0.5$. To measure the heat capacity (and thermal diffusivity), this range is bounded from below by the value $\kappa \sim 0.05$. Thus, the optimal range is $0.05 < \kappa < 0.5$ for the measurement of a series of thermal properties λ, $C_p\rho$, a, and b. Assuming the lower frequency limit of radio facilities to be about 20 Hz, the optimal diameters of wire probes for this frequency is between 2 and 20 μm (the characteristic values of the thermal diffusivity of organic liquids being $a \sim 10^{-7}$ m^2 s^{-1}). In our experiments a platinum wire, 5 to 10 μm in diameter, was used for the probes.

3. ABSOLUTE AND RELATIVE VARIANTS OF THE METHOD

Independent measurement of $|\theta_0|$, tan φ, and $W_0/(2C'_p m'\omega)$ make it possible to calculate dimensionless parameters κ and η. Obviously, the probe radius r_0 and value of $C'_p\rho'$ should be known in order to determine the thermal properties a, $C_p\rho$, and $\lambda = aC_p\rho$.

Direct measurement of r_0 and $C'_p\rho'$ takes much time and is rather tedious, so the realization on this basis of an absolute way of measuring λ and a is expedient only for a metrological purpose. This measurement was exemplified[4] where a probe diameter was determined by the diffraction pattern of a laser beam by a wire.

A relative method of measurement is much simpler. By such a technique, unknown values of r_0 and $C'_p\rho'$ are excluded by calibrating in a substance whose thermal properties are well known. Usually, toluene was used for this purpose. Toluene is a substance whose thermophysical properties have been investigated in different ways by many investigators. A significant amount of data is known for conditions close to normal (it suffices to conduct calibration at only one temperature and one pressure).

Available heat-capacity data on toluene exceed the accuracy of our method, and undoubtedly can be used for probe calibration. Recently, some papers have been published in which measurements of the molecular thermal conductivity (not complicated by radiation heat transfer) of toluene are in good agreement. The recommended values of the thermal conductivity of toluene based on these works are presented elsewhere.[5]

One more variant of measurements is possible and produces absolute values of thermal conductivity. This variant uses relatively small values of κ. Equations (8) and (9) yield

$$\lambda \cong \frac{W_0}{L} \frac{1}{8|\theta_0|} \sqrt{1 + \left(\frac{4}{\pi} \ln \frac{\gamma\kappa}{2}\right)^2} = \frac{W_0}{L} \frac{1}{8|\theta_0| \sin \varphi} \qquad (12)$$

where γ is Euler's constant. Evidently, only the probe length L and the temperature coefficient of resistance (necessary for the determination of $|\theta_0|$) need be known of all the probe characteristics. The variant of the technique was tested—absolute values of the thermal conductivity of toluene were measured at a temperature of $T = 300$ K.

4. THE EFFECT OF RADIATION HEAT TRANSFER

The radiation heat transfer in a medium semitransparent to infrared radiation is one of the most important problems related to the study of the thermal conductivity of liquids and compressed gases.

It is common knowledge that radiation heat transfer might significantly distort measured values of liquid thermal conductivity even at room temperatures, the influence largely growing with temperature. This primarily influences traditional steady-state techniques involving a plane layer and coaxial cylinders. The analysis of radiation–conductive heat transfer under conditions of the periodic heating technique[6] shows that the role of radiation heat transfer in thermal conductivity does not exceed the measurement errors up to temperatures of about 600 K. This problem was analyzed[6] for plane temperature waves (a foil probe), which form the basis for the measurement technique of thermal activity $b = \sqrt{\lambda C_p \rho}$. The nonsteady-state linearized equation of radiation–conductive heat transfer has been solved by the gray medium approximation.[6]

The following simple considerations lead to the conclusions drawn by Kravchun and Phylippov.[6] If the amplitude of the probe temperature oscillations equals $|\theta_0|$, then the amplitude of radiation flux does not exceed the value $q^r = 4\sigma n^2 \bar{T}_0^3 |\theta_0|$, which corresponds to an absolutely black surface of a probe and a fully transparent medium (where $\bar{T}_0$ is the average probe temperature, n is the refractive index, and σ is the Stefan–Boltzmann constant). The amplitude of conductive flux is evaluated at $q^c \approx \lambda |\theta_0| / l$ where l is the length of the temperature wave absorption, equal to $l = \sqrt{a/\omega}$. Thus, the comparative influence of radiation heat transfer is evaluated as

$$\frac{q^r}{q^c} = \frac{4\sigma n^2 \bar{T}_0^3}{\lambda} l \qquad (13)$$

In the general case, this relation should also depend on the reflectance R of the probe surface and nonselective absorption coefficient $\bar{\alpha}$.

Actually, published calculations[6] for the plane case show that the relative difference between the effective thermal conductivity, calculated by the equation without taking into account radiation heat transfer, and the molecular thermal conductivity, calculated by the equation with radiation heat transfer taken into account (if the probe heat capacity is neglected), is equal to

$$\frac{\lambda_{\text{eff}} - \lambda}{\lambda} = \frac{4\sigma n^2 T_0^3}{\lambda} l\Phi(\tau, R) \tag{14}$$

Function $\Phi(\tau, R)$ shows the dependence of radiation heat transfer on the reflectance R and optical thickness $\tau = \bar{\alpha}l$. Function $\Phi(\tau, R)$ as dependent on τ for various values of R is shown in Fig. 2. Because $\Phi(\tau, R) \leq 1$, the value of $4\sigma n^2 \bar{T}_0^3 l/\lambda$ characterizes the upper limit of possible influence of the radiation heat transfer on the values of thermal conductivity being measured. The insignificant effect of the radiation heat transfer in the AC hot-wire (foil) technique is the result of the small pass-length of a temperature wave l.

Thus, for practically a minimal frequency of 23 Hz, at a temperature of 300 K the value of l in toluene is about 25 μm, and at a temperature of 550 K, l decreases down to 17 μm due to a change of thermal diffusivity. As an example, let us estimate the role of radiation heat transfer on measuring the thermal conductivity of toluene ($T = 550$ K) at frequencies of 23 and 100 Hz. We assume that $\bar{\alpha} = 5$–6 mm^{-1} (extrapolation from toluene values of $\bar{\alpha}^{(7)}$ to temperature $T = 550$ K) and $R = 0.8$ (reasonable value for a metal); then the value of $(\lambda_{\text{eff}} - \lambda)/\lambda$ is about 0.35% at 23 Hz and about 0.15% at 100 Hz.

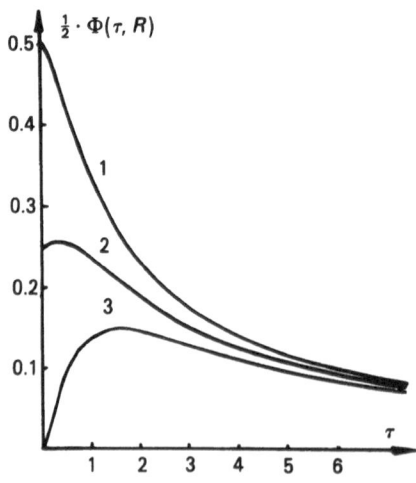

FIGURE 2. Dependence of function $\Phi(\tau, R)$ on optical thickness $\tau = \bar{\alpha}l$ for different values of reflectivity: 1, $R = 0$; 2, $R = 0.5$; 3, $R = 1$.

The study of radiation–conductive heat transfer for a plane probe leads to the conclusion that the radiation effect is practically absent in a real experiment. There is no rigorous solution to the problem for a cylindrical probe, but without doubt, in this case, the effect of radiation heat transfer in a medium is insignificant too.

A very small absorption length $l = \sqrt{a/\omega}$ of a temperature wave in liquids results in a small characteristic thermal disturbance as compared with the absorption length $l_r = 1/\bar{a}$ of thermal radiation (see Table 1).

Under these conditions, the temperature wave field near the probe surface is not complicated with a redistribution of radiant energy in a liquid, and only radiation produced by a probe surface in a medium is important for the energy balance on the probe surface. In other words, to solve the problem it is sufficient to include an appropriate term in the thermal balance equation of a probe,

$$\frac{W_0}{\bar{S}} = \frac{C'_p m'}{\bar{S}} 2i\omega\theta_0 - \lambda \frac{d\theta}{dr}\bigg|_{r=r_0} + 4\sigma n^2 (1 - R) \bar{T}_0^3 \theta_0 \qquad (15)$$

leaving equation (7) unchanged.

The solution of this problem results in the following expressions for the amplitude and phase of probe temperature oscillations (with an accuracy to linear terms of parameter $Bi = 4\sigma n^2 \bar{T}_0^3 l/\lambda$):

$$|\bar{\theta}_0| = |\theta_0| \left[1 + \frac{Bi\,(1 - R)}{\sqrt{2}} \frac{\mathrm{ker}'\,\kappa\,\mathrm{ker}\,\kappa + \mathrm{kei}'\,\kappa\,\mathrm{kei}\,\kappa}{(\mathrm{ker}'\,\kappa + \kappa\eta\,\mathrm{kei}\,\kappa)^2 + (\mathrm{kei}'\,\kappa - \kappa\eta\,\mathrm{ker}\,\kappa)^2} \right] \qquad (16)$$

and

$$\tan \bar{\varphi} = \tan \varphi \left[1 + Bi\,(1 - R) \frac{\mathrm{ker}^2\,\kappa + \mathrm{kei}^2\,\kappa}{\mathrm{ker}\,\kappa\,\mathrm{ker}'\,\kappa + \mathrm{kei}\,\kappa\,\mathrm{kei}'\,\kappa} \right] \qquad (17)$$

where $|\theta_0|$ and $\tan \varphi$ are expressions for the amplitude and phase of temperature oscillations without radiation [see equations (8) and (9)].

TABLE 1

Optical Thickness $\tau = l/l_r = \bar{a}l$ at a Frequency of 23 Hz and a Temperature $T = 300$ K for Various Substances

Substance	Hexane	Decane	Pentadecane	Toluene	Carbon tetrachloride	Cyclohexane	Benzene
τ	0.035	0.05	0.075	0.09	0.04	0.035	0.09

As in the case of a plane probe, the effect of radiation heat transfer is thus determined by parameter Bi and the negligible affect of radiation is explained primarily by the insignificant value of l, which is a factor in this parameter.

When $R = 0$ (absolutely black walls), the correction for radiation is maximum for a transparent medium.

An evaluation of the corrections in line with equations (16) and (17) for $T = 550$ K, $R = 0.7$ to 0.9, and $\kappa = 0.3$ (the probe diameter $2r_0$ is about 10 μm, frequency $\omega/2\pi$ is about 23 Hz, substance is toluene) produces a value of 0.1 to 0.2%, which is less than the measurement errors of $|\theta_0|$ and $\tan \varphi$. Estimates of the radiation heat transfer effect for a cylindrical probe immersed in a low-absorbing medium (as opposed to a transparent medium examined in this work) were considered elsewhere[8] and the effect was found to be insignificant.

5. THE EFFECT OF CONVECTIVE HEAT TRANSFER

Convection is another factor which, along with radiation, complicates the measurement of the molecular thermal conductivity of liquids and gases. To avoid convection in steady-state techniques with a plane layer, coaxial cylinders, etc., one should ensure special conditions such as strict horizontal positioning of plates (in the plane-layer technique), narrow gaps, and small temperature differences. Certain advantages are attributed to the nonsteady-state techniques.

For periodic heating, the temperature oscillation frequency (at least 40 Hz) is rather high, the transient period of convection (1 to 10 s) exceeds manyfold the characteristic time of the process, and therefore temperature oscillations do not cause periodic convection. On the other hand, when a probe is heated with an alternating current, apart from an oscillating temperature with which thermophysical properties are determined, constant overheating of the probe (as compared with the cell walls) occurs. Depending on the experimental conditions, the overheating value varies from several fractions of a degree to several degrees, and this is likely to cause convection.

Experiments with different types of cells (3 to 20 mm in diameter) and diverse differences in average temperature (0.5 to 3 K) have shown that the influence of convection on measurement results does not manifest itself even under conditions where the Rayleigh criterion $Ra = Pr \cdot Gr$ (Pr is the Prandtl number and Gr is the Grashoff number) associated with convection initiation exceeds manyfold the critical value.

A specific experiment[9] was conducted to determine the limits at which hydrodynamic flow starts to appear. A wire probe ($2r_0 \sim 5$ μm) was placed in a liquid flux, and the amplitude and phase of probe temperature oscillations were measured as a function of the flux velocity at various heating frequencies.

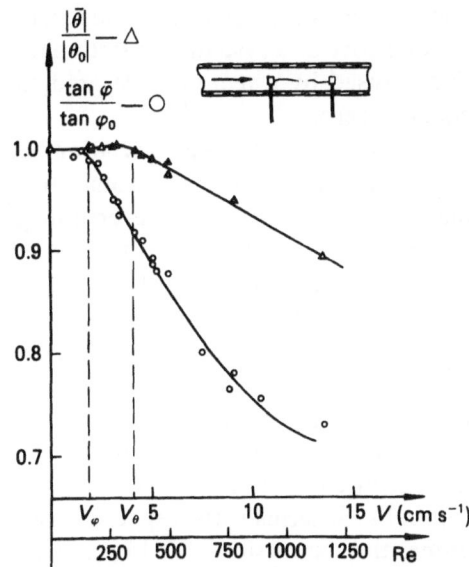

FIGURE 3. Experimental scheme and measurement results of relative values of the temperature oscillation amplitude $|\bar{\theta}|/|\theta_0|$ and phase tangent $\tan\varphi/\tan\varphi_0$ as a function of flow velocity V and Reynolds Number Re, where $|\theta_0|$ and φ_0 are the amplitude and phase of probe temperature oscillations in a stationary medium.

It was found that the flux did not practically influence the measurement results at the lowest frequency $\omega/2\pi \sim 23$ Hz, the velocity ranging from 0 to about 1 cm s^{-1} (see Fig. 3). The velocity range increases with frequency. The solution of the appropriate problem has shown that the existence of this range is associated with the temperature boundary layer whose thickness δ up to a certain flux velocity exceeds the absorption length l of a temperature wave which probes the liquid.

The experimental results[9] have proved not only the insignificant influence of convection on measurement results, but also the possibility of conducting measurements in fluid fluxes.

6. THE EFFECT OF DIFFUSIVE HEAT CONDUCTION IN MIXTURES

The theory of the method for the case of mixtures of liquids and gases has been generalized elsewhere.[10] The generalization is necessitated by the fact that, in mixtures, the temperature gradient causes not only a thermal flux but also a diffusive flux (thermal diffusion, Soret effect). And *vice versa*, the concentration gradient causes not only a diffusive flux but also a thermal flux (diffusive thermal conductivity, Dufour effect). The cross effects complicate the process of heat transfer, therefore two heat conductivity factors were introduced into the description of heat transfer in mixtures: λ_0, the heat conductivity of a system, homogeneous in concentration, and λ_∞, the heat conductivity of a system in a stationary state when the concentration distribution, caused by a temperature gradient, has settled and the diffusive flux is equal to zero.

In order to determine a relation between the amplitude and phase oscillation temperature and the thermal properties of a medium in accordance with the thermodynamics of irreversible processes, a set of equations of heat conductivity and diffusion (with cross effects taken into account) was solved.[10] The main conclusion derived by the solution is that the measured value of thermal conductivity λ^* depends upon the ratio D/a where D is the diffusivity and a is the thermal diffusivity, i.e., $\lambda^* = f(D/a)$.

Thus, in the case of a plane probe, the value of $b^* = \sqrt{\lambda^* C_p \rho}$ is measured, where $\lambda^* = \lambda_\infty[1 + \xi(1 + \sqrt{(D/a_0)})^2]$ with $\xi = (\lambda_0 - \lambda_\infty)/\lambda_\infty$ and $a_0 = \lambda_0/C_p\rho$. It follows from the relationship for λ^* that in liquid mixtures where $D/a \ll 1$ the measured value of b^* is related to $\lambda^* \approx \lambda_0$, i.e., $b^* = \sqrt{\lambda_0 C_p \rho}$ (accurate to within linear terms with respect to the small quantity $\xi \ll 1$). For a cylindrical probe, the expressions for $|\theta_0|$ and $\tan \varphi$ with $D/a \ll 1$ do not differ from equations (8) and (9); however, it should be borne in mind that their conductivity is expressed through λ_0, i.e., the thermal conductivity of a system homogeneous in concentration.

In gaseous mixtures $D/a \sim 1$; the periodic heating method is used to measure the value of λ^*, intermediate between λ_0 and λ_∞.

7. THE LIMITATIONS OF THE METHOD

The method under consideration is similar to the more widely used transient hot-wire (or foil) method. The main difference between the two methods is the following: In the case of the AC hot-wire (foil) method, the response of the system (wire or foil immersed in the substance under investigation) to a sinusoidal disturbance is considered, while in the case of the transient hot-wire (foil) method, the response of the system to a step-like disturbance is considered. This difference leads to differences in the recording of the electrical signals and the methods of calculations for the two cases.

It should be noted that in view of existing fast digital-to-analog and analog-to-digital converters, it may be possible to develop a generalized apparatus which can operate on both transient and AC hot-wire modes. It may also be possible to use more complicated modes, such as heating by currents at two different frequencies.

Similarity of the two methods leads to some common limitations. The main limitation is that the methods are suitable only for dielectric (electrically nonconducting) fluids. There are two cases where this limitation can be overcome. In the first case, the wire should be coated with a dielectric material.

This approach has been used in measurements of the thermal conductivity of some highly conducting liquids.[23,27] The other case of possible application is when the electrical conductivity of the investigated liquid is considerably less than the electrical conductivity of the wire. For some electrolytes this may be the case, but not for liquid metals.[28] Under certain conditions, the effect of the electrical conductivity of the investigated substance on the results of measurements of thermal properties may be negligible, but not for the AC hot-wire method.

The upper temperature limit of both of the methods is near the melting temperature of the wire. Usually, platinum is used as the wire material, thus the upper temperature limit of the apparatus is about 2000 K. However, the upper temperature limit for the measurements of the thermal conductivity and heat capacity of organic liquids is the temperature of the thermal breakdown of these substances, which is about 600 K.

Both the AC and transient hot-wire methods are generally used for measurements on fluids. However, the transient hot-wire method is also used for measurements on solids and loose-fill materials.[29] Attempts have been made to replace platinum with stronger tungsten wire. The connection between the probe and its holders can be achieved either by soldering or mechanical pressure. The probe can also be made by depositing an electrically-conducting material on a nonconducting substrate.

Another limitation of the AC hot-wire method is the assumption of the invariance of thermophysical properties within the temperature oscillation range of the experiment. In the cases of investigations of critical phenomena or second-order phase transformations, this assumption may not be valid. As one approaches critical or phase transition points, the derivatives $(\partial \lambda / \partial T)_p$, $(\partial C_p \rho / \partial T)_p$, and $(\partial \rho / \partial t)_p$ will increase, resulting in nonlinear phenomena. For the AC hot-wire method, this leads to the appearance of harmonics in the spectrum of the probe temperature oscillations (i.e., 4ω, 6ω, ...). The above problems can be minimized by decreasing the amplitude of temperature oscillations. This in turn necessitates the use of more sensitive instrumentation. The increase of the wire length results in an increase of the electrical signal, conserving the value of θ_0. The natural limitation of such an approach is the increase in electric-thermal Nyquist noises. A way of solving the problem would be to develop the appropriate theory which would take nonlinear phenomena into account.

To the first approximation, the amplitude θ_2 of the second harmonic of the plane probe temperature oscillations (frequency 4ω) can be written in the following form:

$$\frac{\theta_2}{\theta_0^2} = \frac{0.104}{\sqrt{1 + \sqrt{2}\xi + \xi^2}} \left[\frac{1}{C_p \rho} \left(\frac{\partial C_p \rho}{\partial T} \right)_p + \frac{0.41}{\rho} \left(\frac{\partial \rho}{\partial T} \right)_p + \frac{1.41}{\lambda} \left(\frac{\partial \lambda}{\partial T} \right)_p \right]$$

When n-hexane is investigated in the vicinity of the critical point, signals caused by the second (4ω) and third (6ω) harmonics of nonlinear temperature oscillations were observed and recorded. The temperature dependence of θ_2/θ_0^2 obtained in our measurements confirm the results predicted by calculations under the conditions of $P =$ constant and $P > P_c$. Thus, it might be said that the AC hot-wire (foil) method can be applied to the regime to yield information on the derivatives mentioned above.

8. THE APPARATUS CIRCUIT

The circuit diagram of the apparatus is presented in Fig. 4. The main part of the circuit is an AC bridge consisting of three branches. In the left-hand branch "A" are probe Z1, immersed into the liquid under investigation, and resistance boxes R1 and R3. The probe Z1 can be switched over to one of the arms of the Wheatstone bridge using the twin switch, to measure temperature dependence $R(t)$ of its resistance (the switch and Wheatstone bridge are not shown in the diagram). In the middle branch "B" are resistance box R2 and probe Z2, which is placed inside the evacuated glass vessel. The probe Z2 is a source of the reference phase signal. The third branch "C" consists of two 100-ohm resistors R4 and R5. All the resistors are of a nonreactive type. The

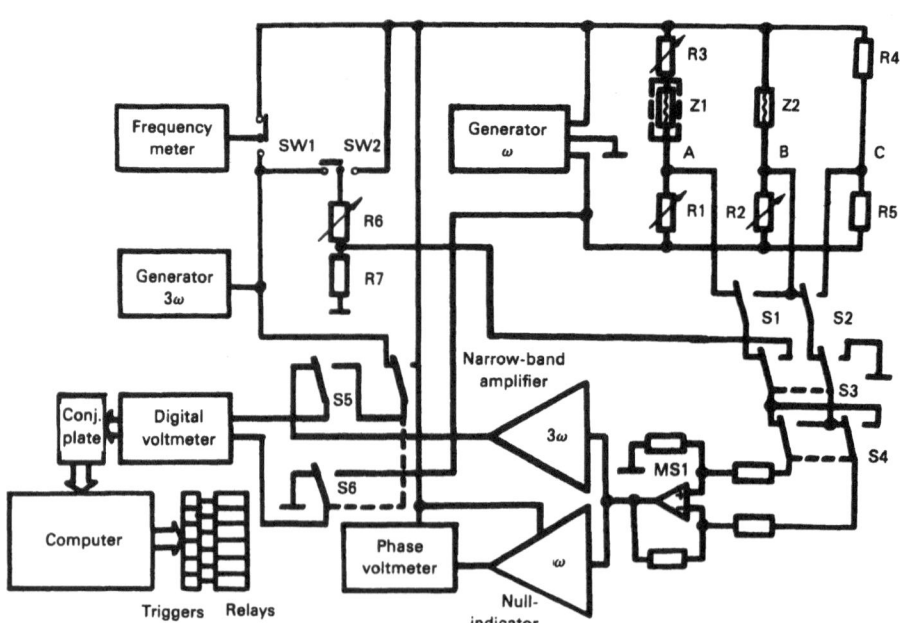

FIGURE 4. Diagram of the electric circuit of the apparatus.

accuracy of the resistance boxes is 0.05%, discreteness being 0.01 ohm. The fixed resistors R4 and R5 are chosen accurate to within 0.01%.

The bridge is powered by a sinusoidal voltage generator whose transformer center point is grounded to reduce noise and AC pickups.

The signal at the output of the preamplifier MS1 (integrated circuit—operational amplifier) is a difference of two signals at its inputs, which are connected in turn to junctions AC, BC, and AB. This imbalance signal of the bridge, a sum of sinusoids of frequencies ω and 3ω, is applied to the inputs of the selective amplifier tuned to frequency 3ω with the null indicator tuned to a frequency ω so as not to shift phases of the signal to be amplified by it. The null indicator output is connected to the input of the phase-sensitive voltmeter. The signal from the ω generator is applied to the phase-sensitive voltmeter as a reference, and to the null indicator as a sweep on its oscilloscope. The size of an ellipse on the null indicator screen is indicated by a disturbance signal value, while the phase is indicated by the tilt angle of the ellipse axes and two scales of the phase-sensitive voltmeter, which shows the "real" and "imaginary" signal components. This allows one to accurately balance the bridge arms.

For accurate measurement of the amplitude of the 3ω signal, which carries the information on the liquid properties, one should know the gain ratio of the 3ω amplifier. It is determined with the aid of the 3ω generator and divider R6–R7. The digital voltmeter measures the signals at the output of the 3ω amplifier and voltages of the ω and 3ω generators. The frequency meter is designed for tuning the generators to frequencies ω and 3ω (frequency-meter changeover button SW1). Actually, to ensure higher accuracy, the frequency meter measures the periods of oscillations T_ω and $T_\omega/3$.

The voltmeter readings are sent to the microcomputer storage through the conjugation plate. The computer automatically switches the preamplifier and digital voltmeter inputs by means of six relays S1 to S6. Table 2 gives the combination of relay states and indicates the spots where the inputs of the voltmeter and preamplifier are connected.

9. THE RELATION BETWEEN ELECTRICAL QUANTITIES BEING MEASURED AND THE AMPLITUDE AND PHASE OF PROBE TEMPERATURE OSCILLATIONS

When a bridge is supplied with sinusoidal voltage $U_1 \cos \omega t$, the probe temperature oscillates at frequency 2ω:

$$T = \bar{T} + |\theta_0| \cos (2\omega t - \varphi)$$

TABLE 2

Combination No.	Relay stage						Digital voltmeter readings	Inputs of		Preparatory operation[a]
								Digital voltmeter	Preamplifier	
								Are connected to		
1	S1	S2	S3	S4	S5	S6	8	9	10	11
	2	3	4	5	6	7				
1	0	0	1	0	1	1	U1 voltage of ω generator	ω generator	Output of divider R6–R7	Setting voltage of ω generator
2	0	0	1	0	1	0	U2 voltage of 3ω generator	3ω generator	Output of divider R6–R7	(a) Checking voltage of 3ω generator (b) Tuning generator frequencies ω, 3ω (c) Adjusting phase-sensitive voltmeter (d) SW2 depressed—adjustment of null indicator (e) SW2 released—adjustment of 3ω amplifier
3	1	1	0	0.1	0	0	E2 signal 3ω from probe Z2	Output of 3ω amplifier	Diagonally opposite pairs of junctions BC and CB	Balancing BC with R2
4	0	1	0	0.1	0	0	E1 signal 3ω from probe Z1	Output of 3ω amplifier	Diagonally opposite pairs of junctions AC and CA	Adjusting E1 ≈ E2 and balancing AC with R1 or R3
5	0	0	1	0.1	0	0	E4 signal 3ω from divider	Output of 3ω amplifier	Output of divider R6–R7	Adjusting E4 ≈ E1 with R6
6	0	0	0	0.1	0	0	E3 vector difference of 3ω signals from probes Z1 and Z2	Output of 3ω amplifier	Diagonally opposite pairs of junctions AB and BA	Starting microcomputer

[a] Button SW2 is depressed only upon adjustment of the null indicator.

where $|\theta_0|$ and φ are the amplitude and phase of temperature oscillations which carry information on the thermal properties of the medium, with $|\theta_0| \ll \bar{T}$. The probe resistances also consist of a fixed $\bar{R}$ and alternating components

$$R = \bar{R} + \frac{dR}{dT}|\theta_0| \cos (2\omega t - \varphi) \tag{18}$$

For the imbalance signal from junctions AC, we have

$$e = U_1 \cos \omega t \left[\frac{R5}{R4 + R5} - \frac{R1}{R1 + R3 + \bar{R} + |\theta_0|(dR/dT) \cos (2\omega t - \varphi)} \right]$$

$$\approx U_1 \left[\left(\frac{R5}{R4 + R5} - \frac{R1}{R1 + R3 + \bar{R}} \right) \cos \omega t + \frac{R1|\theta_0|(dR/dT)}{2(R1 + R3 + \bar{R})^2} \cos (\omega t - \varphi) \right.$$

$$\left. + \frac{R1|\theta_0|(dR/d\theta)}{2(R1 + R3 + \bar{R})^2} \cos (3\omega t - \varphi) \right] \tag{19}$$

Here we have used the fact that the resistance variations are small: $|\theta_0|(dR/dT) \ll \bar{R}$. It follows from equation (19) that a triple-frequency signal is available at the pair of junctions AC, the value of which is proportional to the amplitude, while the phase equals that of the probe temperature oscillations:

$$e_{3\omega} = \frac{U_1 R1}{2(R1 + R3 + \bar{R})^2} \frac{d\bar{R}}{dT} |\theta_0| \cos (3\omega t - \varphi) \tag{20}$$

It also follows from equation (19) that, while balancing the bridge on the frequency ω so that the imbalance signal has phase φ [i.e., the first term of equation (19) equals zero], we obtain the balance of the bridge arms. For a given bridge balancing, the third harmonic of the generator in the supply voltage does not reach the output of the diagonally opposed pair of junctions and does not disturb the signal being recorded.

The amplitude of power developed on a probe is given by

$$W_0 = \frac{U_1^2 \bar{R}}{(R1 + R3 + \bar{R})^2} \tag{21}$$

If this expression and the expression for $|\theta_0|$ derived from equation (20) are substituted into the relationship

$$|\theta_0| = \frac{W_0}{2C_p' m' \omega} S(\kappa, \eta)$$

which relates the amplitude of the probe temperature oscillations to the unknown thermal properties, then we obtain the relationship linking these properties to the electric quantities measured during the experiment:

$$\frac{(R1 + R3 + \bar{R})^4}{U_1^3 T_\omega \bar{R} R1} |e_{3\omega}| = \frac{(dR/dT)}{8\pi C_p' m'} S(\kappa, \eta) \tag{22}$$

where $|e_{3\omega}|$ and U_1 are the effective values, while T_ω is the period of oscillation of the supply voltage and equal to $2\pi/\omega$.

In order to determine the value of signal $e_{3\omega}$, the gain ratio is calibrated with the help of the 3ω generator and divider R6–R7.

The total gain ratio of the preamplifier and 3ω amplifier equals

$$\frac{E1}{e_{3\omega}} = \frac{E4}{U_2 R7} (R6 + R7) \tag{23}$$

where E1 is the amplifier output signal while its input signal is $e_{3\omega}$, U_2 is the voltage of the 3ω generator, E4 is the amplifier output signal while its input signal fed from the divider is $U_2 R7/(R6 + R7)$.

With the junctions AC balanced, the equality $\bar{R} = R1 - R3$ is true. When this is entered into equation (22) and equation (23) employed, we obtain

$$\frac{16 \cdot R1^3 \cdot E1 \cdot U_2 \cdot R7}{U_1^3 \cdot T_\omega (R1 - R3) E4 (R6 + R7)} = \frac{d\bar{R}/dT}{8\pi m' C_p'} S(\kappa, \eta) \tag{24}$$

This relationship is used to determine function $S(\kappa, \eta)$. The left-hand side of equation (24), designated S^0, contains the quantities that are being sought experimentally. On the right-hand side, multiplier $(d\bar{R}/dT)/(8\pi m' C_p')$, here designated B, depends on the parameters of a particular probe and on its temperature. The dependence of function B on the probe resistance (and consequently its temperature) is determined by calibrating in vacuum [where function $S(\kappa, \eta)$ equals 1] or in air at a higher frequency [where function $S'(\kappa, \eta)$ approaches 1] and is calculated on the basis of tabulated data on the thermal properties of air.

Let us examine briefly functions $S(\kappa, \eta)$ and $F(\kappa, \eta)$ for a probe placed in a highly rarefied gas, almost a vacuum. When the asymptotic expressions for the Kelvin functions valid at small values,

$$\text{ker}\,\kappa \approx -\ln\frac{\kappa\gamma}{2}, \qquad \text{kei}\,\kappa \approx -\frac{\pi}{4}$$

$$\text{ker}'\,\kappa \approx -\frac{1}{\kappa}, \qquad \text{kei}'\,\kappa \approx 0 \tag{25}$$

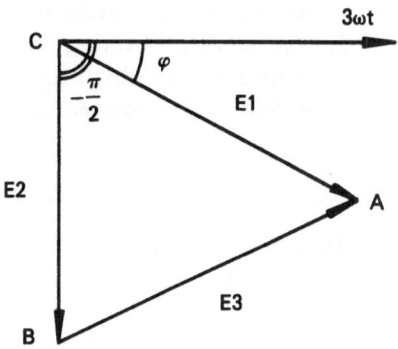

FIGURE 5. Vector diagram of triple-frequency voltages.

(γ is Euler's constant, equal to 1.781...), with parameter η tending to infinity, are substituted into expressions (10) and (11) for functions S and F, then we have

$$S(\kappa, \eta) \to 1, \qquad F(\kappa, \eta) \to -\infty, \qquad \varphi \to -\pi/2$$

This information enables one to measure the function $B(\bar{R}) = (d\bar{R}/dT)(8\pi C'_p m')$, which in vacuum equals function S^0 [see equation (24)], and to use a probe placed in an evacuated vessel as a source of the reference phase.

On the vector diagram (see Fig. 5), vector E2, the signal from the bridge junctions BC, corresponds to the oscillations of probe Z2 placed in a vacuum, while vector E1 (junctions AC) corresponds to the oscillations of probe Z1 immersed in the liquid under investigation. Vector E3 corresponds to the signal from junctions AB and the difference of vectors E1 and E2. By measuring all three signals in succession with the digital voltmeter and using the known geometrical relationship between the sides of a triangle, we obtain an expression for the phase shift angle φ in the form

$$\varphi = \text{arc} \sin \frac{E1^2 + E2^2 - E3^2}{2 \cdot E1 \cdot E2} \qquad (26)$$

10. CIRCUIT ADJUSTMENT, CONTROL AND PROCESSING PROGRAM

Before conducting measurements, a computer program and a file of 64 coefficients used for calculating Kelvin functions are transferred from a tape cassette to the on-line storage of a microcomputer (storage capacity, 16K). The following initial information should also be introduced:

1. Parameters R_0, α, and β, functions of the probe temperature.

2. Results of calibrating functions B (in air or in a vacuum)
3. Results of calibrating the probe in toluene, κ_t, η_t, temperature T_t, and period $T_{\omega t}$ at which the calibration was conducted.
4. Period of the voltage supply to the bridge, T_ω.

10.1. Circuit Adjustment Procedure

The circuit is switched by the computer by means of six similar subroutines, which realize the relay state combinations given in Table 2. This table also shows where the inputs of the digital voltmeter and preamplifier are to be connected. The right-hand column of Table 2 gives preparatory operations performed for a given combination of relay states.

The measurement preparation consists in adjusting the instruments and balancing the bridge, the sequence of which is shown in the table.

1. Slowly increase the voltage of the ω generator up to about 1.5 V (sharp application of the supply voltage may cause damage to the probes).
2. (a) Check the voltage of the 3ω generator;
   (b) use a frequency meter (button SW1) to tune the ω and 3ω generators to the desired frequencies;
   (c) calibrate the phase-sensitive voltmeter by the amplitude and frequency of a reference signal—ω generator signal;
   (d) with button SW2 depressed, adjust the null indicator to zero phase shift: the ellipse on the null-indicator screen shrinks into a line and the "imaginary" component of the signal at the null-indicator output becomes zero, which is shown by the phase-sensitive voltmeter (with button SW2 depressed, the voltmeter displays a somewhat high value of U2);
   (e) with button SW2 released, tune the selective amplifier to the frequency of the 3ω generator.
3. Use resistance box R2 to balance the bridge junctions equal to zero. The value of E2 displayed by the digital voltmeter should be memorized.
4. Use resistance boxes R1 and R3 to adjust the value of signal E1 to the value of E2 (it is necessary for higher measurement accuracy of φ) and balance the bridge junctions AC. As this takes place, the imbalance signal must have phase $\simeq \varphi$ (which is determined by the phase-sensitive voltmeter) to achieve the equality of the bridge arms and exclude the third harmonic of the generator from the signal being measured.
5. Use resistance box R6 to adjust the value of signal E4 to the value of E1 (in order to exclude possible dependence of the gain ratio of the amplifier on the amplitude).
6. The bridge junctions AB become automatically balanced. Now that the preparation work has been completed, the computer is started and executes the program measurements.

10.2. Measuring Part of Program

In the read subroutine, which is accessed after every switching, the computer reads continuously the indications of the digital voltmeter at a frequency of 10 Hz and compares three successive readings $(E_{n-2} \gtrless E_{n-1} \lessgtr E_n)$. When, after a subsequent switching, the voltmeter readings start oscillating about a certain mean value, which is indicative of the cessation of transients, the computer stores 15 successive readings. Then it checks for malfunctions, computes the average value and sends it to the assigned storage cell.

In this way all the required voltages are measured in the following sequence: E3, E1, E2, E4, U2, U1, with E1-E4 measured twice—the preamplifier inputs are reconnected by relay S4—to compensate for possible difference in gain ratios of the inverting $(-)$ and noninverting $(+)$ inputs of the preamplifier.

After the computer has executed the measuring part of the program, it stops and an operator inserts the values of resistance boxes R1, R3, and R6. This done, the data processing is started and proceeds in three stages.

Stage 1. Based on the measured values of voltages E1-E4, U1, U2, and introduced values of R1, R3, and R6, the computer calculates function S^0—left-hand side of equation (24), and phase-shift angle tangent φ—equation (26). The values of functions S^0 and $|\tan \varphi|$ are displayed on the panel, and the computer pauses.

Stage 2. A value of function $B = B_0 (1 + z\bar{R})$, corresponding to probe resistance $\bar{R} = R1 - R3$ and functions $S = S^0/B$, is computed. Then the set of transcendental equations (8) and (9) is resolved, and is reduced to an equation dependent on a variable κ by finding product $\kappa\eta$ from the second equation and substituting it into the first.

As a zero-order approximation for κ (a somewhat low value) we take

$$\kappa_0 = \frac{2}{\gamma} \exp\left(-\frac{\pi}{4|\tan \varphi|}\right)$$

where γ is Euler's constant equal to 1.781..... This expression is obtained by putting the asymptotic formulas of Kelvin functions in equations (25), true for small values of the argument, into equation (9) for $\tan \varphi$.

Further, the computer starts the cycle where it used the secant method, which is chosen because it does not need the derivative computation to compute the next approximations of κ until the desired accuracy is obtained [cycle end condition $(\kappa_n - \kappa_{n-1})/\kappa_n \lesssim 10^{-7}$]. By using the obtained value of κ, η is computed with the help of equation (9).

Kelvin functions ker κ, kei κ, ker$'$ κ, and kei$'$ κ are computed with the help of the coefficients of expansion in terms of the Chebyshev polynomials $T_n(x) = \cos(n \arccos x)$. These coefficients are put in the computer storage from a magnetic tape, with the leading coefficients being accurate to 12 decimal digits. Eight coefficients are put for each of the Kelvin functions, the total number being 64; functions ber κ, bei κ, ber' κ, and bei' κ are also computed. The calculation formulas for the Kelvin functions and the coefficients are taken from Luke's book.[30]

The second stage is completed when the computed values of κ and η are displayed on the panel.

Stage 3. At this last stage of processing, the computer calculates values of thermal diffusivity a_t and heat capacity per unit volume $(C_p\rho)_t$ of toluene, corresponding to the calibration temperature T_t, by the following expressions:

$$a_t(T) = \left(9.49 - \frac{T - 273.2}{36}\right) \times 10^{-8} \qquad (\text{m}^2\,\text{s}^{-1})$$

and

$$(C_p\rho)_t(T) = \left[1.447 + \frac{0.04}{37}(T - 273.2)\right] \times 10^6 \qquad (\text{J}\,\text{m}^{-3}\,\text{K}^{-1})$$

which approximate the data on thermal conductivity,[5] heat capacity,[11] and density[12] of toluene over the room-temperature range.

Based on the obtained toluene properties, the computer calculates the unknown thermal properties a_1, $(C_p\rho)_1$, λ_1, and b_1 of a liquid, using the expressions derived in determining parameters κ and η:

$$a_1(T_1) = a_t(T_t) \frac{\kappa_t^2}{\kappa_1^2} \frac{T_{\omega t}}{T_{\omega 1}} [1 + 0.2 \times 10^{-4}(T_1 - T_t)]$$

$$C_p\rho(T_1) = C_p\rho_t(T_t) \frac{\eta_t}{\eta_1} \frac{1 + 0.304 \times 10^{-3}(T_1 - 273.2) - 0.27 \times 10^{-6}(T_1 - 273.2)}{1 + 0.3 \times 10^{-3}(T_t - 273.2)}$$

$$\lambda_1(T_1) = a_1(C_p\rho)_1 \qquad \text{and} \qquad b_1 = \sqrt{a_1}(C_p\rho)_1$$

Here, the temperature dependence of the specific heat of platinum (probe material) is computed by the formula

$$\frac{C_p'(T)}{C_p'(273.2)} = 1 + 0.332 \times 10^{-3}(T - 273.2) - 0.26 \times 10^{-6}(T - 273.2)^2$$

which is a parabolic approximation of the data on the specific heat of platinum taken from Yokokawa and Takahashi[13] for temperatures ranging from 0 to 400 °C. The dependence of density ρ' on the temperature requires only a small correction to be introduced in the function coefficients. Factor $1 + 0.2 \times 10^{-4}(T_1 - T_t)$ takes account of the increase in wire radius due to temperature rise, which can be neglected as it is too small.

The program is completed when the panel displays computed values of the thermal properties λ_1, $(C_p\rho)_1$, a_1, b_1, and the reference temperature, calculated on the basis of the probe resistance by the following formula:

$$T - 273.2 = \frac{2((R1 - R3)/R_0 - 1)}{\alpha + \sqrt{\alpha^2 + 4\beta\{[(R1 - R3)/R_0] - 1\}}}$$

where R_0, α, and β are the coefficients of the temperature dependence of the probe resistance obtained in the calibration of $\bar{R}$.

11. PROBE CALIBRATION

A newly manufactured probe must be annealed and gradually cooled down, and then calibrated: (1) in vacuum (or in air) to record the function $B(\bar{R})$; (2) in toluene (standard liquid) to obtain parameters κ_t and η_t (when the relative variant of the technique is used); (3) to determine the function $\bar{R}(T)$ of probe resistance as dependent on temperature for obtaining coefficients R_0, α, and β which approximate the dependence.

11.1. Calibration of a Probe in Vacuum or in Air

Equation (24) relating the electrical quantities being measured experimentally and the thermal properties of a medium surrounding a probe may be written as follows:

$$S^0 = B(T)S(\kappa, \eta)$$

where $B(T) = (dR/dT)/(8\pi C_p' m')$ is the function whose dependence on temperature is to be found.

In a vacuum, function $S(\kappa, \eta)$ equals one and so $B = S^0$. In practice, it is convenient to measure B as a function of probe resistance $\bar{R}$ with variation in supply voltage U1, when a probe is overheated to various temperatures. The obtained values of $B_i(\bar{R}_i)$ are approximated by the linear function $B = B_0(1 + z\bar{R})$ using the least-squares technique.

Function $B(\bar{R})$ may also be calibrated in air for increased frequency of the supply voltage (about 500 to 1000 Hz). At this high frequency, function $S(\kappa, \eta)$ is close to 1, therefore (using the approximate wire radius r_0 and reference data on the temperature-dependent relation of the properties of air and platinum) one may calculate the approximate values of κ and η and then calculate function $S(\kappa, \eta)$. The required data file $B_i(\bar{R}_i)$ may be obtained by experimentally measuring S^0 and calculating $S(\kappa, \eta)$ at different temperatures.

For calibration in air, we use a special program which is the slightly modified main program. The electrical quantities needed for computing function S^0 are measured, and function $S(\kappa, \eta)$ is calculated on the basis of this program in accordance with the employed values of the wire diameter and atmospheric pressure p, which proved to only slightly influence the value of $S(\kappa, \eta)$. Reference data on the thermal properties of air[12] and platinum[13] give the following approximations for the thermal diffusivity of air a and parameter η:

$$a(T) = \frac{[24.023 + 0.08115(T - 273.2) - 0.342 \times 10^{-4}(T - 273.2)^2]}{[0.996 + 0.1734 \times 10^{-3}(T - 273.2)]p} 0.2153$$

$$\times 10^{-3} T$$

and

$$\eta(T) = \frac{1 + 0.304 \times 10^{-3}(T - 273.2) - 0.27 \times 10^{-6}(T - 273.2)^2}{[0.996 + 0.1734 \times 10^{-3}(T - 273.2)]p} 3095.8 T$$

where p is the atmospheric pressure (in mm Hg).

Calibration of $B(\bar{R})$ in air is preferable to that in a vacuum. In the latter, at a high temperature of the probe, the end effects might influence the calibration results because the probe ends are exposed to room temperature.

If we use the linear approximation of the temperature dependence of function B and specific heat C'_p of the probe material, namely,

$$B(T) = B(273.2)[1 + z'(T - 273.2)] \qquad \text{and}$$

$$C'_p(T) = C'_p(273.2)[1 + \gamma(T - 273.2)]$$

and bear in mind that

$$\frac{dR}{dT} = R_0\alpha\left(1 + \frac{2\beta(T - 273.2)}{\alpha}\right)$$

then we have

$$B(273.2)[1 + z'(T - 273.2)] = \frac{R_0\alpha}{8\pi C'_p(273.2)m'} \frac{1 + 2\beta(T - 273.2)/\alpha}{1 + \gamma(T - 273.2)}$$

It follows from this relationship that $\gamma = (2\beta/\alpha) - z'$, i.e., using calibration in vacuum (air) and taking account of function $\bar{R}(t)$, one may obtain γ, the temperature coefficient of the heat capacity C'_p of the probe material.

11.2. Calibration of the Probe in Toluene (Standard Liquid)

The expressions for parameters κ and η contain the wire radius r_0 and heat capacity $C'_p\rho'$ of the probe material per unit volume, which are difficult to measure directly. In order to overcome these difficulties, we use a relative modification of the technique where a probe is calibrated in a substance (toluene) whose properties are well known, at room temperature and atmospheric pressure.

The amplitude and phase of the triple-frequency signal of the probe submerged in toluene are measured like those in liquids whose thermal properties are unknown. The data are processed until parameters κ_t and η_t are computed. The calibration temperature T_t, which influences the thermal properties of toluene, is also recorded.

11.3. Calibration of Probe Resistance as a Function of Temperature

Probe Z1 can be switched from an AC bridge to a Wheatstone bridge by means of a twin switch in order to measure its resistance along with the main measurements (the switch and Wheatstone bridge are not illustrated on the diagram of the apparatus). The operating current of the Wheatstone bridge is sufficiently small to ignore probe overheating. The temperature of the copper pig where the measuring cell is positioned is checked with a platinum resistance thermometer.

Probe resistance R_0 is measured when the cell is in a Dewar flask filled with thawing ice. In addition, the temperature in the Dewar flask is checked with a mercury thermometer. By measuring probe resistances at different temperatures, we obtain a number of values of $\bar{R}_i(T_i)$, which are computer-processed by a least-squares technique so as to obtain coefficients α and β in the probe resistance as a function of temperature:

$$\bar{R}(T) = R_0[1 + \alpha(T - 273.2) + \beta(T - 273.2)^2]$$

The root-mean-square deviation σ' is also computed.

The coefficients obtained by the calibration are employed to compute the data reference temperature using the probe resistance. The error in finding the reference temperature is about 0.2 °C, which is quite sufficient as the thermal properties of a liquid are not greatly influenced by temperature (far from the critical point).

11.4. Test of Apparatus Operation

A test of the apparatus operation is required after a new probe has been installed. The test includes the following points.

1. When the probe is calibrated with a standard substance, such as toluene, the value of $\eta = C'_p\rho'/2C_p\rho$ has to be approximately equal to the value of η obtained from tabular data on specific heat and density for toluene and platinum.

2. The best test of the apparatus performance is the measurement of the thermal conductivity and specific heat of one or more substances where properties are well known at room temperature.

3. It is essential to test the reliability of the measurements of the temperature dependence of thermophysical properties in the temperature range 290-370 K. For this purpose, one may use toluene and combine the test with the probe calibration.

12. MEASUREMENT ERRORS

It is convenient to analyze the errors in thermal property measurements by using the asymptotic formulas for $S(\kappa, \eta)$ and $F(\kappa, \eta)$ which correspond to small values of the argument, $\kappa^2 = 2\omega r_0^2/a \ll 1$ and $\kappa^2\eta \ll 1$. In our measurements, κ^2 never exceeded 0.1. The merits of asymptotic formulas are that they enable one to solve the set of equations (8) and (9) in terms of thermal properties. For relative measurements, the expressions governing the thermal properties of a liquid take the following forms:

$$\lambda_1 = \lambda_s \left[\frac{B(T_1)}{B(T_c)} \frac{\omega_1}{\omega_c} \frac{C'_p(T_1)}{C'_p(T_c)} \right] \frac{S_c^0 \sin \varphi_c}{S_1^0 \sin \varphi_1} \tag{27}$$

$$a_1 = a_s \frac{\omega_1}{\omega_c} \exp\left[\frac{\pi}{2}\left(\frac{1}{F_1} - \frac{1}{F_c} \right) \right] \tag{28}$$

$$(C_p\rho)_1 = (C_p\rho)_s \left[\frac{B(T_1)}{B(T_c)} \frac{C'_p(T_1)}{C'_p(T_c)} \right] \frac{S_c^0 \sin \varphi_c}{S_1^0 \sin \varphi_1} \exp\left[\frac{\pi}{2}\left(\frac{1}{F_1} - \frac{1}{F_c} \right) \right] \tag{29}$$

$$b_1 = b_s \left[\frac{B(T_1)}{B(T_c)} \frac{C'_p(T_1)}{C'_p(T_c)} \right] \sqrt{\frac{\omega_1}{\omega_c}} \frac{S_c^0 \sin \varphi_c}{S_1^0 \sin \varphi_1} \exp\left[\frac{\pi}{4}\left(\frac{1}{F_1} - \frac{1}{F_c} \right) \right] \tag{30}$$

In these formulas the subscripts are defined as follows: s indicates the thermal properties of a standard substance, c indicates that the corresponding value was obtained upon calibration in a standard substance, while 1 indicates measurements in the liquid under investigation; $F = \tan \varphi$. It follows from equations (27)-(30) that, for a relative method of measurement, almost all the quantities (except F) are included in the calibration formulas in the form of relationships; this situation is most advantageous.

The total error of a thermal property measurement at confidence level $\alpha' = 0.95$ includes measurement errors in the values of $\delta F/F$ (about 0.3%), $\delta S^0/S^0$ (about 0.2%), $\delta B/B$ (about 0.5%), and $\delta C'_p/C'_p$ (about 1%), as well as known errors in the thermal properties of a standard substance (toluene at $t = 293$ K): $\delta \lambda_t/\lambda_t$ (about 0.5%) and $\delta (c_p\rho)_t/(C_p\rho)_t$ (about 0.5%).

The total error is calculated by the formula

$$\delta\lambda/\lambda = \sqrt{\sum_h \left(K_{\lambda,h} \frac{\delta h}{h} \right)^2}$$

where $K_{\lambda,h}$ are sensitivity coefficients corresponding to S_1^0, S_c^0, F_1, F_c, λ_s, $C'_p(T_1)/C'_p(T_c)$, and $B(T_1)/B(T_c)$. Errors are as follows:

$$\frac{\delta\lambda}{\lambda} \sim 1\text{-}2\%, \qquad \frac{\delta(c_p\rho)}{c_p\rho} \sim 2\text{-}3\%, \qquad \frac{\delta a}{a} \sim 2.5\text{-}3\%, \qquad \frac{\delta b}{b} \sim 1.5\text{-}2\%$$

An idealized mathematical problem forms the basis underlying the theory of the method. The difference between the actual experimental conditions and the mathematical model is another source of errors. The following factors may be identified as causing errors:

1. Nonisothermal medium, i.e., the existence of a radial temperature gradient caused by wire overheating relative to the cell walls.
2. Nonisothermal probe.
3. End effects, i.e., the influence of wire-end length on the amplitude and phase of temperature oscillations.
4. Distribution of temperature oscillation amplitude over the probe radius.
5. Nonideal property of a vacuum probe, i.e., deviation in the reference phase of temperature oscillations from $-\pi/2$ due to the end effects and radiation.
6. Lack of cylindrical shape of the probe, i.e., variable wire diameter over its length and deviation from roundness in the cross section of the wire.

An analysis shows that the overall misrepresentation of $S^0 \sim |\theta_0|$ and $\tan\varphi$ caused by these factors not being taken into account does not exceed 0.2-0.3% and can be neglected in the relative method of measurement.

13. HARDWARE OF SET AND MICROCOMPUTER INPUT/OUTPUT INTERFACE

13.1. Input Device

Information is entered in the microcomputer from the B7-16 digital voltmeter with the aid of the INPUT bus, which consists of eight digits and is capable of receiving one byte per cycle in a parallel code. In order to switch four decimal digits of the voltmeter (or 16 binary digits) to four digits of the INPUT bus, four integrated circuits, type K155KП 1 (multiplexer–selector, analog SN74150), are provided, each of which is capable of connecting one of its 16 information inputs to a single inverted output as a function of a combination of logic zeros and ones at the four control inputs of the integrated circuit and on condition that a gate signal is available (zero to gate input).

All four control inputs of the integrated circuits are connected to four low-order digits of the CONTROL bus, which is used for input/output functions and consists of eight digits. One half-byte of information (i.e., one decimal digit) is determined from the voltmeter by the input instruction INPSd (the instruction code is 1500d, where d is the byte set at the CONTROL bus). To set all four decimal digits, it is necessary to use in succession four input instructions, changing d from 0100 to 0111, while integrated circuits KП 1 are gated by one of the high-order half-bytes of d. After a certain delay (about 1 μs) the computer-available signal shapes the peripheral clock pulse if the "measurement end" signal is supplied by the voltmeter. The voltmeter digits are read in accordance with a special subroutine.

13.2. Output and Control Device

The OUTPUT bus, consisting of eight digits, is used to ensure the output of computer numerical data and to control the set. The output information is stored in the external memory register which is built around six D-flipflops (integrated circuits K155TM7, analog SN7475, each containing four D-flipflops, are used). The information input of each flipflop is connected to one digit of the OUTPUT bus, enabling the input to be connected to the output data accompanying signal on the computer. Each flipflop output is connected to the inputs of the integrated circuits K155Λ A7 (analog SN7422, with open collector output, ensures load current up to 30 mA) and the outputs of each of the integrated circuits ΛA7 are connected to the relays, type PэC-22, each having four contact groups. By writing corresponding combinations of logic ones and zeros into the external register of the memory with the aid of the output instruction OUTSd (code 1501d), it is possible to introduce necessary switchings of the digital voltmeter and preamplifier inputs.

14. MEASURING CELL DESIGN

The thermal properties of liquids at normal pressure and temperatures ranging from 0 to 100 °C were studied using a simple test tube with liquid, into which was immersed probe (1) welded to nickel leads (2) (see Fig. 6). The test tube was put in a liquid thermostat.

The design of a measuring cell for conducting investigations at temperatures up to 600 K and pressures up to 40 MPa is presented in Fig. 6. Each end of probe (1) is put between two strips of nickel foil (2), which are contact-welded. The probe length is 2 to 3 cm. The foil strips are welded to nickel wires passed through two-channel ceramic pipe (4) whose outer diameter is 6.8 mm. The nickel wires are caulked in the contacts of lead-in (5). The lead-in external terminals are connected to the set electric circuits. Probe (1) is put into a hollow thin-walled cylinder (6) whose inner diameter is 7 mm. The joint of lead-in (5) and cylinder (6) is sealed with a shaped teflon washer (9) and union nut (8). A pressure transfer chamber (10) is a steel cylinder whose outer and inner diameters are 50 and 30 mm, respectively. Bellows (11) are welded to one of the cylinder ends. All the cell components which come in contact with a liquid under study are made of stainless steel. Pressure is built by hydraulic press (13) and translated to the liquid under study through pipe (12) and bellows (11). Pressure is measured with manometer (14). After being evacuated, the cell is filled via a hole into which a screw (15) with a sealing teflon washer is driven. The working cell is screwed into a copper

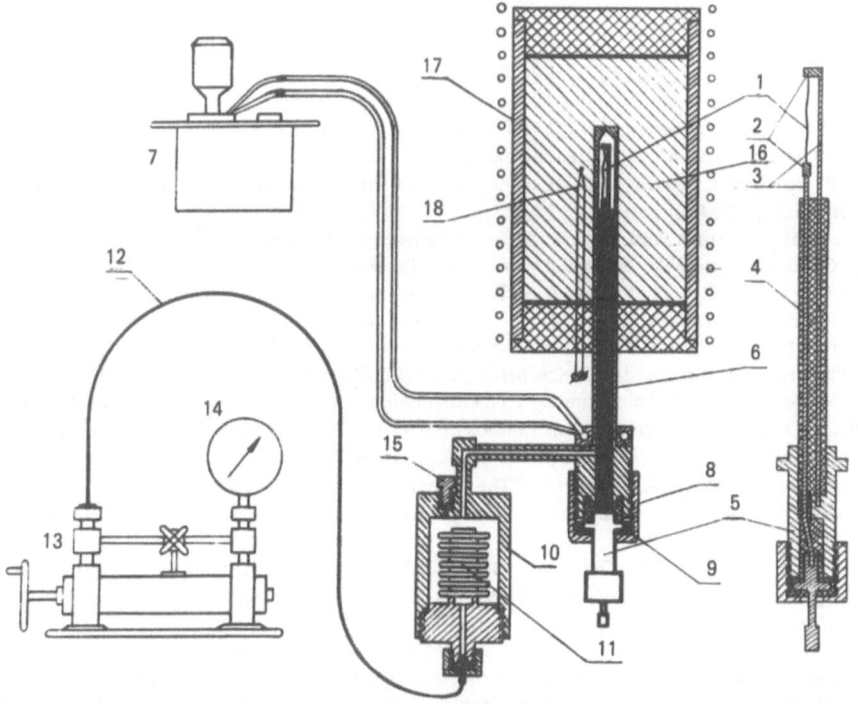

FIGURE 6. Design of measuring cell, heater, and pressure build-up system.

cylinder (16), which is placed inside the heater wound over a thick-walled copper pipe (17). From the outside, the heater is heat-insulated with asbestos. Thermostat (7) serves to cool the cell sealing unit, if necessary. Located near the inner winding of the heater is one of the junctions of a chromel-copel thermocouple used in the temperature regulation circuit. The temperature of the middle part of cylinder (16) is measured by thermocouple (18). The reference temperature of the results is determined by the probe resistance.

15. CONCLUSION

The thermal properties of a large number of liquids, including toluene,[14] cyclohexane,[15] carbon tetrachloride, normal alkanes and their isomers,[15-19] esters, mixtures of normal liquids,[20] and others, have been studied on the described apparatus. It should be noted that the comprehensive method of periodic heating, realized according to somewhat different schemes of recording probe temperature oscillations, has been used elsewhere.[4,21,22]

The experience gained in the investigations proves that the above-described techniques and equipment may be recommended for use in scientific research and engineering laboratories.

REFERENCES

1. L.P. Phylippov, *Measurement of the Thermal Properties of a Substance by the Periodic Heating Method* (in Russian), Energoatomizdat, Moscow (1984).
2. L.P. Phylippov, *Investigation of the Heat Conduction of a Liquid* (in Russian), Moscow University (1970).
3. E. Yanke, F. Emde, and F. Lesh, *Special Functions* (in Russian), Nauka, Moscow (1977).
4. F.G. Eldarov and V.M. Shulga, *Ismeritelnaya Tekhnika*, No. 2, pp. 75-78 (1982).
5. N.B. Vargftik, L.P. Phylippov, A.A. Tarzimanov, and E.E. Totskii, *Thermal Conductivity of Polyatomic Liquids and Gases* (in Russian), p. 68, Isdatelstvo Standartov, Moscow (1981).
6. S.N. Kravchun and L.P. Phylippov, "On Radiation-Conductive Heat Transfer in a Temperature Wave Regime" (in Russian), *Inzhenerno-fizicheskii Zhurnal* **35**(6), 1027-1033 (1978).
7. T.V. Gurenkova, L.L. Suleimanova, T.N. Gorshenina, and C.A. Usmanov, "Study of Radiation Heat Transfer in Semitransparent Liquids of Various Chemical Natures," in: *Teplo-i massobmen v khemicheskoi tekhnologii*, pp. 68-72, Kazan (1981).
8. A.V. Baginskii and A.A. Varchenko, "The Influence of Thermal Radiation on Heat Conduction in Thin Layers of Gray Medium," in: *Teplofizicheskie Svoistva Veschestv i Materialov*, pp. 132-148, Novosibirsk (1979).
9. S.N. Kravchun and A.S. Tleubaev, "On Possibility of Measurement of the Thermophysical Properties of Liquids in Liquid Flows by the AC Heated-Wire Method (in Russian), *Inzhenerno-fizicheskii Zhurnal* **46**(1), 113-118 (1984).
10. S.N. Kravchun, On Use of AC Heated-Wire Method for Measurement of the Thermophysical Properties of Solutions (in Russian), *Inzhenerno-fizicheskii Zhurnal* **42**(6), 949-955 (1982).
11. A.M. Mamedov and T.S. Akhundov, "*Tables of Thermal Properties of Gases and Liquids*" (in Russian), Issue 5, Izdatelstvo Standartov, Moscow (1978).

12. N.B. Vargaftik, *Reference Book of Thermal Properties of Gases and Liquids* (in Russian), p. 720, Nauka, Moscow (1972).

13. Harumi Yokokawa and Yoichi Takahashi, "Laser-Flash Calorimetry. II—Heat Capacity of Platinum from 80 to 1000 K and its Revised Thermodynamic Functions," *J. Chem. Thermodyn.* **11**, 411–420 (1979).

14. S.N. Nefedov and L.P. Phylippov, "Experimental Studies of Thermal Properties of Toluene I. Heat Conduction" (in Russian), *Izvestiya VUZov, Neft i gaz*, No. 11, pp. 65–70 (1979).

15. S.N. Nefedov and L.P. Phylippov, "Experimental Studies of Thermal Properties of Cyclohexane" (in Russian), *Inzhenerno-fizicheskii Zhurnal* **37**(4), 674–676 (1979).

16. S.N. Nefedov and L.P. Phylippov, "Experimental Studies of Thermal Properties of n-heptane" (in Russian), *Zhurnal Fizicheskoi Khimii* **53**(8), 2112–2113 (1979).

17. L.P. Phylippov, S.N. Nefedov, S.N. Kravchun, and E.A. Kolykhalova, "Experimental Studies of Thermal Properties of Liquids" (in Russian), *Inzhenerno-fizicheskii Zhurnal* **38**(4), 664–650 (1980).

18. L.P. Phylippov and S.N. Kravshun, "On Heat Conduction of Normal Alkanes" (in Russian), *Izvestiya VUZov, Neft i gaz*, No. 5, pp. 88–90 (1983).

19. L.P. Phylippov and L.A. Laushkina, "The Study of Heat Conduction and Heat Capacity of Liquids. I: Alkane Isomers" (in Russian), *Zhurnal Physicheskoy Himii* **58**(5), 1068–1071 (1984).

20. L.P. Phylippov and S.N. Kravchoon, "On Heat Conduction of Liquid Solutions," *Zhurnal Physicheskoy Himii* **56**(11), 2753–2756 (1982).

21. S.R. Attala, A.A. El-Sharkawy, and F.A. Gasser, "Measurement of Thermal Properties of Liquids with an AC Heated-Wire Technique," *Int. J. Thermophys.* **2**(2), 155–162 (1981).

22. A.A. Varchenko, "Measurement of Thermal Properties of Liquids by the Method of Radial Temperature Waves," in: *Thermal Conductivity*, Proc. 15th Conference, pp. 255–260, New York–London (1978).

23. W.R. White, R.J. Brunson, R.J. Bearman, and S. Lindenbaum, "Computer Controlled Measurements of Thermal Conductivity of Aqueous Salt Solutions," *J. Solution Chem.* **4**(7), 557–570 (1975).

24. Y. Nagasaka, J. Suzuki, and A. Nagashima, "The Thermal Conductivity of Aqueous KCl Solutions at Pressures up to 40 MPa," in: *Proc. 10th Int. Conf. on Properties of Steam*, Vol. 2, pp. 203–209, Moscow (1984).

25. Y. Nagasaka and A. Nagashima, "Absolute Measurements of Thermal Conductivity of Electrically Conducting Liquids by the Transient Hot-Wire Method," *Thermal Cond.* **17**, 307–314 (1983).

26. M. Hoshi, T. Omotani, and Y. Nagasaka, "Transient Method to Measure the Thermal Conductivity of High Temperature Melts Using a Liquid Metal Probe," *Rev. Sci. Instrum.* **52**, 755 (1981).

27. T. Omotany, Y. Nagasaki, and A. Nagashima, "Thermal Conductivity Measurements of Potassium Nitrate–Sodium Nitrate System Using a Transient Method with a Liquid–Metal Probe," *Thermal Cond.* **17**, 251–256 (1983).

28. M. Takeuchi, S. Katoh, J. Kamoshida, and Y. Kurosaki, "Thermal Conductivity of Aqueous LiCl Measured by Transient Hot-Wire Method," in: *Heat Transfer 1986*: Proc. 8th Int. Conf., San Francisco, Calif., Aug. 17–22, 1986, Vol. 2, pp. 543–548, Washington, D.C. (1986).

29. G.C. Glatzmaier and W.F. Ramires, "Simultaneous Measurement of the Thermal Conductivity and Thermal Diffusivity of Unconsolidated Materials by the Transient Hot-Wire Method," *Rev. Sci. Instrum.* **56**(7), 1394–1398 (1985).

30. Yu.L. Luke, *Mathematical Functions and their Approximations*, Academic Press, New York (1975).

IV

SPECIFIC HEAT

14

Practical Modulation Calorimetry

Ya.A. KRAFTMAKHER

1. INTRODUCTION

Modulation calorimetry consists in creating periodical oscillations of the power which heats the sample and in recording the oscillations of the sample temperature about its mean value. The mean temperature, and the amplitude and phase of the temperature oscillations are assumed to be the same throughout the sample.

The variants of modulation calorimetry differ in the ways of modulating the heating power and in the methods of recording the temperature oscillations. As a rule, measurements are carried out in a regime for which the amplitude of oscillations of the power heating the sample is much larger than the amplitude of oscillations of the heat loss from the sample due to the temperature oscillations. In this regime, called adiabatic, when evaluating the specific heat it suffices to determine the amplitude and frequency of oscillations of the applied power and the amplitude of temperature oscillations.

The applied power is modulated by heating the sample in one of various ways: by passing through it an AC current or a DC current with a small AC component, by periodical electron bombardment or radiation, or with a separate heater. The temperature oscillations are recorded by the electrical resistance of the sample, its emission, or using thermocouples or resistance thermometers.

Modulation calorimetry is valid over a wide temperature range, from fractions of kelvin to melting points of refractory metals. It is applicable for measuring specific heat of metals, alloys, semiconductors, dielectrics, and organic substances. Modulation calorimetry provides high sensitivity (0.01%) and excellent temperature resolution (10^{-3} K), which are unattainable by other methods. It is possible to perform measurements using very small samples

Ya.A. KRAFTMAKHER • Institute of Inorganic Chemistry, USSR Academy of Sciences, Novosibirsk 630090, USSR. Present address: Department of Physics, Bar-Ilan University, 52100 Ramat-Gan, Israel.

$(10^{-3}-10^{-2}\,\text{mm}^3)$. However, modulation calorimetry is in its early stages. The most favorable methods for measuring specific heat at different temperatures have not yet been established, and some suggestions have so far not proved to be sufficiently reliable.

When the high-temperature specific heat of metals was measured, the most important feature of the method was the smallness of corrections for heat losses. In numerous investigations of phase transitions the most important feature was a unique temperature resolution. In many cases the smallness of the samples was important, too.

The basic relations in the modulation calorimetry are as follows[1]:

$$mc = (p/\omega\theta_0) \sin \varphi \qquad (1)$$

and

$$\tan \varphi = mc\omega/P' \qquad (2)$$

where m and c are the mass and specific heat of the sample, p and ω are the amplitude and angular frequency of the oscillations of the power applied to the sample, θ_0 is the amplitude of the temperature oscillations, φ is the phase shift between the power and temperature oscillations, while P' is the heat transfer coefficient assumed to be independent of the modulation frequency.

When $\tan \varphi \geq 10$ one obtains $\sin \varphi \geq 0.995$, so that correction for heat losses is small enough (the correction factor is $\sin \varphi$). In practice one must check the validity of equations (1) and (2). For this purpose, one usually observes the plot of the product $\omega\theta_0$ versus the modulation frequency (Fig. 1). This value should be constant when $\tan \varphi \geq 10$. The decrease at low frequencies

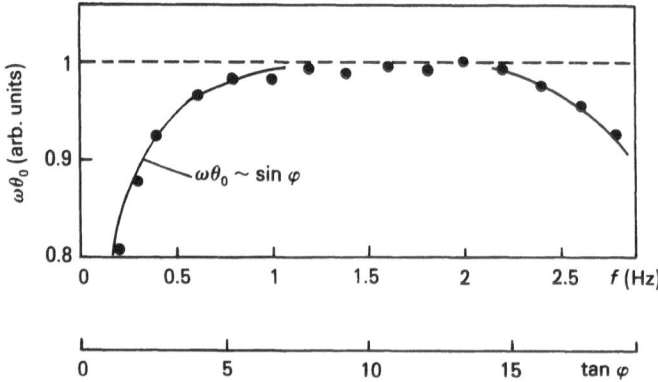

FIGURE 1. An example of the frequency dependence of quantity $\omega\theta_0$. The decrease at low frequencies is caused by nonadiabaticity, while at high frequencies it is related to the inertia of the temperature sensor.

is caused by the nonadiabatic conditions and could be taken into account by the factor $\sin \varphi$. However, in some cases the heat transfer coefficient P' is frequency-dependent (e.g., when the sample is immersed in a liquid or a gas). The decrease in the value of $\omega \theta_0$ at enhanced frequencies is related to inertia, or insufficient contact of the temperature sensor, a thermocouple or a resistance thermometer, with the sample. The modulation frequency is chosen from the interval where the quantity $\omega \theta_0$ remains constant.

The accuracy of modulation measurements depends mainly on the errors in determining the amplitudes of the power and temperature oscillations. In some cases the accuracy of measuring the sample mass may also be important.

Among methods of periodical heating of the sample, the chopped-light heating was employed most often. However, by this method it is difficult to determine exactly the amplitude of oscillations of the power applied to the sample. Therefore, in all these measurements the absolute values of the specific heat have not been evaluated but some precautions were taken to ensure that the absorbed radiation power not be dependent on temperature. For this purpose one must coat the sample surface with a layer of sufficiently high absorptivity which is independent of temperature (such as graphite).

All other methods of heating allow one to determine the amplitude of the oscillations of the applied power and hence to determine the absolute values of the specific heat. Lower accuracy is inherent in induction heating, so this method is no longer employed now practically. For metal and alloy samples (in some cases also for semiconductors) in the form of wires, rods, or strips it is useful to employ direct electric heating. For small samples electron heating is most convenient, while for nonconducting samples it is necessary to use separate electric heaters.

The wide temperature range over which the modulation method can be employed does not allow one experimental setup to be used that is valid for all cases. The methods for measuring temperature oscillations in different temperature regions may be quite different. At low and middle temperatures mainly thermocouples and resistance thermometers are used. In these temperature regions the proper modulation frequency is low and compatible with the inertia of temperature sensors. To avoid corrections for heat losses at high temperatures, one has to increase the modulation frequency. In many cases measurements can be performed in the nonadiabatic regime, allowing one to decrease the modulation frequency. However, for the highest temperatures sufficient thermocouples or resistance thermometers are absent and the temperature oscillations must be recorded by the emission of the sample. Modulation measurements on nonconducting materials were performed at temperatures up to 1500 K.

The relations governing modulation calorimetry are very simple. Results are derived directly and no complicated processing of experimental data is needed.

Since the chapter on modulation calorimetry was prepared for the first volume of the Compendium,[1] many new modulation measurements have been conducted. As earlier, modulation calorimetry was used to study specific heat anomalies at phase transitions, including transitions in cobalt and zirconium,[2] intercalated graphite,[3,4] ferroelectrics,[5,6] quartz,[7,8] and other inorganic substances,[9-16] in liquid crystals and organic substances.[17-20] Two review papers appeared concerning the employment of modulation calorimetry for studying phase transitions.[21,22]

The specific heat of gallium and sodium near the melting points was measured carefully.[23,24] A check was performed[25] of the earlier observed unusual premelting anomaly in the specific heat of platinum.[26] The specific heat of liquid cesium and rubidium at temperatures up to 2000 K and elevated pressures was measured.[27,28] Measurements at pressures up to 0.9 GPa were also performed.[29]

A modulation dilatometer with interferometric recording of the sample-length oscillations was developed.[30] The temperature coefficient of electrical resistivity of platinum and platinum–rhodium alloys was studied using the modulation method.[31] A relaxation effect in the high-temperature specific heat of tungsten was observed caused by point-defect formation.[32] The properties of some rare-earth metals in the range 600–1500 K were investigated.[33] In several papers the technique of modulation calorimetry was described.[34-37] The modulation method for studying thermopower was also used.[38-40]

Some workers use a computer for controlling modulation measurements and data processing. The computer is able to monitor changes in the mean temperature of the sample and in the amplitude of oscillations of the heating power, and to process all the data for evaluating the temperature dependence of the specific heat or another quantity to be measured. In contrast with pulse calorimetry, the use of a computer in the modulation method is unnecessary because the latter is a steady-state method: the amplitude and phase of the temperature oscillations do not depend on time. However, computers save time spent on measurements and their employment will undoubtedly be widespread.

2. MEASUREMENT OF SPECIFIC HEAT BY THE EQUIVALENT-IMPEDANCE METHOD

The equivalent-impedance method is based on recording temperature oscillations by employing oscillations of the sample resistance. The impedance of the sample, in the form of a wire or strip, heated by a DC current with a small AC component is equivalent to the impedance of a resistor and a capacitor connected in parallel (the temperature coefficient of the sample resistance is assumed to be positive). Therefore, the specific heat of the sample can be determined using a bridge[41] or a potentiometer[42] circuit whose balance is independent of the AC component of the heating current. Using the potentiometer circuit one measures the specific heat of the central part of the sample confined by thin potential probes, so any effect of the cold ends of the sample is eliminated. The potential probes are much thinner than the sample, so there are no significant changes in temperature at the points where they connect to the sample. In order to justify this requirement more sufficiently, one can heat the potential probes by passing an additional electric current through them.

The expression for evaluating the specific heat is

$$mc = 2I_0^2 R'/\omega^2 RC \qquad (3)$$

where I_0 is the DC current passing through the sample, $R' = dR/dT$ is the temperature derivative of the sample resistance, while R and C are the parameters of the equivalent impedance. For the potentiometer circuit quantities m, R', R, and C relate to the central part of the sample.

The practice of measurements shows that all quantities measured directly (m, I_0, ω, R, and C) can be determined with a high degree of accuracy. The mass of the sample is determined with an error lower than 1%, while the errors of other mentioned quantities can be made lower than 0.1%. Thus, the accuracy of the measurements is determined mainly by errors in the accepted values of R', which must somehow be evaluated. Most favorable is a case when the temperature dependence of the sample resistance is near-linear, so that the temperature derivative R' depends weakly on the temperature. On the other hand, the equivalent-impedance method should not be used to study specific heat anomalies at phase transitions because they are accompanied by anomalies in resistivity.

Samples of various thickness can be used for measurements, from the thinnest wires or films to samples with thickness of the order of 1 mm. Upon increasing the sample thickness the heating current also increases, and the modulation frequency decreases. The frequency corresponding to adiabatic conditions is inversely proportional to sample thickness. Therefore, for a correct choice of modulation frequency and for a given relation between the DC and AC components of the heating current, the amplitude of the temperature oscillations is independent of sample thickness. At a given temperature, the heating current is proportional to $d^{3/2}$ (d is the diameter of the sample) while the sample resistance and its temperature derivative are inversely proportional to d^2. Therefore, the AC voltage across the sample caused by the temperature oscillations is inversely proportional to $d^{1/2}$. Thus, the decrease in signal due to an increase in sample thickness is not as strong as it appears.

The samples are placed in a vacuum chamber. A moderate vacuum suffices at middle temperatures only to exclude an instability of the sample temperature caused by convection. For high temperatures a perfect vacuum is desirable, and the best way to obtain it is to employ a cryopump.[43] The most difficult case is when an inert gas atmosphere is necessary because requirements pertaining to the vacuum system and gas purity are very strong, particularly when thin samples are used.

The length of the samples is 100 to 200 mm; the central part confined by potential probes is 30 to 60 mm long. The potential probes are connected to the sample by point-welding. The distance between them is determined using a microscope.

The measuring circuit (Fig. 2) provides heating of the sample by a DC current with the addition of a small AC component. The AC voltage is fed to the sample from a low-frequency oscillator through a transformer. An additional winding of the transformer is used to derive an AC current for the compensation circuit. This current must be in direct relation to the AC component of the heating current and have the same phase. The variable resistor R_1 and capacitor C_1 are used for adjusting the amplitude and phase of the compensating current.

The differential input of a selective amplifier is connected first to resistors R_2 and R_3, in order to compare the voltages across them. The resistor R_3 is

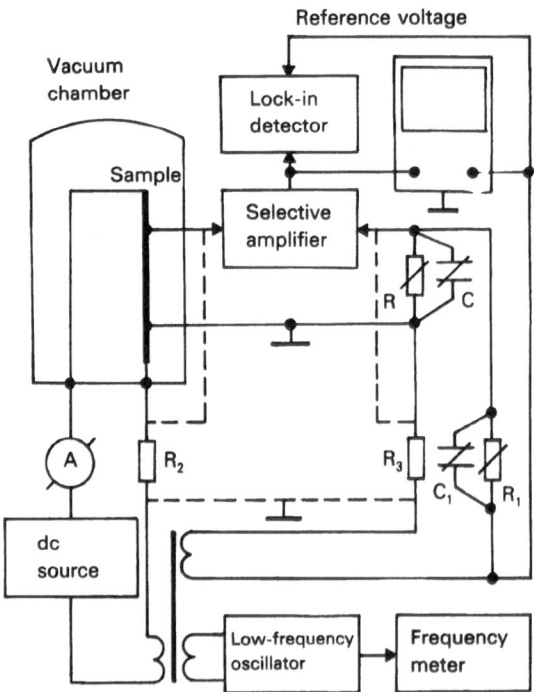

FIGURE 2. Circuit for measuring the specific heat by the equivalent-impedance method.

equal to 100 ohm while the resistor R_2 is 1 ohm for thin samples and 0.1 ohm for thick samples.

After setting the current in the compensation circuit the selective amplifier is switched for comparing AC voltages across the central part of the sample and the circuit RC. By adjusting the resistor R and capacitor C, one has to balance the circuit and determine the parameters of the equivalent impedance that enter equation (3). The resistor R_1 is much greater than the resistor R, so upon adjusting R and C the amplitude and phase of the current in the compensation circuit do not change.

An oscilloscope and a lock-in detector are used to indicate the balance. For convenience, a voltage proportional to the current in the compensation circuit is fed to the X-input of the oscilloscope. A form of the Lissajous figure on the screen allows one to see whether R or C needs to be adjusted for full compensation.

The lock-in detector is employed to enhance the sensitivity. The phase shift in the reference circuit is set to provide the sensitivity of the detector only to the AC component of the voltage, which is in quadrature with the AC current. This component is compensated by adjusting the capacitor C and is much smaller than the component which is in phase with the AC current. If a double lock-in detector is available, it should be adjusted so that one of the detectors is sensitive to the in-phase component of the voltage and the other to the quadrature component. The sign of the output DC voltages of the detectors shows whether increasing or decreasing the values of R and C are necessary for full compensation.

Measurements with the equivalent-impedance method can also be made in the nonadiabatic regime. The expression for evaluating the specific heat has the form[44]

$$mc = 2I_0^2R' \Big/ \left\{ \omega^2 R_0 C \left[1 + \left(\frac{R/R_0 - 1}{\omega RC} \right)^2 \right] \right\} \qquad (4)$$

where R_0 is the DC resistance of the central part of the sample which coincides with the AC resistance at a frequency corresponding to the adiabatic regime, while R and C are parameters of the impedance in the nonadiabatic regime.

When the sample thickness is several tenths of a millimeter, the necessary data on the temperature derivative of the sample resistance, R', can be obtained using the same arrangement. For this purpose one must weld to the sample a thin thermocouple. A DC current with an AC current modulated at an infralow frequency is passed through the sample. In this way temperature oscillations with a period of several seconds and corresponding oscillations of the resistance are created in the sample. The voltage across the central part of the sample contains an infralow component, which is proportional to the amplitude of the temperature oscillations, to the DC current, and to R', the temperature derivative of the sample resistance.[45]

The oscillations of the thermocouple voltage and of the voltage across the central part of the sample are recorded using a two-channel or X-Y recorder. In these measurements adiabatic conditions are not required. Therefore the modulation frequency can be decreased to provide a significant amplitude of the temperature oscillations and to neglect the inertia of the thermocouple. Data on R' derived in this way are used to measure the specific heat by the equivalent-impedance method at higher frequencies corresponding to an adiabatic regime or to a nonadiabatic regime, but with more favorable values of tan φ.

The errors in measurements of the specific heat by the equivalent-impedance method are determined completely by the errors in the values of R'. If this quantity is measured directly, the error depends on the accuracy of the thermocouple used and on the accuracy of recording the temperature oscillations and the oscillations of the sample resistance. Using a Pt-PtRh thermocouple can lead to the total error in the measured values of R' being lower than 3%. Such measurements were carried out at temperatures up to 1900 K.[31,46]

When the temperature derivative of the sample resistance varies with temperature in an anomalous way (phase transitions) or is very small (alloys), the equivalent-impedance method is invalid. In such cases it is better to use equation (2) and determine the specific heat from the heat transfer coefficient P'. For heat transfer by radiation, the value of P' can be expressed in the form[44]

$$P' = \frac{P}{T}\left(4 + \frac{T}{\varepsilon}\frac{d\varepsilon}{dT}\right) = nP/T \qquad (5)$$

where ε is the total emissivity; quantity n is given by $n(T) = d(\ln P)/d(\ln T)$ and is assumed to be independent of the frequency. It is very important that the second term in parentheses is usually several times smaller than the first. Thus, when ε depends weakly on temperature, it suffices to know approximately the derivative $d\varepsilon/dT$.

After deriving the temperature-dependent functions $P(T)$ and $n(T)$, one can find immediately the specific heat of the sample from the phase shift φ between the oscillations of the power applied to the sample and the temperature oscillations. The accuracy of such measurements is best when tan $\varphi \approx 1$. From equations (2) and (5) one obtains

$$mc = nP \tan \varphi / \omega T \qquad (6)$$

When the sample resistance is temperature-dependent, the oscillations of the current heating the sample and those of the voltage across the sample are not in phase. This phase shift is negligible in the adiabatic regime of measurements but becomes important in the nonadiabatic regime. It was shown[44] that when a DC current with a small AC component passes through the sample, equation (6) must be replaced by

$$mc = 2nP/[\omega T(\cot \alpha + \cot \beta)] \tag{7}$$

where α is the phase shift between the temperature oscillations and the AC component of the current, and β is the phase shift between the temperature oscillations and the AC component of the voltage across the sample. The difference between quantities $\cot \alpha$ and $\cot \beta$ is proportional to the temperature

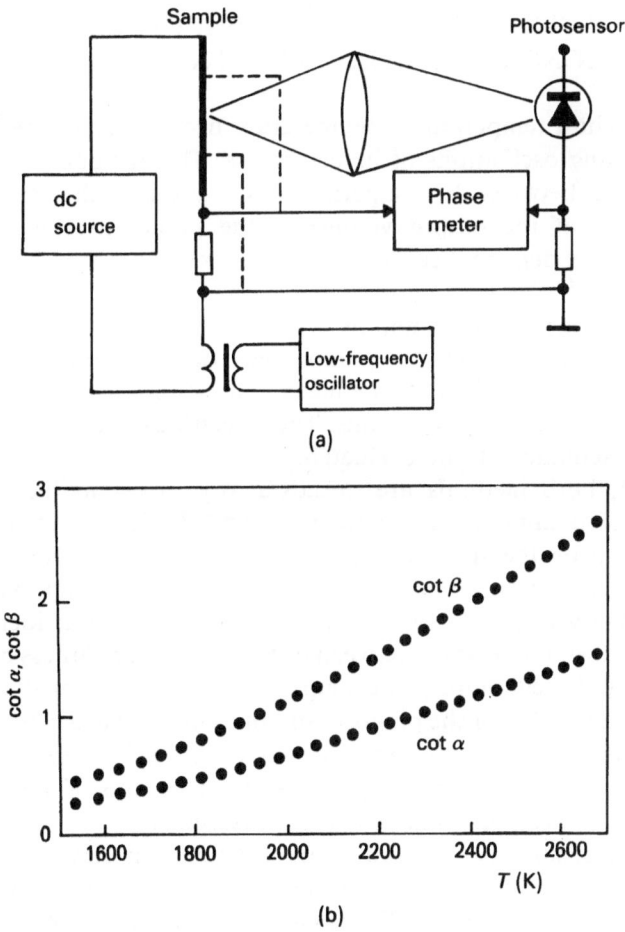

FIGURE 3. (a) Measurement of the phase shifts in modulation calorimetry. (b) An example of the temperature dependence of $\cot \alpha$ and $\cot \beta$.[44]

derivative of the sample resistance and inversely proportional to the modulation frequency.

Thus, when evaluating the specific heat with the aid of the heat transfer coefficient, one must determine two phase shifts, α and β. The simplest approach is to measure them with a phasemeter (Fig. 3). The voltage proportional to the temperature oscillations in the sample is fed to the phasemeter. These oscillations are recorded using a thermocouple or a photosensor and an amplifier. The phase shift in the amplifier must equal zero. The AC voltage across the central part of the sample and the voltage across a standard resistor connected in series with the sample are fed in turn to the second input of the phasemeter.

3. DETERMINATION OF TEMPERATURE OSCILLATIONS BY OSCILLATIONS OF SAMPLE BRIGHTNESS

At the highest temperatures, temperature oscillations in the sample can be detected using oscillations of its brightness. This method is being widely used. A relation between the temperature oscillations in the sample and the AC component of the output voltage of the photosensor is obtained by assuming a certain dependence of the photocurrent on the sample temperature, such as $I = AT^n$ [47] or $I = A \exp(-B/T)$, [48] where A, B, and n are taken to be constant. In this way one obtains $V = nV_0\theta_0/T$ or $V = V_0B\theta_0/T^2$, where V and V_0 are the AC and DC components of the output voltage of the photosensor, while T and θ_0 are the mean temperature of the sample and the amplitude of temperature oscillations. These relations allow the amplitude of temperature oscillations to be evaluated.

However, both methods are unsatisfactory owing to the temperature dependence of quantities A, B, and n (A and B depend on the effective wavelength and on the spectral emissivity of the sample). These difficulties can be excluded and an additional advantage gained by employing samples with a blackbody cavity. If such samples are maintained at identical mean temperatures, then the relation between oscillations of brightness of the blackbody cavities and oscillations of the temperature is the same for all the samples. Without any calibration of the photosensor, this allows one to compare directly the temperature oscillations in different samples. By using a sample of known specific heat one can perform comparative measurements of the specific heat that can be more accurate than absolute measurements.

In such measurements tungsten or platinum can serve as a reference material. Their specific heat has been determined by many workers using all existing methods of calorimetry. Tungsten can be taken as a standard of specific heat at temperatures up to 3000 K and platinum up to 1500 K. The high-temperature specific heat of other metals and alloys is known not so well and comparative measurements of the specific heat in relation to the standard are useful.

Comparative measurements were employed when studying the specific heat of molten metals.[49,50] The molten metal was placed in a niobium capillary heated by an AC current. A comparison of oscillations of brightness of the filled and empty capillaries maintained at identical mean temperatures allowed one to evaluate directly the ratio of the specific heat of the sample to that of niobium.

An important way of increasing the accuracy of modulation measurements is to employ compensation circuits whose balance is independent of oscillations of the heating power.[51] Balancing the compensation circuit gives directly the ratio of the oscillations of the power applied to the sample and the oscillations of the sample temperature. This ratio enters the expression for evaluating the specific heat. The uncertainty of compensation depends on the sensitivity of the zero indicator and on the accuracy of the variable resistors and capacitors, so it can be several times smaller than the error in determining small AC voltages which must be measured in modulation calorimetry.

In the compensation circuit (Fig. 4) a sample (a rod or a capillary) is heated by passing a DC current with a small AC component. In order to exclude any effect of the cold ends of the sample, only its central part confined by thin potential probes is used for the measurements. The mean temperature and the amplitude of the temperature oscillations are constant throughout the

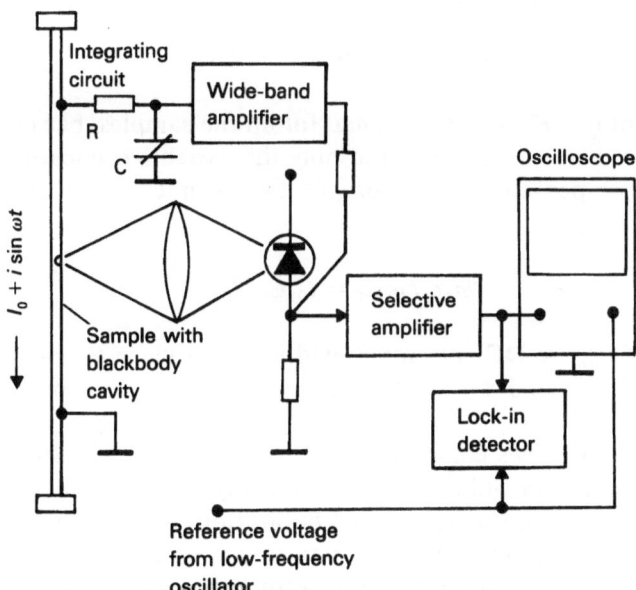

FIGURE 4. Measurement of specific heat using a photoelectric sensor.

central part of the sample. In the adiabatic regime the specific heat can be evaluated from the expression

$$mc = 2I_0 U / \omega \theta_0 \tag{8}$$

where m and U are the mass of the central part of the sample and the AC voltage across it, respectively, while I_0 is the DC current heating the sample.

The radiation from the blackbody cavity is projected onto the photosensor. A selective amplifier tuned to the modulation frequency is connected to the load resistor of the photosensor. The oscillations of the radiation from the blackbody cavity cause an AC voltage at the load resistor proportional to temperature oscillations in the sample: $V_1 = K_1 \theta_0$, where K_1 is a proportionality factor. Due to the blackbody conditions this coefficient depends only on the mean temperature of the sample. It is therefore possible to compare directly temperature oscillations in different samples.

The AC output signal of the photosensor is compensated by the integrating RC circuit, C being a variable capacitor. The potential probes are connected to the input of the circuit. The output of the circuit is connected to the input of a wide-band amplifier and the output signal of the amplifier is fed to the load resistor of the photosensor. The input voltage of the integrating circuit is the AC component U of the voltage across the central part of the sample. The AC voltage on the capacitor C is $U/\omega RC$ ($\omega^2 R^2 C^2 \gg 1$) and the compensation voltage on the load resistor is $V_2 = K_2 U/\omega RC$, K_2 being a proportionality factor. At compensation $V_1 = V_2$, i.e.,

$$mc = 2K_1 I_0 RC / K_2 = KI_0 RC \tag{9}$$

The quantities K and R are equal for all the samples, because the mean temperature remains unchanged and only the variable capacitor C has to be adjusted for compensation. Therefore, for two samples at a given temperature one obtains

$$m_1 c_1 / m_2 c_2 = I_{01} C_1 / I_{02} C_2 \tag{10}$$

where I_{01} and I_{02} are DC currents providing equal mean temperatures of the samples, while C_1 and C_2 are the values of capacitance corresponding to compensation.

This method allows comparative measurements to be performed on any conducting samples in which a blackbody cavity can be made. Calibration of the photosensor is unnecessary. In addition, full radiation of the blackbody cavity can be utilized instead of radiation in a narrow spectral band. This favorable circumstance permits one to recompense a decrease in the radiant flux due to the small dimensions of the blackbody cavity.

When performing measurements, the radiation from the blackbody cavity of both samples and from a standard strip lamp are projected in turn, by the same optical system, onto the photosensor. The standard lamp is employed to determine the mean temperature of the samples. Both samples and the standard lamp are located in a vacuum chamber and can be shifted with a turn-plate. Therefore, no corrections are needed for the reflectance and absorption of radiation in the chamber window.

4. MEASURING THE SPECIFIC HEAT OF NONCONDUCTING MATERIALS

In order to measure the specific heat of nonconducting materials, some additional requirements must be satisfied. In such measurements one employs separate heaters and thermometers. The criterion of adiabatic conditions and a condition of establishing thermal equilibrium in the system including a heater, a sample, and a thermometer pose contradictory requirements for the modulation frequency. At low and middle temperatures when the heat transfer coefficient P' is small enough, it is possible to choose a modulation frequency that quite satisfactorily meets both requirements. However, samples of thickness equal to several tenths of a millimeter should be used in order to decrease the time necessary for establishing thermal equilibrium.

Most modulation measurements on nonconducting samples were carried out using modulated-light heating. As a rule, the absolute magnitude of the heating-power oscillations was not determined. This method is used widely for studying the anomalies in specific heat at phase transition points. The imprecision of such measurements over narrow temperature intervals is of the order of 0.1%, but their inaccuracy amounts to 5-10%.

The first measurements on nonconducting samples were performed long ago.[52] The specific heat of $LiTaO_3$ was measured in the temperature range 300-1000 K. A sample ($3 \times 3 \times 0.1$ mm^3) is placed in a furnace and illuminated by an incandescent lamp (Fig. 5). The modulation frequency is 30 Hz. The mean temperature of the sample is measured with a Pt-PtRh thermocouple. A chromel-alumel thermocouple is employed for measuring temperature oscillations in the sample. The AC voltage of this thermocouple is fed to a lock-in amplifier. The reference voltage for the amplifier is created using an additional light source and a photocell. The output voltage of the amplifier is fed to the Y-input of a plotter while the Pt-PtRh thermocouple is connected to the X-input. Hence, the arrangement was the same as in the first measurements with modulated-light heating.[53]

When measuring the specific heat of nonconducting materials, it is expedient to employ a separate heater as was done for measurements on LiKSO$_4$.[16] One side of a sample ($5 \times 5 \times 0.35$ mm^3) was coated with a thin

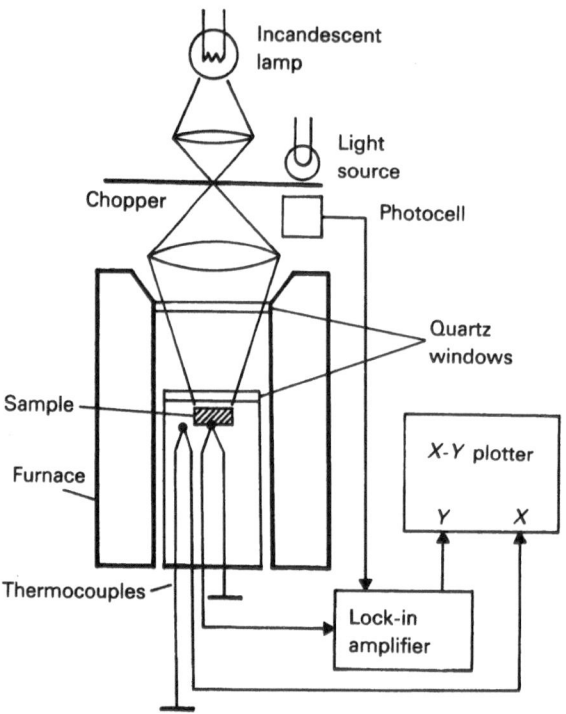

FIGURE 5. Determination of the specific heat of nonconducting materials at high temperatures.[52]

nickel film, which served as a heater. A thermocouple junction was attached to the other side of the sample. Measurements were performed in the 400–750 K range.

Temperatures up to 1500 K were realized while measuring the specific heat of some melts.[54] The liquid under study filled a hollow platinum crucible placed in a furnace (Fig. 6a). A heater with low inertia passes along the axis of the crucible. A thermocouple junction is welded to the crucible. This arrangement serves for creating radial temperature waves in the crucible, but over some frequency range the amplitude of the temperature oscillations depends only on the specific heat of the sample. The uncertainty of specific-heat measurements amounts to 4%.

In order to measure the specific heat of nonconducting samples, one can place a flat resistive heater between two parts of the sample.[55,56] The temperature oscillations are measured with a thermocouple attached to the outer side of the sample.

The heat transfer coefficient grows rapidly with increasing temperature. Its value is proportional to T^3 when heat losses are due to radiation. In fact, the increase is somewhat greater owing to the temperature dependence of the total emissivity of the sample. Therefore, the modulation frequency required to meet the adiabatic conditions also grows with increasing temperature. A low modulation frequency can be maintained at high temperatures by employing the nonadiabatic regime of measurements[44] or by compensating for heat losses.[57] Both approaches are quite practicable if the heat transfer coefficient of the sample is independent of the modulation frequency.

In the nonadiabatic regime, a modulation frequency is used such that the value of tan φ is of the order of unity. For evaluating the specific heat one must measure the phase shift between the oscillations of the heating power and the temperature oscillations in the sample. Measurements in the nonadiabatic regime make it possible to determine the specific heat with the same accuracy as in the adiabatic regime because a necessary correction for heat losses can be defined and taken into account. The nonadiabatic regime allows one to substantially decrease the modulation frequency and to facilitate establishing thermal equilibrium in the heater–sample–thermometer system.

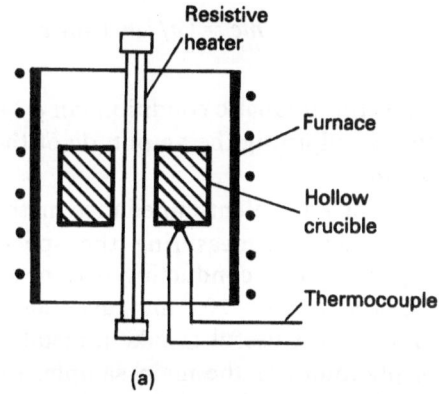

(a)

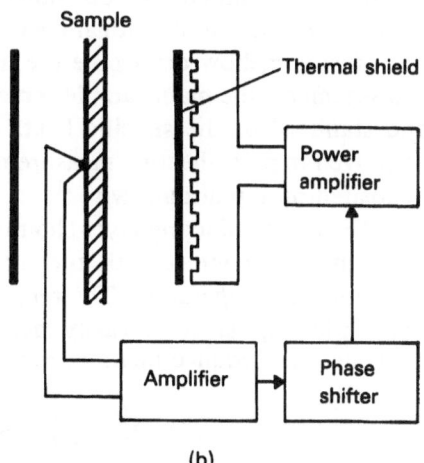

FIGURE 6. (a) Measurement of the specific heat of nonconducting materials.[54] (b) Compensation for heat losses using a thermal shield with an oscillating temperature.[57]

(b)

Another technique consists in surrounding the sample with a thermal shield (Fig. 6b). The temperature oscillations in the shield are of the same frequency and phase as those in the sample and their amplitude can be adjusted to completely exclude the AC component of the heat losses from the sample. The criterion of compensation is a 90° phase shift between the oscillations of the power heating the sample and the temperature oscillations.

The heat balance equation now takes the form

$$mc\theta' + (K_1 - bK_2)\theta = p \sin \omega t \tag{11}$$

where K_1 and K_2 are the temperature derivatives of the mutual heat transfer coefficients in the sample–shield system while b is the ratio of the amplitudes of the temperature oscillations in the shield and in the sample. The solution of this equation is

$$mc = (p/\omega\theta_0) \sin \varphi, \qquad \tan \varphi = mc\omega/(K_1 - bK_2) \tag{12}$$

Thus, the adiabatic condition $\tan \varphi \gg 1$ can be met at low modulation frequencies by adjusting the amplitude of the temperature oscillations in the thermal shield.[57]

The equivalent-impedance method provides high sensitivity, so it can be employed for measuring the specific heat of thin nonconducting layers deposited on a conducting wire or strip. The thickness of the layer, being of the order of 10^{-3}–10^{-1} mm, depends on the thickness of the main (conducting) sample. It is useful to take a metal of known specific heat (such as tungsten or platinum) as the main sample. In this case one can compare directly the specific heats of the coating and of the main sample. In this way it is unnecessary to know the temperature derivative of the sample resistance, because it is identical for the sample with and without the nonconducting coating. However, one must somehow determine the mean temperature of the sample. In such measurements the main sample serves simultaneously as heater, thermometer, and standard of the specific heat. A similar method has been successfully employed in pulse-heating measurements of the specific heat of silver bromide deposited on a platinum wire.[58]

The nonconducting layer should possess constant thickness and maintain good thermal contact with the main sample, not to be disturbed during variations in temperature. The relation for evaluating the specific heat of the nonconducting layer is easily derived from the basic expression for the equivalent-impedance method. For the main sample

$$mc = 2I_0^2 R'/\omega^2 RC \tag{13}$$

and for the composite sample, including the nonconducting layer of heat capacity $m_1 c_1$,

$$mc + m_1 c_1 = 2I_{01}^2 R'/\omega^2 R C_1 \tag{14}$$

When the mean temperatures are identical, the values of R and R' are equal; the modulation frequency also remains the same. Only the DC current heating the sample and the capacitance according to the equivalent impedance of the sample are changed. Hence equations (13) and (14) yield

$$m_1 c_1/mc = I_{01}^2 C/I_0^2 C_1 - 1 \tag{15}$$

It is not difficult to derive a relation for the nonadiabatic regime of measurements.

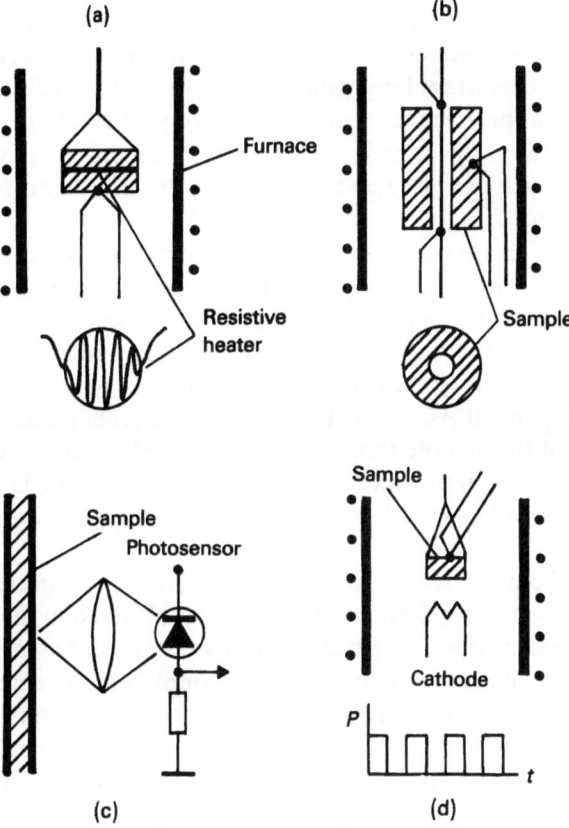

FIGURE 7. Modulation calorimetry of nonconducting materials: (a) a heater between two parts of the sample; (b) the sample in a hollow crucible, radiation heating; (c) the sample in a capillary, comparison of the brightness oscillations of empty and filled capillaries; (d) electron-bombardment heating, the sample in a conducting crucible.

When measuring the specific heat of nonconducting materials, it therefore suffices to fabricate samples of suitable shape compatible with a container or heater at high temperatures (Fig. 7). Probably, this problem cannot be solved satisfactorily in all cases. However, modulation calorimetry is quite applicable to most nonconducting materials.

5. MODULATION DILATOMETER WITH INTERFEROMETRIC RECORDING

The modulation method of measuring thermal expansion consists in creating periodic oscillations of the sample temperature and recording corresponding oscillations of the sample length. Hence the thermal expansion coefficient is measured directly.[59] The advantages of this method become significant at high temperatures owing to the creep of the samples. By using a suitable modulation frequency, selective amplification and lock-in detection allows one to eliminate irregular changes in sample length arising from external disturbances and to perform measurements at very small temperature oscillations. The oscillations in sample length can be recorded in various ways, including those which provide maximum sensitivity. Interferometric recording is one such method.

In a modulation dilatometer with interferometric recording[30] samples take the form of a wire or rod (Fig. 8). The top end is fixed and a flat mirror (1) is attached to the lower end of the sample. The sample is heated by passing a DC current from a stabilized source. In order to eliminate cold-end effects, the temperature oscillations are generated only in the central part of the sample. For this purpose a small AC current from a low-frequency oscillator is fed to the central part of the sample through thin wires welded to it. An offshoot of the AC current in the upper and lower parts of the sample is prevented by linking in series with the sample a coil with a high AC resistance (not shown in the figure).

The beam from a He–Ne laser ($\lambda = 0.63\ \mu$m) is projected through a beam-splitter onto mirror (1) and a second mirror (2) fixed to the piezoelectric transducer. The intensity of the interference pattern is recorded with a photodiode, the output voltage of which is fed to an amplifier. The amplified voltage is applied to the transducer with a polarity such that the oscillations of mirrors (1) and (2) are in phase. Due to a high gain of the amplifier, the displacements of mirror (2) are practically identical to those of mirror (1). Therefore, the AC voltage applied to the transducer is proportional to the oscillations of the sample length.

The selective amplifier measuring this voltage is tuned to the modulation frequency. The output signal of the amplifier is fed to the lock-in detector. The reference voltage for the detector is taken from the low-frequency oscillator

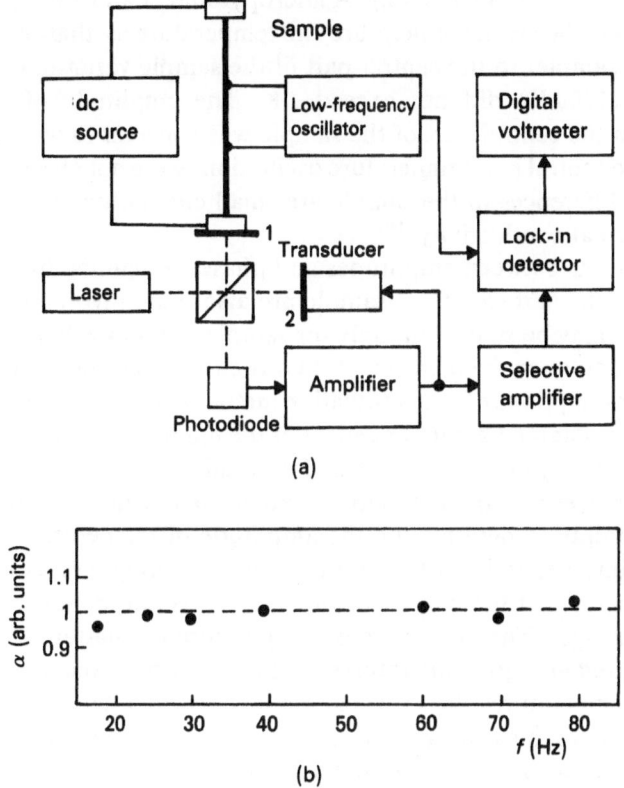

FIGURE 8. (a) Modulation measurement of the thermal expansion coefficient: 1—a mirror attached to the sample, 2—a mirror attached to the piezoelectric transducer. (b) Results of measurements at different modulation frequencies.

supplying the AC current to the central part of the sample. The output voltage of the detector, proportional to the amplitude of the oscillations of the sample length, is measured with a digital voltmeter.

In order to examine the properties of the dilatometer, the output voltage of the lock-in detector was measured for different amplitudes of temperature oscillations in the sample. The sample under study was a piece of platinum wire 0.3 mm in diameter. The results obtained showed linearity of the measuring circuit and the possibility to perform measurements at temperature oscillations of the order of 0.1 K. It turned out that the measurement results are independent of the intensity of the laser beam.

The mean temperature, and the amplitude and phase of temperature oscillations are assumed to be the same throughout the sample. The sample was 150 mm long and temperature oscillations were generated only in its central part, which was 40 mm long. A micropyrometer and a photodiode were used to compare the mean temperature and temperature oscillations at different points of the sample. In the central part of the sample variations of the mean temperature (1500 K) did not exceed 2 K. The amplitude of temperature oscillations in the central part of the sample was constant to within 1%, while away from the central part temperature oscillations were not observable. Radial temperature differences in the sample are small enough due to low thickness and high thermal conductivity.[60]

The relations between amplitudes and phases of temperature oscillations inside and on the surface of the sample are also quite favorable. Differences between them may be significant only for samples with low thermal diffusivity and at low modulation frequencies.[61] In any case, when checking the validity of the basic assumptions of the modulation method it is enough to be convinced that the results obtained are independent of the modulation frequency (Fig. 8).

The uncertainty of measurements of oscillations in sample length is about 1%. The error in evaluating the thermal expansion coefficient depends mainly on the uncertainty in determining the amplitude of temperature oscillations. This uncertainty depends on the accuracy in calibrating temperature sensors and on the features of the measuring devices, but is not identical for different temperature ranges. The amplitude of temperature oscillations in the sample can be measured at high temperatures with an error not exceeding 5%; at low and middle temperatures the error can be lowered to 1–2%.

The setup for measuring thermal expansion was arranged in order to determine simultaneously the specific heat of the sample and the temperature derivative of the sample resistance. When measuring the derivative of the resistance, temperature oscillations of frequency 0.1–0.2 Hz and amplitudes up to 10 K were generated in the sample and measured using a Pt–PtRh thermocouple.

Modulation measurements of the Seebeck coefficient are the simplest. They consist in creating temperature oscillations and measuring them simultaneously with a thermocouple under study and a standard one.[62] The compensation method can be employed to detect small changes in the Seebeck coefficient.

6. ELECTRONIC EQUIPMENT FOR MODULATION MEASUREMENTS

A substantial amount of electronic equipment is necessary to perform modulation measurements. Fortunately, all the required devices are now used extensively and are available. The specific requirements for modulation measurements are moderate and easily satisfied. We list below the necessary equipment and the features important for modulation measurements. This equipment can be employed to construct the apparatus described above and further develop the modulation technique.

Direct-current sources for heating samples to a desired mean temperature. Their main features are upper limits of the voltage and current, smallness of the pulsations of the output voltage, stability and convenience of regulation. Modern stabilized DC sources are quite applicable, so that any available device with corresponding limits of the voltage and current can be used.

Low-frequency *oscillators* for modulating the power applied to the sample. The main requirements are stability of the frequency and amplitude of the output voltage, small distortions, sufficient power, and compatibility with the load, sample or sample heater (maximum power is achieved when the load resistance is equal to the internal resistance of the source). Most commercial oscillators have a suitable frequency range, stability, and a changeable output resistance.

Low-frequency *amplifiers* are necessary for directly heating samples of large cross section. They must provide sufficient power, gain stability, small distortions, and be compatible with the load.

Devices for measurements of the electric current, voltage, and power are necessary for determining the amplitude of oscillations of the power applied to the sample. It is convenient to employ digital devices with an inaccuracy not exceeding 0.1%.

Frequency meters for measuring the modulation frequency. The calibration accuracy of low-frequency oscillators is not usually sufficient, so one must control the frequency with a digital frequency meter. At low frequencies it is preferable to measure the period of the oscillations, because in this case the desired accuracy is achieved in a shorter measurement time. The frequency meter is unnecessary when one employs an oscillator with digital setting of the frequency.

Selective amplifiers for recording temperature oscillations or as zero indicators in compensation circuits. Their main features are sensitivity, selectivity, stability, inherent noise, and convenience of tuning.

We employ the selective nanovoltmeter, type UNIPAN 237 (Poland). It is suitable for selective amplification of signals with a frequency of 1 Hz to 100 kHz. The desired frequency is set with a digital switch (three decades). The selectivity can be set at 25 or 40 dB per octave (Fig. 9). The full-scale deflection is 1 μV to 0.1 V. Preamplifiers and input transformers can be added to the main amplifier, so the sensitivity can be enhanced 100-fold. However, the internal resistance of the signal source should not exceed 10 ohm when using the input transformers.

The output voltages of the amplifier are a DC voltage proportional to the input AC signal (0.1 V for a 100-ohm load) and an amplified AC voltage (1 V for a 600-ohm load). The device can be used also as a wide-band amplifier. The measurement errors are about 10% but the gain stability is better than

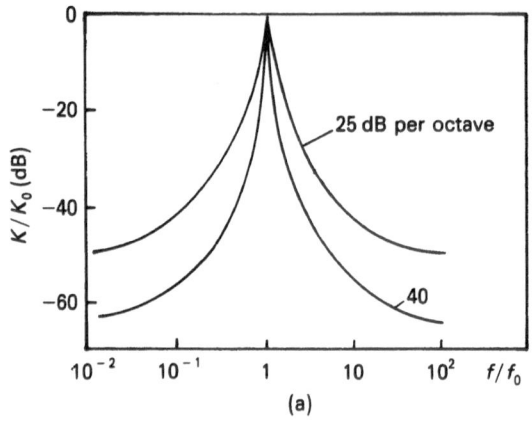

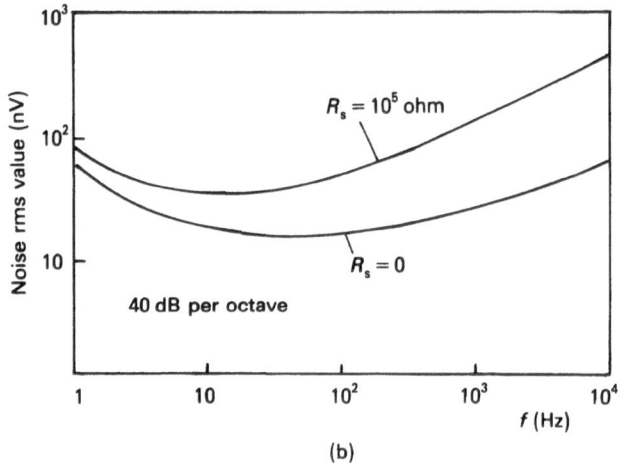

FIGURE 9. Features of the selective nanovoltmeter, type UNIPAN 237: (a) frequency response, f_0 is the tuning frequency; (b) voltage of the inherent noise, R_s is the output resistance of the signal source.

1%. Therefore the amplifier can be employed not only as a zero indicator, but for direct measurements, too.

Oscilloscopes for observing the waveforms of electrical signals and as indicators in compensation circuits. Modulation frequencies are relatively low, and signals after amplification are quite large. Therefore the requirements pertaining to an oscilloscope are quite moderate.

Lock-in detectors for measuring periodical signals buried in a noise. Their main features are sensitivity, stability, and convenience of tuning.

We employ the lock-in nanovoltmeter, type UNIPAN 232 (Poland). In this device an amplifier precedes the lock-in detector (Fig. 10). The frequency bands which can be set with RC-filters are 1.5–5 Hz, 5–15 Hz, and so on to 50–150 kHz. Thus, the selectivity of the amplifier is substantially lower than that of a selective amplifier. The main purpose of the amplifier is to improve the signal-to-noise ratio and to enhance the signal to a level sufficient for lock-in detection.

The full-scale deflection is 300 nV to 30 mV. Using preamplifiers and input transformers, this limit can be lowered to 1 nV. The efficiency of recording signals at a frequency coinciding with a reference frequency depends on the averaging time after detection. This time can be selected as 1 ms to 100 s. The

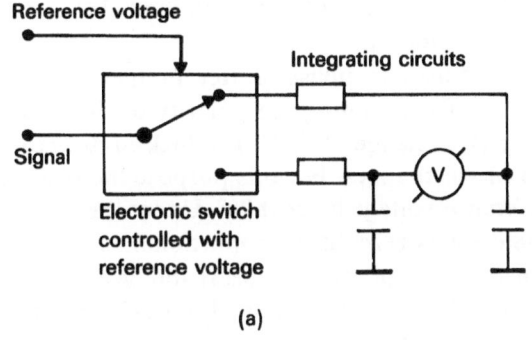

(a)

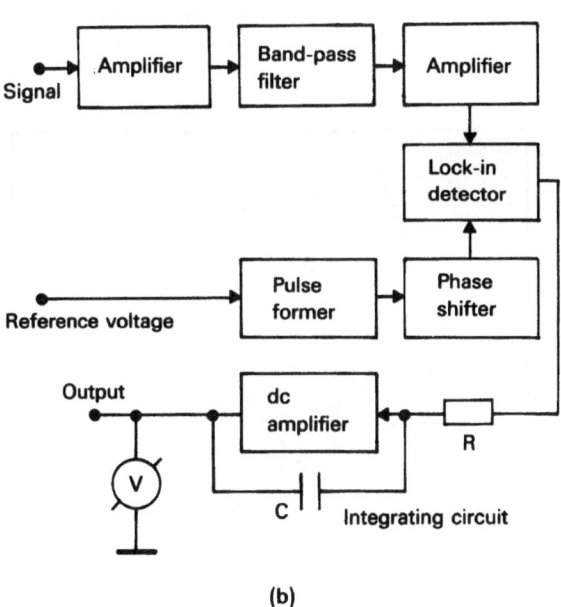

(b)

FIGURE 10. (a) Principle of the lock-in detection: the signal is measured with a frequency equal to a reference frequency. (b) Block diagram of the lock-in nanovoltmeter, type UNIPAN 232.

effective frequency band is related to the time constant τ: $\Delta f = 1/2\pi\tau$. Thus, the lock-in detector is always tuned to the frequency of an expected signal, while the effective frequency band is adjusted by switching the time constant. On increasing the time constant the time of measurements also increases. The time needed to obtain a result with an error of 10% is three times the time constant. Time constants of 1 to 10 s are usually quite sufficient.

The type UNIPAN 232 device allows weak signals to be detected with a signal-to-noise ratio of the order of 1:100. Periodic signals buried in a noise are commonly measured by lock-in detection. The output voltage of the detector is a DC voltage proportional to the input AC voltage (±5 V for a 2-kohm load). There is also an output of the amplified AC signal, which can be controlled with an oscilloscope.

There are an adjustable phase shifter in the reference-voltage circuit ($\pm100°$) and a switch for changing the polarity of the output voltage, which corresponds to a phase change of 180°. The lock-in detector can be also used to measure small phase changes. For this purpose the phase shift between the signal and the reference voltage is set at 90°. Under these conditions the output voltage of the detector is zero (it is proportional to the cosine of the phase shift) but the sensitivity to phase changes is the best.

The inaccuracy of measurements is about 10%, but the gain stability is better than 1%. Therefore the device can be used not only as a zero indicator but for direct measurements.

Recorders are used to record DC signals or periodic signals of infralow frequency. We employ the recorder ENDIM 620.02 (DDR). It can be used as

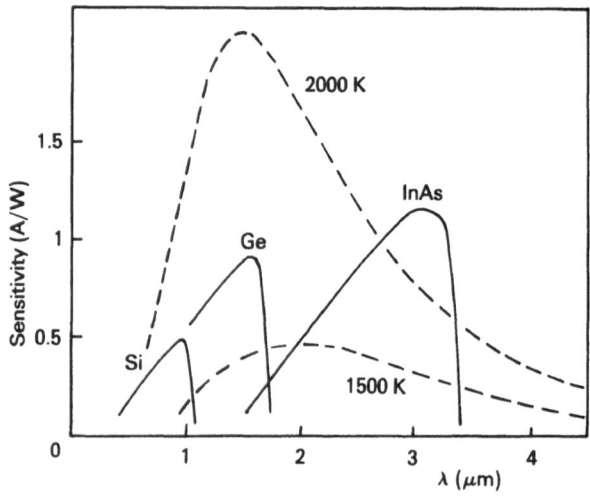

FIGURE 11. Spectral response of photodiodes. The spectra of a perfect blackbody radiation at 1500 and 2000 K are also shown.

a chart recorder or a X-Y plotter. There are two similar DC amplifiers in the X and Y channels providing a sensitivity of 0.1-10^4 cm V^{-1}. The inaccuracy of recording does not exceed 1%; the damping time is lower than 1 s. The rate of scanning can be set at 0.01-5 cm s^{-1}. The dimension of the plot is 270×400 mm^2.

In some variants of modulation measurements, precision variable *resistors and capacitors* are needed for compensation circuits. Their inaccuracy usually does not exceed 0.1%.

Photosensitive sensors for detecting temperature oscillations: photomultipliers, photodiodes, and photoresistors. The main features are spectral response, stability, time constant, and inherent noise. Photomultipliers are very sensitive to visible and near-infrared radiation, but they are not convenient for high feeding voltage. Photodiodes and photoresistors provide a more favorable spectral response and allow one to record the radiation of samples at lower temperatures. The main advantage of the multipliers, the high inherent gain, has no great importance at present owing to the availability of sensitive amplifiers. Most convenient are germanium and silicon photodiodes (Fig. 11).

Modern digital *phase meters* provide a wide frequency band and an inaccuracy of about 0.1°, which is quite sufficient for modulation measurements.

7. CONCLUSION

At present there is no standard measurement technique which can be employed in all cases of modulation calorimetry. In each concrete case one must choose the most appropriate variant according to the temperature range, properties, and form of the sample. However, these difficulties are insignificant in comparison with the important advantage inherent to the modulation method. This conclusion refers also to modulation methods of measuring other thermophysical properties: thermal expansion, electrical resistivity, and the Seebeck coefficient.

NOTATION

m	Mass of the sample
c	Specific heat
p	AC component of power fed to the sample
ω	Frequency ($\omega = 2\pi f$)
θ_0	Amplitude of temperature oscillations
φ, α, β	Phase shift
P'	Heat transfer coefficient
I_0	DC current
R, R_0	Resistance

R'	Temperature derivative of the resistance
C	Capacitance
d	Diameter of the sample
ε	Total emissivity
U, V	AC component of voltage
V_0	DC component of voltage
K, K_1, K_2	Proportionality factor
τ	Time constant

REFERENCES

1. Ya.A. Kraftmakher, in: *Compendium of Thermophysical Property Measurement Methods*, Vol. 1 (K.D. Maglić, A. Cezairliyan, and V.E. Peletsky, eds.), pp. 591-641, Plenum Press, New York (1984).

2. S.V. Boyarskii and I.I. Novikov, "Specific Heat and Some Acoustic Properties of Co and Zr at the Phase Transitions" (in Russian), *Teplofiz. Vys. Temp.* **19**, 201-203 (1981).

3. D.S. Robinson and M.B. Salamon, "Universality, Tricriticality, and the Potts Transition in First-Stage Lithium-Intercalated Graphite," *Phys. Rev. Lett.* **48**, 156-159 (1982).

4. D.N. Bittner and M. Bretz, "Heat Capacity of Antimony Pentachloride-Intercalated Graphite," *Phys. Rev. B* **31**, 1060-1068 (1985).

5. E. Kanda, M. Yoshizawa, T. Yamakami, and T. Fujimura, "Specific Heat Study of Ferroelectric CsH_2PO_4 and CsD_2PO_4," *J. Phys. C* **15**, 6823-6831 (1982).

6. K. Ema, "Critical Behavior in the Heat Capacity of Ferroelectric TGS, TGSe, and TGFB," *J. Phys. Soc. Jpn.* **52**, 2798-2809 (1983).

7. M. Matsuura, H. Yao, K. Gouhara, I. Hatta, and N. Kato, "Heat Capacity in $\alpha-\beta$ Phase Transition of Quartz," *J. Phys. Soc. Jpn.* **54**, 625-629 (1985).

8. I. Hatta, M. Matsuura, H. Yao, K. Gouhara, and N. Kato, "True Behavior of Heat Capacity in α, Incommensurate and β Phases of Quartz," *Thermochim. Acta* **88**, 143-148 (1985).

9. N. Sugimoto, T. Matsuda, and I. Hatta, "Specific Heat Capacity of $Pb_{1-x}Ge_xTe$ at Their Structural Phase Transitions," *J. Phys. Soc. Jpn.* **50**, 1555-1559 (1981).

10. S. Stokka and V. Samulionis, "Specific Heat near Two Phase Transitions in $CsBi(MoO_4)_2$ Crystals," *Phys. Status Solidi A* **67**, K89-K92 (1981).

11. S. Stokka and K. Fossheim, "Specific Heat and Phase Diagrams for Uniaxially Stressed $KMnF_3$," *J. Phys. C* **15**, 1161-1176 (1982).

12. S. Stokka, K. Fossheim, T. Johansen, and J. Feder, "Specific Heat of $CsPbCl_3$ near Three Phase Transitions," *J. Phys. C* **15**, 3053-3058 (1982).

13. T. Goto, M. Yoshizawa, A. Tamaki, and T. Fujimura, "Elastic and Thermal Properties of the Layered Compound $(CH_3NH_3)_2FeCl_4$," *J. Phys. C* **15**, 3041-3051 (1982).

14. M. Yoshizawa, T. Fujimura, T. Goto, and K. Kamiyoshi, "Specific Heat of the $NH_4Cl_{1-x}Br_x$ System," *J. Phys. C* **16**, 131-142 (1983).

15. S. Hirotsu, M. Miyamota, and K. Ema, "Three-State Potts Transition in Sodium Azide: Experimental Study of an Order of the Transition by Means of a Heat Capacity Measurement," *J. Phys. C* **16**, L661-L666 (1983).

16. T. Bręczewski, P. Piskunowicz, and G. Jaroma-Weiland, "Thermal Properties of $LiKSO_4$ Crystals in the Temperature Region from 400 K to 950 K," *Acta Phys. Pol. A* **66**, 555-560 (1984).

17. S. Imaizumi, I. Hatta, and T. Matsuda, "Experimental Study of Dynamic Specific Heat Capacity of Protein Aqueous Solutions," *J. Phys. Soc. Jpn.* **50**, 276-280 (1981).

18. J.M. Viner, D. Lamey, C.C. Huang, R. Pindak, and J.W. Goodby, "Heat Capacity near the Smectic-A-Hexatic-B and Hexatic-B-E Transitions of n-Hexyl-4'-n-pentyloxibiphenil-4-carboxylate (650BC)," *Phys. Rev. A* **28**, 2433-2441 (1983).

19. S.C. Lien, C.C. Huang, and J.W. Goodby, "Heat-Capacity Studies near the Smectic-A–Smectic-C (–Smectic-C*) Transition in a Racemic (Chiral) Smectic Liquid Crystal," *Phys. Rev. A* **29**, 1371–1374 (1984).

20. I. Hatta, K. Suzuki, and S. Imaizumi, "Pseudo-Critical Heat Capacity of Single Lipid Bilayers," *J. Phys. Soc. Jpn.* **52**, 2790–2797 (1983).

21. I. Hatta and A.J. Ikushima, "Studies of Phase Transitions by AC Calorimetry," *Jpn. J. Appl. Phys.* **20**, 1995–2011 (1981).

22. C.W. Garland, "High-Resolution AC Calorimetry and Critical Behavior at Phase Transitions," *Thermochim. Acta* **88**, 127–142 (1985).

23. G. Fritsch, R. Lachner, H. Diletti, and E. Lüscher, "Specific Heat of High-Purity Solid Gallium Close to the Melting Point," *Phil. Mag. A* **46**, 829–839 (1982).

24. G. Fritsch, H. Diletti, and E. Lüscher, "Specific Heat and Surface Melting of Sodium," *Phil. Mag. A* **50**, 545–558 (1984).

25. V.Ya. Fridman, "On a Brightness Anomaly Near the Melting Point of Platinum" (in Russian), *Inzh.-Fiz. Zh.* **44**, 986–988 (1983).

26. Ya.A. Kraftmakher, "Premelting Anomaly in Specific Heat of Platinum," Paper 101, 6th European Conference "Thermophysical Properties—Research and Application," Dubrovnik (1978).

27. L.A. Blagonravov, L.P. Filippov, V.A. Alekseev, and V.N. Shnerko, "Specific Heat of Liquid Cesium at Temperatures up to 2000 K and Pressures up to 12 MPa" (in Russian), *Inzh.-Fiz. Zh.* **44**, 438–444 (1983).

28. L.A. Blagonravov, V.N. Shnerko, L.P. Filippov, and V.A. Alekseev, "Specific Heat of Liquid Rubidium in the Range 1300 to 1900 K and at Pressures up to 16 MPa" (in Russian), *Teplofiz. Vys. Temp.* **22**, 177–179 (1984).

29. O.K. Gulish, I.N. Polandov, and A.A. Kuyumchev, "Pressure Effect on the Thermodynamic State of Solid Solution of TGS–TGSe" (in Russian), *Fiz. Tverd. Tela* **25**, 2115–2119 (1983).

30. S.Yu. Glazkov and Ya.A. Kraftmakher, "High-Temperature Modulation Dilatometer with Interferometric Registration" (in Russian), *Teplofiz. Vys. Temp.* **21**, 769–772 (1983).

31. S.Yu. Glazkov, "Point-Defect Formation and Temperature Coefficient of Electrical Resisitivity of Platinum and Platinum-(10 wt%) Rhodium Alloy in the Range 1100–1900 K", *Int. J. Thermophys.* **6**, 421–426 (1985).

32. Ya.A. Kraftmakher, "Relaxation Effect in the High-Temperature Specific Heat of Tungsten" (in Russian), *Fiz. Tverd. Tela* **27**, 235–237 (1985).

33. A.A. Kurichenko, A.D. Ivliev, and V.E. Zinov'ev, "Investigation of Thermophysical Properties of Rare-Earth Metals Using Modulated Laser Radiation" (in Russian), *Teplofiz. Vys. Temp.* **24**, 493–499 (1986).

34. I.N. Polandov, V.A. Chernenko, and V.K. Novik, "AC Calorimetry of Solids at High Hydrostatic Pressures," *High Temp. High Pressures* **13**, 399–406 (1981).

35. S. Stokka and K. Fossheim, "A Simple System for Automatic Specific Heat Measurements," *J. Phys. E* **15**, 123–127 (1982).

36. S. Imaizumi, K. Suzuki, and I. Hatta, "AC Calorimeter for Liquid Including Suspension of Biological Materials," *Rev. Sci. Instrum.* **54**, 1180–1185 (1983).

37. C.C. Huang, J.M. Viner, and J.C. Novack, "New Experimental Technique for Simultaneously Measuring Thermal Conductivity and Heat Capacity," *Rev. Sci. Instrum.* **56**, 1390–1393 (1985).

38. M. Kawai, T. Miyakawa, and T. Tako, "AC Measurement of Seebeck Coefficient in Disk-Shaped Semiconductors Using cw-Lasers," *Jpn. J. Appl. Phys.* **23**, 1202–1208 (1984).

39. W. Kettler, S.N. Kaul, and M. Rosenberg, "Absolute Thermoelectric Power in Amorphous $Fe_x Ni_{80-x} B_{19} Si_1$ Alloys," *Phys. Rev. B* **29**, 6950–6956 (1984).

40. E. Papp, "The Thermoelectric Power of Ni and Ni-Cu Alloy near the Curie Temperature Measured by an AC Method," *Z. Phys. B* **55**, 17–22 (1984).

41. Ya.A. Kraftmakher, "Modulation Method for Measuring Specific Heat" (in Russian), *Zh. Prikl. Mekh. Tekhn. Fiz.* (5), 176–180 (1962).

42. Ya.A. Kraftmakher, "Potentiometer Circuit for Measuring Specific Heat by Modulation Method" (in Russian), *Zh. Prikl. Mekh. Tekhn. Fiz.* (2), 144 (1966).
43. R.A. Haefer, "Cryogenic Vacuum Techniques," *J. Phys. E.* **14**, 273-288 (1981).
44. A.A. Varchenko and Ya.A. Kraftmakher, "Non-adiabatic Regime in Modulation Calorimetry," *Phys. Status Solidi A* **20**, 387-393 (1973).
45. Ya.A. Kraftmakher, "Electric Conductivity of Nickel near the Curie Point" (in Russian), *Fiz. Tverd. Tela* **9**, 1529-1530 (1967).
46. Ya.A. Kraftmakher and G.G. Sushakova, "Equilibrium Vacancies and Electric Conductivity of Platinum" (in Russian), *Fiz. Tverd. Tela* **16**, 138-142 (1974).
47. G.C. Lowenthal, "The Specific Heat of Metals between 1200 K and 2400 K," *Austral. J. Phys.* **16**, 47-67 (1963).
48. L.P. Filippov and R.P. Yurchak, "Measurement of Specific Heat of Solid and Liquid Metals" (in Russian), *Teplofiz. Vys. Temp.* **3**, 901-909 (1965).
49. I.A. Akhmatova, "Measurement of Specific Heat of Liquid Tin at High Temperatures" (in Russian), *Dokl. Akad. Nauk SSSR* **162**, 127-129 (1965).
50. I.A. Akhmatova, "Specific Heat of Molten Gallium and Copper at High Temperatures" (in Russian), *Izmer. Tekh.* (8), 14-17 (1967).
51. Ya.A. Kraftmakher, in: *Investigations at High Temperatures* (in Russian) (I.I. Novikov and P.G. Strelkov, eds.), pp. 5-54, Nauka, Novosibirsk (1966).
52. A.M. Glass, "Dielectric, Thermal, and Pyroelectric Properties of Ferroelectric $LiTaO_3$," *Phys. Rev.* **172**, 564-571 (1968).
53. P. Handler, D.E. Mapother, and M. Rayl, "AC Measurement of the Heat Capacity of Nickel near Its Critical Point," *Phys. Rev. Lett.* **19**, 356-358 (1967).
54. A.S. Derman and O.V. Bogorodskii, "Complex Measurement of Thermophysical Properties of Melts" (in Russian), *Izv. AN SSSR, Ser. Fiz.* **34**, 1215-1216 (1970).
55. R.P. Yurchak, "An Arrangement for Complex Measurements of Thermophysical Properties of Insulators" (in Russian), *Zavod. Labor.* **37**, 1514-1516 (1971).
56. V.A. Chernenko, I.N. Polandov, and V.K. Novik, "An Arrangement for Investigation of Specific Heat of Ferroelectrics Under High Pressures Using a Dynamical Method" (in Russian), *Prib. Tekhn. Eksper.* (1), 222-225 (1979).
57. Ya.A. Kraftmakher and V.Ya. Cherepanov, "Compensation of Heat Losses in Modulation Measurements of Specific Heat" (in Russian), *Teplofiz. Vys. Temp.* **16**, 647-649 (1978).
58. T.E. Pochapsky, "Heat Capacity and Thermal Diffusivity of Silver Bromide," *J. Chem. Phys.* **21**, 1539-1540 (1953).
59. Ya.A. Kraftmakher and I.M. Cheremisina, "Modulation Method for Measuring Thermal Expansion" (in Russian), *Zh. Prikl. Mekh. Tekhn. Fiz.* (2), 114-115 (1965).
60. L.P. Filippov, *Measurement of Thermal Properties of Solid and Liquid Metals at High Temperatures* (in Russian), Moscow State University, Moscow (1967).
61. L.R. Holland and R.C. Smith, "Analysis of Temperature Fluctuations in AC Heated Filaments," *J. Appl. Phys.* **37**, 4528-4536 (1966).
62. Ya.A. Kraftmakher and T.Yu. Pinegina, "Thermoelectric Power of Iron near the Curie Point," *Phys. Status Solidi* **42**, K151-K152 (1970).

15

Phase-Change Calorimeter for Measuring Relative Enthalpy in the Temperature Range 273.15 to 1200 K

DAVID A. DITMARS

1. INTRODUCTION

The general principles of heat-capacity or enthalpy calorimetry by the method of mixtures have been presented elsewhere together with a copious bibliography.[1] To briefly recapitulate, this experimental approach is well suited to highly accurate and precise measurement of the heat capacity of nonreacting, solid or liquid systems in any condition of subdivision (including massive, crystallites, and powders). It is also of great use in measuring heats of phase changes and their associated temperatures. Usually, the basic datum obtainable through this method is the total enthalpy change of the system under study upon rapid transfer from an initial equilibrium state in a high-temperature furnace to a second equilibrium state in the calorimeter. Thus the chief sources of error are limited to those in temperature measurement and attainment of equilibrium in the furnace, in accurately accounting for heat losses during a sample transfer, in measurement of the heat transferred from the sample to the calorimeter, and in attainment of a reproducible thermodynamic state of the sample in the calorimeter. The equipment required is a mix of custom-fabricated parts, well within the capability of any competent instrument shop, and commercially-obtainable heating, temperature-measurement, and temperature-control components. This chapter presents a specific design and operating criteria for a phase-change calorimetric system using purified water as the working substance and capable of operating at a precision level of 0.01% and an accuracy level of 0.1%. Except for the measurement, control, and recording of temperature, the system is manual in operation.

DAVID A. DITMARS • National Institute of Standards and Technology, Gaithersburg, Maryland 20899, USA.

2. SCHEMATIC OVERVIEW OF APPARATUS

In this section, we present the basic apparatus components and describe their functions. More detailed construction specifications and recommended operating procedures will be covered in the sections below. The three basic assemblies (brackets, Fig. 1) are the resistance-furnace assembly (1) in which the encapsulated measuring sample is suspended to establish the initial sample equilibrium state; the temperature-control and measurement assembly (2), and the calorimeter assembly (3) in which the sample relative enthalpy is measured.

The resistance furnace (4, Fig. 1, and Fig. 2) is mounted so that it may be rotated 90° about an axis (7, Fig. 2) parallel to the calorimeter axis in order to give access to the calorimeter top for the necessary preparations for measurement. It consists of an insulated alundum tube (5) on which are wrapped three Pt heaters. The tube contains a core (see Fig. 2) consisting of coaxially-stacked silver or ceramic segments. Holes drilled through these segments provide sites for temperature-measuring and control sensors. The encapsulated specimen

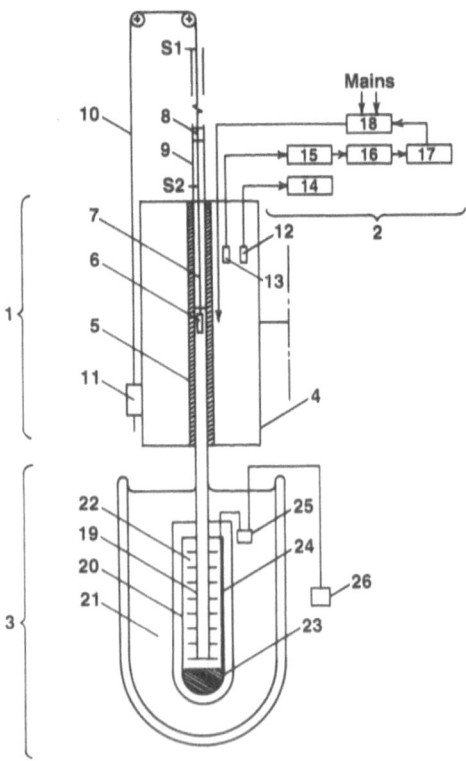

FIGURE 1. Functional assembly of phase-change calorimetric system.

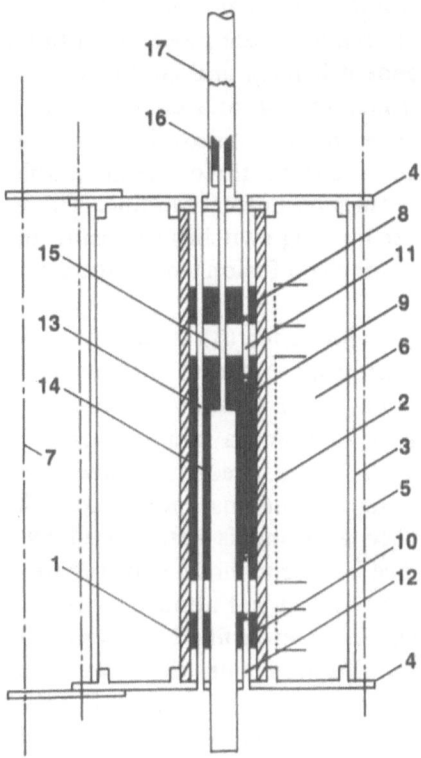

FIGURE 2. Silver-core furnace.

to be measured (6) is suspended from a shield system which is attached by a fine Nichrome wire (7) to an adjustable-mass, cylindrical rider (8). The rider is guided in vertical fall by a thin-walled aluminum tube (9) and is supported from a fine, flexible cord (10) passing over a pulley to a simple friction brake (11). The equipment dimensions together with the constraints set by the upper and lower rider stops (S1, S2) are chosen to ensure that the movable sample capsule resides at the center of the isothermal furnace zone when the rider is positioned against stop S1 and resides very near the calorimeter bottom when the rider rests on stop S2.

Temperature measurement and control in the furnace is affected by PRT or TC elements (12, 13 typical), which are inserted axially into holes in the central core of the furnace (Fig. 2). For temperature measurement, the sensor connects directly to a calibrated potential- or resistance-measuring instrument (14). For temperature control, each sensor (3 required) connects via a stable, adjustable reference emf (15) to a high-sensitivity DC null detector/amplifier (16). The amplifier output is fed to a PID controller (17) configured to control about a zero-input setpoint. The controller output is fed in turn to an SCR control unit (18), which regulates the power level in one of the three furnace heaters. Three independent heaters and control circuits are required: one to control the central furnace section at a constant set temperature and one to control each of the furnace end sections to minimize axial temperature gradients in the silver core. In addition, three thermocouples are installed, each individually next to a furnace heater winding. These are connected to individual external power limiting switches to ensure that temperatures in the furnace never approach the silver fusion temperature closer than approximately 50 K.

The phase-change calorimeter (3) is ideally a system at constant temperature, volume, and pressure, but with variable mass. The change in mass of the system during an experiment is directly proportional to the heat exchanged between the system and a sample introduced into it for measurement. It consists of a thin-walled, reentrant tube (19) provided at its lower end with heat-conducting copper fins and sections. This tube is positioned coaxial with the furnace axis and is surrounded by a double-walled glass enclosure (20), the space between whose walls is kept filled with a gas of low thermal conductivity (Ar or CO_2) at atmospheric pressure. The enclosure is submerged in a strip-silvered Dewar filled with finely divided, solute-free (but air-saturated) ice in equilibrium with clean, air-saturated water (21). The innermost space of the enclosure is completely filled with air-free, solute-free liquid water (22), and triply distilled mercury (23) at the bottom. A tube (24), immersed in this mercury pool, connects it through the enclosure roofs and via a tempering reservoir (25) to a mass-accounting system (26), which is maintained at atmospheric pressure. A fiduciary mark in the mass-accounting system defines the outer limit of the system's "constant volume." A refrigerant such as solid CO_2 or liquid N_2 introduced into the reentrant tube causes a portion of the "working water" within the enclosure to solidify, thereby (due to the greater specific volume of water ice) increasing the pressure and expelling mercury through the exterior accounting system. Subsequent exchange of heat between the working water and a measuring specimen inserted into the reentrant tube will cause a change in the ratio of solid to liquid working water and thus a change in the system mass (water plus mercury) which will be detectable at the external mass-accounting system. We note that although generally used with samples *above* the ice point, this method applies equally well to those samples initially at equilibrium *below* the ice point. The equivalence between an energy exchange within the calorimeter and the mass change detected externally can be established by calibrating with accurately-measured quantities of electrical energy in a suitably-designed heater. We further note that, once measured,[2] this equivalence (expressed as a "calibration factor" in units of energy per unit mass of mercury) will apply equally well to all phase-change calorimeters using H_2O as a working substance.

3. FURNACE FOR TEMPERATURES TO 1200 K

The furnace depicted in Fig. 2 has been used successfully in air to establish isothermal zones for specimen equilibration in the temperature range 320 K to 1200 K. Below about 750 K it can sustain temperature control, using a PRT sensing element, within a 4 to 6 mK range for extended time intervals. Above 750 K the PRT sensors have lacked stability but Pt/Pt10Rh thermocouples have been used successfully for temperature control within a 20–30 mK range to 1200 K. When using thermocouples as control sensors, extreme care must be taken to avoid contacts or circuit configurations which might introduce unpredictable parasitic emfs into the control circuit.

3.1. Furnace Construction

The central alundum tube (1), which carries the three furnace heaters, is of ID 52 mm, OD 65 mm, and length 595 mm. The outer tube surface has cut into it a single spiral groove of pitch 2 mm. The three independent Pt heaters (2) wound on the outer surface correspond in their axial placement to the three silver sections of the furnace core (8, 9, 10) and have room-temperature resistances as follows: top, 6 Ω; middle 23 Ω; bottom, 6 Ω. The furnace core within the alundum tube consists of alternating segments, machined from a vacuum-cast silver ingot or from an alumina rod, which fit closely inside the heater tube. Each core segment has several holes (13) drilled parallel to the segment axis near the outer surface. These holes have identical azimuthal and radial spacing and thus create continuous paths through the assembled core for the insertion of temperature-measurement and control sensors. Two of these holes (11, 12), one starting from each end of the core, terminate blindly in the central silver core segment and are used for emplacement of chromel-alumel differential thermocouple junctions, X. These couples are used for control of the end segment temperatures in order to minimize temperature gradients in the central silver segment. The lower 93% of the center silver segment and all segments below it each have an axial hole (14) bored in them, thus forming, when stacked in place, the chamber in which the measuring sample is equilibrated. The upper 7% of the central silver segment and all segments above it have a 2-mm hole (15) drilled axially. The continuous axial path through the core is provided with a close-fitting liner fabricated from thin-walled inconel tube segments of the two required diameters. The larger diameter segment of this liner extends approximately 11 cm below the furnace base plate; the smaller diameter end terminates in an axial hole in the furnace lid plate. The furnace base and lid plates (4) have holes drilled corresponding to the off-axis holes of the furnace core. Bosses on the plate surfaces facing the furnace serve to position the core/heater-tube assembly coaxial within the 220-mm-OD and 597-mm-long aluminum tube (3) forming the furnace exterior shell. The base and lid plates have projections welded to them to facilitate mounting the furnace so it can be swung about axis 7 away from the calorimeter. Tie rods (5) hold the plates together. The furnace insulation (6), packed uniformly in the annular space between the heater tube and the furnace exterior shell, is finely-divided diatomaceous earth. An adjustable stop (16) permits fixing precisely the terminal position of the dropping rider (8, Fig. 1). A thin-walled aluminum tube (17) is friction-mounted on the projection of the end plate to guide the rider.

3.2. Temperature Measurement and Control

At and below *ca* 800 K, core temperatures are measured with a long-stemmed platinum resistance thermometer positioned with its 25-ohm sensing element at the same level as that at which the sample resides in the furnace core. Also, below *ca* 800 K, a 100-ohm capsule platinum sensor is inserted to reside at the same level. This four-lead sensor sustains a current of 7 mA from a precision constant-current source. The potential drop across this sensor thus becomes the measure of furnace temperature. This potential is opposed by an external reference potential generated by passing current from a second, precision constant-current source through a standard resistor. The sum of sensor and reference potentials serves as the input for a sensitive DC null detector/amplifier.

Above 800 K, single Pt/Pt10 Rh thermocouples referred to the ice point are used for control and measurement purposes with the same control components described below. These thermocouples are carefully annealed and their (hot) junctions mounted in multiple-hole alumina tubes, which fit closely in the holes provided in the furnace core.

The amplifier has four switch-selected ranges to provide flexibility in instances where sensed temperatures are relatively far removed from the desired controlled temperature. The most sensitive amplifier range offers 0.1 μV resolution. Amplifier noise is less than 0.05 μV peak-to-peak; the instrument step-response time with the low-resistance sensors used is less than one second. The amplifier output, $\pm$5 mV, is fed directly to a digital PID controller configured with a zero-input setpoint. The controller has both manual and automatic operating modes with provision in automatic mode for independent variation of the proportional, integral, and derivative (rate) control parameters. The controller output is applied to some control element such as a silicon-controlled rectifier which adjusts the AC furnace power supplied. Three independent control loops are used, one to control the temperature of the central furnace silver-core segment and two to control separately the temperature of each of the two silver guarding segments above and below the central segment. Maximum equilibrium furnace power at 1200 K is about 600 W.

4. PHASE-CHANGE CALORIMETER

The heart of the phase-change calorimeter (Fig. 3a) is an enclosed volume (1) surrounding a reentrant tube (2) and consisting of an "interior" and an "exterior" part. "Interior" and "exterior" in this context means interior to and exterior to the Dewar-contained ice bath (21, Fig. 1). This volume contains *only* solute-free, air-free water (1) and liquid mercury (3). It must be constructed of nominally rigid elements. The (larger) interior part communicates via rigid tabulation (4) to an exterior mass-accounting system (Fig. 4) and terminates there in a capillary tube open to the atmosphere. The total exterior volume must be kept as small as possible, consistent with practical, operational requirements of the calorimeter (to be described later). During operation, all parts of the enclosed volume must be of unvarying temperature, though of necessity the interior and exterior parts will be at different temperatures. A fiduciary index on the capillary tube of the exterior mass-accounting system defines the termination of the "constant-volume" referred to in Section 1. We consider it essential for efficient operation and maintenance that key sections of the enclosed volume be constructed of *glass* to permit viewing the contents.

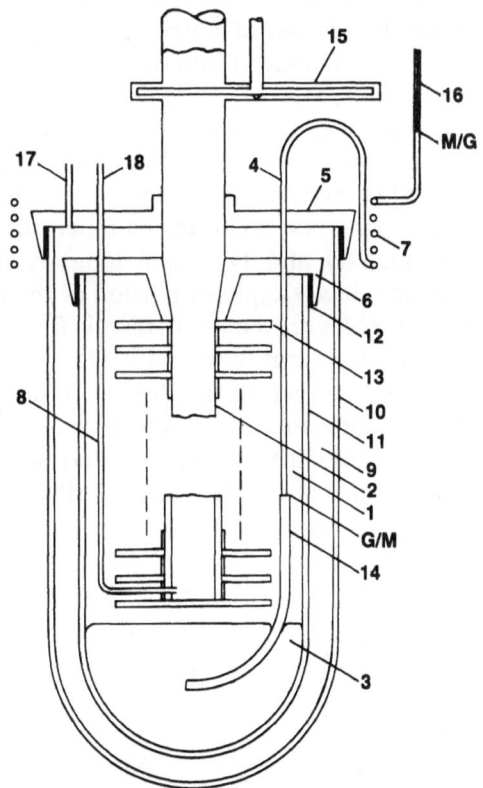

FIGURE 3a. Detailed construction of phase-change calorimeter.

4.1. Calorimeter Construction and Assembly

The calorimeter described in this section (Figs. 3a, 3b, and 3c) will have a maximum continuous energy-measuring capability of *ca* 54 kJ, after which a day or two are required to prepare for any further heat measurements. The seamless reentrant tube (2) is of monel with 25.4-mm OD and 0.6-mm wall. The lower 18.5 cm of this tube is reduced in diameter to 22.2 mm. Onto this tube are first soldered the brass outer (5) and inner (6) calorimeter lids, the latter positioned up against the tube segment of transitional diameter. Each lid has two holes to accommodate the two tubes which must enter the inner space (1). The top lid has, in addition, a third hole through which an environment of dry, low-thermal-conductivity gas (Ar or CO_2) at atmospheric pressure can be created and maintained in the interenvelope space (9). Glass envelopes, 10 (outer) and 11 (inner), with smoothly-rounded, nesting bottoms are fabricated from 90-mm- and 75-mm-OD pyrex tubing of 2-mm wall thickness. The inner envelope is of a length which provides about 45-mm axial clearance below the end of the reentrant tube. The lid inside diameters and envelope outside diameters are selected to provide 0.1-mm radial gaps (12) between the metal lids and the individual glass envelopes, which in the final assembly are attached to the lids with Apiezon "Q" wax. In general, the glass tubing available will be slightly out-of-round, requiring judicious grinding of the outer glass surfaces where they fit into the metal lids. The heat-distribution assembly (13) consists of 21 "thimbles" (Fig. 3b) individually soft-soldered to the reentrant tube, excluding all air gaps. Each thimble consists of a 0.5-mm-thick copper fin hard-soldered to a 8.25-mm-long segment of thin-walled monel tubing whose ID permits snug assembly over the reentrant tube. After 8 thimbles are installed (the first abutting the inner lid), a heavy-wall copper tube segment (29.7-mm OD and 122.7-mm long) is snugly fit with soft solder over the remaining, lower exposed portion of the reentrant well tubing. Thirteen additional thimbles are then individually fitted over and soft-soldered to this copper tube. The reentrant tube end is closed off using a thimble having no central hole. We note that all thimbles should be mounted with their grooves at the same azimuth above one another. A 2-mm thin-walled inconel tube (8) is installed through both lids, securely soldered wherever tube and lids are in contact. This tube enters a hole drilled through to the reentrant tube interior

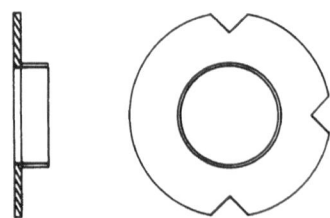

FIGURE 3b. Heat-distribution thimble.

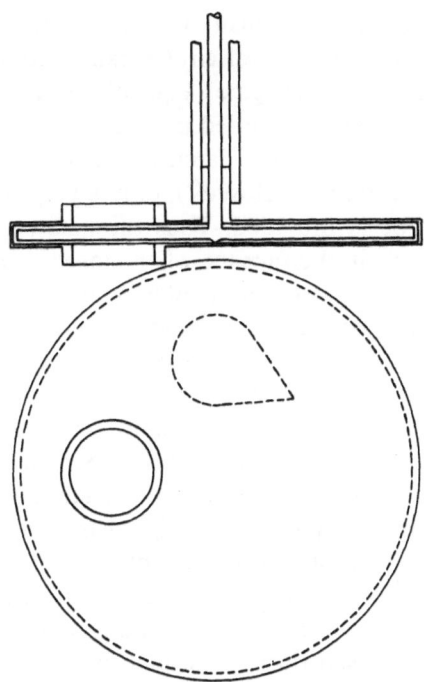

FIGURE 3c. Calorimeter gate.

between the lowest two fins and is soldered there. The purpose of this tube is to introduce helium gas into the calorimeter during operations. A second (long) piece of the thin-walled inconel tubing of 3-mm OD is formed into a five-turn helix (7) of 110-mm diameter, which in the final assembly will be positioned to reside at the outer lid level. The excess tubing at the bottom of the helix is formed to enter through the lids (where it also is soldered) into the inner space, terminating about 10 cm above the lowest fin. To protect the metal surfaces against corrosion, the entire top lid and the outside of the bottom lid are tin-coated and, to preserve the purity of the calorimeter working water, all metal surfaces which will be in contact with the working water in the inner space are either tin or silver-plated. A short pyrex glass tube (14) is waxed onto the end of the 3-mm tube at joint G/M and curves at its lower end to terminate so that its end is very close to the bottom center of the inner envelope after assembly.

The top end of the reentrant tube is soldered into the calorimeter gate case (15, and Fig. 3c) 100 mm above the top lid. Another section of reentrant tube stock is soldered into the top of the calorimeter gate case. It is onto this last-named section that mechanical supports for the calorimeter are attached. All metal surfaces above the upper lid are tinned as well. It should be noted that our experience with tin kept at 0 °C (below the solid–solid tin transition temperature at 13 °C), for almost 20 years, has shown no tendency of the tin to transform. The transition is a notoriously sluggish one and apparently requires special physical circumstances to initiate it at all.

The calorimeter gate (Fig. 3c) consists of a rotating disk with a special cut-out section and limited angular freedom of movement. At one extreme of the movement ("open"), a clear path down the reentrant tube is provided. At the other extreme ("closed"), only a very small area at a radius equal to that of the tube is provided. Thus the gate serves multiple functions: it blocks radiation from the furnace when closed; it pulls the drop wire and hence the sample capsule over into contact with the tube wall when closed; and it helps to contain possible convection currents of warm gas within the calorimeter while the heat is being transferred to the calorimeter.

After mechanically mounting the calorimeter, the clean glass envelopes are waxed in place onto the calorimeter lids. Care must be exercised to avoid thermal relaxation of the inner lid seal while working on the outer lid seal. The calorimeter is mounted at a height sufficient to permit later moving the large strip-silvered Dewar (Fig. 1) in below the calorimeter. At this time, a pyrex glass capillary tube (16) of 1.5-mm ID is joined onto the free tube end at the top of the tube coil (joint M/G, Fig. 3a). This can be accomplished satisfactorily with epoxy cement, but we note that since this joint may be severely stressed during the placement and packing of ice in the Dewar to cover the calorimeter, it should be adequately strengthened and protected. The capillary tube rises to the level of the reentrant tube top end, is then led horizontally the minimum distance required to clear the Dewar outside, and is brought down to be joined to the mercury mass-accounting system described below.

4.2. Mercury Mass-Accounting System

The accuracy and precision of each enthalpy measurement made with the calorimeter depend to a great extent upon the correct design and operation of the mercury mass-accounting system (Fig. 4). This is used to measure the mercury forced into the calorimeter by atmospheric pressure during each experiment. The mass-accounting system and the tubulation connecting it to the calorimeter provide the sole and only path connecting the inner calorimeter space to the outside. It is the important external component of the "constant volume" referenced in Section 1. Since it is filled with mercury and more or less exposed to ambient temperature, mass changes of the contained mercury due to ambient temperature change will be inextricably included in the mercury mass changes caused by heat release within the calorimeter. Therefore, every attempt must be made during design and fabrication to minimize the contained mercury volume of the system, consistent with operational requirements.

The mass-accounting system is connected directly to the glass capillary (1, Fig. 4; 16, Fig. 3a) issuing from the calorimeter. It is situated so that the

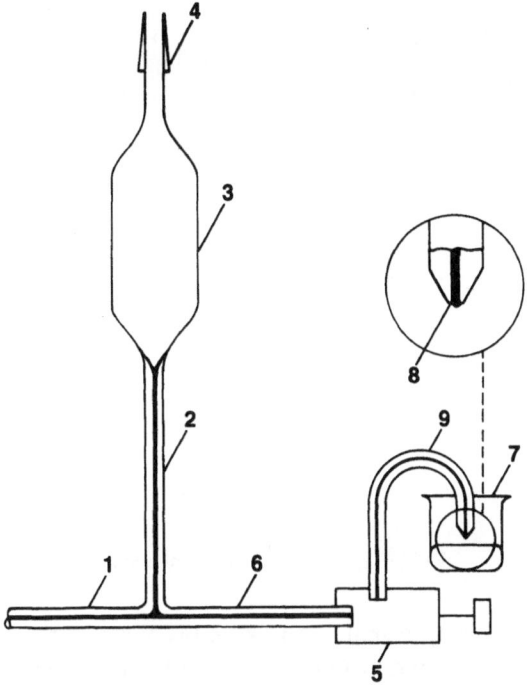

FIGURE 4. Mercury mass-accounting system.

needle valve (5) is approximately level with the middle of the calorimeter fin assemblage. It consists of a capillary "T" constructed from 0.6-mm-ID tubing, the vertical leg (2) being a 10-cm section of *precision-bore* capillary tubing. It must be emphasized that great care exercised in fabricating smooth capillary joints with no sudden or large diametral changes in the capillary duct will prevent much grief in the later maintenance of the calorimeter. A reservoir (3) of *ca* 50 ml capacity, terminating in a tapered glass joint (4), is attached atop the vertical leg. The needle valve (5), shown here schematically and in detail in Fig. 5, has a stainless-steel body. Its seat port is waxed onto the short capillary arm (6) and a curved capillary tip (9) is waxed into its packing port. Certain precautions must be observed in the fabrication of this tip in order that the maximum mercury mass resolution becomes achievable. (This resolution should be equal to the mass of mercury contained in a 0.01-mm length of the precision capillary, *ca* 0.06 mg.) The inset illustrates the required, reproducible meniscus appearance after each immersion in the mercury in beaker 7. Each immersion must without fail result in the exclusion of any air pockets at the tip end; otherwise, those may be forced into the accounting system at some time, thereby falsifying a mass accounting. This is best realized by optical grinding and lapping techniques which produce a smooth, flat tip end showing no raggedness or broken edges at the capillary wall termination

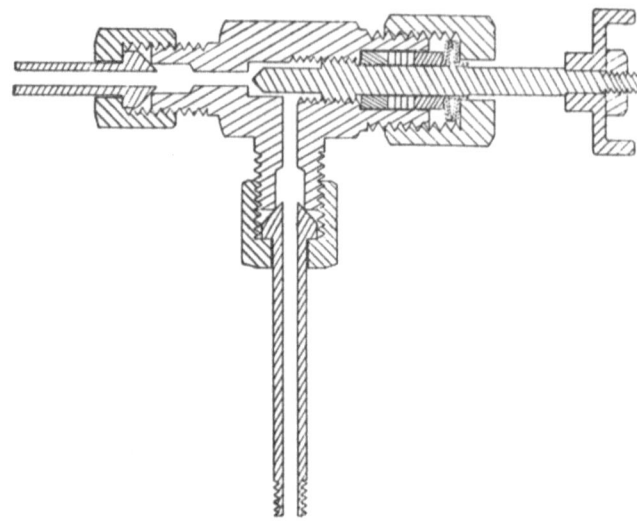

FIGURE 5. Calorimeter needle valve.

(8), even on a microscopic scale. An assembly view of the miniature needle valve used at NBS is shown in Fig. 5. It is made of stainless steel, uses teflon packing, and offers a very small incremental volume between the ends of the glass capillary tubing sections connected by it. The accounting system is securely attached to a plate, which in turn is mounted on the main calorimeter support. Due to inevitable dimensional variances during connection of the calorimeter to the mass-accounting system, the glass capillary tube which connects them may be strained. It is desirable after assembly to heat a small length of this tube to near the softening temperature, in order to relieve this strain.

4.3. Filling the Calorimeter with the Working Fluids

After physical assembly, the next step is to fill the calorimeter with the correct proportion of working fluids: solute-free, air-free water and triply-distilled mercury.

4.3.1. Production of Working Water

We have found that the most efficient procedure is to vacuum-distill ordinary laboratory distilled water. This is done in the apparatus illustrated in Fig. 6. In this still (1), constructed of metal except for the sight tube (2), all metal surfaces which contact water are either of silver, or are tin-coated. Distilled water is admitted to the still through valve 4 as a water aspirator

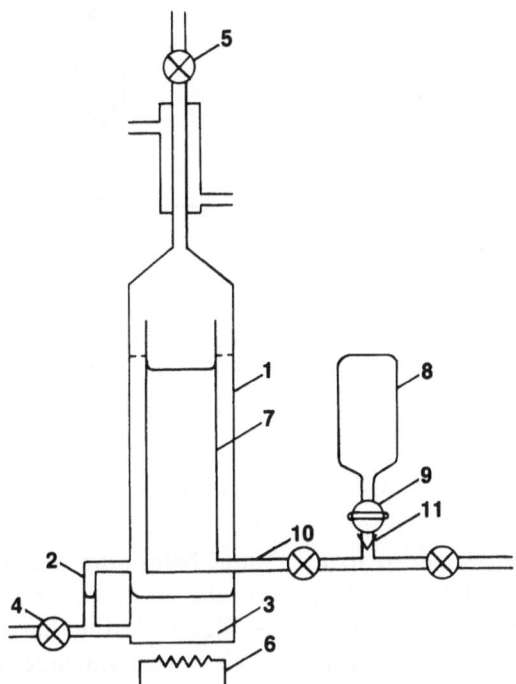

FIGURE 6. Vacuum still for preparation of calorimeter working water.

maintains the still at a low pressure through valve 5. Heater 6 vaporizes the charge, part of which condenses in the still neck and is collected in container 7. After sufficient distillate has collected, it is drained under gravity and its own vapor pressure into the previously evacuated, two-liter glass container (8). This container has a greaseless stopcock closure (9) and is attached to the distillate line (10) with a tapered, vacuum-tight seal (11). A very good test of the efficiency of the distillation is to observe whether or not the flask (8) can be completely filled with water or whether an air pocket remains at the inverted flask bottom. After a flask has been completely filled, it is drained of perhaps 50 ml water (providing a space to accommodate thermal expansion), the stopcock (9) closed, and the flask removed from the system. It takes less than a day to produce a complete flask full of calorimeter working water in this manner.

4.3.2. Filling the Calorimeter with Working Water and Mercury

The last assembly step is to introduce the working water and mercury into the calorimeter. This is accomplished through the external mass-accounting system—water first, followed by the necessary amount of mercury—using the apparatus arrangement shown schematically in Fig. 7 and described below. The necessary amount of mercury is that amount which fills the inner

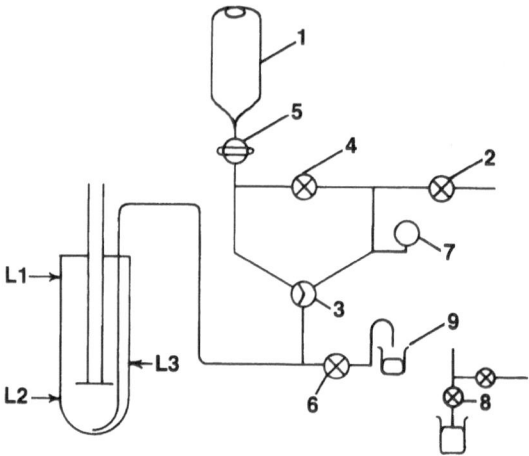

FIGURE 7. Apparatus arrangement for filling calorimeter with water and mercury

envelope up to *ca* 15 mm below the lowest calorimeter fin plus the entire tubulation leading out to the mass-accounting system. It is presumed that the calorimeter space to be filled with fluid has, prior to this, been vacuum-tested (helium leak detector use is recommended) and is now vacuum-tight. The steps necessary to fill the calorimeter are now outlined (cf. Fig. 7):

1. Attach flask of working water (1), a "high-vacuum" source (2), through a three-way stopcock (3) to the tapered joint at the top of the mass-accounting system (4, Fig. 4).

2. With the cock 3 in a blocked position, evacuate through 4. As soon as high vacuum is attained, close 4 and immediately open 5, filling the tube with water.

3. Close calorimeter needle valve 6 and adjust 3 so that calorimeter inner space is now evacuated through the capillary tubing. This will obviously be a lengthy process and it is well to monitor progress with a vacuum gage (7), such as a Pirani gage under nondynamic conditions. We find it adequate to pump for 7–10 days to achieve a vacuum in the range 10^{-3} to 10^{-4} torr.

4. While pumping on the calorimeter, simultaneously fill the entire space from the capillary tip back to the valve seat with mercury. This is easily accomplished after temporarily connecting to the tip a vacuum and mercury source (shown schematically as 8 in Fig. 7). Now raise a full breaker of clean mercury (9) in which to immerse the tip.

5. When the calorimeter vacuum is satisfactory, adjust 3 to connect the filling flask (1) to the calorimeter. Water will now flow into the calorimeter inner space under gravity and its own vapor pressure. When level L1 has been attained, switch 3 back to the blocked position. The

remaining free volume above L1, now filled with water vapor, is equal to the envelope volume available below L2 (the desired final mercury level, *ca* 15 mm below the lowest fin).

6. Open the needle valve and *slowly* admit mercury to the calorimeter through the capillary line (the lowest calorimeter fin is coated with Apiezon Q wax on its underside during assembly to prevent mercury from splashing onto the metal surface) while adding fresh mercury to 9 as required.

The entire filling apparatus is now removed from the mass-accounting system and mercury is drained out the capillary tip till it falls in the precision bore "T" section to its equilibrium level, approximately the level of mercury in 9. *Hereafter, the tip must always be kept immersed in mercury except for very brief periods while changing breakers.* The tube leading to the interenvelope space (17, Fig. 3a) is connected to a vacuum-or-gas source and the helium tube (18, Fig. 3a) is connected through a needle valve to a helium source. The calorimeter is now ready for operation.

5. OPERATING THE CALORIMETER

At the outset, the operator must always bear in mind that *each and every thermal exchange within the calorimeter will now result in a corresponding mass exchange at the accounting system. The mercury must never be allowed to recede out of the precision-bore capillary section.*

The large strip-silvered Dewar is positioned under the calorimeter and raised till its upper edge is about 2 cm below the top of the reentrant tube. It should be deep enough to allow 15 to 20 cm space beneath the bottom of the outer envelope. The Dewar is filled with a clean ice–water slush and the calorimeter allowed to equilibrate to the ice point. This takes about 24 hours. The interenvelope space is then evacuated and filled with dry Ar or CO_2 gas at atmospheric pressure. Provision should be made that this condition is maintained continuously regardless of atmospheric pressure changes. Also, a very low flow of dry He gas should be maintained into the bottom of the reentrant tube and its top should be lightly capped when not connected to the furnace to prevent moisture diffusion and condensation in the calorimeter. *The interior of the calorimeter must be kept free of all traces of water during operation.*

5.1. Calibration of Calorimeter

The phase-change calorimeter has two calibration figures of note. The first, referred to as the "calibration constant," expresses the equivalence of heat exchange within the calorimeter and mercury mass exchange at the mercury-accounting system. As noted in Section 1, this is identical for *all* phase-change calorimeters using the same working substance. It has been accurately measured[2] and has the value 270.48 ± 0.02 J per gram of mercury, when H_2O is the working substance.

The second calibration figure varies from calorimeter to calorimeter and expresses the mass of mercury required to fill a 1.00-mm length of the precision-bore capillary section of the mass-accounting system. This can only be measured when the calorimeter working water is ice-free and is at thermal equilibrium. A typical value for this would be around 6 mg mm^{-1}.

5.2. Production of Ice Mantle

In order to make enthalpy measurements, the first requirement is to create a cylindrical mantle of ice surrounding the fin assembly and extending out nearly to the inner envelope wall. Due to the lens effect of the water-filled envelope, it is difficult to judge how far out the mantle really extends. It is certainly possible to misjudge and fracture the inner envelope; we have done so on one occasion. We have adopted the convention of halting ice formation when the image of the exterior ice surface ceases to express the undulations which are caused by the fins, and becomes smoothly cylindrical.

The refrigerant sources for mantle formation are usually solid CO_2 or liquid N_2. A few ml of liquid N_2 poured directly into the reentrant tube (hereafter called the "well") will rapidly initiate ice formation on the surfaces of the lowest few fins. As soon as this has been accomplished, a closed-end copper tube of slightly lesser OD than the well ID is inserted and finely crushed dry ice is added to 5–10 cm depth. This depth is maintained by periodic additions during the mantle formation process, which takes about 1.5 hours. As soon as the outer ice surface below the last three or so fins appears smoothly cylindrical, the copper tube is raised so its bottom is at the second fin level. As the smoothly cylindrical surface progresses upward, the tube is progressively raised until the mantle now envelops even the topmost fin. (It is entirely feasible to carry out the entire process with some type of capsule-shaped cooler fed with liquid N_2 and perhaps save a little time at the expense of some complication.) Of course, during mantle formation one must collect the mercury forced out the accounting system. As soon as the ice formation process is completed, the copper tube is filled with warm water and inserted to full depth (first making sure that mercury is available to feed the accounting system). This melts the ice immediately in contact with the fin assembly and provides the narrow, tortuous water path (around and over each fin in turn) which, in later heat measurements, the working water from *outside* the mantle must follow to reach the mantle *interior* to equalize the pressure.

As a result of the temperature and pressure fluctuations attending mantle formation, the calorimeter will be far from thermal equilibrium. Therefore, the ice bath is drained of water, repacked, and additional ice and a portion of the cooled drain water added, after which it is allowed to equilibrate at least till the next working day.

5.3. Heat-Measurement Sequence

5.3.1. Heat Leak

In general, there is likely to be a small but finite heat exchange ("heat leak") between the calorimeter and its ice-bath environment. This is manifested in a slow rise or fall of the mercury column in the precision-bore capillary tube and is best observed with a precision cathetometer. It is important to understand that we are dealing with heat leaks in the tens of microwatts range or less and that temperature differences of millidegrees between parts of the calorimeter or between the calorimeter and the ice bath are *very* significant. Brief reflection will show that the calorimeter interior will be at the water triple point minus a few mK depression due to the mercury pressure (if the accounting-system level is higher than the level of the mercury pool in the calorimeter). The exterior ice bath, however, is at the air-saturated ice point (273.15 K) or perhaps a bit below if one has been careless in handling the ice. Therefore, the normal circumstance should be the slow *formation* of fresh ice and the mercury column will *rise*. Depending on the thermal resistance of the calorimeter support, there may be a compensating heat lead *into* the calorimeter. In any case, it is the *net* heat leak which is observed as the mercury column rises or falls, and this should be 100 μW or less. It may vary from day to day and, if the right circumstances obtain, may be essentially unobservable on a time scale of several hours (zero heat leak).

5.3.2. Sequence of Experiment

The furnace is first swung into position over the calorimeter and held in the coaxial position. The sample is raised into the furnace and resides at the furnace center for a time sufficient to bring it to thermal equilibrium. The time required depends on the temperature and on the particular system studied and must be estimated knowledgeably or measured from case to case. It can vary from half an hour or less to several days (if, for instance, one is carrying out measurements very near to a phase transition). Furnace temperature should be recorded periodically to ensure that the temperature variance is acceptable. Approximately half an hour before one schedules a "drop" (sample transfer from furnace to calorimeter) the mercury needle valve is closed, the accounting-system tip is immersed in a *weighed* beaker of mercury, and observations at roughly 5 to 10 minute intervals are made of the height of the mercury column. If the heat leak is satisfactorily small and is linear, the valve may be opened, the calorimeter gate swung open, and the sample dropped into the calorimeter.

Then the gate is closed. The gate need be open for no longer than two seconds. *It is important to note the exact time when the sample entered the calorimeter.* If a large amount of heat transfer is anticipated within the calorimeter, care must be exercised that the level of mercury does not fall to the level of the "T" joint in the accounting system. Otherwise air could enter the mercury line. After the sample has equilibrated in the calorimeter (typically 30 to 40 minutes but, again, varying widely from case to case) the mercury needle valve is closed with the mercury column near to the last height observed before the experiment. The column level is again observed periodically to establish the post-experiment heat leak. The mercury beaker is removed for weighing and an unweighed beaker put in its place.

5.3.3. Calculating the Sample Enthalpy

Every enthalpy measurement results in two fundamental data: the gross mass change M_g in the beaker of mercury in which the accounting-system tip is immersed, and a positive or negative correction to this which is derived from observations on the mercury column in the accounting system. This correction is made in order to account for the calorimeter heat leak while the sample is equilibrating in the calorimeter and for any change in the (arbitrary) fiduciary indices assigned to the vertical mercury column in the "pre-experimental" and "post-experimental" intervals (Fig. 8).

The accounting-system correction can be understood with reference to Fig. 8, in which the bold lines represent the actual readings of mercury column height in the accounting system. Times before τ_0 (the actual time the sample entered the calorimeter) are considered pre-experimental; times after τ_0 are considered post-experimental. Heat transfer to the calorimeter from the sample starts at τ_0. At τ_1, the needle valve is closed. The entire sample heat has been transferred by τ_2. The accounting-system mass correction to be added to the gross beaker mass change cited above is given by $(h_i - h_f) \cdot b$, where b is the accounting-system calibration.

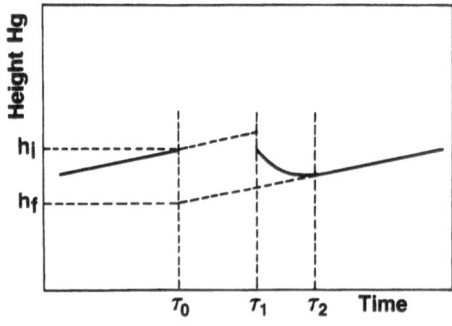

FIGURE 8. Interpretation of observations at the mercury accounting system.

Following the above procedure, any individual enthalpy measurement is calculated using the equation

$$(H - H_{273.15}) = [M_g + (h_i - h_f) \cdot b] \cdot 270.48 \quad J$$

The *sample* enthalpy is derived by subtracting *at the same temperature* from the measured enthalpy of an encapsulated sample, the measured enthalpy of the encapsulation. Generally, the enthalpy of the encapsulation is measured using a separate (empty) capsule, identical in physical composition. In practice, however, it is rarely possible to achieve *exact* mass equality between the actual sample encapsulating materials and those of the empty capsule. It is convenient to adjust the measured enthalpy of the encapsulated sample to correct for this mass deficit or excess. Since with care in encapsulation such mass differences can be held below, say, 100 mg, it is quite acceptable to use literature enthalpy values in calculating this correction.

6. GENERAL OBSERVATIONS AND RECOMMENDED OPERATING PROCEDURES

As with all experiments, there are both pitfalls for the unwary and shortcuts to success. We offer the following comments from our experience:

1. The rate of data production will be somewhat dependent on the particular material and investigation, however, three enthalpy measurements per working day may be considered a reasonable, long-time average.
2. The assumption that one can derive the *sample* enthalpy from the difference of two measurements as outlined in Section 5.3.3 above depends implicitly on the assumption that the sample itself loses a negligible amount of heat during the physical transfer to the calorimeter. This demands that the thermal resistance between the sample and its encapsulation be made large through appropriate encapsulation design (though not *so* large that the sample takes excessively long to equilibrate in either furnace or calorimeter).
3. *Any* water in the mercury transit line, including both glass and metal sections, is inadmissible. If the calorimeter filling process has not swept this line free of water, one must partially melt a mantle and allow the terminating meniscus of the mercury in the transit line to retreat till visible in the short glass tube section at its end. The line is then evacuated to remove all traces of water and refilled under vacuum by re-forming an additional portion of mantle ice.

4. After experiments have added heat (consumed ice) equivalent to about 200 grams of mercury, it can happen that a hole is created in the mantle. This is fatal to the experiment for warm (and, hence, denser) water will escape through this hole, mix with the water outside the mantle, and an indeterminate portion of its heat will never be registered as a phase transformation. At this point, the remaining ice must be melted (usually, by an electric heater inserted into the calorimeter well) and a new mantle formed. *However*, this melting process warms the mercury, since the warmer water will gravitate to the bottom. If one goes immediately to new mantle formation after melting an old one, the bottom of the new mantle will then be quite thin as one will then be freezing in competition with heat from the warm mercury pool. We allow the calorimeter to equilibrate overnight after having totally melted a mantle.

5. Despite the best precautions, a small amount of air may get into the needle valve, be it through worn packing, failure to keep the capillary lip clean, or some gross operational error. If this happens, the valve stem is removed, the valve cleared of mercury, the mercury in the transit line is drawn back toward the calorimeter, the valve is re-assembled, and finally valve and line evacuated and refilled with mercury using the techniques outlined in Section 4.3.2.

REFERENCES

1. D.A. Ditmars, "Heat-Capacity Calorimetry by the Method of Mixtures," in: *Compendium of Thermophysical Property Measurement Methods* Vol. 1, pp. 527–553, Plenum Press, New York and London (1984).
2. D.C. Ginnings, A.F. Ball, and D.T. Vier, *J. Res. Natl. Bur. Stand.* **50**(2), 2392 (1953).
3. R. Hultgren, P.D. Desai, D.T. Hawkins, M. Gleiser, K.K. Kelley, and D.D. Wagman, *Selected Values of the Thermodynamic Properties of the Elements*, American Society for Metals, Metals Park, Ohio 1953.

16

Apparatus for Investigation of Thermodynamic Properties of Metals by Levitation Calorimetry

VITALIY YA. CHEKHOVSKOI

1. INTRODUCTION

The major advantage offered by levitation calorimetry is that this technique enables one to substantially extend[1] the high-temperature limits for the application of the method of mixing for the purpose of experimental investigation of the calorimetric properties of refractory metals and their alloys in solid and liquid states, as well as other electrically conducting substances. The result is achieved, since one has an electromagnetic crucible instead of a ceramic crucible or a crucible fabricated from some other materials. Crucibles inevitably start reacting with the investigated substance or its melt at temperatures higher than 2000 K, thus causing misrepresentation of the experimental results. With an electromagnetic crucible the specimen being investigated is held in a suspended state (levitated), while being heated up to the required temperature through the interaction between the high-frequency electromagnetic fields of the inductor and the specimen. Such heating processes eliminate any contact between the specimen under examination and the structural elements of the experimental apparatus, thus preventing any chemical interaction.

The levitation heating technique and drop calorimetry have been employed in investigating the heats of fusion, melting points, entropies of fusion, spectral emissivities of melts in the vicinity of the melting points, as well as enthalpies and specific heats of the following metals in the solid and liquid states: molybdenum,[2-8] niobium,[5,7-10] vanadium,[5,7-9,11-13] ruthenium,[5,8,14-16] platinum,[17] copper,[7,17] titanium,[5,7,8,13,15,18,19] iron,[7,13] cobalt,[7,20] pal-

VITALIY YA. CHEKHOVSKOI ● Institute for High Temperatures, USSR Academy of Sciences, Moscow 127412, USSR.

ladium,[7,20] nickel,[7,21] silver, chromium, and zirconium,[7] rhodium,[15,22,24] iridium,[23-25] tantalum,[7,26] and gallium.[27] Most recently levitation calorimetry has been employed to achieve much higher temperatures for measuring the caloric properties of molybdenum[28] at 3382 K, niobium[29] at 3159 K, and tungsten[1] at 4017 K. The above experimental data and available references enabled the semiempirical relationships[29] to be derived, from which it appeared possible to predict the heats and entropies of fusion for metals not yet studied, such as rhenium, osmium, technetium, lutecium, americium, actinium, hafnium, chromium, thorium, holmium, manganese, rhodium, and francium.

However, along with the above advantages, levitation calorimetry exhibits a number of problems. In particular, there are no methods so far for computing the optimum geometrical dimensions and configurations of inductors and specimens being studied as applied to specific metals for keeping these suspended upon reaching high temperatures. These problems are attacked by trial-and-error methods.

As mentioned elsewhere in the literature,[31] the major portion of the error in measuring the thermodynamic properties by levitation calorimetry is accounted for by the error in measuring the temperature of the specimens in the solid and liquid states. This is further enhanced by the fact that the heated solid specimens are in no way isothermic with height because of the nonuniformity of their interaction with the high-frequency electromagnetic field of the inductor. That is why, when using a blackbody cavity for measuring the temperature of a specimen, one should know the temperature distribution along the height of the specimen.

Intensive electromagnetic mixing of a liquid metal takes place within a molten specimen, thus rendering the specimen isothermic. Bearing in mind that it is impossible to create a blackbody cavity within a molten specimen, one should know in advance the relationship between the spectral emissivity and temperature of the melt in order to measure the latter. However, these data are not always available. Incidentally, a promising method to measure the true or proper temperature of the surface of a body makes use of a multi-wavelength pyrometer.[32,33]

Analysis procedures and specific features involved in applying levitation calorimetry are described in detail elsewhere.[31] These prerequisites underlay the design and operation of the apparatus intended for making experimental analysis of the thermodynamic properties of refractory metals and their melts in the Institute of High Temperatures of the USSR Academy of Sciences. An appreciable contribution to these efforts was made by the scientific workers of the Institute, B.Ya. Berezin and S.A. Katas. This chapter is designed to present a description of the apparatus, its individual units and assemblies, their interaction, as well as certain aspects of the method as practical in High Temperature Institutes.

2. MAJOR UNITS OF EXPERIMENTAL APPARATUS

Used as a basis for levitation calorimetry[2,34,35] is the method of drop calorimeters and heating a specimen under study by the levitation technique, whereby the specimen is heated up to a high temperature in a suspended state within a high-frequency electromagnetic field produced by an inductor. To realize a measurement of the enthalpy, use is made of a massive copper drop calorimeter with an isothermic jacket.

As shown in Fig. 1, the major components of the apparatus operating as a levitation calorimeter are the high-frequency generator 1, a hermetically sealed chamber 2, the calorimeter 3, and a vacuum system consisting of a vapor-oil-diffusion pump 4 and roughing-down pump 5.

The pressure-tight volume of chamber 2 and calorimeter 3 can be evacuated down to a pressure of 10^{-3}–10^{-4} Pa with the aid of the vacuum system (4 and 5). To obtain a higher vacuum use is made of a liquid-nitrogen trap mounted within the branch pipe of the oil-diffusion pump 4, and a getter in the form of a titanium foil strip heated by electric current. The getter is placed within chamber 2.[18,19] The pressure-tight volume of the apparatus can be filled with a noble gas from cylinder 6. Gas pressure or vacuum are measured

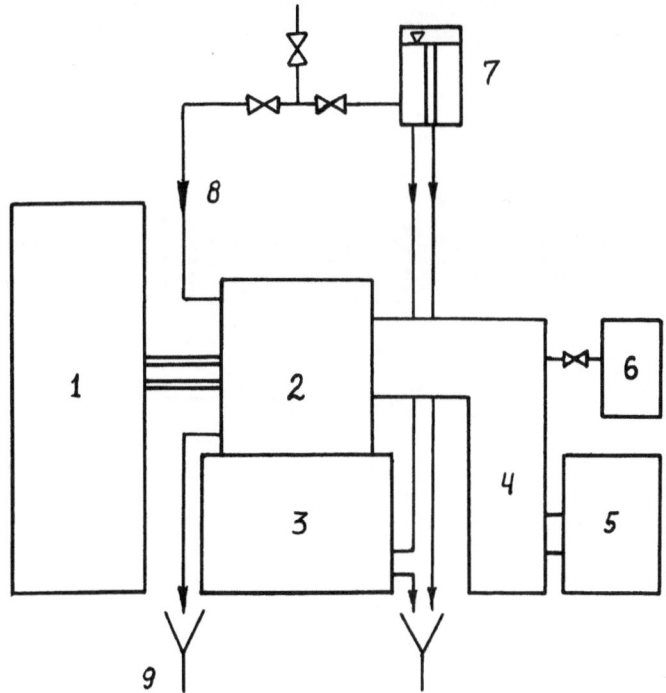

FIGURE 1. Block diagram of the major units of the apparatus operating on the principle of levitation calorimetry.

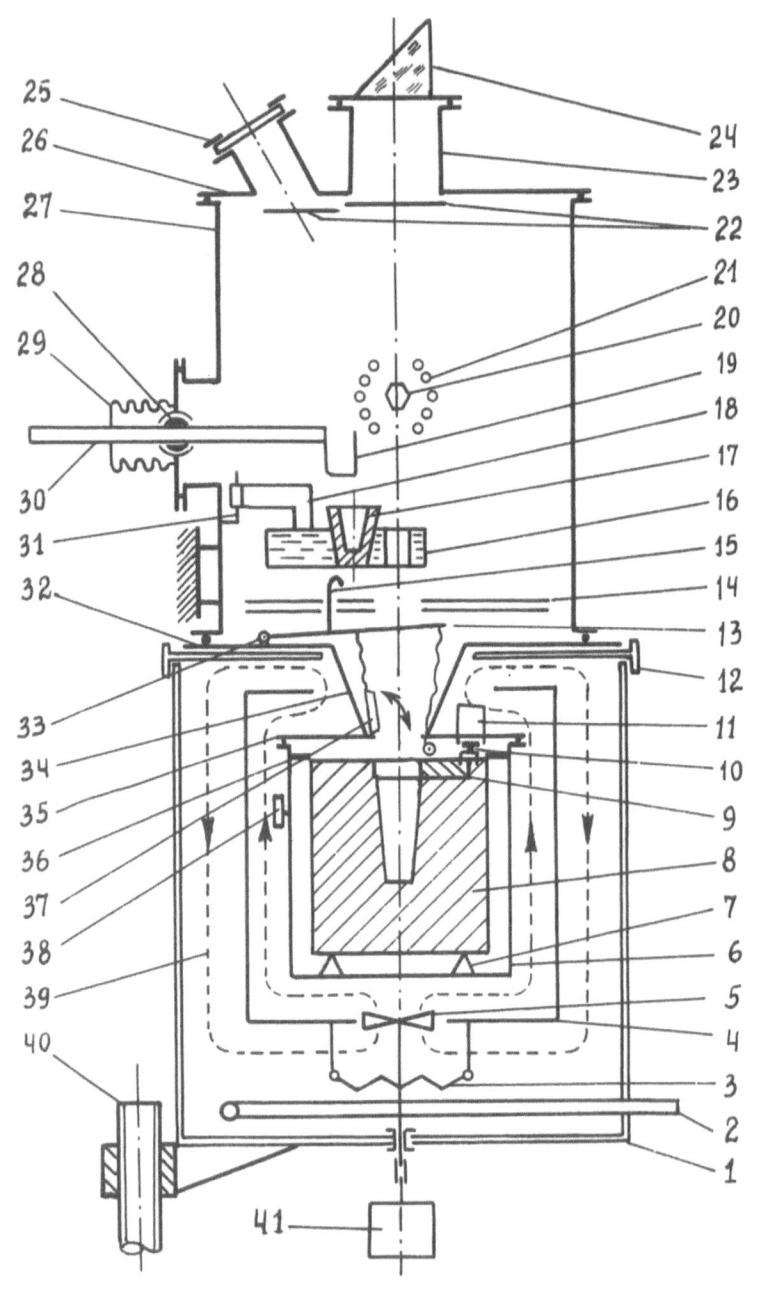

FIGURE 2. Design of heating chamber and calorimeter with thermostat.

with the aid of a compound pressure and vacuum gage, thermocouple, and manometer valves.

The body of chamber 2 is cooled with tap water 8, which is discharged into drain 9. Chamber 2 is designed to accommodate an inductor arranged to heat the specimen under investigation up to the specified temperature while in the suspended state. The induction coil is connected to a high-frequency generator 1 rated at 60 kW nominal; the frequency of electromagnetic oscillations is 66 kHz. A specimen heated up to the specified temperature within chamber 2 is dropped into calorimeter 3, which measures the amount of heat brought by the specimen. The calorimeter consists of a calorimetric system and its jacket, which is placed within a liquid thermostat. The temperature of the calorimeter jacket is maintained constant within the thermostat. The constant temperature of the liquid within the thermostat is ensured by a cooler and electric heater, the power capacity of which is adjusted automatically by a high-accuracy controller. A constant flow of cooling water through the cooler of the thermostat is maintained by means of a pressure water tank 7 having a constant level of water inside due to the use of an overflow tube mounted at the center of the tank.

Now we consider in greater detail the design and operation of individual units of the apparatus, which are shown schematically in Fig. 2.

3. HERMETICALLY SEALED CHAMBER

The body 27 of the chamber[2,35] (Fig. 2) rests upon a stationary support. A demountable brass cover 26 covers the chamber from above. The cover is made to carry two branch pipes (23 and 25), as well as a vacuum gage, thermocouple, and manometer valves arranged to measure gas pressure or vacuum; these three members are not shown in Fig. 2. On top of branch pipe 23 there is a total internal reflection prism 24 used to sight on the specimen 20 when measuring its temperature with an optical pyrometer. Branch pipe 25 is hermetically sealed with a window arranged to observe the inner components and parts of the chamber. Shutters 22 are employed to cover prism 24 and window 25 in order to protect them from being sprayed in the course of experiments; the shutters can be moved aside when necessary.

Body 27 of the chamber is fabricated from a copper seamless pipe, 232 mm in diameter and 3.5 mm wall thickness. Three copper branch pipes, which end in flanges, are welded to the cylindrical wall of the chamber body. One of the flanges is used for making connection to the oil-diffusion pump, the second flange is arranged to introduce inductor 21. The third flange which is shown in Fig. 2 serves for inserting mechanical arm or manipulator 30. The inductor is made of copper tubes $\frac{3}{4}$ mm in diameter. The most preferable configurations of the inductor[11] and specimen[18] are illustrated in Figs. 3 and 4. An inductor

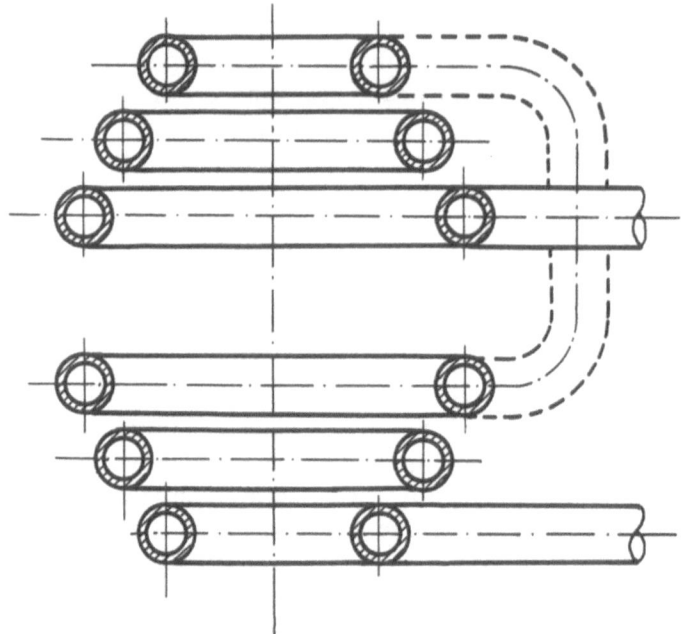

FIGURE 3. Inductor.

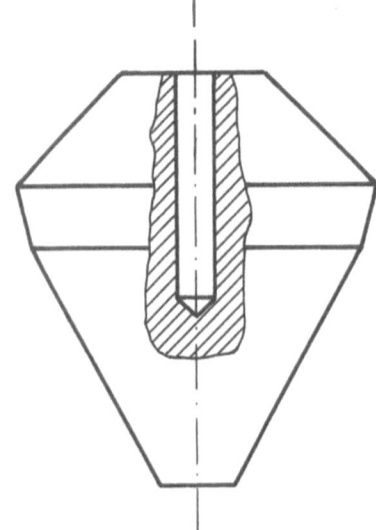

FIGURE 4. Specimen.

of such design has proved itself to be a good device for measuring the enthalpies of metals in the solid and liquid states. Dimensions of the inductor and specimen have been slightly varied depending upon the properties of the metal under investigation. This configuration of the inductor which has been employed ensured the stable position of a solid specimen in the course of heating. This feature is of importance in measuring temperature of a specimen with a pyrometer directed at the blackbody. Furthermore, such configurations of the specimen and inductor ensure but slight variations in their mutual position while electric power applied to the inductor is being changed. This, in turn, results in only negligible changes in the distribution of temperature with height of the specimen and, as mentioned elsewhere,[31] in a practically constant correction factor to the nonisothermal regime of the specimen.

Due to the ball-and-socket joint 28 and bellows-type seal 29, bracket 19 of the manipulator can be moved along its axis and circumference with the aid of handle 30, allowing one to make the following operations. Prior to actuating the high-frequency generator, specimen 20 under examination is held within the inductor 21 at the end of the manipulator bracket 19. Once the generator is energized, the specimen will interact with the high-frequency electromagnetic field of the inductor becoming suspended and the manipulator can be withdrawn.

The second function of the manipulator is to control the position of the movable stainless-steel shutter 16, which is in the form of a disk. The disk shutter 16 is cooled by running water, because the shutter is adjacent to the inductor and can be exposed to induction heating. Besides, the cooling of disk 16 is further required because the disk must shield calorimeter 8 from heat radiated by the melted specimen. There is a pocket in disk 16 arranged to accommodate casting mould 17 machined from a piece of copper, with a joint in the plane passing through the axis of the casting mould. Casting mould 17 is intended for catching specimens in case one falls from the inductor, as well as for accepting specimens in the course of start-up and adjustment. In addition to casting mould 17 there is an aperture made through disk 16 to allow a specimen to fall into the calorimeter. Disk 16 is suspended from lever 18 so as to be capable of turning around the vertical axis 31 in the horizontal plane, whereby the disk can arrive at one of the two extreme fixed positions. In the first position casting mould 17 is just under a specimen, being stationed there while an experiment is being prepared, when a specimen is being brought into the inductor 21, and while heating the specimen up to the specified temperature. Prior to releasing a specimen into the calorimeter, disk 16 is turned into the second extreme fixed position where the aperture in disk 16 is just under the specimen as shown in Fig. 2. The disk is turned into that second position with the aid of manipulator 19 or a tie-rod extending from a special-purpose motor with a gear. Once a specimen is dropped into the calorimeter, disk 16 will be brought back into the initial position with the aid of the manipulator or motor, so that casting mould 17 is positioned just under the inductor.

Several shields 14 are positioned under disk 16, and are fabricated from molybdenum foil to protect the calorimeter 8 and the resistance thermometer wires from radiation and splattering of liquid specimen 20. Resistance thermometers are used to measure the temperature increase in the calorimetric system. Thermometer wires pass from copper block 8 through truncated cone 34, resting on flange 32. These wires are brought out between two gaskets made of vacuum rubber, which serves to seal the joint between chamber body 27 and flange 32. Truncated cone 34 and flange 32 serve as a link between chamber body 27 and jacket 6 of the calorimetric system, the principal portion of which is copper block 8. Flange 32, truncated cone 34, and flange 35 of the calorimetric system jacket are welded together by argon-arc welding. These parts and jacket 6 are machined from stainless steel.

It can be seen from Fig. 2 that the calorimeter, which includes a thermostat 1, calorimetric system* 8, and calorimeter jacket 6, can be disconnected from chamber 27 at the point where the chamber is tightly joined to flange 32. When disconnected, the calorimeter can be lowered and turned around the support screw 40 for performing installation work.

4. THERMOSTAT

The jacket 6 of the calorimeter is placed within a thermostat[2,35] 1 (see Fig. 2) which has the form of a cylinder, 300 mm in diameter and 220 mm high. The walls of the thermostat are of a double-layer design made of stainless-steel sheets so as to provide for good thermal insulation. The spacing between the double walls is 10 mm. Cover 12 of the thermostat consists of two halves, where the joint plane passes through the axial line. When assembled, the thermostat cover 12 is fastened to the body of the thermostat 1 with the aid of coupling locks.

About 15 liters of transformer oil is used as a working fluid within the thermostat. The thermostat includes a cooler 2, electric heater 3, temperature sensor 38, and propeller mixer 5. The shaft of the mixer passes via a stuffing box seal, being rotated by motor 41 at 1400 rpm. In order to reduce to a minimum transmission of vibration from motor 41 to the thermostat, the shafts of the mixer and motor are coupled through a resilient coupling, while the motor itself is mounted on a separate insulated support.

The design of partition 4 made of stainless-steel sheet is such that the thermostatic liquid flows along the path shown in Fig. 2 by dotted line 39.

---

*A calorimetric system is understood to be the totality of pieces involved in the determination of the amount of heat being measured.

Partition 4 separates the liquid into two flows—the internal and external flows. The latter flow plays the role of a thermal screen, reducing the influence of the ambient temperature variations on jacket 6 of the calorimetric system. The internal flow of oil heated by electric heater 4 and pumped by propeller 5 exhibits a higher velocity as compared with the external flow. This feature facilitates quick temperature regulation in the liquid by varying the power capacity of electric heater 3. The use of the transformer oil has substantially reduced the difficulties associated with the problem of ensuring electric insulation of electric heater 3, having at the same time minimized its thermal lag, which feature is of considerable importance from the standpoint of ensuring efficient thermal control in the thermostat. The heater was made in the form of a 0.5-mm Ni–Cr alloy wire spiral, which is fastened at four points to insulators positioned on partition 4.

A type KMT-4 thermistor is used as temperature-sensitive element within the thermostat. The nominal resistance of the thermistor is 100 kΩ; the temperature coefficient of resistance is $4\,k\Omega\,\mathrm{deg}^{-1}$.

The thermostatic system of the calorimeter jacket[23] appears in Fig. 5. Thermistor 1 is fitted into the measuring arm of the DC resistance bridge circuit 2. A type P 325 self-compensation microvoltmeter–nanoammeter 3, used as a voltage amplifier, is connected to the bridge circuit instead of a galvanometer. Once bridge unbalance occurs due to variations in temperature within the thermostat, an error signal is amplified by the type P 325 instrument, arriving at the matching resistance box 4. A type BPT-3 high-accuracy temperature controller 5 is used as an actuator in the temperature-control system of the thermostat. Voltage brought from resistance box 4 is applied to the temperature controller. The type BPT-3 controller is adapted to vary the electric current flowing through heater circuit 6 in the thermostat in such a way as to reduce to zero the unbalance signal delivered by the resistance bridge circuit. Variations in voltage in the course of a calorimetric experiment are recorded by a type H-39 recording millivoltmeter 7. The readings taken from the type

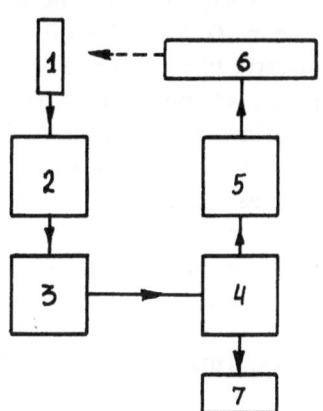

FIGURE 5. Block diagram of temperature control for the calorimeter jacket within a thermostat.

H-39 instrument have been converted to degrees kelvin. To that end, use was made of an artificial unbalance of the bridge circuit corresponding to a known number of degrees. In this way, the extent of temperature stability within the thermostat in the course of experiments has been estimated as 2×10^{-4} K.

The use of the thermostatic liquid circulation system shown in Fig. 2, as well as of the low inertial heater 3, and the optimum mutual position of electric heater 3, mixer 5 and cooler 2, enabled one to obtain a very short response time in controlling temperature within the thermostat. In other words, we have managed to reduce to a minimum the time lag between the signal delivered by temperature sensor 38 and the respective change in temperature of the thermostatic liquid against the wall of jacket 6 as a result of the variation in electric power capacity of heater 3.

For each experimental enthalpy measurement, the level of the stabilized temperature within the thermostat was selected and maintained so as to achieve a minimal correction factor for the heat exchange between the calorimetric system 8 and its jacket 6, not exceeding 3–5% of the total amount of heat introduced into the calorimeter by a dropping specimen.

5. JACKET OF THE CALORIMETRIC SYSTEM

As can be seen from Fig. 2, the massive block 8 of the calorimetric system is surrounded by a thin-walled jacket 6 made of stainless steel.[2] The block is supported by sharp ribs of three fluoroplastic support members 7 mounted at the bottom of jacket 6. The fixed position of copper block 8 within jacket 6 is ensured by spacer ring 36 made of acrylic plastic. The ring touches the upper end of the massive block 8 only at those points where it is fastened with two bolts. The barrel of jacket 6 is covered by flange 35 from above. The joint between these elements is hermetically sealed with an oil-resistant rubber gasket.

A gap between the massive block 8 and jacket 6 of the calorimeter is 10 mm. To lower heat transfer by radiation within the gap, the surface of the massive block 8 was nickel-plated, and the inner surface of the stainless-steel jacket was polished. It has been demonstrated in the course of calibrating the calorimeter that within such a gap the heat exchange factor remained constant even in those cases where the temperature drop between the surfaces was as high as 7 degrees.

Cover 37 of the calorimeter jacket 6 is mounted on a horizontal shaft within truncated cone 34. Cover 37 is connected by a flexible thread with lever 13, which can rotate around horizontal shaft 33. Still another flexible thread is on lever 13, the end of this thread being fastened to cover 9 of the massive block. This thread is passed through two rollers, thus changing its direction.

The shaft of one of the rollers is mounted horizontally at the inner surface of flange 35 carried by the calorimeter jacket, as shown in Fig. 2, while the shaft of the second roller is mounted vertically at the upper end of the massive block 8. If connecting rod 15 is lifted by bracket 19 of the manipulator together with lever 13, then one of the threads would lift and open cover 37 on the calorimeter jacket, while the second thread would slide over the guiding rollers thus opening cover 9 on the massive block by shifting it to the right into a fixed position as shown in Fig. 2.

Flange 35 of the jacket is made to carry electromagnet 11. When electromagnet 11 is energized, lock 10 on cover 9 of the massive block will be lifted, thus releasing cover 9, and the latter would shut the receiving cavity of the copper block 8.

6. CALORIMETRIC SYSTEM

Figure 6 shows an axial section and top view of the calorimetric system. The major components of the system are as follows: body of the massive block 1, cone insert member 2, specimen receiver 3, massive block cover 10, and resistance thermometers 13. The mass of the copper block together with other components is about 5 kg. Dimensions of the block are shown in Fig. 6. The mass of the block has been selected proceeding from considerations of the optimum increases in temperature of the calorimetric system when measuring enthalpies within a wide temperature range, as applied to specimens of refractory transition metals in the solid and liquid states; the mass of typical specimens is about 10 g.

Copper cone-shaped insert member 2 is composed of the upper cylindrical part and the lower conical part. The conical surface of insert member 2 is provided with a double-start square screw 16.8-mm pitch thread, with a manganin 0.5-mm wire of the calibration heater being double-wound into the thread. The heater wire is electrically insulated with a fluoroplastic cambric, 0.05-mm thick. There is a gap 12 between the ring-shaped horizontal surfaces of insert member 2 and body of block 1, which is employed for effecting a special type of fastening of the calibration heater terminals on the ring-shaped surface of insert member 2. The fastening of these terminals is made in the following manner. Copper strips, 0.5-mm thick, are soldered to the ends of the calibration heater wires which are located at the upper end of the screw thread on insert member 2. The copper strips are glued on the lower ring-shaped surface of conical insert member 2 through a thin insulating mica sheet. The length of each strip is somewhat less than that of the semicircle of the ring-shaped surface. The opposite ends of the copper strips are soldered to current and potential copper wires passing outside through holes 16 made in

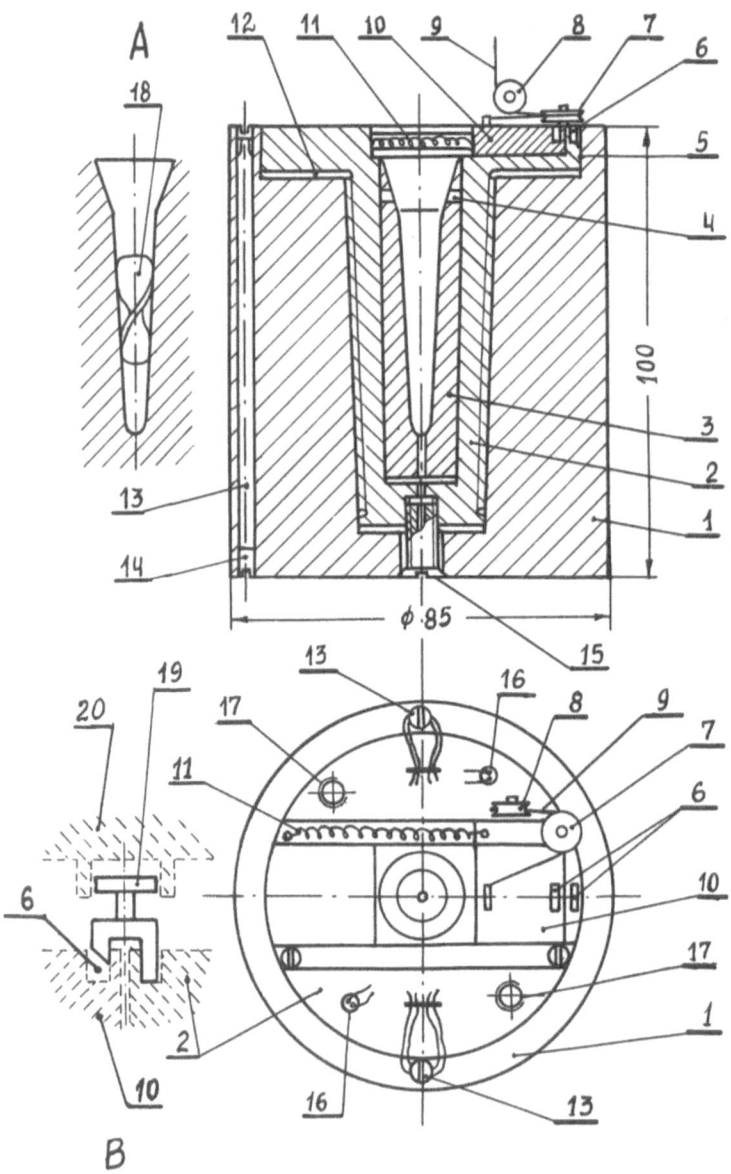

FIGURE 6. Calorimetric system: elevation and plan view.

insert member 2. The purpose of the copper strips is to eliminate heat dissipation through the potential and current wires of the heater when calibrating the calorimeter.

The cone-shaped surfaces of block 1 and insert member 2 are snug lapped to tightly fit one another, being tightened by screw 15 so as to provide for a sure thermal contact between these parts. It is for this same purpose that the annular gap 5 between the cylindrical surfaces of parts 1 and 2 is filled with a fusible Wood's alloy.

Two diametrically opposite apertures 17 having a screw thread are used in those cases where one has to remove the copper block from the jacket, or take insert member 2 from block 1.

The cone-shaped insert member 2 is made to accommodate a copper part 3 designed to receive a dropping specimen. The receiver has two diametrically opposite apertures 4 adapted for removing receiver 3 with a special grip. The touching conical surfaces of receiver 3 and insert member 2 are thoroughly lapped, thus providing for good thermal contact with one another. A hole, 0.2 mm in diameter, is machined through the bottoms of receiver 3, insert member 2, and screw 15, with a noble gas passing via the hole when receiver 3 is being fitted into insert member 2 or a molten metal specimen being dropped into receiver 3.

Figure 6 shows cover 10 of specimen receiver 3 opened, wherein two springs 11 are stretched, and bosses of lock 19 rest within the recesses in insert member 2 and cover 10; an enlarged view of lock 19 appears in Fig. 6 in a separate sketch. Figure 6 is helpful in enabling one to imagine more graphically the interaction between the parts upon the opening of cover 10 of the massive block. With the aid of a flexible thread 9 passing through the guiding rollers (7 and 8), cover 10 is drawn from the extreme left position to the right-hand side. At the same time, springs 11 are being stretched, and cover 10 is pressing upon the bevel of the left boss of lock 19, thus lifting it upward along the guides made in grooves 6 and in the calorimeter jacket cover 20. In the extreme right-hand position of cover 10, the left boss of the lock is forced by gravity down into groove 6 in cover 10, holding the lock open. On energizing the electromagnet (see 11 in Fig. 2), lock 19 will be drawn upward releasing cover 10, and the latter will be forced by springs 11 to shift to the left, shutting the receiver in the copper block.

The dropping of a molten specimen into the calorimeter had been observed to cause splattering of metal upon hitting the bottom of receiver 3. To eliminate the trouble, use was made of a helical copper insert 18 shown in Fig. 6, section A. Insert 18 imparts spin to the falling liquid metal, pushing the melt toward the side surfaces in receiver 3. This gadget is helpful in avoiding undesirable direct impacts of a metal melt upon the bottom of receiver 3 which caused the splattering of metal, while the increase in the area of contact between the spinning metal and the walls of receiver 3 facilitated metal cooling.

The side surfaces of insert 18 were painstakingly fitted with those of receiver 3 to ensure reliable thermal contact between them. No interaction between copper parts of the calorimetric system and melts of molybdenum, niobium, vanadium, rhodium, ruthenium, or other metals had been observed.

7. RESISTANCE THERMOMETERS

Copper resistance thermometers[36] were employed to measure temperature variations in the calorimetric system. The major advantage offered by the new design of these temperature sensors is that one can easily achieve a combination of high accuracy and low inertia in measurements. Furthermore, the electrical resistance of copper is a linear function of temperature, and the cost of a copper resistance thermometer is substantially lower than that for platinum resistance thermometers or other types of temperature transducers.

In deciding on what type of thermometer should be used—portable or stationary—preference was given to a portable thermometer because it can be directly calibrated at 0 °C and 100 °C, to check on the stability of readings with time.

An important parameter of temperature sensors is their thermal lag, which can account for an error of method in measuring temperature variations in the calorimeter. To avoid such inaccuracy one tries to design calorimetric temperature sensors with the lowest thermal lag possible. However, in each particular case this problem is being attacked from different sides. Incidentally, we have elaborated an original design of a low lag resistance thermometer intended for use in metal-block calorimeters. The design of such a thermometer is illustrated in Fig. 7. The body 1 of the copper resistance thermometer is in fact a copper screw 95-mm long with a metric M4 screw thread over the entire length thereof. A copper wire 3, 0.07 mm in diameter, having an enamel insulation, is layed into the recesses of the screw thread in a bifilar manner. Screw 1 is screwed into the threaded hole made in copper block 2. In order not to damage wire 3 when screwing in screw 1, the crests of the thread were trimmed by 0.15 mm with the aid of a drill. Section A of the resistance thermometer, shown on an enlarged scale in Fig. 7, presents the mutual position of the massive block 2, body 1, and wire 3 of the thermometer.

At the upper part of body 1 there are two flats ending in a cylindrical neck portion 6 and then in screw head 5 made to receive a screwdriver. Screw head 5 is arranged to screw the thermometer into the copper block 2. Junctions 7 between the thermometric wires and the potential and current terminals of the resistance thermometer are insulated with mica and glued onto the flats on body 1. The terminals are made of 0.3-mm copper wire, being insulated with fluoroplastic cembric tubes and fastened onto the neck portion 6 of the

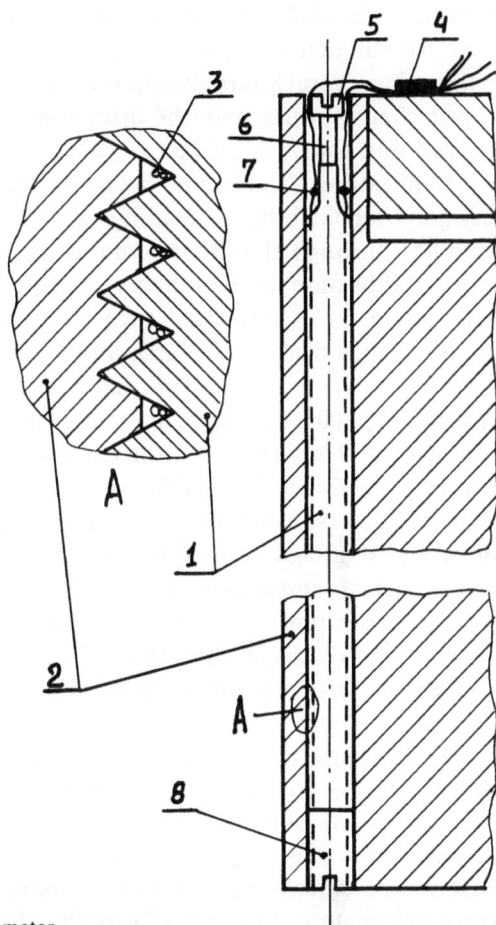

FIGURE 7. Calorimetric resistance thermometer.

screw with the aid of a kapron thread. In addition, the wires are pressed to the end surface of the copper block 2 by means of fixture 4 so as to reduce heat dissipation through the wires. Thermometer body 1 is fixed from below with a tightening screw 8 arranged to fix the position of the thermometer body within block 2, reducing to a minimum the contact thermal resistance between them. Assuming good thermal contact between the massive block and the body of the thermometer, the thermal lag of the thermometer will be mainly determined by the thermal resistance between the thermometric wire 3 and body 1 of the thermometer, since the heat capacity of the thermometer wire *per se* is low.

A number of copper resistance thermometers have been fabricated with the temperature-sensitive elements taken from a single batch of wire. The electric resistance of each of the thermometers at 0 °C was roughly about 11 Ω.

Prior to being calibrated, these thermometers were exposed to thermal treatment: for 300 hours they were kept at 100 °C, after which they were immersed 15 times for several hours alternately in ice and boiling water. Upon completion of such thermal treatment, the insulation resistance of the thermometers were checked and found to exceed 200 MΩ.

Calibration of the thermometers made at 0 °C and 100 °C has demonstrated very good repeatability of readings. A measuring electric current of 5 mA was set to flow through all of the thermometers, being maintained at this level both during calibration and enthalpy measurements. The electric resistance of thermometers was measured with the aid of a DC potentiometer having a measurement error of ±0.002%, and 10-ohm standard resistance coil having a certified accuracy of ±0.01%. Each resistance thermometer was placed into a glass tube, 300 mm long and 5 mm inside diameter. Ten measurements were conducted at 0 °C for each of the ten test runs. Each test run was accompanied by filling the Dewar flask with a new batch of ice, and changing the position of thermometers inside the flask. A single measurement in each test run lasted for no less than 3–4 hours. The random root-mean-square deviation given by a single measurement in each test run or series for all the thermometers was found to be almost the same and equal to $S = [\sum (R - \bar{R})^2/(10 - 1)]^{0.5} = 8 \times 10^{-6}$ ohm corresponding to 2×10^{-4} °C. The confidence interval of a random error for a root-mean-square value of nine test runs for each of the thermometers was equal to $\bar{S} = t_s[(R - \bar{R}_{ser})^2/9.8]^{0.5} = \pm 7 \times 10^{-5}$ ohm, which is almost consistent with the reproducibility of the melting point for ice, i.e., ±0.001 °C (t_s is Student's coefficient for a 95% confidence level).

The boiling point of water was established in five test runs, wherein five measurements of electric resistance were made in each test run. As a result of these measurements we calculated temperature coefficients of resistance for five thermometers; their values were found to be within the range 0.4267×10^{-2} to 0.4269×10^{-2} K^{-1}.

Two resistance thermometers were mounted in the copper block of the calorimeter. These were placed at diametrically opposite positions at equal distances from the axis of the copper block, being spaced several millimeters from its external cylindrical surface. The small distance from the thermometers to the outer surface of the copper block enabled one to measure the temperature on this surface, the knowledge of which was required for computing heat exchange with the jacket of the calorimeter. The use of two thermometers in order to calibrate the calorimeter and measure the enthalpy dramatically increases the "survivability" of the calorimeter, since in the case of failure of one of the thermometers the other can be used to continue measurements. Moreover, all of the results obtained with two thermometers enable one to check, at regular intervals, the functioning of the calorimeter in the course of experiments.

8. CALORIMETER CALIBRATION

Calibration of a calorimeter requires one to determine its thermal value

$$H = (Q/\Delta t)_{\text{grad}} \tag{1}$$

where Q is the known quantity of heat introduced into the calorimetric system with the aid of a calibration electric heater; while Δt is the temperature increment within the calorimetric system. If we count the temperature increase in the calorimetric system from the same initial temperature of the main period of a calorimetric experiment, for example, 22 °C in the calibration and enthalpy-measurement runs, then the heat quantity Q_{spec} introduced by a specimen under investigation will be determined from the measured increase in temperature Δt_{spec} in the calorimetric system:

$$Q_{\text{spec}} = H \cdot \Delta t_{\text{spec}} = (Q/\Delta t)_{\text{grad}} \cdot \Delta t_{\text{spec}} \tag{2}$$

Bearing in mind that expression (2) includes the ratio $\Delta t_{\text{spec}}/\Delta t_{\text{grad}}$, temperature variations in the calorimetric system can be measured in units of electric resistance R, since the latter is directly proportional to temperature in the case of copper resistance thermometers. Then expressions (1) and (2) can be replaced by

$$Q_{\text{spec}} = (Q/\Delta R)_{\text{grad}} \cdot \Delta R_{\text{spec}} = H \cdot \Delta R_{\text{spec}} \tag{3}$$

Accordingly, electric resistance was substituted for temperature in all the formulas[37] employed for calculating the results of a calorimetric experimental run shown schematically in Fig. 8:

$$\Delta R = (R_n - R_0) + \delta R = (R_n + \delta R) - R_0 = R'_n - R_0 \tag{4}$$

$$(dR/d\tau)_0 = (R_0 - R_1)/(\tau_0 - \tau_1) \qquad (dR/d\tau)_n = (R_2 - R_n)/(\tau_2 - \tau_1) \tag{5}$$

$$k = -\frac{(dR/d\tau)_n - (dR/d\tau)_0}{R_n - R_0} \tag{6}$$

$$R_k = \frac{(dR/d\tau)_n + (dR/d\tau)_0}{2k} + \tfrac{1}{2}(R_n + R_0) \tag{7}$$

$$\delta R = k\left[\int_{\tau_0}^{\tau_n} R \, d\tau - R_k(\tau_n - \tau_0)\right] \tag{8}$$

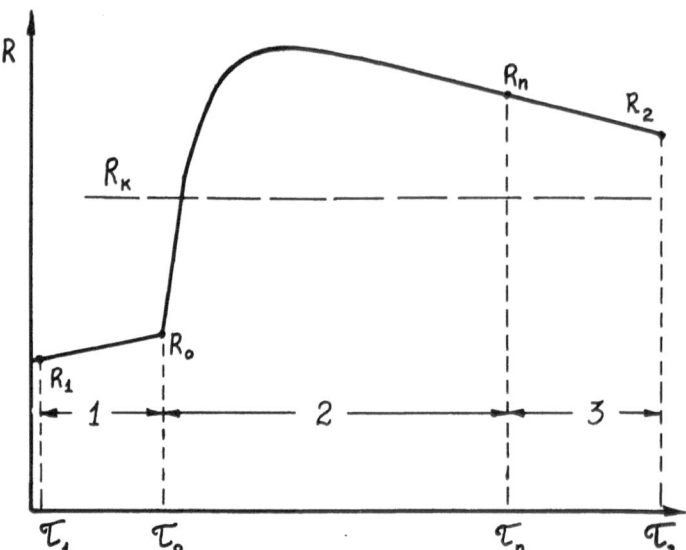

FIGURE 8. Diagram of calorimetric experiment.

where R_0 and R_n are electric resistances of the thermometer at the beginning (τ_0) and end (τ_n) of the main period of the calorimetric experiment, R_k is the electric resistance corresponding to the convergence temperature, k is the rate of cooling of the calorimetric system, δR is the correction factor for the heat exchange of the calorimetric system with a jacket, ΔR is the variation in electric resistance of the thermometer with due account of heat exchange between the calorimetric system and the jacket, while $(dR/d\tau)_0$ and $(dR/d\tau)_n$ are the rates of change of electric resistance at the initial and final periods of the calorimetric experiment.

In order to simulate the conditions of experiments related to measurements of enthalpy in calibrating the calorimeter, the conditions of heat introduction have been varied substantially with the aid of a calibrating electric heater. Thus, for example, electric power released in the calibrating heater was varied from 4.8 to 36 W and temperature increments in the calorimetric system were changed from 2 to 8 degrees; operating times of the calibrating heater varied from 5 to 42 min; and the rates of heating of the calorimetric system were changed within the range of 0.2-1.5 deg min^{-1}. In carrying out calibrations, and then in measuring the enthalpies, we have made provision in advance for such a convergence temperature $t_k(R_k)$ that correction for heat exchange between the calorimetric system and its jacket according to equation (8) would not exceed $\delta R/\Delta R = 3$-5%. All in all we conducted 24 calibration experiments directed toward establishing the heat value of the calorimeter. From the experimental results obtained we have calculated through a least-squares adjustment the coefficients of the approximating equation for the relationship between the heat value of the calorimeter and variation in electric resistance for the two thermometers:

$$H_1 = 40634 + 217 \, \Delta R \qquad J \, \Omega^{-1} \qquad (9)$$

$$H_2 = 40748 + 221 \, \Delta R \qquad J \, \Omega^{-1} \qquad (10)$$

The maximum departure of experimental quantities from the straight lines calculated with equations (9) and (10) did not exceed 0.1%.

As was mentioned above, the temperature of the beginning of the main period t_0 must be the same in the calibration experiments and in the enthalpy-measurement experiments, equal to 22 °C, since the heat value of the calorimeter is the value of a mean heat capacity of the massive copper block, which can be counted depending on the initial temperature. To violate this condition would cause experimental errors. It has been recommended[38] that t_0 be maintained stable with an accuracy of no less than ±0.01 °C without giving any grounds for this quantity. In practice, to meet the above recommendation would often involve certain difficulties calling for a much longer duration of the calorimetric experiment. In this connection it is of interest to discuss[39] how the violation of condition $t_0 = idem$ would affect the accuracy in determining heat in a calorimetric experiment, and how to estimate a correction factor associated with the departure from $t_0 = idem$ when carrying out high-precision investigations. To that end, we express the heat value of the calorimetric system through its heat capacity C_p, mass m, and temperature variation with due account of the correction factor for the heat exchange, $\Delta t = t'_n - t_0$:

$$H = \frac{Q}{\Delta t} = \frac{m \int_{t_0}^{t'_n} C_p \, dT}{\Delta t} \qquad (11)$$

Bearing in mind that in calorimetric experiments it is common that $\Delta t \leqslant 5\text{-}6$ °C, it is clear that within such a small temperature range taken at room temperature, the true heat capacity of the calorimetric system (copper, aluminum and the like) versus temperature can be regarded as a linear function:

$$C_p = A + bt \qquad (12)$$

Then it can be inferred from equations (11) and (12) that

$$H = \frac{m}{\Delta t} \int_{t_0}^{t'_n} (a + b)\, dt = m\left[(a + bt_0) + \frac{b}{2}(t'_n - t_0)\right] = A + B\,\Delta t \quad (13)$$

This latter expression shows that the heat value of a calorimeter is a linear function of temperature t_0 for fixed Δt. If the temperature at the beginning of the main period t_{01} in the experiment differs from temperature t_0, which was maintained when calibrating the calorimeter, than for an identical $\Delta t = t'_n - t_0 = t'_{n1} - t_{01}$ one can find a correction to the heat value of the calorimeter:

$$\Delta H = mb(t_{01} - t_0) = 2B(t_{01} - t_0) \tag{14}$$

If the value of C_p for the calorimetric system is growing with temperature, this correction is positive for $t_{01} > t_0$, but negative for $t_{01} < t_0$. Equation (14) enables one to estimate the relative error resulting from the violation of condition $t_0 = idem$:

$$\Delta H/H = mb(t_{01} - t_0)/m\bar{C}_p|_{t_0}^{t'_n} = b(t_{01} - t_0)/\bar{C}_p|_{t_0}^{t'_n} \tag{15}$$

where $\bar{C}_p|_{t_0}^{t'_n}$ is a mean specific heat capacity of the calorimetric system within the temperature range $t'_n - t_0$. Assuming that the temperature deviation at the beginning of the main period is $t_{01} - t_0 = 1\,°C$, then from available reference data given elsewhere in the literature on heat capacity one can estimate a relative error in calculating the heat value $\Delta H/H$ for copper, aluminum, and silver calorimeters as 0.037%, 0.075%, and 0.024%, respectively.

9. SEQUENCE OF EXPERIMENTAL STEPS

Interaction between separate units of the chamber and calorimeter is of great importance while carrying out an experiment. Let us consider the sequence of steps in conducting an experiment. With chamber cover 26 (Fig. 2) removed, cover 9 of the copper block 8 is closed as well as cover 37 of the calorimeter jacket. Then specimen 20 of the metal under investigation is placed in inductor 21, in which the specimen is supported from below by bracket 19 of manipulator 30. Prior to starting an experiment, specimen 20 is weighed. Disk 16 is set aside in an extreme position where casting mold 17 is positioned under inductor 21. The temperature of copper block 8 is preset somewhat lower than 22 °C, from which temperature we have measured the beginning of the main period of the calorimetric experiment when calibrating the calorimeter. The temperature level of liquid in thermostat 1 is selected so that the correction factor for heat exchange calculated with formula 8 would not exceed 3-5% of the heat quantity introduced into the calorimeter by the specimen.

Having finished these preparatory operations, cover 26 is returned to its original place, the pressure-tight volume of the apparatus is pumped down to $10^{-3}-10^{-4}$ Pa, then filled with a noble gas to a pressure somewhat higher than atmospheric. When investigating metals[18,19] which actively absorb gaseous impurities, the noble gas is purified with a getter in the form of a strip of titanium heated by an electric current. Recording of the initial period of the calorimetric experiment is started and the high-frequency generator is switched on. The inductor starts interacting with the specimen to heat it and levitate it. Bracket 19 of the manipulator is drawn aside and this step is monitored through window 25. The temperature of the specimen is raised over several minutes up to the required level by changing the power applied to the inductor from the high-frequency generator. The temperature of specimen 20 is measured by an optical pyrometer through a total internal reflection prism 24. Several seconds before dropping specimen 20 into the calorimeter, cover 37 of the calorimeter jacket and cover 9 of the copper block 8 are opened with the aid of two flexible threads, which are pulled upward by bracket 19 of the manipulator, by link 15, and lever 13. Covers 9 and 37 are fixed in an open position with appropriate mechanical locks. In order to eliminate the possibility of an emergency should a red-hot specimen or, which is more dangerous, a drop of melt fall onto covers 37 or 9, which had failed for some reason to be opened, provisions are made for an optical check device shown in Fig. 9. Continuous monitoring of the covers is effected in the following manner. Light from lamp 2 is passed through a total internal reflection prism 4 fastened on disk 1 further onto a miniature mirror 7 on cover 8 of jacket 9 of the calorimeter or a mirror on cover 5 of copper block 6, being reflected from one of the mirrors back to photoresistor 3. The positions of covers 7 and 8 can be judged from the level of a signal as read from the indicator.

With completion of the last measurement of the temperature of a specimen, disk 16 (Fig. 2) is brought to the second extreme position where the aperture in the disk is positioned right below inductor 21; this step is done with the aid of bracket 19 of the manipulator or a micromotor with a gear. At the same time the high-frequency generator is switched off, the specimen starts dropping into the calorimeter, and a system of time delay relays and electric switches is energized. The time taken for a specimen to drop from the inductor into the calorimeter is 0.19 s, and the time delay relay being energized operates in 0.30 s, closing the electric circuit of electromagnet 11 for a period of about one second, thus lifting lock 10. Having been released from lock 10, cover 9 of the copper block 8 is forced by the spring to immediately close the receiver

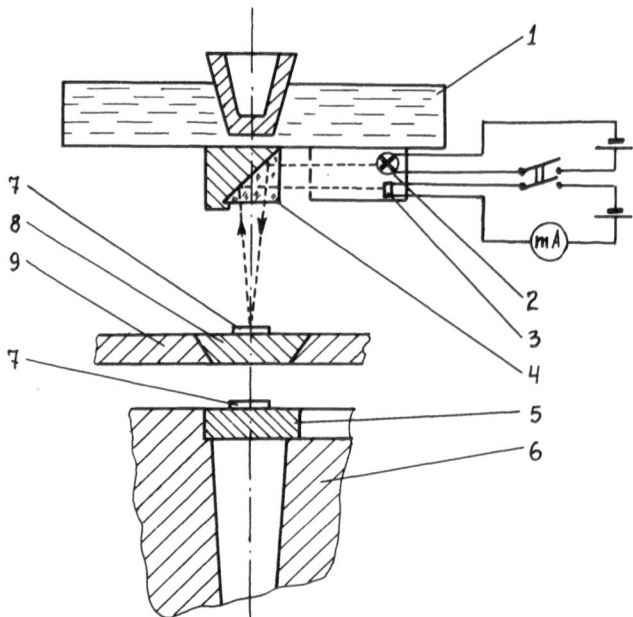

FIGURE 9. System for controlling the position of covers on the copper block and on the jacket of the calorimetric system.

in the copper block once a specimen has dropped into the receiver. At the end of its travel cover 9 pushes by its boss the stop at cover 37 of the calorimeter jacket, the cover of which is thus also closed. Covers 9 and 37 must be closed during the entire experiment when measuring enthalpy, except for a few seconds when a specimen is ready to be introduced into the calorimeter. This feature eliminates uncertainty in calculating heat exchange between the calorimetric system and the jacket, preventing the loss of heat from the specimen dropped into the copper block. Recording of the main and final periods of the calorimetric experiment is started from the instant the specimen falls into the calorimeter.

Upon completion of the experiment, cover 26 of the chamber, cover 37 of the calorimeter jacket, and cover 9 of the copper block are again opened. To remove a specimen dropped into the copper block, a special-purpose device in the form of a tube is lowered along the axis of the apparatus through inductor 21, and reaches receiver 3 with the specimen (see Fig. 6). There is a slot at the end of the tube to which a movable cone member is fitted. By pulling the cone member upward, the ends of the split tube are pressed against the inner walls of receiver 3 thus enabling one to remove the receiver together with the specimen from the copper block and to measure the mass of the specimen after the experiment.

NOTATION

K	Degrees Kelvin
°C	Degrees Celsius
$R, \bar{R}, \bar{R}_{ser}$	Electric resistance and its mean value as measured in a single series or several series, or test runs
S	Random root-mean-square deviation of individual measurement
$\bar{S}$	Random root-mean-square deviation of arithmetic mean
t_s	Student's coefficient
$t_0 (R_0)$	Temperature (electric resistance) at the beginning of the main period of a calorimetric experiment
$t_n (R_n)$	Temperature (electric resistance) at the end of the main period of a calorimetric experiment
τ_0, τ_n	Time of the beginning and end of the main calorimetric experiment
$t_k (R_k)$	Convergence temperature (resistance)
k	Rate of cooling of a calorimetric system
$\delta t (\delta R)$	Correction factor for heat exchange between the calorimetric system and jacket
$t'_n = t_n + \delta t,$ $R'_n = R_n + \delta R$	Final temperature (electric resistance) of the calorimetric system with due account of heat exchange
$(dt/d\tau)_0, (dR/d\tau)_0$	Rate of temperature change, or of electric resistance, during the initial period of a calorimetric experiment
$(dt/d\tau)_n, (dR/d\tau)_n$	Rate of temperature change, or of electric resistance, during the final period of a calorimetric experiment
H	Calorimeter heat value
Q_{grad}, Q_{spec}	Amount of heat introduced into a calorimetric system by a calibration heater or by a specimen under investigation
$t_{grad} (R_{grad})$	Increment of temperature (electric resistance) in a calorimetric system with due account of correction for heat exchange in calibration experiments
$\Delta t_{spec} (\Delta R_{spec})$	Increment of temperature (electric resistance) in a calorimetric system with due account of correction for heat exchange in specimen-enthalpy measurement experiments
m	Mass of a calorimetric system
C_p	Heat capacity of a calorimetric system

REFERENCES

1. E. Arpaci and M.G. Frohberg, "Enthalpy Measurements on Solid and Liquid Tungsten by Levitation Calorimetry," *Z. Metallkd.* **75**(8), 614–618 (1984).
2. V.Ya. Chekhovskoi, A.E. Sheindlin, and B.Ya. Berezin, "Enthalpy Measurements by Drop Calorimetry Using Electromagnetic Levitation," *High Temp. High Pressures* **2**, 301–307 (1970).
3. B.Ya. Berezin, V.Ya. Chekhovskoi, and A.E. Sheindlin, "Enthalpy of Solid and Liquid Molybdenum by Levitation Calorimetry. Heat of Fusion of Molybdenum," *High Temp. High Pressures* **3**, 287–297 (1971).
4. V.Ya. Chekhovskoi and B.Ya. Berezin, "Experimental Investigation of Enthalpy of Molybdenum in Solid and Liquid States," in: *Thermophysical Properties of Solid Substances*, pp. 123–134, "Nauka" Publishers, Moscow (1973).
5. B.Ya. Berezin, V.Ya. Chekhovskoi, and S.A. Kats, "The Experimental Results of Heats of Fusion of Refractory Metals," in: *Proc. 4th Int. Conference on Chemical Thermodynamics*, Vol. 9, pp. 77–84, Montpellier, France (1975).

6. J.A. Treverton and J.L. Margrave, "Thermodynamic Properties of Liquid Molybdenum by Levitation Calorimetry," in: *Proc. of 5th Symposium on Thermophysical Properties*, pp. 489-494, ASME, New York (1970).

7. D.W. Bonnel, J.A. Treverton, A.J. Valerga, and J.L. Margrave, "The Emissivities of Liquid Metals at Their Fusion Temperatures," in: *Temperature, Its Measurements and Control in Science and Industry*, Vol. 4, Part 1, pp. 483-487, Pittsburgh Instrument Society of America (1972).

8. B.Ya. Berezin, S.A. Kats, and V.Ya. Chekhovskoi, "Spectral Emissivity of Liquid-State Refractory Metals," (in Russian), *Teplofiz. Vys. Temp.* **14**, 497-502 (1976).

9. B.Ya. Berezin and V.Ya. Chekhovskoi, "Enthalpy and Heat Capacity of Niobium and Vanadium Over Temperature Range from 298.15 K to Temperatures of Fusion" (in Russian), *Teplofiz. Vys. Temper.* **14**, 772-778 (1977).

10. A.E. Sheindlin, B.Ya. Berezin, and V.Ya. Chekhovskoi, "Enthalpy of Niobium in the Solid and Liquid State," *High Temp. High Pressures* **4**, 611-619 (1972).

11. B.Ya. Berezin, V.Ya. Chekhovskoi, and A.E. Sheindlin, "Enthalpy and Specific Heat of Molten Vanadium," *High Temp. Sci.* **4**, 478-485 (1972).

12. B.Ya. Berezin, V.Ya. Chekhovskoi, and A.E. Sheindlin, "Heat of Melting of Vanadium" (in Russian), *Dokl. Akad. Nauk SSSR, Tech. Fiz.* **201**(3), 583-586 (1971).

13. J.A. Treverton and J.L. Margrave, "Thermodynamic Properties by Levitation Calorimetry. III. The Enthalpies of Fusion and Heat Capacities for the Liquid Phase of Iron, Titanium and Vanadium," *J. Chem. Thermodyn.* **3**, 473-481 (1971).

14. A.E. Sheindlin, S.A. Kats, B.Ya. Berezin, V.Ya. Chekhovskoi, and M.M. Kenisarin, "Some Thermophysical Properties of Ruthenium in the Neighbourhood of the Melting Point," *Rev. Int. Hautes Temp. Refract.* **12**, 12-15 (1975).

15. S.A. Kats, B.Ya. Berezin, N.B. Gorina, V.P. Polyakova, E.M. Savitski, and V.Ya. Chekhovskoi, "Enthalpy and Capacity of Liquid Ruthenium" (in Russian), *Izv. Akad. Nauk SSSR, Metally* (6), 87-89 (1974).

16. S.A. Kats, V.Ya. Chekhovskoi, and A.E. Sheindlin, "Investigation of Temperature Dependence of Emissivity of Refractory Metals," in: *Thermophysical Properties of Substances at High Temperatures*, pp. 81-84, Institute for High Temperatures, USSR Academy of Sciences, Moscow (1978).

17. A.K. Chaundhuri, D.W. Bonnell, L.A. Ford, and J.L. Margrave, "Thermodynamic Properties by Levitation Calorimetry. I. Enthalpy Increments and Heats of Fusion for Copper and Platinum," *High Temp. Sci.* **2**, 203-212 (1970).

18. B.Ya. Berezin, V.Ya. Chekhovskoi, S.A. Kats, and M.M. Kenisarin, "Some Thermophysical Properties of Titanium in the Neighbourhood of the Melting Point," in: *Proc. of the 6th Symposium on Thermophysical Properties*, August 5-8, 1973, Atlanta, pp. 263-273, ASME (1973).

19. B.Ya. Berezin, S.A. Kats, M.M. Kenisarin, and V.Ya. Chekhovskoi, "Heat and Temperature Melting of Titanium" (in Russian), *Teplofiz. Vys. Temp.* **12**, 524-529 (1974).

20. J.A. Treverton and J.L. Margrave, "Levitation Calorimetry. IV. Thermodynamic Properties of Liquid Cobalt and Palladium," *J. Phys. Chem.* **75**, 3737-3740 (1971).

21. D.W. Bonnell and J.L. Margrave, "High Temperature Thermodynamic Data for Liquid Metals by Levitation Calorimetry," in: *Proc. 3rd Int. Conference on Chem. Thermodyn, Jointly with the Symposium Phys.-Chem. Tech. at High Temp.*, Austria, September 3-7, 1973, THT 4/5, pp. 105-111 (1973).

22. S.A. Kats, V.Ya. Chekhovskoi, N.B. Gorina, V.P. Polyakova, and E.M. Savitski, "Thermodynamic Properties of Rhodium in the Neighbourhood of the Melting Point," in: *Trans. 7th All-Union Sci. Conference on Calorimetry*, January 31-February 3, 1977, Moscow, Chernogolovka, pp. 351-357 (1977).

23. S.A. Kats, "Thermophysical Properties of Some Refractory Metals in Solid and Liquid States in the Neighbourhood of Their Melting Points," Diss. Abstr., Institute for High Temp., USSR Academy of Sciences, Moscow (1978).

24. S.A. Kats, V.Ya. Chekhovskoi, N.L. Korenovski, V.P. Polyakova, and E.M. Savitski, "Some Thermophysical Properties of Iridium in the Neighbourhood of the Melting Point," in: *Rare Metals and Alloys with Monocrystalline Structure*, pp. 203–210, "Nauka" Publishers, Moscow (1981).

25. V.Ya. Chekhovskoi and S.A. Kats, "Investigation of the Thermophysical Properties of Refractory Metals Near the Melting Point," *High Temp. High Pressures* 13, 611–616 (1981).

26. B.Ya. Berezin and V.Ya. Chekhovskoi, "Enthalpy of Tantalum in the Temperature Range from 2400 K to the Melting Point" (in Russian), *Izv. Akad. Nauk SSSR, Metally* (3), 63–65 (1977).

27. R.L. Montgomery, P.C. Sundareswaren, D.W. Ball, and J.L. Margrave, "Thermodynamic Properties by Levitation Calorimetry—V. High Temperature Heat Content of Liquid Gallium," *Int. J. Thermophys.* 5, 161–175 (1984).

28. G. Betz and M.G. Frohberg, "Enthalpy Measurements on Solid and Liquid Molybdenum by Levitation Calorimetry," *High Temp. High Pressures* 12, 169–178 (1980).

29. G. Betz and M.G. Frohberg, "Enthalpy Measurements on Solid and Liquid Niobium by Means of Levitation Calorimetry," *Z. Metallkd* 71, 451–455 (1980).

30. S.A. Kats and V.Ya. Chekhovskoi, "Enthalpies of Fusion of Metallic Elements," *High Temp. High Pressures* 11, 629–634 (1979).

31. V.Ya. Chekhovskoi, "Levitation Calorimetry," in: *Compendium of Thermophysical Property Measurement Methods*, Vol. 1, pp. 555–589, Plenum Press, New York and London (1984).

32. L.N. Latyev, V.Ya. Chekhovskoi, and E.N. Shestakov, "Determination of the True Temperature of Substances Having Continuous Radiation Spectrum" (in Russian), *Metrologiya* (1), 35–41 (1982).

33. L.N. Latyev, V.Ya. Chekhovskoi, and E.N. Shestakov, "Determination of Temperature by Contactless Method," in: *Methods and Apparatuses in Optical Pyrometry*, pp. 14–20, "Nauka" Publishers, Moscow (1983).

34. V.Ya. Chekhovskoi and B.Ya. Berezin, "Experimental Investigation of Melting Heat Refractory Metals," in: *Collect. of Addresses of Sections "Thermodynamics of Phase Transitions, Flow and Irreversible Processes" and "Thermophysical Properties of Substances"* (All-Union Sci. Conference on Thermodynamics, July 4–8, 1968, Leningrad), pp. 379–382, Leningrad (1970).

35. V.Ya. Chekhovskoi and B.Ya. Berezin, "Experimental Unit for Measurement of Enthalpy Heat Capacity of Refractory Metals" (in Russian), *Teplofiz. Vys. Temp.* 8, 1320–1322 (1970).

36. B.Ya. Berezin and V.Ya. Chekhovskoi, "Quick-Response Resistance Thermometer for Massive Calorimeter" (in Russian), *Metrologiya* (3), 69–70 (1971).

37. V.A. Kirillin, A.E. Sheindlin, and V.Ya. Chekhovskoi, "Experimental Determination of Enthalpy of Corundum at Temperatures of 500 to 2000 °C" (in Russian), *Inzh.-Fiz. Zh.* 4(2), 3–17 (1961).

38. M.M. Popov, *Thermometry and Calorimetry*, Moscow University Press, Moscow (1954).

39. B.Ya. Berezin, L.A. Reshetov, V.D. Tarasov, and V.Ya. Chekhovskoi, "Effect of Initial Temperature of Main Period on Accuracy of Calorimetric Experiment by Method of Mixing" (in Russian), *Metrologiya* (7), 74–77 (1971).

17

A Millisecond-Resolution Pulse Heating System for Specific-Heat Measurements at High Temperatures

ARED CEZAIRLIYAN

1. INTRODUCTION

In the first volume of this Compendium,[1] a general description and a brief survey of the pulse techniques, utilizing resistive self-heating of the specimen, for the measurement of specific heat at high temperatures were given. The presentation covered techniques over a wide regime of time response: from nearly a second to submicrosecond resolution. Most of the techniques were of an exploratory nature and were developed for specific immediate applications. However, as a result of extensive research performed during the last 25 years, the millisecond-resolution pulse heating technique for specific-heat measurements has reached a mature stage.

The objective of the present chapter is to give details of the millisecond-resolution technique developed in the Dynamic Measurements Laboratory of the U.S. National Institute of Standards and Technology (NIST), formerly the U.S. National Bureau of Standards (NBS). The presentation includes a description of the measurement system, measurement procedure, consideration of various phenomena that affect the design and operation of the system, estimate of errors, and a discussion summarizing recent improvements and additions to the overall system.

The development of the millisecond-resolution system at the NIST began in the early 1960s. The system became operational in the late 1960s and has been improved continuously since then. The original system and its operation are described elsewhere.[2,3] A summary of the measurements performed with this system up to 1984 may also be found in the literature.[4]

ARED CEZAIRLIYAN • Thermophysics Division, National Institute of Standards and Technology, Gaithersburg, Maryland 20899, USA.

2. MEASUREMENT SYSTEM

The millisecond-resolution pulse heating system for specific-heat measurements consists of an electric pulse power circuit, associated measuring and control circuits, and various components and instruments. A functional diagram of the complete system and a photograph of a portion of the system are presented in Figs. 1 and 2, respectively. The major items of the system are described in the following subsections.

2.1. Pulse Power Circuit

The pulse power circuit includes the specimen in series with a battery bank, a standard resistor, an adjustable resistor, and a switching system. The battery bank consists of 14 series-connected 2-V batteries each having approximately 1100 A-h capacity. A standard resistor (1 mΩ) made of manganin strip is used for the measurement of the pulse current through the specimen. An adjustable resistor, made of water-cooled Inconel tubes (total resistance, 30 mΩ), enables control of the heating rate of the specimen and the shape of the current pulse. The switching system consists of two series-connected, fast-acting switches. The second switch is used as a backup in the event that

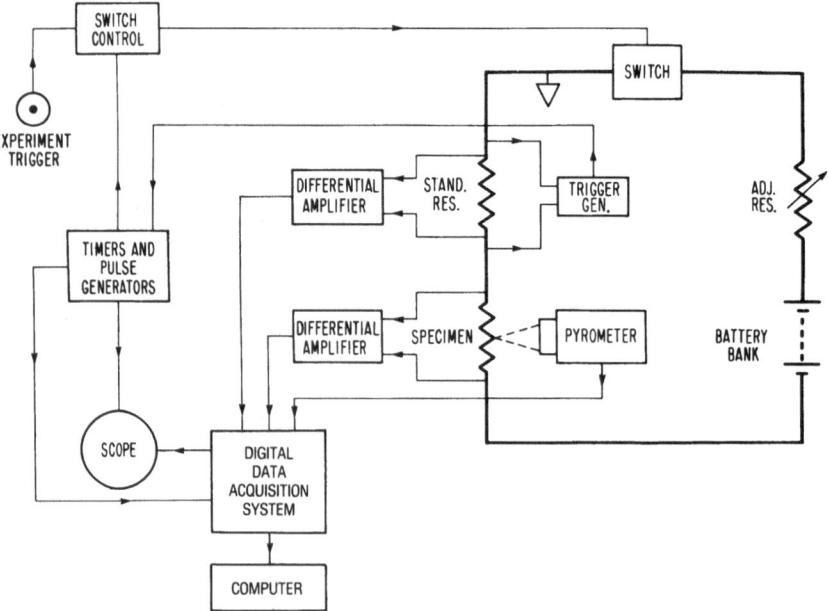

FIGURE 1. Functional diagram of the complete millisecond-resolution pulse heating system for specific-heat measurements.

FIGURE 2. Photograph of the experimental chamber and the high-speed pyrometer.

the first one fails to open at the end of the heating period. The switches are operated by pulse generators.

2.2. Measuring and Control Circuits

Voltage signals from both the standard resistor and the specimen are sent to the data acquisition system via differential amplifiers. A potentiometric system is used to calibrate pulse-current and pulse-voltage measuring circuits including the differential amplifiers and the data acquisition system. A 400-Hz synchronous motor used in the high-speed pyrometer provides the time base for the data acquisition system. Timing of various events, such as closing of the switches and triggering of various electronic equipment, during a pulse experiment is achieved by a series of pulse generators. Opening of the switches at the end of the desired heating period is automatically activated when the specimen temperature reaches the preadjusted value. A Kelvin bridge is used to measure the resistance of the specimen at ambient temperature before and after a pulse experiment. Triaxial cables are used for signals that are recorded, and coaxial cables for all the controls.

2.3. Specimen

The specimen is a tube of the following nominal dimensions: length, 76 mm; outside diameter, 6.3 mm; and wall thickness, 0.5 mm. During the earlier period of this project, somewhat longer (102 mm) specimens were used. In general, for maximum cross-sectional uniformity, the specimen is fabricated from a solid cylindrical stock by removing the center portion by an electro-erosion technique. The outer surface of the specimen is polished to reduce heat loss due to thermal radiation. About 12.7-mm-long portions at each end of the specimen are used for clamping the specimen to the current leads. Therefore, the portion of the specimen that undergoes heating is 51 mm in length. A small rectangular hole (1 × 0.5 mm) is fabricated in the wall at the middle of the specimen to approximate blackbody conditions for optical temperature measurements. In order to compensate for the cross-sectional nonuniformity created by the hole, a portion from the rest of the specimen is removed by grinding its surface. The criterion is that the cross section of the removed portion is equal to that of the opening created by the hole. From geometrical considerations this may be expressed as

$$ab = \frac{R^2}{2}\left(\frac{\pi\theta}{180} - \sin\theta\right) \tag{1}$$

where a is the width of the hole, b the depth of the hole (thickness of the tubular specimen), R the outer radius of the specimen, and θ is the angle subtended by the arc corresponding to the ground flat.

After solving the above transcendental equation for angle θ by iteration, the pertinent dimensions as shown in Fig. 3 may be obtained from the following relations:

$$L = 2R \sin\frac{\theta}{2} \tag{2}$$

and

$$C = R \cos\frac{\theta}{2} \tag{3}$$

A photograph of the specimen is shown in Fig. 4. The rectangular sighting hole in the specimen wall is fabricated 0.8 mm off-center to improve the blackbody quality.

Two narrow and shallow grooves (referred to as voltage probe marks) are fabricated on the specimen approximately 25 mm from its ends and on the opposite side of the blackbody hole for locating the voltage probes.

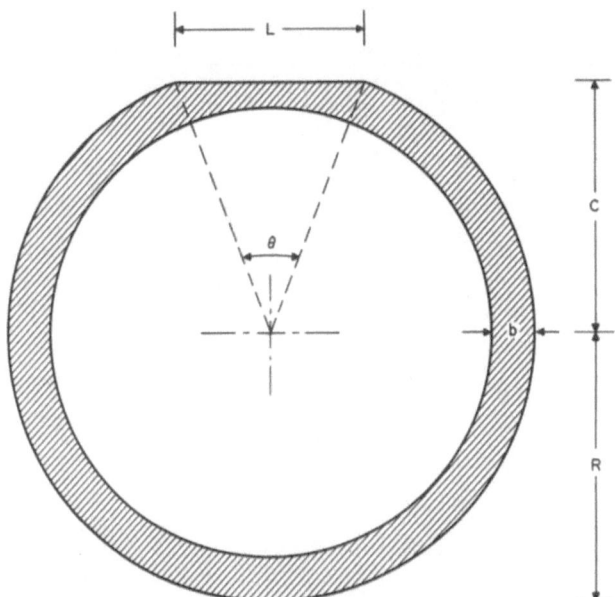

FIGURE 3. Specimen cross section at the plane of the flat.

2.4. Experimental Chamber

The experimental chamber, shown schematically in Fig. 5, contains the specimen, the clamping electrodes, an expansion joint, voltage probes, and other auxiliary components.

The specimen is mounted vertically about 6 mm off-center with respect to the axis of the chamber to reduce the effect of internal reflections. The inner wall of the chamber is coated with nonreflecting paint. The chamber wall, as well as the specimen clamps, are water-cooled. Thermocouples are connected (electrically insulated) to the two end clamps to measure the specimen temperature before each pulse experiment. An expansion joint allows expansion of the specimen in the downward direction.

The voltage probes are spring-loaded knife-edges made of the specimen material and are placed at a distance approximately 13 mm from the end

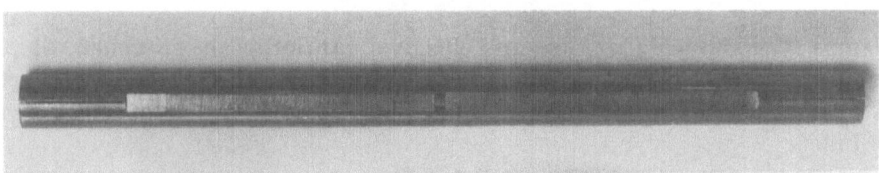

FIGURE 4. Photograph of the specimen.

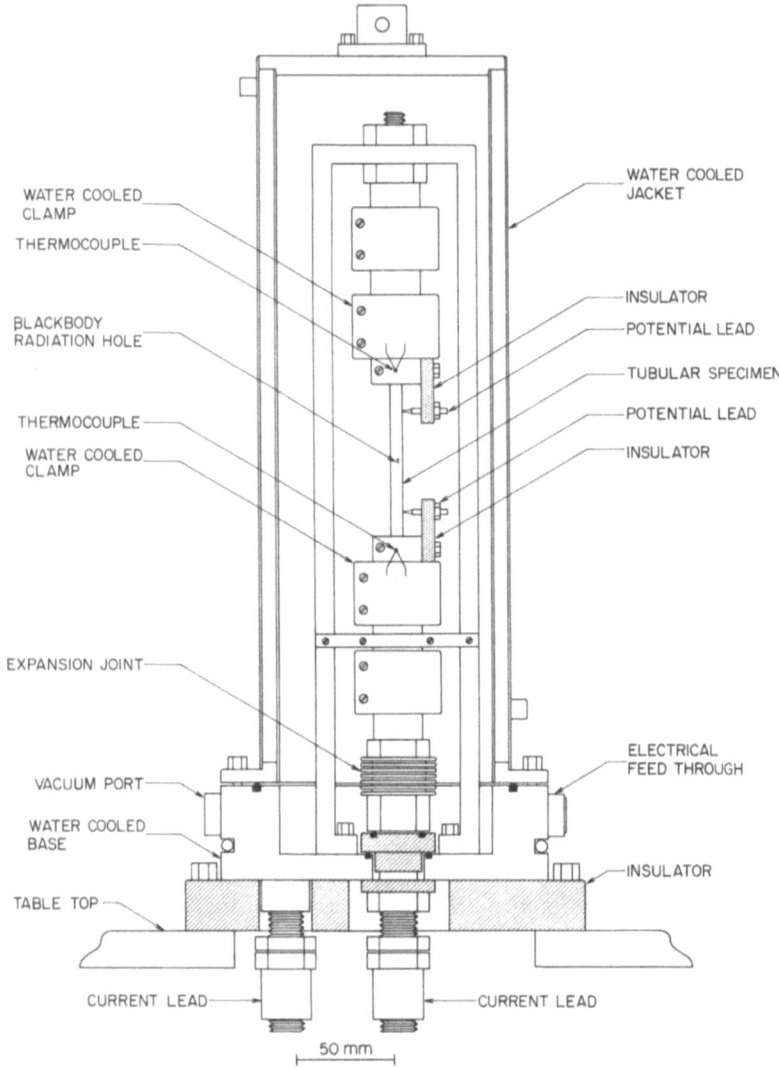

WATER COOLED CLAMP

THERMOCOUPLE

BLACKBODY RADIATION HOLE

THERMOCOUPLE

WATER COOLED CLAMP

EXPANSION JOINT

VACUUM PORT

WATER COOLED BASE

TABLE TOP

CURRENT LEAD

WATER COOLED JACKET

INSULATOR

POTENTIAL LEAD

TUBULAR SPECIMEN

POTENTIAL LEAD

INSULATOR

ELECTRICAL FEED THROUGH

INSULATOR

CURRENT LEAD

50 mm

FIGURE 5. Experimental chamber.

clamps. The knife-edges define an "effective" portion of the specimen, which should be free of axial temperature gradients for the duration of the experiment. A photograph of the specimen with the clamps and voltage probes is shown in Fig. 6.

The chamber is designed for conducting experiments with the specimen either in vacuum or in a controlled atmosphere.

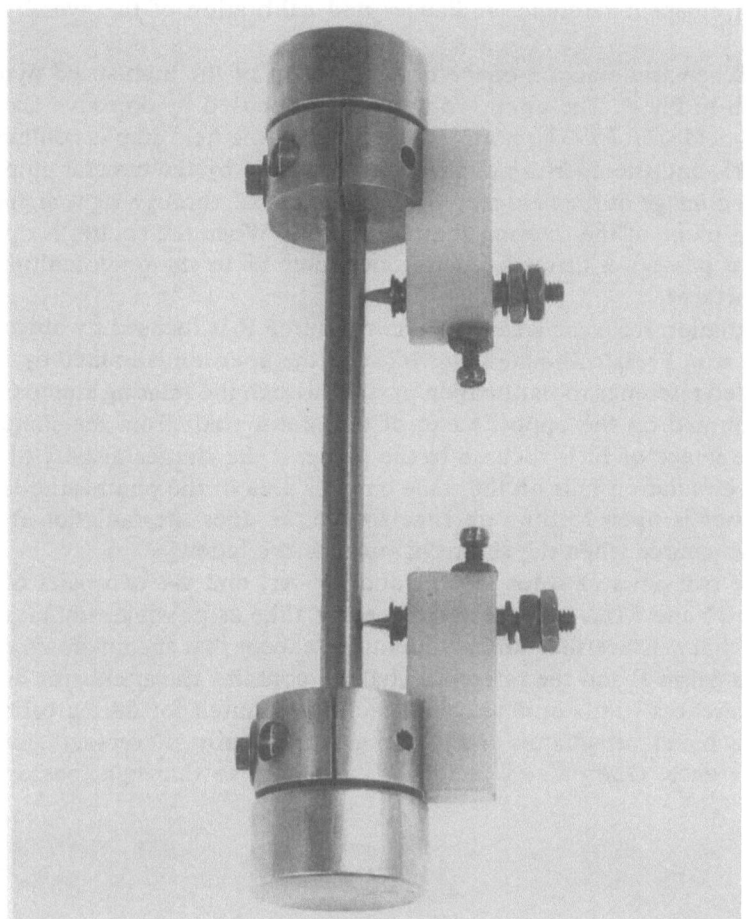

FIGURE 6. Photograph of the specimen, clamps, and the voltage probes.

2.5. High-Speed Pyrometer

The temperature of the specimen during a pulse experiment is measured with a high-speed photoelectric pyrometer designed and constructed specifically for this system.[5] The pyrometer permits specimen temperature evaluations at the rate of 1200 per second. The pyrometer alternately passes precisely timed samples of radiance from the specimen and a reference source through an interference filter to a photomultiplier. During each exposure, the photomultiplier output is integrated and recorded. Successive exposures to the reference source are taken through a sequence of three different optical attenuators mounted on a rotating disk, resulting in a staircase of reference exposures. The specimen temperature is determined from interpolation between adjacent reference radiances and calibration of the overall optical system.

A schematic diagram of the optical system of the high-speed pyrometer is shown in Fig. 7. The unknown target X is focused by objective Ox onto a circular field stop Fx. The portion which passes the field stop is collimated by lens Mx1, and the aperture of the system is fixed by the circular stop Ax. A magnified image of the field stop is formed by Mx2, through right-angle prism P, in the plane of the rotating shutter disk DS. When the shutter is open, the radiation passes on through interference filter IF to the photomultiplier PC (S-20 surface).

Radiation from the steady reference source R is focused by objective Or on field stop Fr, is collimated by Mr1, and the aperture is limited by Ar. The collimated reference radiation then passes through the rotating attenuator disk DA, mounted on the opposite end of the motor shaft from the shutter disk DS. The image of Fr is focused in the plane of the shutter disk by Mr2. The reference radiation falls on the same circular area of the photocathode, when the shutter is open to the reference source, as does the radiation from the unknown source when the shutter is open to the latter.

The two pairs of stops (Fx–Fr and Ax–Ar) and the two pairs of lenses (Mx1–Mr1 and Mx2–Mr2) are all as nearly alike as possible, so that the two optical channels are substantially identical, except that the unknown channel contains prism P and the reference channel contains the attenuator disk DA. The differences in the optical channels are accounted for during calibration.

The metal attenuator disk DA has three pairs of sectors around its circumference. One pair of sectors is cut away, so that light passes freely.

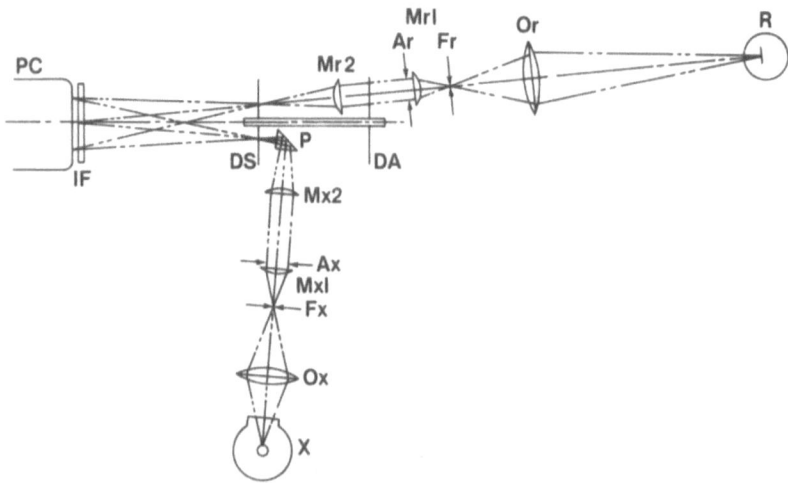

FIGURE 7. Schematic diagram of the optical system of the high-speed photoelectric pyrometer.

Another pair of sectors is perforated in a pattern of small holes, so as to transmit nominally half of the radiation which falls on the sector. A third pair of sectors is similarly perforated with smaller holes, so as to transmit nominally one-quarter of the radiation. The number of holes is made very large, so that any fluctuation in transmission during the rotation of the disk owing to the finite number of holes is negligible by comparison with the noise due to the finite number of photons in each sample. The effective transmittance of the attenuator is determined experimentally in a calibration procedure, accounting for any diffraction effects which occur. As the disk rotates, the reference illumination thus changes in three steps during each half-revolution. A film of transparent resin covers the perforated areas to prevent dirt from changing the effective size of the openings.

The shutter disk has six large apertures each subtending 15° at the center of the disk. It and the attenuator disk are rotated at 200 rps by a synchronous motor. The shutter disk opens the reference radiation path for 208 μs, then closes for 208 μs, then opens the unknown radiation path for 208 μs, closes for 208 μs, and then opens the reference path again. The reference pulses form two "staircases" per revolution of the sampling motor; the unknown pulses are interlaced with the reference pulses. The shutter disk also has a circle of timing apertures, which is used to generate photoelectrically the pulses to control the electronic system.

The optical system was designed to view a target circle 0.2 mm in diameter on the unknown (specimen) from a distance of 10 cm. The nominal wavelength passed by the interference filter is 650 nm, and the nominal bandwidth of the filter is 10 nm.

The pyrometer is calibrated with a gas-filled tungsten-strip lamp for measurements up to about 2500 K, the limit of reliable operation for such lamps. For measurements at higher temperatures, calibrated optical attenuators are placed in the path of the radiation from the specimen.

2.6. Digital Data Acquisition System

Electrical signals from the pyrometer as well as those corresponding to the current through and the voltage across the specimen are recorded as functions of time. This is accomplished through the use of a high-speed digital data acquisition system which consists of a mainframe and two plug-ins. The plug-ins are a high-speed digital voltmeter and a high-speed FET multiplexer. The data acquisition system is in turn controlled by a host computer. The scan and measure sequence can be triggered either with the internal source of the mainframe or external pulses. The maximum data acquisition rate of the system is 100,000 readings per second corresponding to a sample period (i.e., the time between two successive measurements) of 10 μs. The sample period is adjustable to permit slower acquisition of data. The voltmeter has four scale ranges, namely, ± 10.24 V, ± 2.56 V, ± 0.32 V, and ± 0.04 V, full scale with autoranging capability. Each range has a full scale resolution of 12 bits plus a sign bit. Moreover, there is an on-board memory buffer in the voltmeter with a capacity for storing over 64,000 readings. The large memory capacity eliminates the need for time-outs for transferring data from the voltmeter to the host computer during an experiment, avoiding breaks in the data as a function of time. At the end of an experiment, the data stored in the memory buffer are transferred to the host computer for analysis.

3. MEASUREMENT PROCEDURE

3.1. Directly Measured Quantities

3.1.1. Temperature

The specimen temperature is determined with the high-speed photoelectric pyrometer, which is used as a device to compare the radiation from a standard lamp and the radiation from the specimen.

3.1.1a. Calibration. A calibration is first made under steady-state conditions to determine parameters inherent to the pyrometer, by comparing the radiation of the standard lamp to that of an equivalent reference lamp. These parameters were then used in the determination to compare the reference lamp radiation with the specimen radiation under conditions of rapidly changing temperature. The Wien radiation equation, modified for use with this pyrometer, serves as the transfer mechanism:

$$\frac{1}{T_X} = \frac{1}{T_R} + A_R - A_X + \frac{\lambda}{c_2} \ln \left(\frac{R + C_4}{X + C_4} \right) \tag{4}$$

where T_X is the temperature of the standard lamp (or specimen), T_R the temperature of the reference lamp, A_R the optical attenuation of the channel through which the reference lamp radiation is received by the pyrometer, A_X the optical attenuation of the channel for the standard lamp (or specimen) radiation, λ the mean effective wavelength of the interference filter, c_2 the second radiation constant, R the output of the data acquisition system for the radiance received in the reference lamp channel, X the corresponding voltage for the radiance in the standard lamp (or specimen) channel, and C_4 is the constant inherent in the electronic circuit of the pyrometer.

The quantities A_R, A_X, and C_4 are properties of the pyrometer. They are determined in calibration experiments, in which a standard lamp is substituted for the specimen with the window of the chamber interposed between the lamp and the pyrometer.

The attenuator disk of the pyrometer produces three levels of radiance in the reference channel for each half-revolution of the attenuator, and each unknown or standard radiance is sandwiched between two reference radiances. There are, therefore, twelve radiances per revolution, six reference and six unknown. By making three exposures per calibration, one with the standard lamp adjusted to produce a response in the unknown channel nearly equal to that obtained in the reference channel with the largest attenuator opening (smallest attenuation), one with the standard lamp adjusted to a response near that of the smallest attenuator opening, and one near that of the medium attenuator opening, it is possible to calculate all six A_X values and all six A_R values, and the C_4 value for each exposure. The redundancy of three separate exposures (36 equations for 15 parameters to be determined) makes it possible to choose those ratios of R and X which give the most accurate values of A and C_4. Because of the random variation of the radiances in any one exposure, the data from a large number of revolutions (usually 30) were averaged before calculating the A values. The standard deviation of an individual radiance from its mean is normally about 0.2%, and the standard deviation of the mean is normally about 0.04%.

3.1.1b. Specimen Temperature. The A_R and A_X values obtained from the calibration make it possible to calculate the temperature of an unknown specimen placed in the same position that the standard lamp occupied in the calibration, the reference lamp remaining in the reference channel. Equation (4) is now used to obtain T_X directly. The quantity C_4 is calculated separately for each half-revolution by comparing the high-reference radiance (largest attenuation opening) with the low-reference radiance and inserting the corresponding A_R values; the temperature terms drop out because "T_X" = T_R. Because of possible pyrometer drift, calculations of a given T_X are confined within a half-revolution (2.5 ms); it is also found that, within this time period, temperature determinations are most accurate if unknown radiances are calculated against reference radiances nearest in level rather than closest in time. Each unknown temperature is calculated separately from each of the bracketing reference radiances, and the two T_X values thus obtained are weighted linearly for their nearness to the respective references and averaged. Hence

$$T_X = \frac{T_{X_1}(X - R_2) + T_{X_2}(R_1 - X)}{R_1 - R_2} \tag{5}$$

where T_{X_1} is the temperature calculated by equation (4) from reference radiance R_1, T_{X_2} is similarly obtained from R_2, X is the unknown radiance, while R_1, R_2, and X are corrected for C_4.

In calculating all unknown (specimen) temperatures, corrections are made for (1) scattered light and (2) departure from blackbody conditions (geometrical); they are discussed in Section 3.1.1c.

A typical oscilloscope trace photograph of the radiance of rapidly heating specimen is shown in Fig. 8.

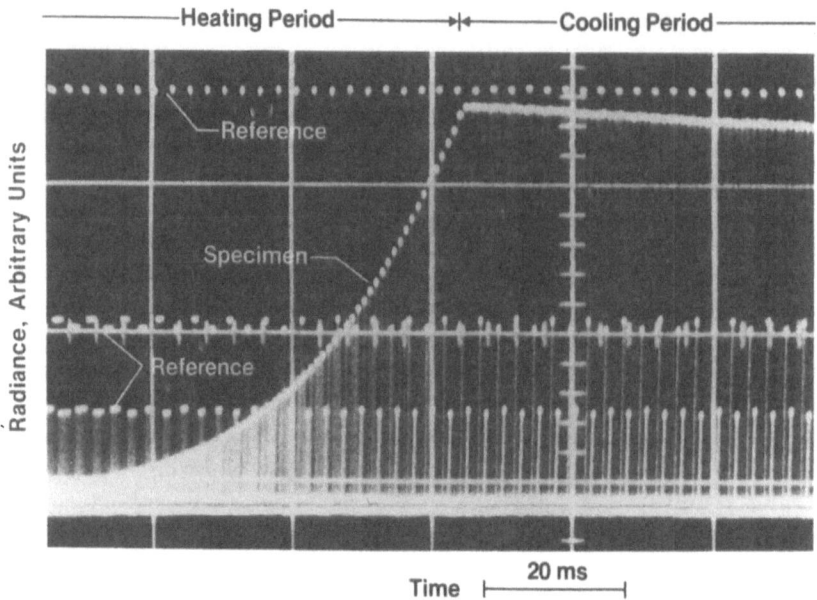

FIGURE 8. Oscilloscope trace photograph of radiance of rapidly heating specimen. Dots forming the long horizontal lines correspond to radiances from the reference source.

Temperature determinations beyond the upper limit of the standard lamps are made by inserting an attenuator in the optical path of the specimen radiation and performing two calibrations, one without and one with the external attenuator. From the A values of the first calibration without the external attenuator and the apparent temperature of the standard lamp through the external attenuator in the second calibration, the attenuation of the external attenuator is determined and added to all the A_X values, so that specimen temperatures higher than the standard lamp range can be calculated.

3.1.1c. Corrections. The most significant corrections made in the computations of the temperature of the specimen are due to the scattered-light effect and departure from blackbody conditions.

Scattered-Light Effect. The optics of the pyrometer scatter light from regions immediately surrounding the target area into the pyrometer. Since the radiance distribution around the target on the tubular specimen differs from that around the target area on the calibrating lamp, a correction is required.

Temperature correction due to the scattered-light effect may be expressed by

$$\frac{1}{T_c} = \frac{1}{T_o} - \frac{\lambda}{c_2} \ln\left(\frac{1}{S}\right) \tag{6}$$

where T_o and T_c are observed and corrected temperatures, respectively, and S is the scattered-light correction factor, which is given by

$$S = \frac{1 + \varepsilon_{N,\lambda}K}{1 + K} \tag{7}$$

where $\varepsilon_{N,\lambda}$ is normal spectral emittance and K is the scattering ratio.

The scattering ratio, K, is the fraction of radiation that the pyrometer senses and that comes from an area outside the defined target. The parameter K may be expressed by

$$K = K_1 - K_2 \tag{8}$$

where K_1 is the total scattered radiation from outside the target and K_2 is the scattered radiation from the background.

The quantity K_1 can be obtained experimentally by employing a specimen similar to the one used in dynamic experiments, except that the portion of the tube behind the sighting hole is removed. Then, K_1 is the ratio of radiation from the hole to that from the surface. The required values were obtained by conducting two experiments, one with the pyrometer aimed at the hole, and another with the pyrometer aimed at the surface.

In the above case, radiation from the hole includes scattered radiation from adjacent surfaces, as well as scattered radiation from the background (chamber wall) as seen through the hole. In order to determine the magnitude of the background radiation, separate experiments were conducted on a specimen composed of two strips (3.2 mm wide) placed parallel to each other with a separation of approximately 4 mm. Since in this arrangement the opening between the strips was large, the contribution of surface scattering was negligible and thus one was able to obtain the scattered radiation from the background. This quantity was then used to correct K_1 according to equation (8). All of the above-described experiments were performed under dynamic heating conditions, and the pyrometer outputs for each pair of experiments were related to each other on the basis of times corresponding to equal resistance values of the specimen.

The initial optical arrangement of the pyrometer required a correction of about 6–8% for the scattered-light effect. Subsequent improvements in the optical components reduced this correction to about 1%.

Blackbody Quality. Temperature correction due to departure of the specimen from blackbody conditions may be expressed by

$$\frac{1}{T_c} = \frac{1}{T_o} - \frac{\lambda}{c_2} \ln\left(\frac{1}{Q}\right) \tag{9}$$

where T_o and T_c are observed and corrected temperatures, respectively, and Q is the blackbody quality.

Departure from ideal blackbody conditions (geometrical factor) for the specimen was estimated using De Vos's[6] method. Computations were made assuming perfectly diffuse, perfectly specular, and two intermediate conditions for the internal surface of the tubular specimen. Computations were made for two different specimen lengths, 51 and 76 mm for the heated portion, which correspond to 76 and 102 mm total length, respectively. The sighting hole $(1 \times 0.5$ mm) is 0.8 mm off-center to improve blackbody conditions. All the results were in the range 0.994 to 0.997, indicating a very small dependence of the blackbody quality on specimen length and internal surface conditions. A value of 0.995 is used for Q in all the computations related to the temperature of the specimen.

3.1.2. Voltage and Current

Voltage across and current through the specimen during a pulse experiment are determined by multiplying the output of the digital data acquisition system by calibration factors. The calibration factors are obtained by supplying accurately known stable voltages to the entire voltage and current measuring circuits (including the differential amplifiers) under steady-state conditions before a series of pulse experiments. Oscilloscope trace photographs of typical voltage and current pulses are shown in Fig. 9.

3.1.3. Dimensions, Mass, and Density

The total length, "effective" length (length between voltage probes), outer diameter, and sighting hole dimensions of the specimen are measured at room temperature with a micrometer microscope to the nearest 0.02 mm. The wall thickness is calculated from the mass, density, and length.

The "effective" mass of the specimen is determined from the measurements of the mass of the entire specimen and the appropriate dimensions (total and "effective" lengths).

The density of the specimen material is determined either by the water displacement method in a pycnometer, or by mass and volume measurements of a representative material of solid cylindrical geometry.

3.2. Determination of Specific Heat

3.2.1. Formulation of Relations

The power balance for the "effective" specimen during the heating period

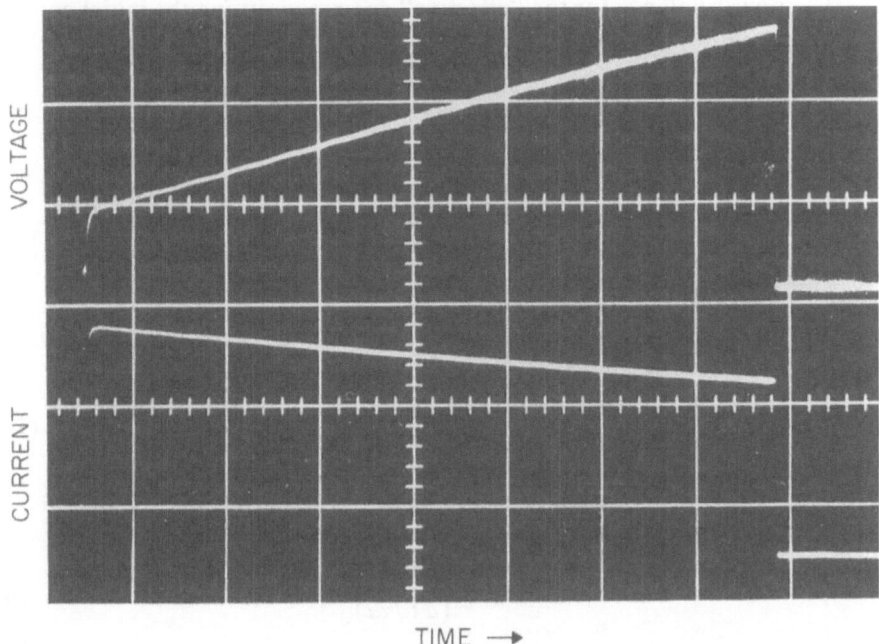

FIGURE 9. Oscilloscope trace photograph of typical voltage and current pulses. Equivalence of each major division is: time, 50 ms; voltage, 2 V; and current, 1000 A.

may be expressed as

$$\text{Power imparted} = \text{power absorbed} + \text{power loss}$$

In the present experiments, the major source of power loss is that due to thermal radiation. Using the proper quantities, the above relation becomes

$$ei = C_p m (dT/dt)_h + \varepsilon \sigma A_s (T^4 - T_e^4) \tag{10}$$

where e is the voltage across the "effective" specimen, i the current through the specimen, C_p the specific heat at constant pressure, m the quantity of "effective" specimen, ε the hemispherical total emittance, σ the Stefan–Boltzmann constant, A_s the effective surface area, T the specimen temperature, T_e the ambient temperature, and $(dT/dt)_h$ is the heating rate.

The solution of equation (10) for C_p yields

$$C_p = \frac{ei - \varepsilon \sigma A_s (T^4 - T_e^4)}{m (dT/dt)_h} \tag{11}$$

The power balance for the "effective" specimen during the initial cooling period may be written as

$$\text{Power loss} = \text{power radiated}$$

which can be expressed as

$$-C_p m (dT/dt)_c = \varepsilon \sigma A_s (T - T_e^4) \tag{12}$$

where $(dT/dt)_c$ is the cooling rate.

When equations (11) and (12) are combined and solved for ε, one obtains

$$\varepsilon = \frac{ei}{\sigma A_s (T^4 - T_e^4)(1 + M)} \tag{13}$$

where

$$M = -\frac{(dT/dt)_h}{(dT/dt)_c} \tag{14}$$

3.2.2. Procedure in Experiments

Usually, measurements are performed from 1500 K to near the melting temperature of the specimen. The lower temperature is set by the limitation of the pyrometer's operation, and the upper limit is due to the disintegration of the specimen upon melting.

In order to optimize operation of the pyrometer, the temperature region above 1500 K is divided into several ranges. This is accomplished by inserting calibrated optical attenuators at appropriate places in the optical path of the pyrometer. Typically, the ranges are:

Range 1:	1500–1700 K
Range 2:	1700–1900 K
Range 3:	1900–2100 K
Range 4:	2100–2350 K
Range 5:	2350–2700 K
Range 6:	2700–3200 K
Range 7:	3200–4000 K

Measurements at higher temperatures can be performed by using additional optical attenuators.

For a given specimen, experiments are performed starting with Range 1 and up to the range which corresponds to melting of the specimen. At least one and usually two experiments are performed on a given specimen in each temperature range, except for the highest temperature range where only one final experiment is conducted that results in melting of the specimen. Generally, measurements are conducted on at least three different specimens of the same material.

3.2.3. Calculation of Specific Heat

Experimental data on voltage, current, and temperature for each experiment are recorded as functions of time and used to obtain polynomial (quadratic or cubic) functions for each quantity in terms of time using the least-squares method. These functions are used in the calculations given in Section 3.2.1.

Owing to the relatively small temperature range covered during the initial cooling period of each experiment, equation (13) is used to compute an average value for ε corresponding to an average temperature near the maximum temperature for that range. The values of ε determined for all the experiments for a given specimen are then used to obtain a function (linear or quadratic), by the least-squares method, to represent ε over the entire temperature range of the experiments. Then, ε values from this function are substituted in equation (11) to compute the specific heat at equal temperature intervals (usually 20 or 50 K) for all the ranges.

The calculated specific-heat results for all the temperature ranges and for all the specimens are then combined and fitted by a polynomial function using the least-squares method. The resultant expression represents the specific heat of the material of interest over the temperature range of interest.

4. CONSIDERATION OF VARIOUS PHENOMENA

In this section various physical phenomena that affect the design and operation of the millisecond-resolution pulse heating system and the measurement of the experimental quantities are presented.

4.1. Axial Temperature Distribution

Heat transfer from a specimen to the clamps causes the establishment of temperature gradients in the specimen. However, since the voltage probes are about 13 mm away from the clamps, sharp temperature gradients near the clamps do not affect the measurements. The magnitude of the axial temperature gradient in the "effective" specimen is estimated by solving the transient heat-conduction equation assuming constant properties and zero radial heat transfer.[7] The series solution is given by the following equation:

$$T = \left(\frac{i^2\rho l^2}{2\lambda A^2}\right)\left\{1 - \frac{x^2}{l^2} - \frac{32}{\pi^3}\sum_{n=0}^{\infty}\frac{(-1)^n}{(2n+1)^3}\right.$$

$$\left.\times \cos\left[\frac{(2n+1)\pi x}{2l}\right]\exp\left[-\frac{a(2n+1)^2\pi^2 t}{4l^2}\right]\right\} \quad (15)$$

where i is the electric current, ρ the electrical resistivity, A the cross-sectional area of the specimen, l the half-length of the "effective" specimen, x the axial distance from the midpoint of the specimen, λ the thermal conductivity, and a is the thermal diffusivity. If radial heat transfer is considered, the actual axial temperature gradient will be somewhat smaller than that given by equation (15).

As an example, the temperature gradient between the midpoint of a molybdenum specimen and various points along the specimen is calculated using equation (15) (with $n = 30$) for several heating pulses. The normalized results are shown in Fig. 10. Computations for heating pulses of 350 and

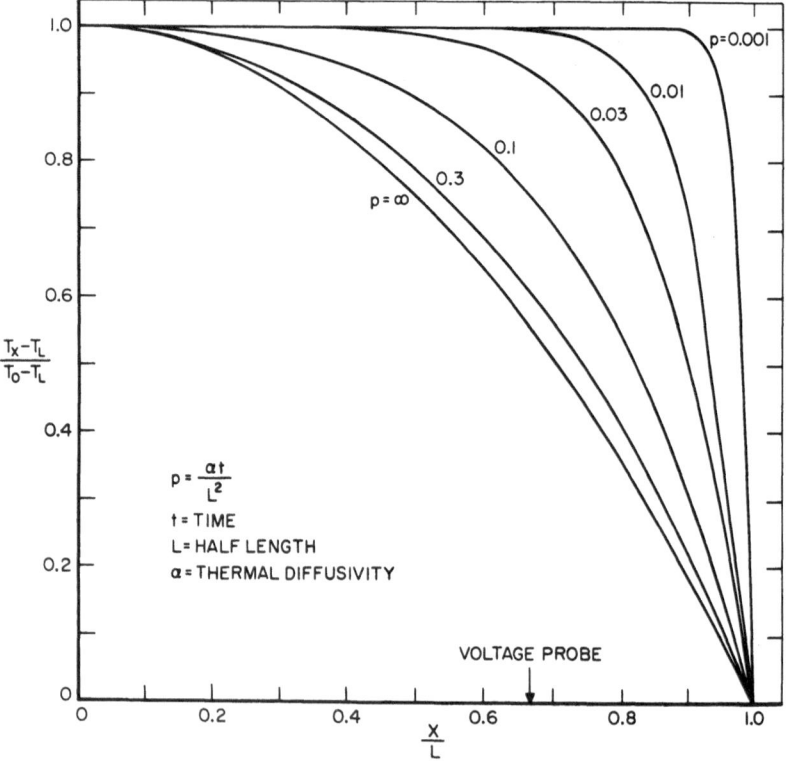

FIGURE 10. Axial temperature distribution in a resistively self-heated specimen under transient conditions.

700 ms duration indicate that the specimen temperature at the plane of the voltage probes is approximately 99.9 and 99% of its midpoint value, respectively. The respective average temperature of the "effective" specimen is 99.99 and 99.9% of its midpoint value.

4.2. Radial Temperature Distribution

There is no exact solution for the time-dependent radial temperature distribution in conductors carrying pulse currents with radiative heat transfer from their surfaces. Even the numerical methods become complicated owing to the nonlinearity of the differential equation.

In pulse experiments where heat loss from the surface is small compared with the power imparted to the specimen, the radial temperature drop (temperature difference between inner and outer surfaces for tubular conductors) may be estimated using the following approximate relation:

$$\Delta T_t = N \Delta T_s \qquad (16)$$

where ΔT_t is the temperature drop under transient conditions, ΔT_s the temperature drop under steady-state conditions, and N is the ratio of the heat loss from the surface to the imparted power under pulse conditions.

The solution of the steady-state radial heat-conduction equation, assuming constant properties, for a radial temperature drop is

$$\Delta T_s = \frac{r_o \varepsilon \sigma T^4}{2\lambda(r_o^2 - r_i^2)} \left[(r_o^2 - r_i^2) - 2r_i^2 \ln\left(\frac{r_o}{r_i}\right) \right] \qquad (17)$$

where r_o is the outer radius, r_i the inner radius, σ the Stefan–Boltzmann constant, ε the hemispherical total emittance, and λ is the thermal conductivity.

Computations on a molybdenum specimen indicate that the radial temperature drop at 2800 K may be approximately 0.3 K for a heating pulse of 0.4 s.

4.3. Cooling Rate of Specimen by Radiation

The cooling rate of the specimen from a high temperature following the opening of the switch depends, in addition to temperature, on the specimen material and its geometry.

Power balance during cooling yields the following expression for the cooling rate of the specimen:

$$\left(\frac{dT}{dt}\right)_c = -BT^4 \qquad (18)$$

where

$$B = \varepsilon \sigma A / C_p m \tag{19}$$

A being the surface area of the "effective" specimen, C_p the specific heat, m the mass of the "effective" specimen, T the specimen temperature, ε the hemispherical total emittance, σ the Stefan–Boltzmann constant, and $(dT/dt)_c$ the cooling rate.

In order to obtain an expression for the specimen temperature during cooling as a function of time, equation (18) is rearranged and integrated. The resultant relation is

$$T = \left(\frac{1}{3Bt + 1/T_m^3}\right)^{1/3} \tag{20}$$

where t is time and T_m is the specimen temperature at the start of cooling (at $t = 0$).

The variation in the temperature of a molybdenum specimen during cooling as a function of time corresponding to different initial temperatures is shown in Fig. 11.

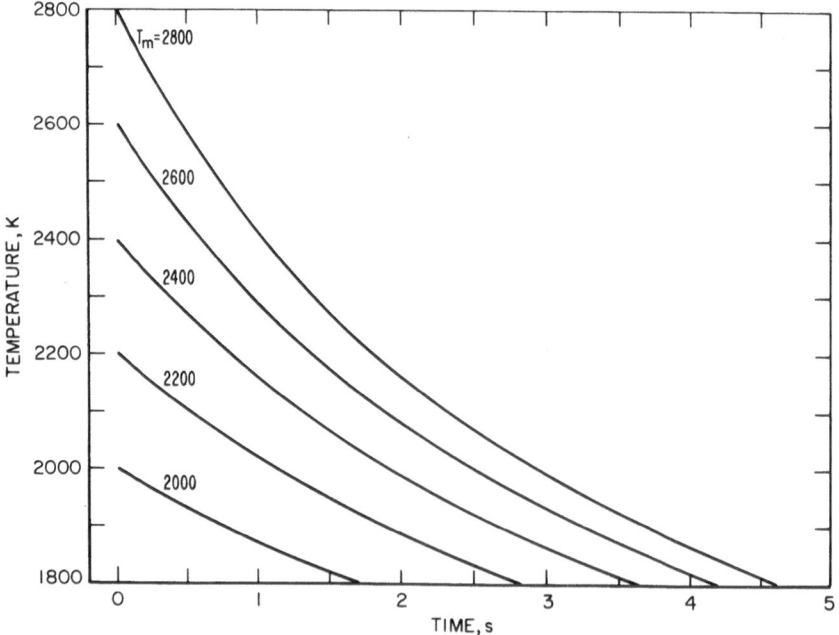

FIGURE 11. Variation in the temperature of a molybdenum specimen during cooling from various initial temperatures.

In preliminary calculations and also in computations related to error analysis, it is necessary to know the quantity P_r/P_i, where P_r is power loss from the specimen due to thermal radiation and P_i is power input to the specimen. From power balances for heating and cooling periods the following relation is obtained:

$$\frac{P_r}{P_i} = \frac{1}{1+M} \tag{21}$$

where

$$M = -\frac{(dT/dt)_h}{(dT/dt)_c} \tag{22}$$

It may be seen that for the steady-state case $(dT/dt)_h = 0$, which implies that $M = 0$, and thus $P_r = P_i$.

4.4. Circuit Characteristics

The pulse power circuit is a simple RL circuit whose dynamic characteristics, after closing of the switch, may be expressed by

$$E - L\frac{di}{dt} = iR \tag{23}$$

where R is the resistance of the complete circuit, L the self-inductance of the complete circuit, E the battery voltage, i the instantaneous current, and t is time. The solution of equation (23) is

$$i = \frac{E}{R}(1 - e^{-Rt/L}) \tag{24}$$

This equation, which gives the increase in current as a function of time, is obtained on the assumption that quantities E, L, and R are constant. In actual experiment, this assumption is satisfied approximately for E and L, however, R increases with time as the specimen is heated. Therefore, the current versus time relationship is of a distorted exponential form. The circuit's "time constant," which is expressed by the quantity L/R, decreases with increasing R as the specimen is heated.

Energy stored in the magnetic field of an inductor, with no ferromagnetic material in its vicinity, is given by

$$W = Li^2/2 \tag{25}$$

It is noteworthy that stored energy depends on the instantaneous value of the current and not on the current history. In a typical millisecond-resolution experiment where current is of the order of 10^3 A, magnetic energy stored in the "effective" specimen may be about 0.01 J. This is very small compared to imparted energy, which is of the order of 10^3 J. For a complete cycle of operation, consisting of closing the switch and, after a finite period, opening the switch, the stored energy in the specimen is zero.

It is important to consider self-inductances of the standard resistor and the "effective" specimen, and the mutual inductances between the pulse power circuit and the measuring circuits.

Using the equations given in the literature,[8] the self-inductances of the standard resistor and the "effective" specimen are computed to be 0.13 and 0.006 μH, respectively. For the maximum rate of change of current of about 3000 A s^{-1} during measurements in a typical pulse experiment, the induced voltages (due to self-inductance) in the standard resistor and the "effective" specimen are about 0.4 and 0.02 mV, respectively.

The contribution of the mutual inductance between the pulse power circuit and the measuring circuits is difficult to calculate. However, its effect can be determined by performing the following experiment. The voltage connections to the standard resistor and the specimen are disconnected and are shorted while duplicating, as much as possible, the geometrical configuration that the leads have during an actual pulse experiment. Also, the objective lens of the pyrometer is covered, allowing no radiation from the specimen to reach the detector. Then, a heavy current pulse is sent through the main circuit and recordings of the experimental quantities are made with the digital data acquisition system, just like in a regular pulse experiment. The lack of voltage signals indicates that there is no significant mutual inductive coupling between the pulse power circuit and the measuring circuits.

4.5. Inductive Effects

Inductive effects should be considered in the design and operation of a system where the rate of change of current (dI/dt) is high. Their contribution is usually small and often negligible in millisecond-resolution systems, provided that the components are designed carefully and proper operational precautions are taken.

4.6. Skin Effect

The skin effect, which is caused by changing currents in a circuit, alters the temperature distribution in the specimen. For cylindrical conductors the skin effect is a function of the following parameter:

$$x = 2\pi a \sqrt{2\mu f / 10^9 \rho} \tag{26}$$

where a is the radius of the conductor, f the current frequency, ρ the electrical resistivity of the specimen, while μ is the permeability of the specimen. For nonmagnetic materials, $\mu = 1$.

In the case of tubular (thin-wall) specimens, the skin effect is a function of the following parameter:

$$\beta = x\tau/a\sqrt{2} \tag{27}$$

where x is the quantity defined by equation (26), a the outer radius of the specimen, and τ is the thickness of the specimen.

In general, the contribution of the skin effect in a specimen is expressed in terms of the ratio of its electrical resistance at the frequency in question to that under DC conditions. Results on the resistance ratio in terms of parameters x and β have been tabulated elsewhere.[8] Computations indicate that for the specimen geometry and the type of pulses used in the present measurement system, the contribution of the skin effect is much less than 0.01% for nonmagnetic materials.

4.7. Magnetic Force

A high electric current through a conductor creates a magnetic force which acts toward the center of the conductor. The effect of this force is especially critical for thin-wall tubular specimens.

The magnetic force (per unit length) on a tubular specimen resulting from current through it is given by

$$F = \left(\frac{4i^2}{3b}\right)\left[\frac{1 + 2m}{(1 + m)^2}\right] \tag{28}$$

where i is the current and $m = a/b$, b being the outside radius and a the inside radius. The corresponding magnetic pressure is given as

$$P_m = \left(\frac{2i^2}{3\pi b^2}\right)\left(\frac{1 + 2m}{(1 + m)^2}\right) \tag{29}$$

Excessive magnetic forces, which act toward the center of the specimen, tend to collapse it. The collapsing pressure for thin-wall tubes is

$$P_c = \left(\frac{2E}{1 - \mu^2}\right)(1 - m)^3 \tag{30}$$

where E is the modulus of elasticity and μ is Poisson's ratio.

The maximum allowable current is obtained by equating equations (29) and (30), which yields

$$i = \sqrt{\left(\frac{3\pi b^2 E}{1 - \mu^2}\right)\left[\left(\frac{1 - m}{1 + 2m}\right)(1 - m^2)^2\right]} \tag{31}$$

4.8. Evaporation

The rate of evaporation, $m(T)$, from a metallic surface in vacuum may be expressed by

$$m(T) = A \, e^{-E/T} \tag{32}$$

where T is temperature (in K), A is a constant for a given surface, and E is equal to the energy of evaporation in the units of the Boltzmann constant.

Since in high-speed experiments the heating period is short compared to the cooling period, one may assume that most of the evaporation takes place during cooling of the specimen. Total evaporation during cooling becomes

$$M = \int_{t_1}^{t_2} m(T) \, dt = \int_{t_1}^{t_2} A \, e^{-E/T} \, dt \tag{33}$$

After performing a time-to-temperature transformation using equation (20) and integrating equation (33), one obtains

$$M = \frac{A}{BE^3} [e^{-\theta_m}(2 + 2\theta_m + \theta_m^2) - e^{-\theta_f}(2 + 2\theta_f + \theta_f^2)] \tag{34}$$

where $\theta_m = E/T_m$, $\theta_f = E/T_f$, B is the quantity defined in equation (19), while T_m and T_f are initial and final temperatures, respectively.

Computations on a molybdenum specimen indicate that the total weight loss due to evaporation during cooling from 2800 K is approximately 0.003% of the specimen weight.

4.9. Thermionic Emission

The contribution of thermionic emission is manifested in the following two forms: (1) energy loss from the specimen as a result of energy carried away by emitted electrons, and (2) establishment of undesirable current paths around the specimen.

The dependence of thermionic emission current on temperature is given by

$$I = AT^2 \exp\left(-e\phi/kT\right) \tag{35}$$

where A is the thermionic emission constant, ϕ the thermionic work function, e the electronic charge, and k is the Boltzmann constant.

For a molybdenum specimen, the thermionic emission current is computed to be approximately 7×10^{-5} A mm^{-2} at 2000 K and 0.15 A mm^{-2} at 2800 K.

It should be noted that computed values are for the case where electrons are continuously removed from the surface, which requires a collector plate at positive potential. In the present system, the chamber wall is electrically neutral and the only electrical potential gradient that exists is that across the specimen, which is less than 10 V. Thus, in experiments with the present system, thermionic emission is most probably space charge limited; its contribution therefore would be much less than the above-computed values.

4.10. Thermoelectric Effects

Thermoelectric effects play an important role in experiments where temperature gradients exist and where currents are through interfaces of dissimilar conductors. The Peltier effect between the specimen and clamps and the Thomson effect in the specimen where sharp temperature gradients exist may be appreciable. However, they do not affect the measurements in the present experiments since their contribution is negligible in the "effective" specimen. To further reduce the possible contribution of thermoelectric effects, the knife edges used for voltage measurements are made of the same material as that of the specimen. In the pulse experiments, the record of voltage readings at high temperatures after opening the main switch shows "zero" for the voltage between the probes. This indicates that there are no measurable erroneous signals due to the interaction (in a thermoelectric sense) of the probes with the specimen.

4.11. Thermal Expansion

A material expands (due to thermal effects) approximately with the speed of sound in that material. Since the speed of sound in most solid conductors is of the order of a few millimeters per microsecond, thermal expansion in millisecond-resolution experiments does not present a problem. It is important, however, to make provision (such as an expansion joint) for the free expansion of the specimen at any time during the rapid heating period.

4.12. Thermodynamic Equilibrium

The conventional definition of specific heat is satisfied only when it is obtained from measurements where the specimen is under thermodynamic equilibrium at any given instant. Whether a specimen is under thermodynamic equilibrium while measurements are taken during rapid heating may be determined by considering the relaxation times of various rate processes.

The total relaxation time that must be considered for thermodynamic equilibrium calculations may be expressed in the form

$$\frac{1}{\tau} = \sum_i \frac{1}{\tau_i} \tag{36}$$

where each τ_i represents a different mechanism of relaxation. The various relaxation mechanisms are the results of different scattering processes. The main three relaxation processes are due to electron-phonon, electron-imperfection, and phonon-phonon interactions.

In electrical conductors, electron-phonon interaction is the predominant one in determining the overall relaxation time. For estimative purposes, the corresponding relaxation time may be expressed as

$$\tau_{ep} = \sigma m / ne^2 \tag{37}$$

where e is the electronic charge, m the electronic mass, n the number of electrons per unit volume, and σ is the electrical conductivity.

The relaxation time due to electron-phonon interactions for metals at room temperature is of the order of 10^{-12} s. Since relaxation time is directly proportional to electrical conductivity, and since the latter decreases with increasing temperature, it takes longer to establish thermal equilibrium at high temperatures. The variation of electrical conductivity in metallic elements between room temperature and their melting points is within a factor of 100; therefore, it can be seen that thermal equilibrium in metals takes place in times of the order of 10^{-10} s.

Information concerning relaxation times for vacancy equilibrium is not readily obtainable; but since the mechanism is related to that of diffusion, the relaxation time is an exponential function of temperature. The magnitude of vacancy relaxation time, as observed in quenching experiments at lower temperatures and extrapolated to higher temperatures, may be of the order of a millisecond for refractory metals above 2000 K.

5. ESTIMATE OF ERRORS

5.1. Errors in Directly Measured Quantities

The estimated total error (which includes random as well as systematic errors) in directly measured quantities, such as temperature, voltage, current, dimensions, and mass, is obtained from analyses of sources and magnitudes of errors, experimental results, and auxiliary tests. The details are given below.

5.1.1. Errors in Temperature Measurements

The total error in temperature measurements results from several contributions. Their estimates are summarized in the following subsections.

5.1.1a. Standard Lamp. The most recent report of calibration by the National Bureau of Standards for the tungsten filament lamp used as temperature standard in this work states that the total estimated uncertainty of the reported values corresponds to ±0.8 °C at 1400 °C, ±1.3 °C at 1800 °C, and ±2.0 °C at 2300 °C.

5.1.1b. Instability of Standard Lamp. Calibrations of the standard lamp before and after several groups of experiments (corresponding to the use of the lamp for a cumulative period of over 20 hours) indicate a change in the reported values of approximately 1 °C at 1800 °C and 1.5 °C at 2300 °C.

5.1.1c. Attenuator Calibration. The calibration of the optical attenuator, placed in the optical path of the specimen radiation for temperature determinations beyond the upper limit of the standard lamp, has an uncertainty of about 0.5% in radiance, which corresponds to about 2 K at 3000 K.

5.1.1d. Radiation Source Alignment. Since the pyrometer had to be moved between a calibration and a series of pulse experiments and again for a calibration after the experiments, an additional uncertainty in the calibration of the pyrometer develops as a result of the uncertainties in the visual alignment of the lamp with respect to the pyrometer.

Based on an analysis of experimental results with approximately 20 tests using steady-state radiation sources, it is estimated that the error arising from this uncertainty is no more than 1 K at 2000 K, which corresponds to an uncertainty of about 2 K at 3000 K.

5.1.1e. Pyrometer Reproducibility. Based on calibration results, reproducibility of the pyrometer (between two calibrations), after operating for over 20 hours, is estimated to be about 1 K at 2000 K and 2 K at 3000 K. The primary cause for this is the changes in the optical attenuation in the pyrometer that result from the collection of dust, etc.

5.1.1f. Scattered Light Correction. Although a correction is made to temperature to allow for the scattering effect in the optics of the pyrometer, an uncertainty still exists as to the actual magnitude of this correction. The uncertainty in the scattering factor S given by equation (7) is obtained by considering the uncertainties in the scattering ratio and in the normal spectral emittance. The results indicate an uncertainty of less than 0.2% in radiance, which corresponds to an uncertainty of less than 0.5 K at 2000 K and less than 1 K at 3000 K.

5.1.1g. Window Attenuation. Uncertainty in the measurement of the attenuation of the experimental chamber window is estimated to be no more than 0.2% in radiance, which corresponds to an uncertainty of less than 0.5 K at 2000 K and less than 1 K at 3000 K.

5.1.1h. Electric and Magnetic Fields. The effect on the pyrometer of varying electric and magnetic fields associated with the pulse current was tested and found to be negligible (less than 0.5 K, the resolution of temperature measurements).

5.1.1i. Use of Wien's Law. For simplicity, in the computations of temperature, Wien's instead of Planck's Law was used. The difference in temperature obtained by these two functions for the mean effective wavelength of 0.65 μm is approximately 0.003 K at 2000 K and about 0.2 K at 3000 K. Since the same procedure was used for both calibration and pulse experiments, the error introduced as the result of using Wien's Law is completely negligible.

5.1.1j. Blackbody Quality. The computed blackbody quality (0.995), which is based on geometrical considerations only, is estimated to have a maximum uncertainty of 0.5% in radiance, which corresponds to approximately 1 K at 2000 K and 2 K at 3000 K.

5.1.1k. Specimen Temperature Nonuniformity. Based on the solution of the transient heat-transfer equation in the axial direction, temperature gradients between the midpoint of the specimen and the voltage probes can cause the average "effective" specimen temperature to be about 0.1% lower than the midpoint value. This corresponds to a value of 2 K at 2000 K. Radial temperature gradients are estimated to be less than 1 K at 2000 K. From potentiometric measurements of the resistance of the specimen along its length at room temperature, random variations in the thickness of the tube averaged over lengths of approximately 3 mm are estimated. Temperature nonuniformity corresponding to variations in the thickness of the specimen is approximately 2 K at 2000 K and 3 K at 3000 K. The total temperature uncertainty resulting from the items discussed in this subsection is estimated to be approximately 3 K at 2000 K and 4 K at 3000 K.

5.1.1l. Resultant Error in Temperature Measurements. From a consideration of various errors and their magnitudes listed above, it may be concluded that the uncertainty in temperature measurements is approximately 4 K at 2000 K and 7 K at 3000 K. A summary of estimated uncertainty in temperature measurements is given in Table 1.

5.1.2. Errors in Electrical Measurements

Identifiable sources of errors that affect the measurement of electrical quantities, such as voltage and current, are given below. Errors that contribute less than 0.01% to the measurements are considered negligible, since this limit is significantly less than the resolution of the digital data acquisition system (about 0.025% of the full scale).

5.1.2a. Skin Effect. Calculations for tubular specimens (nonmagnetic material) and trapezoidal-shape current pulses indicate that the contribution of the skin effect on the electrical measurement is much less than 0.01%.

5.1.2b. Inductive Effects. Consideration of self and mutual inductances of pertinent components indicate that their contribution to the electrical measurements is no more than 0.01%. This is due to the fact that current changes relatively slowly during the period of the measurements.

5.1.2c. Thermoelectric Effects. Both computations and experimental checks indicate that the contribution of the thermoelectric effects on the electrical measurements is less than 0.01%.

TABLE 1

Estimated Uncertainty in Temperature Measurements

Item No.	Source	Uncertainty (K)	
		at 2000 K	at 3000 K
1.	Standard lamp	1.3	2[a]
2.	Instability of standard lamp	1	1.5[a]
3.	Attenuator calibration	—	2
4.	Radiation source alignment	1	2
5.	Pyrometer reproducibility	1	2
6.	Scattered light correction	0.5	1
7.	Window attenuation	0.5	1
8.	Blackbody quality	1	2
9.	Specimen temperature nonuniformity	3	4
	Total uncertainty in temperature[b]	4	7

[a] Uncertainty at 2300 °C, the highest temperature at which the standard lamp is calibrated.
[b] Square root of the sum of the squares of the individual uncertainties given in items 1 through 9.

5.1.2d. Errors Due to Calibration and Stability. Use of a high-precision voltage source (0.001% stability) and a potentiometric system assured calibration of the electrical measuring circuits (for each experimental quantity) with an uncertainty of about 0.02%. Stability of the electrical components, including amplifiers, between successive calibrations (usually before and after a set of pulse experiments over a period of a few weeks) were determined to be no more than 0.01%.

5.1.2e. Resultant Error in Electrical Measurements. Based on the above considerations, total uncertainty in the measurements of an electrical quantity (current or voltage) is estimated to be no more than 0.05%. It is noteworthy that this value is comparable to the resolution of the digital data acquisition system. In order to have a more direct check on the accuracy of the electrical measurements, experiments were conducted to determine the value of a test resistor by measuring the current through it and the voltage drop across its terminals under transient conditions, and comparing this measurement with its resistance under steady-state conditions as determined by the potentiometric method. The average of the results of nine experiments indicates that resistance measurements by pulse and steady-state techniques are in agreement within 0.03%.

5.1.3. Errors in Other Measurements

5.1.3a. Length. The uncertainty in an individual length measurement is about 0.02 mm which, when related to the dimension of the "effective" specimen, corresponds to an uncertainty of about 0.1%.

5.1.3b. Mass. The uncertainty in an individual mass measurement is about 0.2 mg, producing an uncertainty of less than 0.01% for the specimen. However, the uncertainty in determining the mass of the "effective" specimen is greater than this. Based on the combined uncertainties of mass and length measurements, the uncertainty in the mass of the "effective" specimen is estimated to be about 0.2%.

5.1.3c. Density. The uncertainty in density measurements is about 0.1%. This uncertainty is reflected in the computation of the wall thickness of the tubular specimen.

5.1.3d. Time. The time base of the experiments is a synchronous motor in the pyrometer that was driven by a 400-Hz regulated power supply. Measurements have indicated that the uncertainty in time due to frequency instability is much less than 0.01%.

5.2. Errors Due to Departure from Assumed Conditions

5.2.1. Errors in Heat-Loss Correction

Since most of the pulse heating experiments are conducted with the specimen in a vacuum environment, heat loss from the specimen takes place by thermal radiation and by axial conduction to the end clamps. In Section 4.1, it was shown that axial temperature gradients in the "effective" specimen are very small; thus heat loss by conduction in the axial direction is negligible. Heat conduction to the voltage probes is also negligible, since the cross-sectional area of the knife-edges is small and the conduction path is long. Thus, thermal radiation is the only major source of heat loss from the "effective" specimen. To correct for this, a term representing thermal radiation is included in the relation for specific heat, equation (11). However, any uncertainty in this term affects the value computed for specific heat. Radiation heat loss, which is obtained from data during the initial cooling period, has an uncertainty not exceeding 5%. The contribution of this uncertainty to the uncertainty of specific heat is obtained by multiplying the uncertainty in the heat loss (due to thermal radiation) by the ratio of radiated power to input power. The latter ratio is usually of the order of a few percent and seldom exceeds 10%. Thus, under the worst conditions, the contribution of the uncertainty in the radiation heat-loss correction to specific heat is of the order of 0.5%.

If the experiments are performed with the specimen in an inert gas environment, additional heat losses take place due to convection and conduction to the gas. However, these losses are negligibly small compared to the radiation heat loss; moreover, they are lumped with the radiation heat loss in the procedure used to determine the heat loss from the specimen. Thus, the heat-loss term obtained from data during the initial cooling period accurately represents the total combined heat loss from the specimen. In order to have a more direct check of the effect of the gas environment on measured specific heat, experiments were conducted with the specimen in a vacuum and in an argon gas environment at atmospheric pressure. The results did not indicate any significant difference in the measured property.

5.2.2. Errors Due to Improper (Low) Heating Rate

The validity of the determination of specific heat is based on the assumption that the "effective" specimen is under uniform temperature at any given time during the experiment. However, this is not completely realized due to axial and radial temperature gradients and due to variations in the cross section of the specimen. Their contributions to the uncertainty in temperature measurements were estimated in Section 5.1.1k. In the present section, the effect of heating rate, as a contributor to axial gradients, on computation of specific heat is discussed.

When a specimen heats at a low rate, temperature gradients are established in the axial direction due to heat conduction from the specimen to the clamps. As a result, the average temperature of the "effective" specimen will be lower than the measured temperature (based on thermal radiation from the blackbody hole at the specimen midpoint). This will tend to lower the computed value of specific heat (since it has positive temperature coefficient) with respect to that determined under conditions of uniform temperature. However, in the case of specific heat, a competing effect arises. The presence of axial gradients will yield a temperature lower than that measured in the absence of gradients. Since this difference in measured temperature will increase with temperature and since, in measurements with pulse heating techniques, specific heat is inversely proportional to dT/dt [see equation (11)], a second effect of a low heating rate will be to increase the computed value of specific heat. Because of the nature of this term (derivative), the magnitude of the second effect will usually be greater than that of the first, implying that for low heating rates computations based on experimental data will yield high values for specific heat. This may be seen in the example given for molybdenum (Fig. 12). It may be noted from the figure that the results of the measurements corresponding to heating rates lower than about 4000 K s^{-1} are dependent on heating rate, while those greater than about 4000 K s^{-1} are relatively insensitive to the changes in heating rate. The above suggests that selection of the proper heating rates is essential for reliable determination of specific heat. The proper heating rates will be different for different materials, mainly because of the differences in thermal conductivity values, and thus a careful investigation (both analytical and experimental) is essential before the final experimental work on a given material.

Another example for the heating-rate effect is given in Fig. 13 for graphite. It may be seen that, because of the low thermal conductivity of graphite in

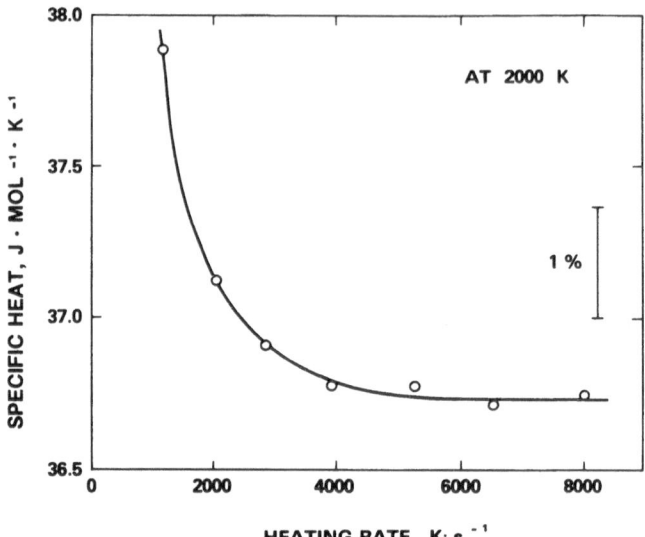

FIGURE 12. Dependence of specific heat (at 2000 K) of molybdenum on the specimen heating rate.

comparison to molybdenum, specific heat does not show any appreciable dependence on heating rate even at rates as low as $600 \, \mathrm{K \, s^{-1}}$.

The results shown in Fig. 13 also strengthen the basis for the calculation of radiative heat-loss correction in measurements by pulse heating techniques. It may be seen that there is no significant trend in specific heat with changing heating rate even though the heat-loss correction (as a fraction of input power) increases by almost one order of magnitude, from 6.5% at $6500 \, \mathrm{K \, s^{-1}}$ to 44% at $600 \, \mathrm{K \, s^{-1}}$. An uncertainty of about 1% in heat-loss correction would yield a corresponding uncertainty in specific heat of about 0.1% at $6500 \, \mathrm{K \, s^{-1}}$ and

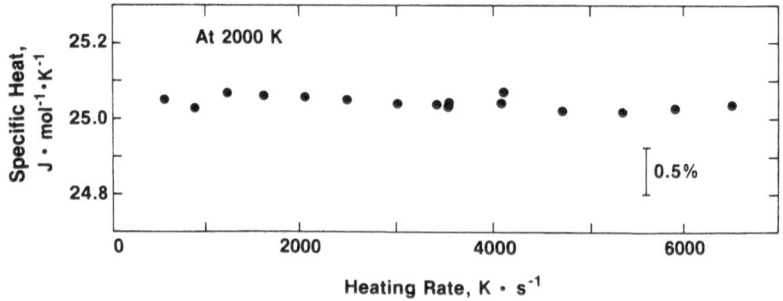

FIGURE 13. Dependence of specific heat (at 2000 K) of graphite (POCO AXM-5Q1) on the specimen heating rate.

0.5% at 600 K s^{-1}. The results in Fig. 13 suggest that the uncertainty in heat-loss correction is no more than 1%, which is well within the estimate reported in Section 5.2.1.

5.2.3. Errors in Realization of the "Effective" Specimen

The "effective" specimen is defined as the portion of the specimen between the voltage probe marks where the knife-edges touch the specimen. The mass of the "effective" specimen is obtained from measurements of the mass of the entire specimen, the total length of the specimen, and the length between the voltage probe marks. As discussed in Section 5.1.3b, the "effective" mass can be determined with an uncertainty of about 0.2%. This assumes that voltage probes touch the specimen exactly at the voltage probe marks on the specimen. In actual experiments, additional uncertainties (up to 0.5%) in specific heat may arise due to the improper contact between the voltage probes and the specimen. The improper contacts may be the result of poor fabrication of the grooves, poor maintenance of the voltage probes including dull knife-edges, improper alignment of the voltage probes, relative movement of the voltage probes and the grooves during an experiment, etc.

In order to minimize the uncertainties in the measured specific heat due to the above problems, it is very important to pay the utmost care to the fabrication of the voltage probes and their alignment with respect to the voltage probe marks on the specimen.

5.3. Summary of Error Estimates

Based on the detailed analysis of all identifiable errors in each measured quantity given in the previous sections, uncertainty in specific heat, determined

TABLE 2

Estimated Uncertainty and Imprecision of Measured Quantities and Specific Heat

Source	Uncertainty		Imprecision	
	at 2000 K	at 3000 K	at 2000 K	at 3000 K
Temperature	4 K	7 K	0.5 K	1 K
Voltage	0.05%	0.05%	0.02%	0.02%
Current	0.05%	0.05%	0.03%	0.03%
Length	0.1%	0.1%	0.05%	0.05%
Mass	0.2%	0.2%	0.1%	0.1%
Density	0.1%	0.1%	0.05%	0.05%
Time	0.01%	0.01%	0.005%	0.005%
Specific heat	2%	3%	0.5%	0.5%

from measurements with the millisecond-resolution pulse heating system, is estimated to be no more than 2% at 2000 K and 3% at 3000 K. Imprecision (often referred to as reproducibility) of the specific-heat results over this temperature range is usually less than 0.5%. A summary of the estimated uncertainty and imprecision of measured quantities and specific heat for representative experiments is given in Table 2.

The uncertainty limits of the measurements with the pulse heating system is further verified by the satisfactory agreement of the specific-heat values obtained using the pulse heating technique with those obtained by other reliable calorimetric methods in the overlapping temperature regions, usually in the range 1500–2000 K, for several refractory metals.

6. DISCUSSION

Among the various pulse heating techniques that utilize the resistive self-heating method, the millisecond-resolution technique has reached a mature stage of development. This technique is now capable of providing reliable data on the specific heat of electrically conducting solids in the temperature range 1500 K to near the melting point of the specimen. The advances made in developing this technique may be attributed primarily to two factors: (1) the need for high-temperature properties of materials in various applications, such as aerospace, nuclear energy, etc., and (2) the emergence of highly sophisticated electronics primarily in the area of fast digital data acquisition systems.

The original millisecond-resolution system developed in the Dynamic Measurements Laboratory of the U.S. National Bureau of Standards during the 1960s has undergone continued changes, additions, and improvements during the 1970s and the early part of the 1980s. The system described in this chapter includes most of the refinements introduced during these years. The key areas where changes took place are summarized in the following paragraphs.

The capacity of the battery bank has been almost doubled (increased to 48 V) to permit faster heating rates for the specimen and also to enable performance of the measurements on nonmetallic conductors, such as graphitic materials.

The digital data acquisition system was one of the major components of the overall system that continuously evolved to the present stage taking advantage of very rapid developments in digital electronics. Other electronic equipment and components, such as timers, pulse generators, amplifiers, and electrical calibration equipment, also underwent continuous modernization.

An area where continued major effort has been placed is fast optical pyrometry. A new generation of solid-state pyrometers has been and is being developed that utilize silicon photodetectors. One such pyrometer[9] was constructed for microsecond response. Another pyrometer,[10] operating at six wavelengths, has been developed and is presently being tested. Solid-state pyrometers are demonstrated to be reliable and have the advantage of being simpler to construct and operate than the original pyrometer that utilized a photomultiplier.

It should be noted that the pulse heating technique described in this chapter has the added advantage that it permits accurate measurement of several other properties (electrical resistivity, thermal emittance, thermal expansion, etc.) in addition to specific heat. The technique also has the potential of extending the measurements to other related properties, such as velocity of sound, mechanical properties, etc., to temperatures beyond the limit of accurate conventional techniques.

At this time, in addition to the system at the U.S. National Institute of Standards and Technology (NIST) described in this chapter, the only other fully operational millisecond-resolution system is that at the Istituto di Metrologia "G. Colonnetti" (IMGC) in Italy. The IMGC system is very similar to that of the NIST and is described in the literature.[11,12]

REFERENCES

1. A. Cezairliyan, in: *Compendium of Thermophysical Property Measurement Methods*, Vol. 1 (K.D. Maglić, A. Cezairliyan, and V.E. Peletsky, eds.), pp. 643–668, Plenum Press, New York (1982).
2. A. Cezairliyan, M.S. Morse, H.A. Berman, and C.W. Beckett, *J. Res. Natl. Bur. Stand.* **74 A**, 65–92 (1970).
3. A. Cezairliyan, *J. Res. Natl. Bur. Stand.* **75C**, 7–18 (1971).
4. A. Cezairliyan, *Int. J. Thermophys.* **5**, 177–193 (1984).
5. G.M. Foley, *Rev. Sci. Instrum.* **41**, 827–834 (1970).
6. J.C. De Vos, *Physica* **20**, 669–689 (1954).
7. H.S. Carslaw and J.C. Jaeger, *Conduction of Heat in Solids*, Oxford University Press, Oxford (1959).
8. F.W. Grover, *Inductance Calculations*, Van Nostrand, New York (1946).
9. G.M. Foley, M.S. Morse, and A. Cezairliyan, in: *Temperature: Its Measurement and Control in Science and Industry* (J.F. Schooley, ed.), Vol. 5, pp. 447–452, ISA, Pittsburgh (1982).
10. A. Cezairliyan, G.M. Foley, M.S. Moise, and A.P. Miiller, in preparation.
11. F. Righini, A. Rosso, and G. Ruffino, *High Temp. High Pressures* **4**, 597–603 (1972).
12. F. Righini and A. Rosso, *Measurement* **1**, 79–84 (1983).

18

The Application of Differential Scanning Calorimetry to the Measurement of Specific Heat

M.J. RICHARDSON

1. DIFFERENTIAL SCANNING CALORIMETRY

Differential scanning calorimetry (DSC) is a simple and rapid method for determining the heat capacities of small samples over a wide range of temperature. It is applicable to materials in general and a variety of forms (bulk, powder, film, granular, and liquid). Although easy to operate, the relevant instrumentation is of complex construction and it is normal to use commercial, rather than home-made, equipment. In this chapter, following a brief general introduction, the remarks will refer specifically to Perkin-Elmer power-compensation calorimeters—although they can usually be generalized to include other makes. The performance of most currently available DSCs is such that an accuracy of ±1-2% should be routine but this can be undermined by unsuitable samples, incorrect calibration, or by inadequate data treatment. Particular emphasis will therefore be placed on these aspects rather than on the setting-up and basic operation which are fully described in the manufacturers' manuals.

In any DSC experiment the response of the sample (subscript s) relative to that of an inert reference (subscript r) is monitored as the two are heated, or cooled, at a constant rate—typically 5–40 K min^{-1} for heat capacity work. In power-compensation DSC (Fig. 1) the ordinate signal is the differential power $\Delta P = P_s - P_r$ that is needed to keep both sample and reference at the same programmed temperature. In heat-flux DSC (Fig. 2) there is a common heat source and the signal is the differential temperature between sample and

M.J. RICHARDSON • Division of Materials Metrology, National Physical Laboratory, Teddington, Middlesex TW11 0LW, England.

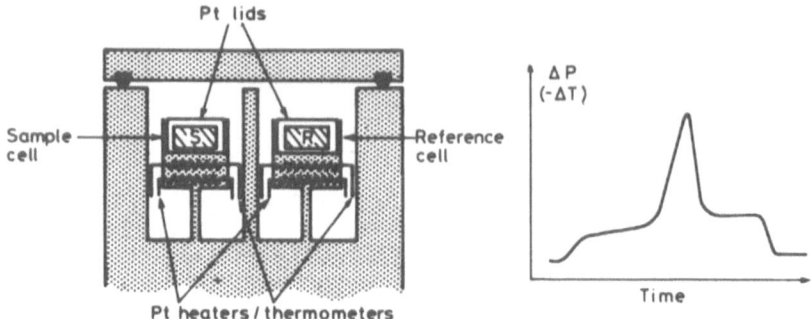

FIGURE 1. (a) Power-compensation DSC. (b) DSC signal through a transition.

reference cells; specific arrangements of thermocouples or thermopiles charac-
terize the many heat-flux systems that are currently available. The reference
cell usually contains only an empty pan and the signal ΔX ($X = P$ or T) is
of opposite sign for the two forms of DSC because in the heating mode, for
example, a loaded sample cell can only be maintained in the $\Delta T = 0$ condition
of power-compensation DSC when $P_s > P_r$; conversely, with a common heat
source the sample cell temperature invariably lags behind that of the reference
cell and $T_s < T_r$ for heat-flux DSC. Conditions are reversed in cooling experiments. The abscissa is time in both forms of DSC, but it is usual to transform
this to temperature using the known linear heating or cooling rate.

2. SPECIFIC HEAT

A DSC determination of heat capacity (c_p) is based upon a comparison
of signals in the scanning mode from the sample and a calibrant (subscript c)
of known c_p. It should be emphasized that the calibrant is always associated
with the sample cell: it should not be confused with "reference," which
throughout this chapter is used only to describe the reference cell (Figs. 1a
and 2). The *isothermal* signal (I, Fig. 3) is due only to heat transfer effects

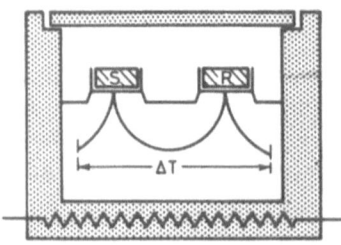

FIGURE 2. Schematic of heat-flux DSC. Many
configurations are available.

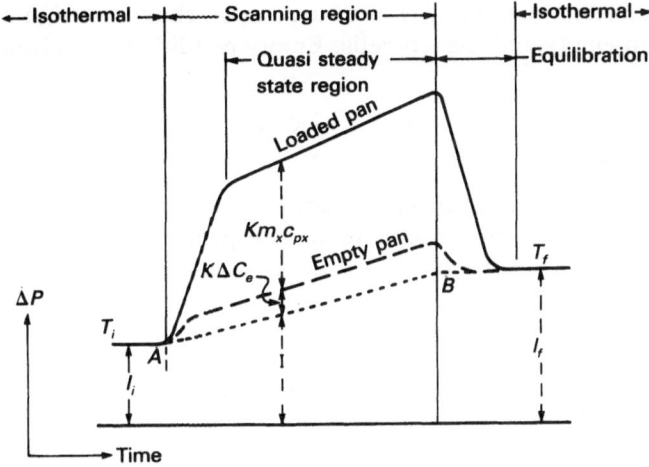

FIGURE 3. Schematic curves showing instrumental response as empty and loaded pans ($x =$ calibrant or sample) are heated from one isothermal temperature (T_i) to another (T_f). AB is the isothermal baseline that is reached should the programmed temperature rise be stopped between T_i and T_f.

between the cells and their surroundings. The absolute value of I is unimportant, because in any DSC experiment interest is always in how the signal *changes* in response to a programmed change of temperature and/or some thermal "event." One property of I that should be noted is its temperature dependence so that, in general, two isothermal regions are characterized by different values of I—although the slope $I(T_f) - I(T_i)$ can be minimized by appropriate use of the instrumental controls.

When the DSC undergoes a programmed change of temperature there is an additional contribution ($K\Delta C$) to the signal due to the difference between the heat capacities (ΔC) of the two cells (Fig. 3) and the scanning signal S is given by

$$S = I + K\Delta C = I + K(\Delta C_e + m_x c_{px}) \tag{1}$$

Equation (7) below will show that the reciprocal of the constant K is an ordinate-to-heat capacity conversion factor, ΔC_e is the difference between the heat capacities of the empty cells, m_x is the additional mass of sample ($x = $ s) or calibrant ($x = $ c) of specific heat c_{px}. "Empty" (subscript e) normally refers to conditions with an empty pan in both sample and reference cells—that in the latter is to "back off" the contribution of the sample pan, it ensures that ΔC_e remains small, and it is undisturbed throughout successive experiments; the ΔC_e term may be positive or negative, or may change sign during an experiment. Determinations of c_{ps} are based upon changes of signal in sequential runs with empty and loaded cells. Empty conditions are given by

$$S_e = I + K\Delta C_e \tag{2}$$

where all terms may be functions of temperature. The behavior of K, which in principle is temperature-independent, is especially important and is considered in more detail in Section 4.2.1. Successive runs with ΔC_e enhanced by calibrant or sample give

$$S_c = I + K(\Delta C_e + m_c c_{pc}) \tag{3}$$

and

$$S_s = I + K(\Delta C_e + m_s c_{ps}) \tag{4}$$

Subtraction of equation (2) from equations (3) and (4) gives

$$S_c - S_e = K m_c c_{pc} \tag{5}$$

and

$$S_s - S_e = K m_s c_{ps} \tag{6}$$

Quantities $S_x - S_e$ ($x = c$ or s) of equations (5) and (6) will be referred to as dynamic signals; they are the basic quantities needed in a DSC determination of heat capacity. Although the heating or cooling rate $\dot{T}$ does not specifically appear in equation (5) or (6), it may be introduced upon replacing K by $K_1\dot{T}$, where K_1 is the value when $\dot{T} = 1$, to show that the dynamic signal depends on sample mass, heat capacity, and rate. Division of equation (6) by (5), followed by rearrangement, gives

$$c_{ps} = \frac{S_s - S_e}{S_c - S_e} \cdot \frac{m_c}{m_s} \cdot c_{pc} = \frac{S_s - S_e}{K m_s} \tag{7}$$

Equations (1)–(4) are deliberately cast in a somewhat different form to normal (when $I + K\Delta C_e$ are usually combined to give the "empty" signal, ΔC, of

Chapter 17 in Vol. 1) in order to emphasize the basic assumptions of this method of determining c_p. These are that $I(T)$ and $\Delta C_e(T)$ are reproducible from one run to another. The value of ΔC_e will only change if a pan of different mass is used in any (or all) of the operations corresponding to equations (2)–(4). Ideally, the same pan is used throughout any set of measurements but this is impossible after crimping or sealing, or for a strongly adhering sample, and here pans of matched mass must be used. Aluminum pans are easily matched but the process can be difficult when only a few specially made containers are available. In this case an additional run must be made with the empty pan (mass m_e, specific heat c_{pe}) removed from the sample cell so that the "empty" contribution is reduced by $Km_e c_{pe}$ and the signal $S_{e'}$ becomes

$$S_{e'} = I + K(\Delta C_e - m_e c_{pe}) \tag{8}$$

Subtraction of equation (8) from (2) gives the heat capacity c_{pe}

$$S_e - S_{e'} = Km_e c_{pe} \tag{9}$$

If the mass of the pan used to contain the calibrant, for example, is M_e, then equation (5) must be modified by the addition of a term, $(S_e - S_{e'}) \cdot (1 - M_e/m_e)$, to the left-hand side.

Reproducibility of I demands identical surface conditions in successive runs, and in principle these are obtained by careful attention to the position of the platinum lid (Figs. 1 and 9). It is, however, also important to ensure that the "package" of pan + sample also presents a reproducible environment to the cell. In practice, this means that lidded pans should always be used. This is because there is always a vertical temperature gradient, even under isothermal conditions, since the heater is located in the base of the cell. If work is carried out with, for example, graphite on the one hand or titanium dioxide on the other and no lid to the pan, then the temperature gradient through the cell will change (especially at high temperatures) and some dependence on the nature of the sample will be shown.

Although the constant K need not specifically appear in equation (7), its role is emphasized because its temperature response is a very useful indication of instrumental performance. Ideally it should be independent of temperature for most DSCs—for power-compensation instruments because it is an electrical conversion factor, unrelated to temperature, and for heat-flux instruments because the inherent temperature dependence is nowadays almost always nullified by some form of electronic compensation. The observed behavior of K is discussed later (Section 4.2.1, Fig. 13). Relation (7) is the basic DSC heat-capacity equation. It may also be applied, with qualifications, to such thermal events as melting or chemical reactions which are taken to be regions of unusual heat-capacity requirements. Qualifications are needed because, in theory, such regions require a more complex treatment to account for the disturbance of quasi-steady-state conditions due to the absorption of latent heat, for example. In practice, use of the simple equation (7) means that, although the apparent *shape* of a melting peak is incorrect, the integrated area beneath the peak still gives the correct total enthalpy change through the event. Because of the peak distortion there will be errors, for example, if attempts are made to calculate entropy changes through the event by summation of incremental values of $(c_p/T)\Delta T$.

Experimental uncertainties mean that the application of equation (7) is not straightforward, and a variety of problems must be recognized and overcome before the full potential of the technique can be realized. Problems are associated with baseline balancing, calibration, and sample preparation and stability; these are not necessarily independent—sublimation, for example, can dramatically affect baseline reproducibility—but form coherent topics for discussion.

3. BASELINE BALANCING

The idealized isothermal baselines of Fig. 3 only occur fortuitously in practice; normally (Fig. 4) all isothermal regions differ—a point that is made very clearly when digital methods of data recording are used. Some degree of curvature of the instrumental isothermal baseline is common, but in principle any pair of curves for empty and loaded conditions could still be made to coincide by a vertical shift of one curve on to the other. In practice, minor baseline differences are inevitable consequences of the uncertainty of any experimental measurement; major differences imply additional sources of error. Whatever the source or magnitude of the baseline imbalances, a mathematical formalism must exist to deal with the situation; the final results can then be used to define what "minor" and "major" differences mean in practice.

Two methods of baseline balancing are available. In one an arbitrary linear base is drawn as shown in Fig. 4a and the scanning signals (S'_s or S'_e, the latter is a negative quantity as shown in Fig. 4a) are as illustrated. The fact that the true baseline may be curved is immaterial, as any error should be the same for loaded and empty pan conditions—and will vanish on subtraction. If it is *not* the same, errors will arise and these are discussed in Section 6 (Fig. 17). The other method of baseline balancing forces the two curves into coincidence in one isothermal region, T_f in Fig. 4b, by adding $I_{fs} - I_{fe}$ to all points for the empty pan curve. The resultant curve i)' (Fig. 4b) remains unbalanced by an amount Δ (defined in Fig. 1b) at T_i; Δ is assumed to be proportional to the temperature difference so that at any temperature T

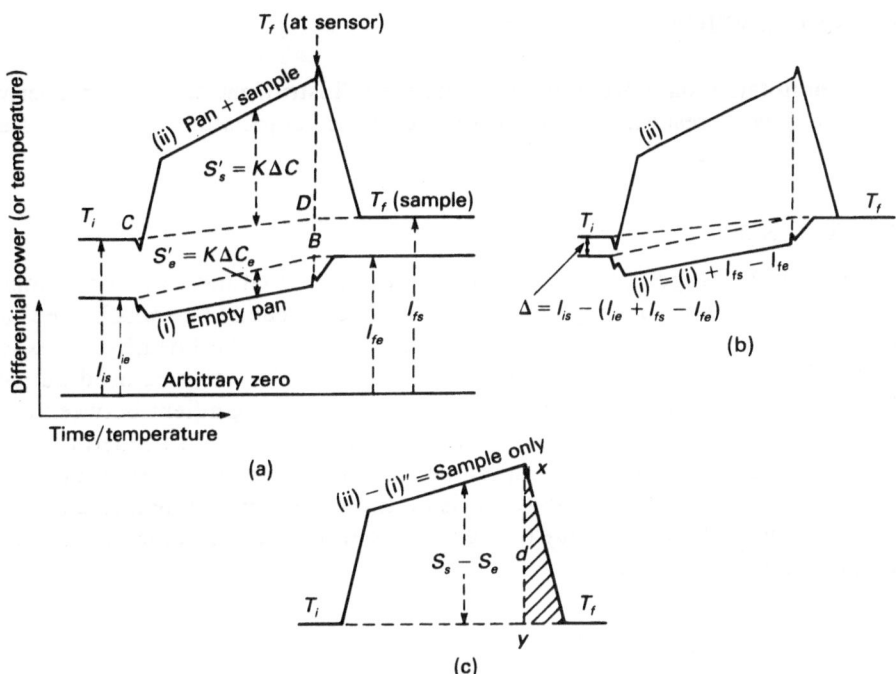

FIGURE 4. Baseline balancing to overcome differences in slope, $I_f - I_i$, for successive runs. See text (Section 3) for discussion. The shaded area in (c) is used to calculate thermal lag (Section 4.1.2).

$(T_i \leqslant T \leqslant T_f)$ the imbalance is $\alpha\Delta$, where $\alpha = (T_f - T)/(T_f - T_i)$ and $0 \leqslant \alpha \leqslant 1$. Appropriately scaled corrections $\alpha\Delta$ are then added to all data points for the empty curve i)' (for $T \leqslant T_f$) to give i)" (not shown), which closes the loop. Subtraction of i)" from ii) gives the final curve (Fig. 4c), which is usefully normalized to unit mass or molarity to allow direct comparison of materials. Whatever method of baseline balancing is chosen, it is always useful to record Δ as a routine measure of the goodness, or otherwise, of the match between the isothermal baselines.

It is helpful to display a DSC curve as it is being produced (chart recorder or electronic display) to give immediate warning of any unexpected behavior but data should always be stored and subsequently analyzed in digital form. (The need to measure ordinate displacements reproducibly from a chart recorder to fractions of a millimeter places unrealistic demands on both the recorder and the operator, as does the work involved in generating manually a "continuous" heat-capacity curve covering a span of perhaps 100–200 K.) In our own experience c_p can be routinely obtained from digital data with an overall accuracy of $\pm 1\%$ to $\pm 2\%$ (depending on the temperature) for suitable samples (requirements are discussed later), but this is impossible if reliance must be placed on recorder traces alone.

4. CALIBRATION

Quantitative data are only available if the instrument has been properly calibrated with respect to both ordinate (time/temperature) and abscissa (differential power).

4.1. Temperature

In any DSC experiment that features both isothermal and scanning regions the ordinate is time, but where necessary this can readily be transformed to temperature by using the known linear scanning rate (the linearity is easily checked). Naturally, the temperature sensors themselves may be used but as these are not located within the sample there will be a gradient between the two, even under isothermal conditions, and calibration is still needed to give the sample temperature. The temperature calibration is conveniently divided into two: an isothermal contribution (is an indicated constant temperature of T_I really correct?) and one due to thermal lag when a programmed change of temperature is occurring.

4.1.1. Isothermal Calibration

This is readily carried out using slow, stepwise, temperature increments through the melting point T_m or transition temperature T_t of a well-characterized reference material. A typical set of stepwise melting curves is shown in Fig. 5. Away from T_m only heat capacity must be supplied in each increment and equilibration takes only one or two minutes. As T_m is approached, premelting increases the amplitude of the perturbation for each increment Z while at T_m itself the transmission of enthalpy of fusion may take many minutes—the rate of heat transfer is slow because sample and heat source are almost in equilibrium. It is always good practice to perform two or three further increments beyond T_m to check that melting is indeed complete. The isothermal temperature correction is $\delta T_I = T_m - T_I$ (T_I is the indicated temperature of the instrument at T_m) and it may be reduced to fractions of a degree over the whole range of operating temperatures by following procedures

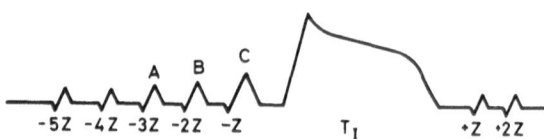

FIGURE 5. Isothermal temperature calibration using a material of known melting point T_m. Premelting shown at A, B, C; when correctly calibrated $T_I = T_m$. Temperature increments Z can vary from 5 K (rough survey) to 0.1 K (final definition).

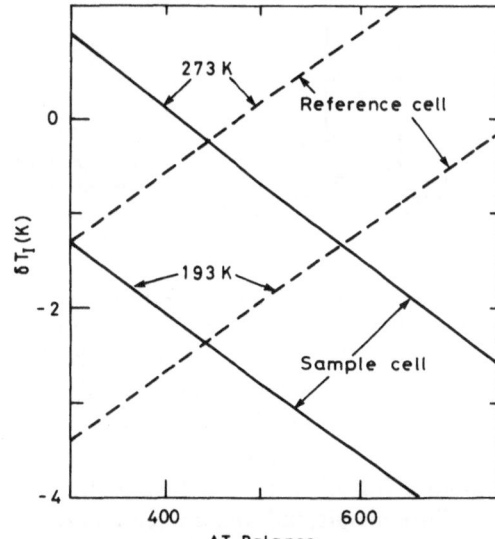

FIGURE 6. Quantity δT_I (Section 4.1.1) as a function of the "ΔT Balance" setting and of the temperature (193 K or 273 K) ambient to the DSC cells.

detailed in the instrument manual. However, provided the form of δT_I is known (a quadratic in T_I is generally quite sufficient) there is little to be gained by spending excessive amounts of time on instrumental manipulation to minimize δT_I; the appropriate equation can always be incorporated in the final data treatment.

The quantitative response of a DSC is influenced by instrumental settings. Obviously, changes of those specific controls that affect temperature and power must be followed by fresh calibrations, but other factors also influence quantitative behavior. Particularly relevant are the "ΔT Balance" control and the ambient temperature. Baseline curvature is minimized by appropriate use of ΔT Balance but, as Fig. 6 shows, both sample and reference temperatures are affected—and in opposite directions. Careful calibration is particularly important in this case if the *reference* cell contains a temperature standard, because this kind of "internal" calibration is sometimes used to monitor "accurately" the temperature at which an unknown event occurs (Fig. 7).

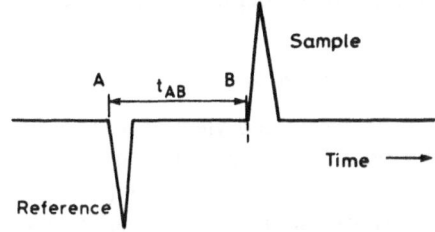

FIGURE 7. A melting point calibrant is sometimes used as an "internal standard" in the reference cell to define the melting behavior of an "unknown" sample. For this procedure to be successful, the time t_{AB} (and hence temperature difference) must be supplemented by the information of Fig. 6.

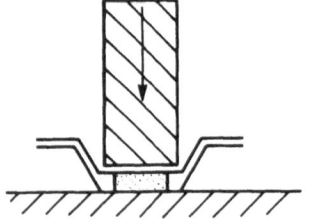

FIGURE 8. Use of two sealable pan bases to minimize the vapor space when using a volatile material as a temperature calibrant. The two are pushed firmly together on a solid surface prior to sealing.

Various methods are available to reduce the temperature ambient to the DSC cells for low-temperature operation. Figure 6 shows how δT_I changes as ambient is reduced from 273 K (ice bath) to 193 K (Intracooler 2). If frequent changes of ΔT Balance and/or ambient temperature are made (the latter is increasingly likely with newer cooling systems, which allow "ambient" temperature to be preselected) it is worth incorporating their values into an equation for δT_I which, in turn, is built into the general DSC data treatment.

Pure metals, indium, tin, lead, and zinc are ideal as temperature calibrants, melting over only 0.1–0.2 K, but they cover only the range 430–700 K. Outside this range organic standards (low temperatures) or inorganic materials must be used. Mercury is a useful additional low-temperature ($T_m = 234.3$ K) standard but gallium ($T_m = 302.9$ K) must be contained in nonaluminum pans. The range of materials for which melting and/or transition temperatures are thermodynamically well defined to ±0.1 K is surprisingly limited. DTA temperature calibration standards are supplied by the US National Institute of Standards and Technology (NIST) on behalf of the International Confederation for Thermal Analysis (ICTA) and organic materials with well-defined melting points are available from the UK Laboratory of the Government Chemist (LGC). It is always wise to seal organic compounds in pans intended for use with volatile materials, because vapor pressures may be such that there is a steady loss of material—often with unfortunate consequences for subsequent instrumental stability. Since only small amounts of material are needed, a good sample for temperature calibration can be made by sandwiching it between two "volatile" pan bases (Fig. 8). In this way the vapor space is negligible and it becomes practicable to use benzoic acid, for example, as a temperature calibrant. Whatever melting point standard is used it should always have been premelted to ensure good sample/pan contact.

4.1.2. Thermal Lag

An expanded view of a loaded DSC 2 sample holder is shown in Fig. 9. During a heating run the sample temperature lags behind that of the sensor, the magnitude of the lag depending on the heating rate (especially) and on the sample size and geometry. The two interfaces LM and PQ are major barriers

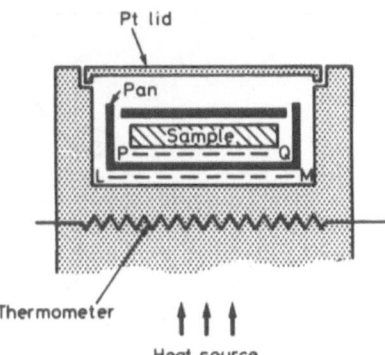

FIGURE 9. Loaded sample cell, Perkin-Elmer DSC 2, showing the major sources LM, PQ of thermal lag.

to efficient heat transfer. The effect of PQ may be minimized by crimping, sample compaction, or prior melting to give good contact, but for many materials none of these may be feasible and large lags will be inevitable. Any estimate of thermal lag δT should therefore be specific to an individual experiment and a simple procedure is to base it on the enthalpy lag δH at the end of a run. In the absence of lag there would be an immediate return to the baseline (XY, Fig. 4c), but in practice a finite time is required and the shaded area δA traced out is related to δH and δT via

$$W\delta A = \delta H = m_s c_{ps}\delta T = \frac{d}{K}\,\delta T \qquad (10)$$

where W is the area-to-enthalpy calibration factor (discussed in more detail in Section 4.2.2) and d is as shown in Fig. 4c. Although equation (10) shows how c_{ps} is calculated [as $d/(m_s K)$], it is only after the derivation of δT that it becomes clear to what temperature the calculation refers.

This experimental derivation of δT is very useful in showing what factors are important in day-to-day operation. Figure 10 shows how δT varies with heating rate for two alumina disks of the same diameter but different thickness. Extrapolation of a family of similar curves to zero thickness (or mass) gives the line $\delta T_0'$ and this is the value of δT at the interface PQ. If samples are placed directly in the DSC cell (this must only be done using thermally inert materials), then the equivalent extrapolation, to δT_0, gives the lag at the face defined by LM (Fig. 9). From Fig. 10 the ratio $\delta T_0': \delta T_0$ is roughly 2:1, the ratio of the number of interfaces. If equivalent dynamic signals are considered—that from a sample of mass $2m$ heated at a rate $0.5\dot{T}$ is the same as that from mass m heated at $\dot{T}$—Fig. 10 shows that lag is less for the large sample heated slowly. If the treatment for thermal lag described here *cannot* be applied, temperature errors for equivalent signals are clearly minimized by the slow heating of large samples.

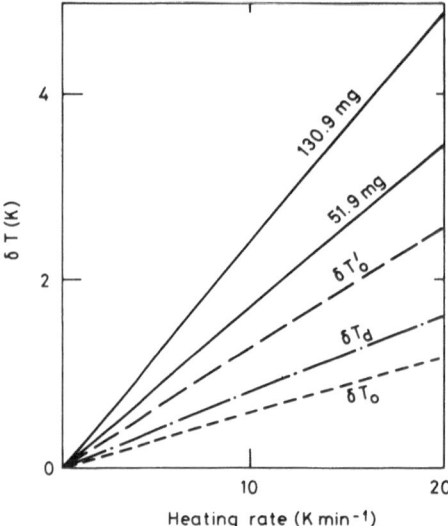

FIGURE 10. Thermal lag (δT) for 6-mm-diameter disks of α-alumina with the mass shown. δT_0, $\delta T_0'$ are the derived values of thermal lag at the interfaces LM, PQ (Fig. 9); δT_d is the conventional dynamic temperature calibration.

4.1.3. Dynamic Temperature Calibration

It was emphasized above that samples should normally always be in pans for any type of DSC work. The "panless" operations leading to δT_0 were only carried out to show the link between the conventional, dynamic temperature calibration δT_d, described in most instrument manuals, and the quantity δT discussed above. The value of δT_d is obtained from the shape of the melting curve of a very pure material which has already been through a melting and crystallization cycle so that it is in intimate contact with the pan. The major thermal resistance in the system is therefore the remaining interface LM. Normal scanning conditions are perturbed at a phase change and it can be shown theoretically that the melting point T_m is represented by the point of intersection T_d of the "leading edge" with the scanning baseline (Fig. 11), so that $\delta T_d = T_m - T_d$. Quantity δT_d, like δT_1, can be minimized over a wide

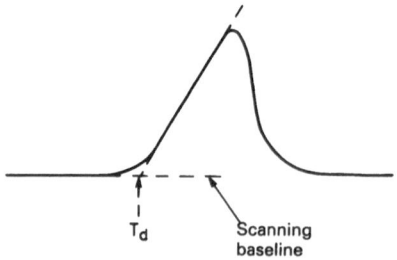

FIGURE 11. Definition of T_d in a dynamic temperature calibration.

range of temperature by manipulation of the instrumental controls. If, for simplicity, $\delta T_I = 0$, then the relationships between the various lags are easily demonstrated. Figure 10 shows that δT_d is always a few tenths of a degree greater than the corresponding value of δT_0—the lag at the interface LM. In fact, in an idealized experiment with a very small sample perfectly in contact with the sample pan, the two would be equal. Some idea of the error in temperature $\delta T - \delta T_d$ that occurs if δT_d alone is used may be obtained from Fig. 10. An obvious way to bring δT_d more into line with conditions in a typical heat-capacity experiment on granular material, for example (this will certainly involve the additional interface PQ), is to simulate the extra barrier by introducing another lid (Fig. 12), when δT_d will increase to a value slightly greater than $\delta T_0'$.

4.1.4. General Comments on Temperature and Thermal Lag

The conventional temperature calibration δT_d is only valid for a small sample in good thermal contact with its pan (conditions that are the reverse of those expected in heat capacity work, especially on solids). It is naturally only valid for the particular heating rate used. It cannot be used for cooling experiments because all materials supercool (often by tens of degrees) and the degree of supercooling is impossible to predict for the conditions of a DSC experiment. The only reversible phase changes on the time scale implied by scanning rates of up to 40 deg min^{-1} are liquid crystal (lc) $\rightleftharpoons$ lc, or lc $\rightleftharpoons$ isotropic liquid transitions. These, however, cluster over the region 400 ± 100 K while cooling experiments are of particular value in defining the high-temperature performance of a DSC (Section 6).

By contrast there is never any ambiguity about the isothermal temperature correction δT_I (Section 4.1.1) and the lag term δT can obviously be derived from cooling as well as heating curves. Any DSC determination of heat capacity requires measurements on both sample and calibrant and the two will only fortuitously have identical values of δT. Data can be interpolated to common temperatures for the most careful work or a mean value can be used—this is usually sufficiently accurate, especially when some experience has been gained, for it is not difficult to match sample and calibrant dimensions to give similar

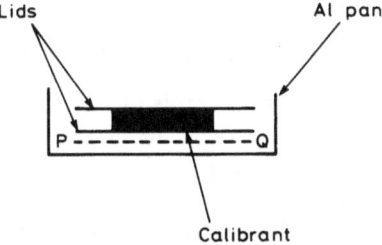

FIGURE 12. Insertion of an additional lid underneath the temperature calibrant restores the interface PQ and gives conditions for the dynamic calibration that better approximate those in heat-capacity work.

values of δT. An additional complication is the variation of δT with temperature. This may be due to monotonous changes in the thermal properties of the complete sample/cell assembly or, more important, events such as melting of the sample when the interface PQ may effectively be eliminated. The first effect is small and may be neglected for a run spanning as much as 200 K but, where there is melting at some intermediate temperature, use of δT calculated at the end (molten state) of the experiment will mean an error in temperature for the phase present initially. In this case the run should be split, to end first at a temperature in the solid state before a second run through the melt.

Comments on thermal lag refer specifically to thermally inert regions where phase changes, etc. are absent. The special conditions that obtain in a melting region, for example, are irrelevant for the purposes of this chapter except for the definition of T_d as the melting point (Fig. 11). However, even here, as mentioned in Section 2, the apparatus records the *total* enthalpy change through T_m correctly—the directly recorded curve is only incorrect in showing the *distribution* of the enthalpy change with temperature.

Thermal lag has been reduced by coupling the pan to the sample holder by a drop of silicone oil. This effectively removes the δT_0 term, but the technique has little application in heat-capacity work because it is impossible to keep a constant mass of oil as the pan is repeatedly inserted in the DSC. Additional complications are thermal volatilization/degradation at higher temperatures or glass temperature or melting phenomena at low temperatures.

4.2. Heat Capacity Calibration

Equation (7) shows how c_p is related to the dynamic signal—the ordinate scale of the DSC—through K and m_s. Formally, since the ordinate scale is a differential power, it must be divided by the heating or cooling rate to give the required units of heat capacity (cal or, preferably, $J \deg^{-1}$). However, specific heats are always *ratios* of the dynamic signals [equation (7)] from experiments using a common rate, so scale conversion is unnecessary. It is appropriate to work in terms of K, which can be obtained over a wide range of temperature by measurements [equation (5)] on a stable calibrant, such as α-alumina.

4.2.1. Ordinate to Heat Capacity Conversion

A 50-mg sample of alumina heated at 20 deg min^{-1} gives a dynamic signal of some 40 mV at 400 K on DSC range 1. Since $c_{pc} \sim 1.00 \, J \, g^{-1} \, deg^{-1}$, K is about 0.8 V deg J^{-1}. Routine data treatment should always record K so that its stability, both with respect to temperature and from day to day, can be checked (comparisons must always be made using the same instrument settings

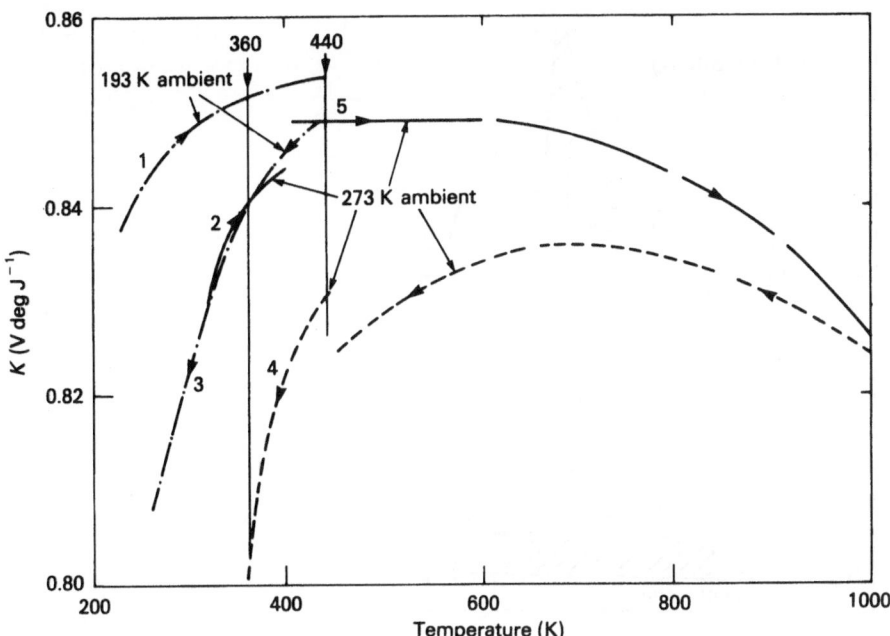

FIGURE 13. The factor K [from equation (5), alumina calibrant] as a function of DSC temperature and ambient conditions. Heating or cooling is indicated by direction of arrows. Results at 360 K and 440 K are shown in Table 2.

and experimental conditions). Figure 13 shows K values for a Perkin-Elmer DSC 2 in both heating (+) and cooling (−) modes and using low-temperature refrigeration (Intracooler 2) or ice as the DSC cell coolant. Clearly K is *not* independent of temperature and $K(+) = (1.01 \rightarrow 1.03) \cdot K(-)$. The general pattern shown in Fig. 13 is reproducible to ±1-2% (2% in the range 800–1000 K) over a period of many months. Other calibrants with very different thermal properties (metals, solid and liquid organics) give K values that agree to within ±1% with those of Fig. 13. Obviously for calibrants there is no distinction between this and the reverse process, the calculation of c_p using K values based on alumina, and ±1% defines the overall accuracy of specific heats determined by DSC—this is considered in more detail in Section 7.2.

4.2.2. Area to Enthalpy Conversion

In this method of calibration an area A_c, defined in some way by DSC curves (Fig. 14a-c), is related to a known enthalpy (H) change to give the conversion factor W:

$$WA_c = m_c[H_c(T_2) - H_c(T_1)]$$

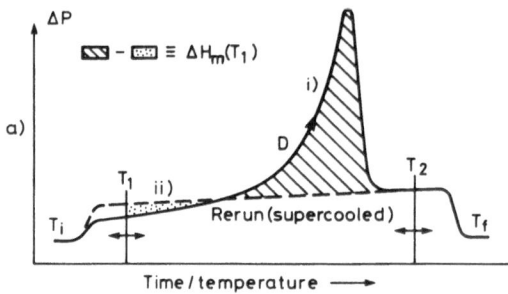

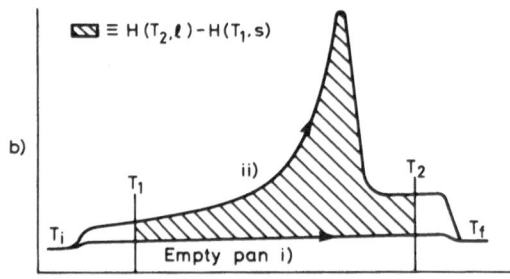

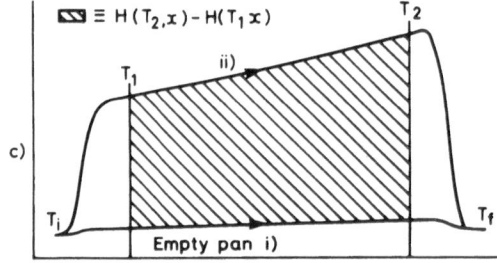

FIGURE 14. Definition of areas used for the calibration of enthalpy changes according to Section 4.2.2. (a) and (b) illustrate melting (solid, s → liquid, l) but any well-defined phase change may be used; in (c) the phase x remains unchanged from $T_1 → T_2$.

and W is then in turn used to convert an unknown area to an enthalpy change; division by $T_2 - T_1$ gives the average heat capacity over the range $T_1 → T_2$—the same procedure that is used in adiabatic calorimetry. (T_1 and T_2 represent any two temperatures in either the dynamic or isothermal regions.) The calibrant area usually corresponds to a heat of melting (ΔH_m) or transition (ΔH_t) (Figs. 14a, b) but a straightforward enthalpy change due to heat capacity alone is equally valid (Fig. 14c). An "area" type of calibration was the first to demonstrate the quantitative potential of DSC, but certain material characteristics must be recognized before results are compatible with instrumental capabilities. Although the detailed geometry of a DSC melting curve is mainly dominated by thermal resistance across the DSC cell/sample interface, premelting prior to the T_m region can be appreciable on the scale of normal solid or liquid heat capacities. There may also be large differences between c_{ps} and c_{pl}. Both these effects must be considered when deriving an area that represents "ΔH_m." By using the fraction transformed it is possible to construct a baseline that accounts for the varying power demands through a melting (or phase-change) region. There is, however, a much simpler technique which only requires runs in i) melting, followed by cooling and ii) rerunning the super-cooled liquid (Fig. 14a). The sample remains undisturbed throughout and any errors will tend to be similar and therefore cancel in subtracting the two curves to give

$$\Delta H_m(T_1) = [H(T_2, l) - H(T_1, s)] - [H(T_2, l) - H(T_1, l)] \qquad (11)$$

Provided that T_1 is some 10 K or more below T_m, "premelting" will be small—and most materials will supercool to this extent on the time scale of a DSC experiment. Of course, calibration data in the literature refer to $\Delta H_m(T_m)$, not $\Delta H_m(T_1)$, but for a well-characterized calibrant c_p data should be available to allow the correction from T_m to T_1 to be made. The original literature often gives $H(T_2, l) - H(T_1, s)$ [or $H(T) - H(0)$, which leads to this quantity] and this may be obtained via DSC by running the empty pan as well as the pan + sample over the range $T_1 \rightarrow T_2$ (Fig. 14b). Three variants of the "area" calibration are therefore possible, each involving two runs:

(a)	i) sample melting ii) supercooled liquid	area equivalent to $\Delta H_m(T_1)$	$T_1 < T_m < T_2$
(b)	i) empty pan ii) sample melting/ crystallizing	area equivalent to $H(T_2, l) - H(T_1, s)$	$T_1 < T_m < T_2$
(c)	i) empty pan ii) sample	area equivalent to $H(T_2, x) - H(T_1, x)$ x = solid, s or liquid, l	$T_1 < T_2 < T_m$ $T_m < T_1 < T_2$

(b) and (c) may be carried out in cooling as well as heating. Emphasis is placed on a variety of calibration procedures because it is then possible to monitor instrumental performance under a range of conditions. Thus (c) above is perhaps the least demanding with the signal (from heat capacity effects alone) changing only slowly with temperature. In both (a) and (b) there is a rapid change which may be extreme for (b) in cooling when, given sufficient supercooling, some materials crystallize in only a few seconds. Since ΔH_m must be a function of temperature it is extremely unlikely that independent data will be available for the crystallization step many degrees below T_m. This is not serious, however, because for the temperature cycle $T_2 \rightarrow T_1 \rightarrow T_2$ the overall enthalpy change should be zero—that is, $H(T_2, l) - H(T_1, s)$ may be used from data obtained in heating.

Area calibrations confirm the results of Section 4.2.1 with respect to temperature dependence and 1–2% differences between factors for heating and cooling experiments.

Although initially disturbing, the changes in K (Fig. 13) and W can be turned to advantage if it can be demonstrated that, at a given temperature, a range of K values always gives a consistent specific heat. It will be seen in Section 7 that this is indeed the case.

4.2.3. Calibration Materials

Calibrants should be pure, stable, widely available materials for which reliable thermodynamic data exist—preferably obtained by adiabatic or drop calorimetry. The most generally used heat capacity calibrant is α-alumina, which is available in very high purity billets that may be machined to disks that fit snugly into the various types of DSC pan. Below room temperature the heat capacity of alumina drops rapidly and a better calibrant for temperatures up to about 320 K is probably benzoic acid. This, however, sublimes well below its melting point of 395.5 K and vaporization can be troublesome unless special precautions are taken (Fig. 8). A calibrant with very different thermal properties is molybdenum, which is a useful alternative to alumina at high temperatures.

Heats of fusion of indium, tin, and lead are potential enthalpy calibrants in addition to integrated heat capacities of the calibrants above. A whole range of organic materials including diphenyl ether and many n-alkanes are potential candidates for the cooling/heating cycles of the type described in Section 4.2.2. Water, of course, has been widely investigated and it is especially interesting for the 1% decrease in c_p over the range 273–303 K followed by a 1% increase to 373 K—DSC should show this effect clearly but, as with benzoic acid, effective sealing is essential.

5. CONTAINMENT AND SAMPLES

5.1. Pans

Aluminum pans, 1 mm × 6 mm diameter, are used for routine work up to 800 K. They are closed by a lid which rests on the sample and may be crimped (lightly, at three points on the circumference, or more securely using a special tool) if required—for example, if the sample stress relaxes, melts, or otherwise changes shape during a run so that the lid, unless restrained, may tilt. An important reason for using a lid is to maintain reproducible conditions throughout a set of measurements (see Section 2).

Traces of oil from the forming operation may remain on aluminum pans and lids, and batches should be cleaned in an organic solvent prior to use. A routine operation should be to flatten the base of every pan before use by means of a metal rod whose diameter is equal to the internal diameter of the pan. Such a tool is also of service in reforming a lightly crimped pan which has contained an easily removed material.

A variety of other metals (especially gold or platinum) or nonmetals (graphite, silica, alumina) can be used in place of aluminum for containing reactive samples or to extend the operating temperature range to 1000 K.

If a material *consistently* loses mass on repeated runs over a given temperature range, it is a sign that it has a high vapor pressure and should be contained in one of the sealable pans that most manufacturers supply— otherwise, the apparent "heat capacity" will contain an additional, unknown component due to sublimation or evaporation (*consistent* mass loss is emphasized; many samples lose moisture or other volatiles on the first heating). Du Pont (now TA Instruments) sealable pans are especially useful, being substantial enough to contain internal pressures of a few atmospheres. Liquids are generally more easily dealt with using sealed pans, although the conventional type should not be ruled out—the lid is generally held in place by surface tension and careful handling should avoid spillage. Special pans are available for containing pressures of the order of tens of atmospheres, but these are generally not relevant to heat capacity work.

5.2. Samples

A great advantage of DSC is its ability to accept samples having a variety of physical forms—although the small mass needed means that thought must be given as to how representative of the bulk is the sample chosen. The most convenient form to deal with is a disk, which can simply be laid in a pan with an uncrimped lid on top. At the end of a run it is easily removed and may be replaced by a fresh sample, so that sequential operations are particularly easy—the mass of the pan + lid remaining constant throughout.

Any surface burrs can generally be removed from a disk by rubbing on an appropriate grade of abrasive paper. Irregular chunks may be similarly treated to give one flat face for good thermal contact. Powders and films generally present no problems, but the pan-shaping tool discussed in Section 5.1 is also useful for gentle tamping, especially of light, fluffy material such as some freeze-dried organics. It is tempting to consolidate such difficult-to-pack specimens by compacting them in a die, but great care is needed—the pressure intensification from a 75-mm- to a 5-mm-diameter ram, for example, is large and the sample can be left with stored energy, which may be released on subsequent heating to give an apparent minimum in the c_p-T curve (Fig. 15). If this is suspected, a rerun will produce a monotonous curve that may

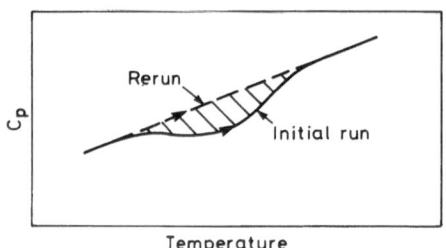

Temperature

FIGURE 15. An apparent decrease in heat capacity may be due to the release of stored energy. The amount (shaded area) is defined by a second run.

be used to define the energy released (shaded area, Fig. 15). Similar effects are also found when a material anneals and this is quite common after crystallization by cooling in the DSC.

When working with materials for the first time, especially at high temperatures, it is wise to preheat the sample/pan combination in an independent furnace to check whether there is any reaction between the two. If a separate furnace is not available, an initial screening run should be made with the DSC cell protected either by a layer of mica or by enclosing the whole sample/pan assembly in a larger pan of some inert material. Even if there is no reaction the surface tension of some molten materials is such that they can creep out of their containers, and if there is any possibility of this it should be checked by preliminary experiments because cleaning a contaminated cell can be very difficult.

For high-temperature work (above about 800 K, although much depends on the condition of the DSC) it is good practice to give the system (including sample) a preliminary bake at the maximum operating temperature. Without this there tends to be extensive drift (Fig. 18) in isothermal regions. The prebake may occasionally destroy features of interest (annealing, phase changes) and obviously cannot be used in such circumstances, but for the heat capacity of a thermally inert material it is very useful and considerably reduces the

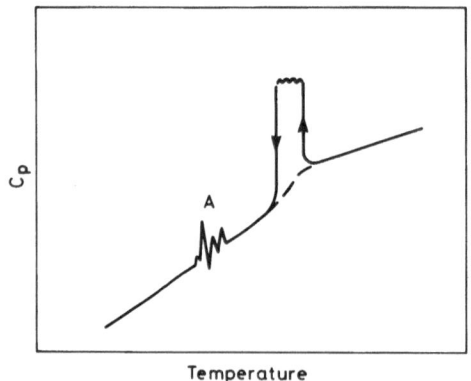

Temperature

FIGURE 16. Differential contraction on cooling may lead to a sample pulling away from the pan. A locally noisy signal (A) is often found at this point.

uncertainty and scatter of results. A very different problem is sometimes encountered in cooling runs some way below a vitrification or crystallization temperature. Differential contraction is such that the sample suddenly pulls away from the pan—often quite violently—and the event is marked by seemingly random noise in the DSC output (Fig. 16).

6. OPERATION AND SOURCES OF ERROR

A heat capacity-temperature curve requires three sets of data, exemplified by equations (2)-(4). Several samples may be run in conjunction with a common empty/calibrant pair (bearing in mind the comments of Section 2 concerning the sample pan mass) but these latter should be run midway through the total number. In this way the empty pan and calibrant baselines correspond to some "average" value even if the overall curvature is slowly changing, perhaps because of a slow buildup of ice in low-temperature operation: in principle, the working area of a DSC cell is protected from icing by enclosing it in a dry box but, in practice, it is difficult to prevent *some* ice from forming. Because of this, the normally matt black surroundings of the cell become white and the heat transfer term I [equations (1)-(4)] is changed. Fortunately, quantitative work is still possible once an initial layer of ice has formed as subsequent changes are slow (a timescale of hours).

If measurements are required over a wide range of temperature, those using low-temperature refrigeration should not be extended to more than 400-500 K because icing uncertainties become increasingly important at high temperatures. Evolution of volatiles compounds the problem because these immediately condense on the surroundings, again changing I. Measurements in the range 300-1000 K are therefore best made using ice or a thermostatted water supply as "ambient." In this way volatiles tend to be carried well clear of the cell environment or, if they condense, it is often as a liquid which has little practical effect on I and hence on the baseline curvature. When covering a large range of temperature, runs should be split into 100-200 K increments—the lower value is recommended for the region 700 or 800 K to 1000 K where errors are larger. An advantage of the incremental procedure is that it gives some idea of experimental reproducibility from the smoothness with which successive curves join—especially when comparing results using cryogenic and room-temperature conditions for ambient, because these require separate calibrations.

Where possible, measurements should always be made in both heating (+) and cooling (−) modes, because baseline errors tend to be of opposite sign. Figure 17a shows idealized curves for empty and loaded pans heated from one isothermal region (at T_i) to another (at T_f) and then cooled back to T_i. Thermal lag is neglected as are the small differences between results obtained in heating and cooling. With these simplifications the dynamic signals

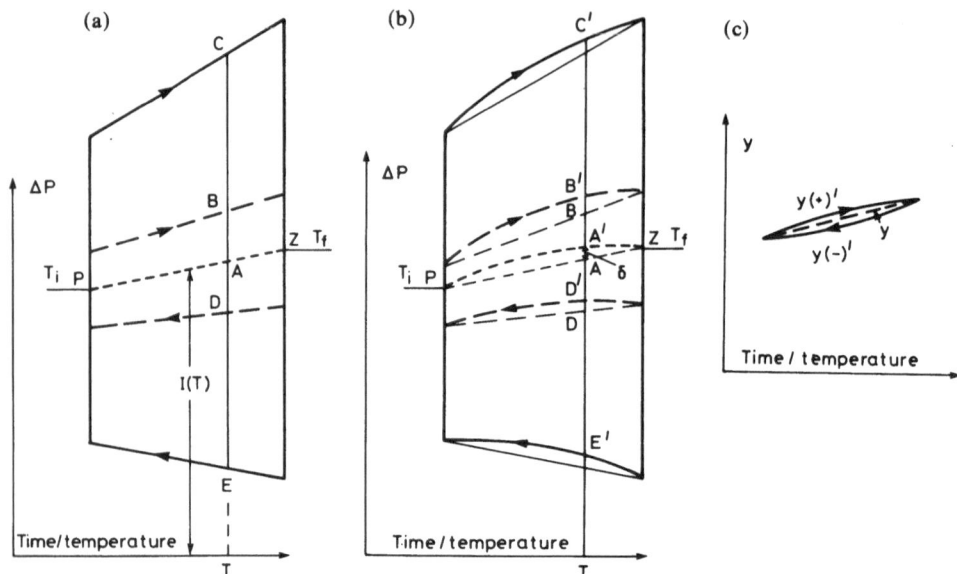

FIGURE 17. Simplified DSC curves to show the effect of errors in baseline curvature. See text for discussion.

of equations (5) and (6) are given by $y(+) = AC - AB$, and $y(-) = AE - AD$, and $y(+) = y(-)$. The same curves are shown in outline in Fig. 17b, but the bold curves here refer to conditions when the isothermal baseline, PA'Z, has some curvature. Locations corresponding to equivalent points on the previous curve (Fig. 17a) are denoted by primes and, again,

$$y(+) = A'C' - A'B' = y(-) = A'E' - A'D'$$

These relationships recognize that $I(T)$ is reproducible (PAZ or PA'Z) for each pair of runs in Fig. 17b. However, if the isothermal baseline is different for some reason (incorrect location of platinum lids, sublimation, or ice formation) for one of the runs in the empty/calibrant/reference cycle this will become immediately obvious from a comparison of results in heating and cooling. Suppose that the empty pan has been correctly run but conditions have changed for the sample: in the latter case signals will still be taken relative to the now incorrect linear baseline PAZ. The values of y are $y(+)' = AC' - AB$ and $y(-)' = AE' - AD$ and $y(+)' = y(-)' + 2\delta$ where $\delta = AA'$ (the mismatch of the isothermal baseline at the particular temperature T). The variation of y' with temperature is given in Fig. 17c, which shows clearly how baseline errors are reflected in heating and cooling experiments. Maximum errors tend to occur around the middle of the temperature increment, which is why it was recommended above that this should not normally cover more than about

200 K. Good "point" estimates can always be made at the end of a run by noting the signals at points M,Z(sample) and N,Z(empty pan) (Fig. 18) to obtain the quantity $h - h_e$ whence $c_p = (h - h_e)/(Km_s)$. The constant K [equation (5)] may be known from previous work or it can be obtained by repeating the above procedure for the calibrant. The final reading Z should correspond to the attainment of equilibrium but, even when this is not reached, due to some endo- or exothermic event (Fig. 18) it may still be possible to estimate the underlying sample heat capacity. For sluggish events there is often a nearly-linear region of drift which may be back-extrapolated to the point (Z') where the programmed maximum temperature is first reached; MZ is then replaced by M_xZ' or M_nZ' in the calculation of h. This procedure is very effective in the examination of such materials as coals, which show a complex series of physical and chemical effects in the region 300–600 K. An advantage is that complete sets of data going from one isothermal region to another are not required; it is sufficient that quasi-steady-state conditions have been attained at M—in practice, this means that at least a minute of programmed heating has elapsed before reaching M.

For simplicity, the two different isothermal baselines of Fig. 17 were shown with equal slopes, $I(T_f) - I(T_i)$, but different curvatures. In practice, the latter would be signalled by changes in the former and for this reason values of $I(T_i)$ and $I(T_f)$ should be routinely recorded so that unusual slopes

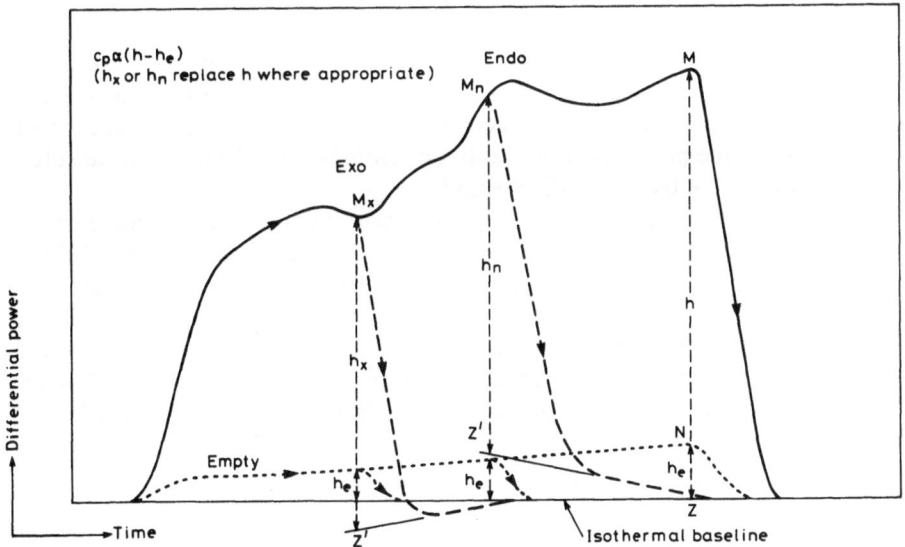

FIGURE 18. Schematic DSC curve showing exo- (x) and endothermic (n) events. The programmed temperature rise stops at M. If this is in an exo- or endothermic region (at M_x or M_n, for example) c_p at this point can be estimated using the construction shown.

are immediately obvious. Remedial action can then be taken—relocation or reshaping the platinum lid of the cell, or removal of ice or sublimed material—before rerunning.

Discussion of errors has so far concentrated on the reproducibility of the $I(T)$ term of equations (1)–(4). In principle there should be little difficulty in avoiding errors in the $\Delta C = \Delta C_e + m_x c_{px}$ terms. Quantity ΔC_e requires either pans with masses identical to ± 0.01 mg for samples of 10–100 mg (typical values for c_p work in, respectively, organic materials and metals) or the application of the mass corrections detailed in equations (8) and (9). Problems can still arise with composite pans, but these are unusual for heat capacity work. Modern microbalances are capable of routine weighing to much better than the ± 0.01 mg accuracy needed for m_x. Errors associated with this term are mainly due to loss of material during a run. If the volatile component is important, then the problem can be minimized by using sealed pans (although the measured c_p may contain a small additional component due to heat of vaporization). More usually, volatilization implies the presence of an undesired impurity (moisture, machine oil, finger grease) which should be removed by prebaking or by rerunning if its presence was only discovered by a loss of mass at the end of the experiment (reweighing should be routine, if only to confirm the original value). Loss of moisture is particularly troublesome because the DSC curve can be perturbed over a wide range of temperature (largely determined by the sample geometry); ice will form if ambient conditions are cryogenic and the change of I inevitably leads to a degradation in the instrumental performance.

The heat capacity calibrants recommended in this chapter, benzoic acid and, especially, α-alumina (Section 4.2.3), have been very carefully investigated using adiabatic calorimetry and uncertainties in c_{pc} are thought to be, at most, a few tenths of one percent—a figure that obviously sets a limit on the absolute accuracy attainable by the DSC method.

From the preceding discussion gross errors in the final c_p data may be expected for certain combinations of samples and experimental conditions. Such combinations must obviously be eliminated before the full instrumental potential can be realized, but the final requirements are not particularly stringent; samples must be stable and the environments of the sample and reference cells should remain unchanged throughout a series of experiments. Results obtained under these conditions are discussed in the final section.

7. INSTRUMENTAL PERFORMANCE

7.1. Consistency

Performance can be judged at several stages along the route to the final figure for the overall accuracy. Reproducibility of K values over a period of

TABLE 1
Heat Capacity ($J g^{-1} deg^{-1}$) via Heating (+) or Cooling (−) Experiments

Temperature (K)	350		550		750		950	
Material	+	−	+	−	+	−	+	−
High nickel superalloy	0.450	0.452	0.499	0.505	0.527	0.524	0.604	0.608
Calcium silicate	0.801	0.801	0.960	0.954	1.035	1.025	1.049	1.046
Coke	0.843	0.840	1.292	1.275	1.521	1.548	—	—

time has already been discussed (Section 4.2.1). The limits of ±1% to ±2% (depending on temperature) refer to entirely independent experiments covering many months. The unexpected features of Fig. 13—the regions where $K = f(T)$ and the differences between $K(+)$ and $K(-)$—cannot be explained by errors in the temperature calibration; these would need to be unbelievably large to restore the theoretical expectation of a unique value for K. This point is stressed to emphasize the importance of the comparative procedure [equation (7)] that is used to determine c_p. This gives excellent reproducibility and agreement from one set of experimental conditions to another. Table 1 shows $c_p(+)$ and $c_p(-)$ for three very different types of material. The *consistency* demonstrated in Table 1 is also shown for other changes in experimental conditions. Figure 13 shows that, although K is a function of the temperature ambient to the DSC cells, whatever conditions are used (differences in ambient, heating/cooling) overlapping heat capacity data always agree to within ±1% (Table 2). This production of consistent data gives confidence in both the instrumental performance and the data treatment.

An additional check on the consistency of results is the summation of enthalpy changes over the temperature cycle $T_2 \to T_1 \to T_2$. If the physical state at T_2 is well defined (a liquid, for example) the overall enthalpy change should be zero, and if this condition is met it again gives confidence in the calorimetric response of the DSC. The test is especially demanding if there is

TABLE 2
Heat Capacity of UO_2 ($J g^{-1} deg^{-1}$)[a]

Data set	Heating (+) or cooling (−)	360 K		440 K	
		c_p	(factor K)	c_p	(factor K)
1	+	0.255	(0.851)	0.270	(0.854)
2/5	+	0.253	(0.841)	0.270	(0.849)
3	−	0.252	(0.840)	0.269	(0.849)
4	−	0.254	(0.799)	0.270	(0.830)

[a]Conditions as in Fig. 13.

a phase change between T_1 and T_2. Experience in this laboratory is that there are random errors (i.e., not biased toward heating or cooling) of, at most, 1% of the overall value of $H(T_2) - H(T_1)$ for T_2 up to about 700 K. Thus the imbalance at the end of a total cycle of 800 J g^{-1} (400 J g^{-1} in cooling and heating) would be expected to be less than 4 J g^{-1}.

7.2. Overall Accuracy

The *consistency* discussed in Section 7.1 is an essential step on the path to calorimetric data, but the latter can only be quantified by comparison with independent measurements. Many organic and inorganic materials are available that have been characterized in this laboratory using adiabatic calorimetry. This has an overall uncertainty of ±0.1% and DSC data always agree with adiabatic results to within ±1%—this figure also applies to overall enthalpy changes. No independent NPL data are available to test the high-temperature performance of the DSC, but the NIST figure for the heat capacity of molybdenum at 1000 K can be reproduced to ±2%.

All tests made in this laboratory can be readily summarized by the statement that the overall accuracy of DSC determinations of heat capacity is ±1-2% with the higher figure referring to the 800-1000 K region. This accuracy is routinely obtained with stable materials, but inappropriate samples (e.g., having appreciable vapor pressures, adsorbed water or other contaminants, or in metastable states that anneal in the measuring range) can lead to inaccurate or even meaningless results. However, such problems are not relevant to a discussion of the basic performance of an instrument which may be considered to be a very versatile calorimeter—one that makes possible the routine characterization of materials on the basis of their thermodynamic properties.

BIBLIOGRAPHY

Specific references are not given in the text but sources of calibrant data and suggestions for further reading follow here.

Theory

E.S. Watson, M.J. O'Neill, J. Justin, and N. Brenner, *Anal. Chem.* **36**, 1233-1238 (1964).
M.J. O'Neill, *Anal. Chem.* **36**, 1238-1245 (1964).
A.P. Gray, in: *Analytical Calorimetry* (R.S. Porter and J.F. Johnson, eds.), Vol. I, pp. 209-218, Plenum Press, New York (1968).
R.A. Baxter, *Thermal Analysis*, Vol. I, pp. 65-84, Academic Press, New York (1969).
J.H. Flynn, *Thermochim. Acta* **8**, 69-81 (1974).
S.C. Mraw, *Rev. Sci. Instrum.* **53**(2), 228-231 (1982).

Temperature Calibration

M.J. Richardson and N.G. Savill, *Thermochim. Acta* **12**, 213-220 (1975).
J.M. Barton, *Thermochim. Acta* **20**, 249-252 (1977).
G. Vallebona, *J. Thermal Anal.* **16**, 49-58 (1979).
G.W.H. Höhne, H.K. Cammenga, W. Eysel, E. Gmelin, and W. Homminger, *Thermochim. Acta* **160**, 1-12 (1990).

Heat Calibration

G.W.H. Höhne, *Thermochim. Acta* **69**, 175-197 (1983).
G.W.H. Höhne, W. Eysel, and K.-H. Breuer, *Thermochim. Acta* **94**, 199-204 (1985).
G.W.H. Höhne, *J. Thermal Anal.* (in press).

Calibrant Materials

H.G. McAdie, P.D. Garn, and O. Menis, NBS Special Publication 260-40 (1972) (ICTA temperature standards).
P.D. Garn and O. Menis, *Thermochim. Acta* **42**, 125-134 (1980) (ICTA low temperature standards).
G.T. Furukawa, R.E. McKoskey, and G.J. King, J. Res. Natl. Bur. Stand. **47**, 256-261 (1951) (heat capacity, benzoic acid).
D.C. Ginnings and G.T. Furukawa, *J. Am. Chem. Soc.* **75**, 522-527 (1953) (heat capacity, alumina, benzoic acid, diphenyl ether, water).
D.A. Ditmars and T.B. Douglas, *J. Res. Natl. Bur. Stand.* **75A**, 401-420, (1971) (heat capacity, alumina).
J.F. Messerley, G.B. Guthrie, S.S. Todd, and H.L. Fincke, *J. Chem. Eng. Data* **12**, 338-346 (1967) (transition and melting temperatures, heat capacities, enthalpy changes, etc., C5-C18 n-alkanes).
R.J.L. Andon and J.E. Connett, *Thermochim. Acta* **42**, 241-247 (1980) (triple point temperatures and heats of fusion, organic standards for DSC available from LGC).
D.A. Ditmars, A. Cezairliyan, S. Ishihara, and T.B. Douglas, NBS Special Publication 260-55 (1977) (heat capacity, molybdenum).
K.N. Marsh (ed.), *Materials for the Realization of Physicochemical Properties*, Blackwell, Oxford, 1987 (general source of information on calibration materials—not only for DSC—published on behalf of the International Union of Pure and Applied Chemistry).

V

THERMAL EXPANSION

19

Methods of Measuring Thermal Expansion

R.K. KIRBY

1. INTRODUCTION

When heat is added to or removed from a solid material there is a change in temperature, ΔT, and a change in dimensions, ΔL. If the material is isotropic, then the change in dimensions is the same in all directions and the mean coefficient of linear thermal expansion is defined as

$$\alpha_m = \frac{1}{L_0} \frac{\Delta L}{\Delta T} \tag{1}$$

where L_0 is the length at some reference temperature, preferably 293 K. The limiting value of this definition (at constant pressure P) for a differential change in temperature is defined as the coefficient of linear thermal expansion or as the expansivity

$$\alpha = \frac{1}{L_0} \left(\frac{\partial L}{\partial T} \right)_P \tag{2}$$

The relative change in dimensions, or the thermal expansion $(\Delta L)/L_0$, is usually expressed in parts per million (μm/m) or as a percent.

Ordinarily, the coefficient of linear thermal expansion is not measured directly but is either calculated from consecutive determinations of the mean coefficient or by differentiating an equation that represents the expansion. If

$$\frac{L_T - L_0}{L_0} = a_0 + a_1 T + a_2 T^2 + a_3 T^3 + \cdots \tag{3}$$

R.K. KIRBY ● Retired from National Institute of Standards and Technology, Gaithersburg, Maryland 20899, USA.

then

$$\alpha = a_1 + 2a_2 T + 3a_3 T^2 + \cdots \tag{4}$$

If the mean coefficient has been determined over a temperature range ΔT, a curvature correction[1] may be needed to obtain the true coefficient at the mean temperature. For instance, if the expansion can be represented by a third-power polynominal [see equation (3)], then

$$\alpha = \alpha_m - \frac{a_3}{4}(\Delta T)^2 \tag{5}$$

If the solid is not isotropic, as many as six coefficients[2] may be needed to describe its change in volume and shape as follows:

Cubic	α_1	α_1	α_1	0	0	0
Hexagonal	α_1	α_1	α_3	0	0	0
Tetragonal	α_1	α_1	α_3	0	0	0
Trigonal	α_1	α_1	α_3	0	0	0
Orthorhombic	α_1	α_2	α_3	0	0	0
Monoclinic	α_1	α_2	α_3	0	α_5	0
Triclinic	α_1	α_2	α_3	α_4	α_5	α_6

The expansivity of a single crystal in a given direction is given by

$$\alpha = \sum_{i=1}^{3} \alpha_i \cos^2 \omega_i \tag{6}$$

where ω_i are the angles between the direction in question and the principal crystallographic axes. For hexagonal, tetragonal, and trigonal symmetries this relationship reduces to

$$\alpha = \alpha_1 + (\alpha_3 - \alpha_1) \cos^2 \omega_3 \tag{7}$$

The coefficient of thermal expansion is nearly proportional to the heat capacity, C_p, at all temperatures[1]; i.e., at high temperatures the expansivity is slowly increasing and at low temperatures it approaches zero as the temperature approaches 0 K (see Fig. 1). The total expansion from 0 K to the

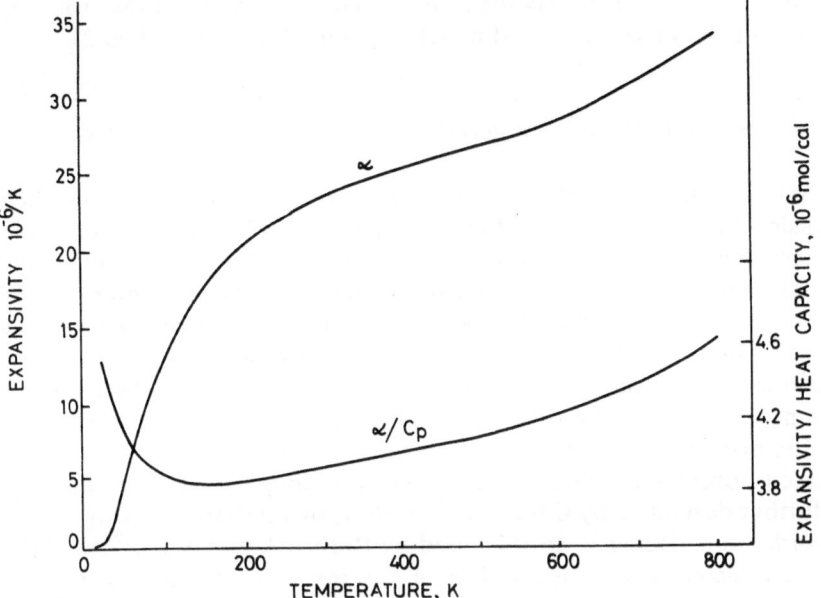

FIGURE 1. Expansivity and ratio of expansivity to heat capacity of aluminum.

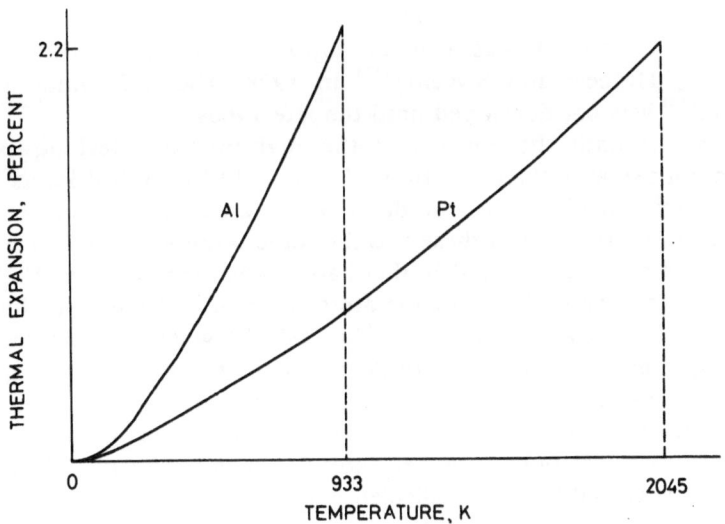

FIGURE 2. Thermal expansion of aluminum and platinum from 0 K to their melting points.

melting temperature is nearly the same for similar materials; for instance, the total expansion of close-packed metals is about 2.2%[1] (see Fig. 2).

2. METHODS FOR MEASURING THERMAL EXPANSION

Many different methods and their variants for measuring thermal expansion have been developed to meet the needs of measurement problems. For instance, an industrial application may call for automation and fast response at moderate accuracy, while academic research may require extremely high sensitivity at cryogenic temperatures. The choice of methods may also depend upon the material to be measured (the expected value of its thermal expansion, for instance), the amount of material available, the temperature range of interest, and the precision and/or accuracy that is needed.

The first measurement methods generally used one or two microscopes[3] to directly observe the expansion of relatively long specimens. This method was further developed by Callendar[4] in 1887, by Holborn and Day[5] in 1900, by Souder and Hidnert[6] in 1926, and by Rothrock and Kirby[7] in 1967.

Interferometry was used by Fizeau[8] in the mid-1800s to measure a large group of materials while a Fabry–Perot interferometer was used by Fraser and Hollis-Hallet[9] in 1965 to measure the thermal expansion of several metals at low temperatures. In 1976 Bennett[10] used a double-passed polarizing interferometer to measure thermal expansion with great accuracy.

An X-ray camera was used by Becker[11] in the 1920s to measure thermal expansion at high temperatures while others, including Mauer and Bolz,[12] used X-ray diffractometer techniques.

Pushrod dilatometry was also developed around the early 1900s and improved by Hidnert and Sweeney[13] in 1928. The differential pushrod dilatometer[14] was not developed until the late 1960s.

It was not until the 1960s that the high-precision techniques were developed for use at low temperatures. These included optical levers[15] and grids,[16] a high-sensitivity variable differential transformer,[17] and electrical capacitance cells. The best of these was the three-terminal capacitance cell as used by White and Carr[18] in 1964 that had a sensitivity of about 10^{-10}.

At the other end of the temperature scale a fast-pulse method was developed by Miiller and Cezairliyan[19] in 1982 in which a polarizing interferometer system is used to measure thermal expansion to the melting point of electrically conducting materials. In this system, temperatures above 2000 K can be obtained in less than one second by resistively heating the specimen with a large electrical pulse. Temperatures are measured with a high-speed optical pyrometer and the data collected with a digital acquisition system.

The methods of measuring thermal expansion that are most widely used now include interferometry, pushrod dilatometry, X-ray diffraction, and the high-precision three-terminal capacitance cell. The various techniques that use interferometry were described by Ruffino[20] in Volume 1 of this Compendium so will not be described here. In this chapter brief descriptions will be provided of X-ray diffraction methods and the three-terminal capacitance cell along with detailed information on pushrod dilatometry.

3. X-RAY METHODS

These methods are based on the diffraction of a beam of monochromatic X-rays that are scattered by atoms in a crystal lattice. The Bragg law

$$\lambda = 2d(hkl)\sin\theta \tag{8}$$

gives the condition for first-order constructive reflection of the incident radiation. Here d is the spacing of the lattice planes as defined by the Miller indices h, k, and l, while θ is the angle between the incident beam and those lattice planes. Except for a small correction due to refraction, the expansion is independent of wavelength:

$$\frac{\Delta d}{d} = -\cot\theta(\Delta\theta) = \frac{\sin\theta_1 - \sin\theta_2}{\sin\theta_2} \tag{9}$$

The relationship between the separation of lattice planes and the symmetry of the crystal lattice is given by

$$d^2 = [1 - \cos^2\alpha - \cos^2\beta - \cos^2\gamma + 2\cos\alpha\cos\beta\cos\gamma]$$

$$\times \left[\left(\frac{h}{a}\right)^2\sin^2\alpha + \left(\frac{k}{b}\right)^2\sin^2\beta + \left(\frac{l}{c}\right)^2\sin^2\gamma + \frac{2hk}{ab}(\cos\alpha\cos\beta - \cos\gamma) \right.$$

$$\left. + \frac{2hl}{ac}(\cos\alpha\cos\gamma - \cos\beta) + \frac{2kl}{bc}(\cos\beta\cos\gamma - \cos\alpha) \right]^{-1} \tag{10}$$

Using this equation it can be shown that the expansion of cubic crystals (where $a = b = c$ and $\alpha = \beta = \gamma = 90°$) can be obtained from any set of lattice planes (hkl):

$$\frac{\Delta a}{a} = \frac{\Delta d}{d} = -\cot\theta(\Delta\theta) \tag{11}$$

Not all the angles possible are diffracted, however, because of the structure factor.[21] For example, in the case of an fcc crystal, reflections can occur only from those planes for which the Miller indices (hkl) are all even or all odd. In the case of hexagonal crystals $(a = b \neq c$ and $\alpha = \beta = \gamma = 90°)$

$$\frac{\Delta d}{d} = \left[\frac{4c}{3a}(h + hk + k)\frac{\Delta a}{a} + l\frac{\Delta c}{c} \right]\left[\frac{4c}{3a}(h + hk + k) + l \right]^{-1} \tag{12}$$

the expansion in the a and c directions must be determined from simultaneous equations except for special planes such as (100) and (010) where

$$\frac{\Delta d}{d} = \frac{\Delta a}{a} \tag{13}$$

and (001) where

$$\frac{\Delta d}{d} = \frac{\Delta c}{c} \tag{14}$$

X-ray methods can be employed to measure the expansion of crystalline materials under conditions that preclude the use of any other method. These conditions include the situation when the specimens are very small, weak, and irregular in shape. These methods are also unique in that they can determine the principal coefficients of thermal expansion of anisotropic crystals, permit direct observation and identification of phase changes, and determination of the change in specific volume resulting from a phase change. There is a further advantage in that measurements with X-rays do not include effects that are observed in measurements on bulk specimens, such as the effect of impurities and the generation of thermal vacancies.[22]

When using the camera technique the specimen can be either a fine-grained polycrystalline wire, a fine powder, or a single crystal that is rotated during the exposure of the film. The Debye–Scherrer powder X-ray camera technique,[23] which has been used extensively over a wide range of temperatures, is illustrated in Fig. 3. The specimen is located at the center of the cylindrical camera and the film placed on the inside wall. Filtered X-rays enter through a collimator and either pass directly through the exit, are scattered, or are diffracted by the crystal planes. The diffracted rays are recorded on the film as sharp lines. The values of θ for the diffracted rays are determined from the position of the lines on the film. For increased accuracy, the value of d is obtained using large values of θ. The accuracy also depends upon the centering of the specimen, the axial divergence of the collimated beam, corrections for absorption and refraction within the specimen, knowledge of the camera diameter, and the shrinkage of the film during development.

X-ray diffractometers,[12] in which a diffracted beam is detected with an ionization counter mounted on a wide-angle goniometer, are also widely used for expansion measurements. In both techniques it must be possible to align and maintain the specimen on the focusing circle. Any tilt of the specimen surface about the axis of the goniometer or the plate holder results in broadened X-ray lines while a displacement of the specimen in the x-direction will cause an error in the measurement of d. In the diffractometer technique there is a 10 to 25% loss of intensity due to absorption in the furnace windows and a

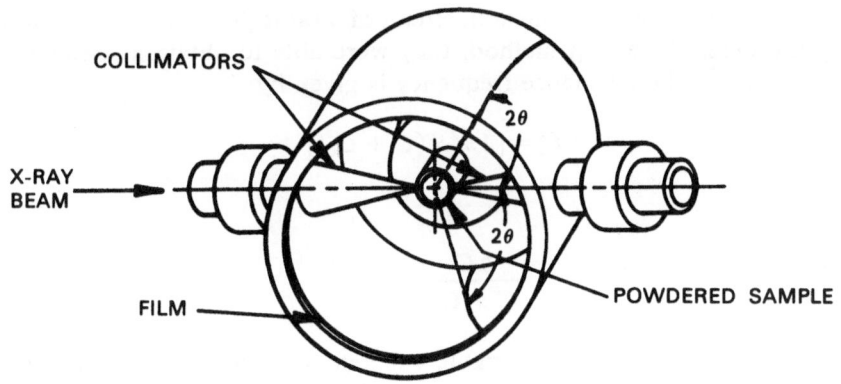

FIGURE 3. Schematic diagram of a Debye–Scherrer X-ray camera.

reduction of from 5 to 40° (2θ) in useful angular range due to limiting windows, radiation shields, etc. Because of this reduction some diffractometers are not able to measure the more complex crystals such as the orthorhombic, mono-clinic, and triclinic systems, although Mauer and Hahn[24] have used the Bond technique to measure the expansion of a monoclinic crystal. In this technique the specimen is rotated between equivalent diffraction positions on either side of the incident beam. The value of the diffraction angle is thereby unaffected by any specimen eccentricity, absorption, or tilt. A sensitivity of 10^{-7} in measurements of expansion has been obtained with this technique.

The major problem with using X-ray methods is in the determination of the specimen temperature. Owing to the tight space requirements around the specimen and the unique material problems of the film or the furnace (in the diffraction technique the windows must be transparent to the X-rays) it is not unusual to encounter very large temperature gradients close to the specimen and therefore inaccurate determinations of the temperature. Temperature gradients in the specimen also cause the X-ray lines to be broadened. In powder methods, however, it is possible to mix a reference material with the specimen and measure both while in intimate contact. Both gold and MgO have been well characterized by measuring temperature this way.[23]

4. CAPACITANCE METHODS

Capacitor techniques have been used for many years in which the expansion of a specimen causes a change in the separation of the capacitor plates. The first applications were by Prytherch[25] in 1932 and by Haughton and Adcock[26] in 1933. The earliest application of this technique to measure-ments at low temperatures was by Bijl and Pullan[27] in 1955. By incorporating a two-terminal capacitor in the tank circuit of a radio-frequency oscillator and using the heterodyne beat method, they were able to obtain a sensitivity of about 6×10^{-7}. The resonance frequency is given by

$$f^2 = [4\pi^2 \mathscr{L}(C + C_c)]^{-1} \tag{15}$$

so that

$$\frac{\Delta C}{C + C_c} = -2\frac{\Delta f}{f} \tag{16}$$

where C is the value of the variable capacitance cell, C_c is the distributed capacitance of the circuit, and $\mathscr{L}$ is the inductance of the tank coil. The variable capacitor is in series with a standard capacitor as part of a circuit that oscillates at radio frequencies. This technique was further developed by Kos and Lamarche[28] in 1969 to obtain a sensitivity of about 10^{-11}.

The three-terminal parallel-plate capacitor technique that was developed by White and co-workers[18] in 1961 utilizes a capacitance bridge operating at 1000 Hz to compare the value of the variable capacitance cell (about 10 pF) to that of reference capacitors kept at constant temperature. Figure 4 indicates the basic design for the differential cell which, except for the specimen, is made completely of copper. The temperature of the cell can be held at any temperature between the boiling point of helium and 300 K. At any measurement point the specimen (1) and the guard electrode (3) are at the same temperature while the capacitance change between the upper face of the specimen and the lower face of electrode (2) is a measure of the difference in expansion between the specimen and the copper cell. The guard electrode completely surrounds the specimen and the reference electrode to form an

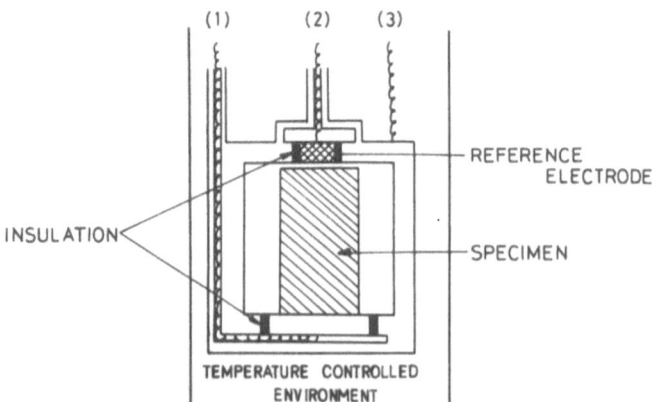

FIGURE 4. Schematic diagram of the three-terminal parallel-plate capacitor dilatometer.

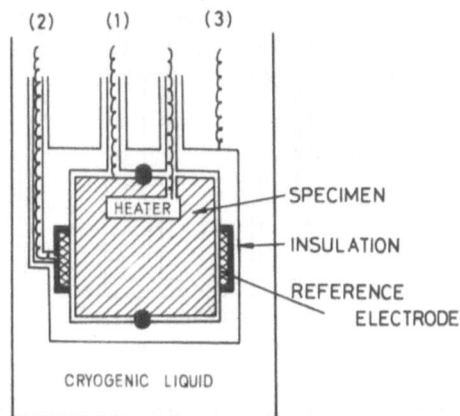

FIGURE 5. Schematic diagram of the absolute three-terminal capacitor cell.

earth shield, so that the capacitance between electrodes (1) and (2) does not include the lead wires. The differential expansion is given by

$$\left(\frac{\Delta L}{L}\right)_S - \left(\frac{\Delta L}{L}\right)_R = \left(\frac{L_R - L_S}{L_S}\right)\left(\frac{\Delta c}{c}\right) \tag{17}$$

where the value of $(L_R - L_S)/L_S$ is usually about 10^{-3}. A change of 10^{-7} pF will therefore give an expansion of about 10^{-10}.

White[29] also developed an absolute cell to measure the expansion of copper at low temperatures. This cell is illustrated in Fig. 5. In this cell the specimen is thermally isolated, so that its temperature can be varied by a small heater imbedded in it while the reference electrode is held at the temperature of the cryogenic bath. The change in the diameter of the specimen on heating above the bath temperature is therefore measured directly by the change in capacitance between it and the inner surface of the reference electrode.

5. PUSHROD DILATOMETRY

5.1. General Considerations

The pushrod dilatometer method for measuring thermal expansion is experimentally simple, reliable, and easy to automate. With this method the relative expansion of the specimen is transmitted out of the cooled or heated zone to a measuring device (an extensometer) by means of tubes and/or rods of some stable reference material. Three variations[30] of this method are used; see Fig. 6. In the first technique the specimen is placed in the end of a tube and a smaller rod (or tube with closed ends) is placed in the tube in contact

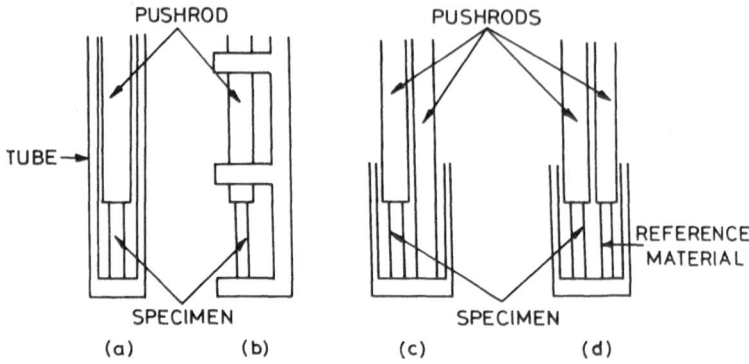

FIGURE 6. Schematic diagram of dilatometer methods: (a) tube and pushrod; (b) rod with fused base and pushrod guides; (c) double pushrod; (d) differential dilatometer.

with the specimen. An extensometer is then able to detect the difference in expansion between the specimen and an equal length of the tube. It is assumed that the expansion of the rest of the tube and the smaller rod will cancel each other. The major problems with this arrangement are that there is bound to be a temperature difference between the tube and the rod, since they are in different thermal environments, and that they have different coefficients of expansion since they would probably come from different lots of material. A somewhat better arrangement is to replace the tube with a rod of the same diameter as the pushrod and to fuse a foot (or base) to it to support the specimen. In this technique there is a better chance that the expansion of the two equal-diameter rods will cancel each other. In a third arrangement, the support for the specimen is independent of the measurement system except for supplying a common reference base. Two techniques are possible with this arrangement. One uses two pushrods of unequal length and the specimen, while the other uses two pushrods of equal length and test and reference specimens of equal length. The latter technique is considered to be the best for two reasons:

1. The system can be easily calibrated to provide accurate values of expansion.
2. The reference specimen can be selected to match the expansion of the test specimen and thereby provide a small difference to be measured at high sensitivity.

The ability to closely match the thermal expansion, thermal mass, and thermal diffusivity of the test specimen also results in a reduction of errors due to temperature and extensometer uncertainties and allows for higher heating and cooling rates without sacrificing accuracy. This technique is properly called a differential dilatometer.

In the range from 80 to 1100 K dilatometers are usually made from fused silica or from some low-expansion glass ceramic. The fused silica parts should be fine annealed* after they have been fabricated and should thereafter not be heated above 1200 K. Even temperatures as low as 800 K can cause fused silica to crystallize if it has alkali compounds on its surface from an operator's hands. Cleaning the parts can be accomplished by immersion in an aqueous solution containing 10% hydrofluoric acid and thoroughly rinsing with distilled water. Handling the parts with plastic gloves is recommended before heating to high temperatures.

The use of low-expansion materials is recommended as it reduces the errors caused by temperature gradients. Other materials must be used, however, when measuring thermal expansion at temperatures above 1100 K. High-purity dense alumina[31] or single-crystal synthetic sapphire rods are recommended materials for use in air or vacuum up to 2000 K. Although tantalum has been used in the past, tungsten is recommended for use in high vacuum up to 2500 K and graphite in carbon-rich or inert atmosphere up to 2800 K.[32]

5.2. Extensometers

The simplest device and, up to a few years ago, the most widely used for measuring the change of length with a dilatometer is a dial gage. Dial gages can readily be obtained that will indicate a length change of 0.25 μm. With a 10-cm specimen, therefore, the sensitivity is 2.5×10^{-6}. Dial gages have long measuring ranges and they are stable. Once they have been calibrated with end standards or with accurate screw micrometers they will stay calibrated unless mistreated, i.e., damaged by dropping or by being placed in a corrosive atmosphere. One of their problems, however, is that the dial movement tends to stick and so needs a light tap or two to bring to a correct reading. Dial gages are of course manual devices that cannot be used for automatic recording.

At the other end of the scale is the use of an interferometer to measure the change in length. A Fizeau interferometer was used for this purpose by Ruffino[33] in 1961 and by Meyerhoff and Smith[34] in 1962. In this system the two optical flats are positioned close together so that a simple light source can be used, they are held at room temperature, and the detection of the fringe movement can be easily automated. This type of extensometer is stable and does not need to be calibrated. With a 2.5-cm specimen the sensitivity can be 10^{-6} or better.

The most widely used extensometer is the LVDT (Linear-Variable-Differential Transformer). This is available in all commercially made dilatometers because its high sensitivity can be varied (it is a function of the

*Fine anneal of fused silica: (1) Heat at 100 K h^{-1} to 1500 K. (2) Hold at 1500 K for 2 h. (3) Cool at 60 K h^{-1} to 1300 K. (4) Cool at 120 K h^{-1} to 900 K. (5) Cool at 200 K h^{-1} to room temperature.

applied voltage) and its electrical output easily recorded. The highest sensitivity with a 2.5-cm specimen is about 0.2×10^{-6}. The problem with LVDTs is that they must be calibrated with end standards, screw micrometers, or some equally accurate device, their voltage supply must be very stable and repeatable, and their temperature must be maintained constant to within less than 1 K.

Unless an extensometer is protected from temperature changes, a low-expansion material such as Invar should be used for attaching it to the tube and/or pushrods of the dilatometer.

5.3. Temperature Sensors

In dilatometers that are used at temperatures below 1700 K, thermocouples are the preferred sensors for measuring thermal expansion. Types E, K, and T can be used at temperatures down to 80 K. Type T can be used at temperatures up to 650 K, Type E up to 1150 K, and Type K up to 1500 K. Types R and S can be used at temperatures up to 1700 K and, because of their stability, are recommended for measurements above 700 K.

Photoelectric optical pyrometers can be used for temperatures above 1100 K. When using blackbody conditions the measured temperatures are accurate to 2 degrees at 1100 K and 8 degrees at 2800 K.

When using a thermocouple it should be referenced to the ice point or to some other known constant temperature. For best results the measuring junction should be in close contact with the specimen and the leads near it should not be exposed to a temperature gradient. For accurate results the thermocouple or the spool of wire from which it was taken must be calibrated according to appropriate national standards and not used in conditions where contamination can occur. The accuracy of these calibrations is generally considered to be as indicated below:

Type	Accuracy of calibration
T	0.2 K
E	0.5 to 1 K at 1150 K
K	0.5 to 1 K at 1500 K
R	0.5 to 2 K at 1700 K
S	0.5 to 2 K at 1700 K

Analog or digital recording systems for both expansion and temperature can contribute errors because they are not linear; possess enough sensitivity; the hardware or software includes assumptions that involve approximations, i.e., an ice-point correction is added to the thermocouple output but the reference junction is allowed to drift; the thermocouple and extensometer responses are assumed to be linear when in fact they never are; or corrections are made for the expansion of the reference material but they are not accurate.

5.4. Error Analysis

The precision and accuracy of determining both thermal expansion and the coefficient of thermal expansion depend upon the simultaneous measurement of temperature and relative length. Random error is usually associated with the precision of repeated temperature and length measurements, but other variables may affect the results. For instance, the thermocouple may lose its thermal contact or the specimen may change its position. Systematic error is usually larger and can result from many sources. These include the accuracy of the length and temperature measurements; the deviation of the mean temperature of the specimen from that indicated by the thermocouple; the temperature gradients between and along the specimen, the tube, and/or the pushrods; the effect of surface contacts between the specimen and the extensometer; the accuracy with which the expansion of the tube, pushrod, or reference specimen is known; the effect of room temperature variations; and the influence of the temperature and length recorders on the data. Little can be done to improve the random errors once the equipment has been selected, except to follow good experimental practice. Systematic errors, however, can be reduced by careful calibration of the individual components and of the total system with reference materials.

Repeat measurements have confirmed that the precision with which linear thermal expansion is measured can be estimated from the precision of the length and temperature measurements. This estimate is obtained from

$$\delta\left(\frac{\Delta L}{L_0}\right) = 2\left\{\left[\frac{1}{L_0}\delta(\Delta L)\right]^2 + [\alpha\delta T]^2\right\}^{1/2} \tag{18}$$

where $\delta(\Delta L)$ and δT are the precisions of the single measurements of the length and temperature. (The error caused by determining the initial length, L_0, is usually so small that it does not contribute to this estimate.) An example is

$$\delta\left(\frac{\Delta L}{L_0}\right) = 9 \ \mu\text{m/m}$$

when $L_0 = 0.05$ m, $\delta(\Delta L) = 0.1$ μm, $\alpha = 20$ μm/mK, and $\delta T = 0.2$ K. It is obvious that increasing the length of the specimen will improve the precision of the measurement and that more care must be taken in measuring the temperature of the specimen if its coefficient is large.

5.5. Calibration with Reference Materials

Calibrations of the total system can be made with certified reference materials supplied by government agencies* or by materials for which good reference data are available. This latter group includes platinum,[35-37] copper,[38,39] aluminum,[40,41] silicon,[38,42,43] sapphire,[44-47] and graphite,[48] while the certified reference materials include vitreous silica, borosilicate glass, tungsten, stainless steel, molybdenum, Invar, copper, pyros, zircon, and magnesia.

The use of reference materials to calibrate the total system is especially important if measurements are made during heating and cooling conditions rather than by taking measurements at a series of constant temperatures. The best results are obtained when the thermal properties of one of the reference materials closely matches those same properties of the specimen. When this is not possible, the properties of the reference materials should at least bracket those of the specimen.

The expansion of a specimen in a dilatometer can be represented by

$$\frac{\Delta L}{L_0} = A\left(\frac{\Delta L}{L_0}\right)_a + B \tag{19}$$

where A is a calibration constant, $(\Delta L/L_0)_a$ is the apparent expansion as indicated by the extensometer, and B is the expansion of the material from which the dilatometer is made (or the reference specimen), i.e., fused silica, alumina, tungsten, graphite, etc. While both the extensometer and the temperature sensors, including the indicating and recording devices, should be calibrated independently, the dilatometer system as a whole should also be calibrated. If the system is working perfectly, the value of A should be equal to one and B should be the expansion of the dilatometer material or the reference specimen. Temperature-dependent variations from the ideal, however, always occur. The dilatometer can be calibrated with reference materials in any of three ways:

---

*See the chapter in this Compendium on certified reference materials.

1. The usual method is to assume that $A = 1$ and use a reference material that has an expansion equal to or greater than the test specimen to determine the temperature variation of B.
2. Assume the temperature variation of B from reference data and use a reference material to determine the temperature variation of A.
3. Use two or more reference materials to determine the temperature variation of both A and B. (A least-squares technique will be needed when more than two tests are made.)

While Plummer[14] has suggested that a baseline correction for a differential dilatometer can be determined by reversing the positions of two nearly identical specimens, it is not considered necessary to do so since the values of B will include that correction. It is obvious, of course, that a calibration must be made for every reference specimen that is used in the system.

5.6. Specimen Shape

While the vertical position of the dilatometer has been used the longest, and in some high-temperature furnaces it is the only practical position, the horizontal position is considered to be the most reliable because the specimen and pushrod are resting in stable positions. In the vertical position the specimen must sit on a stable base and the pushrods must be constrained so that they will not move except in the direction of the measured expansion. For this reason the preferred shapes of the specimen and contacting dilatometer parts are shown in Fig. 7. The contacting dilatometer surfaces should be slightly rounded, since pointed contacts could indent the specimen surface.

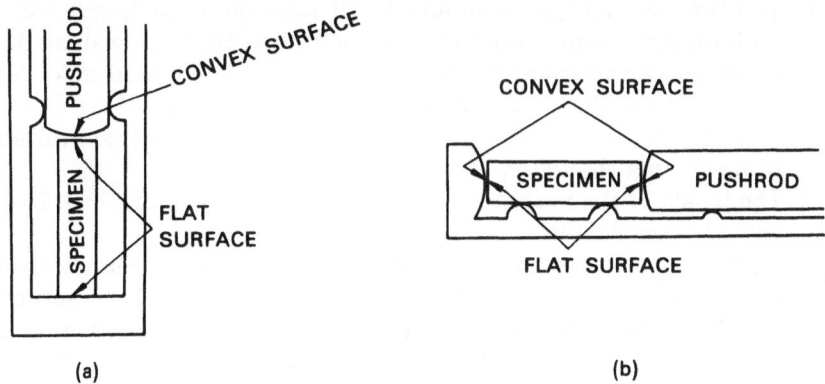

FIGURE 7. Shape of specimen and dilatometer contacts for vertical (a) and horizontal (b) positions.

5.7. Recommendations

Recommendations for the design and operation of a research dilatometer are as follows:

1. Use a horizontal differential dilatometer.
2. The dilatometer should be made from high-purity fused silica (it should have a negligible hydroxyl content and be annealed) for temperatures up to 1100 K and single-crystal synthetic sapphire for temperatures up to 2000 K.
3. Use a high-quality LVDT for measuring the expansion. It should be supported with a spring hinge that incorporates an adjustment for specimen length and means for calibration. The LVDT should be shielded from the furnace and maintained at a constant temperature.
4. A highly stable voltage supply should be used for the LVDT that has a set of fixed (nonadjustable) voltage levels.
5. Use an appropriate thermocouple made from supplies of calibrated wires to measure the temperature. Prepare new thermocouples whenever a previously used one is suspected of having been contaminated. Place the measuring junction in thermal contact with the specimen at its midpoint. The thermocouple leads near the junction should be thermally tempered to an isothermal region near the specimen to prevent any temperature gradient in the leads near the junction. The reference junction should be held at a known constant temperature.
6. Use a flat-ended specimen that has a length of 5 cm and a diameter of at least 5 mm.
7. The furnace or cryostat should be at least 60 cm in length with a uniform temperature zone of at least 20 cm. High-thermal-conductivity shields and guard heaters will help provide this uniformity. The thermal mass of the furnace should be kept as low as possible for easy heating and cooling. Full automatic temperature control and programming should be used.
8. The length of the pushrods should be at least 40 cm, of which 25 cm should be in the furnace or cryostat.
9. A desk-top computer can be used to store the information necessary to convert the data from the LVDT and thermocouple to accurate length changes and temperatures. Programs can be written so that the results of calibration runs, thermocouple tables, LVDT corrections, etc., are placed into memory for this purpose.
10. An $X-Y$ plotter is useful in following the progress of the test, but digital readout and recording are preferred for obtaining the data.

11. Only for a few materials, and then only over limited temperature ranges, can the thermal expansion be accurately represented by a straight line. It is not unusual that a fourth- or fifth-degree polynomial must be used for temperature ranges as small as 100 K.

12. For best results in determining the thermal expansion of a material, several tests should be conducted using a series of constant temperatures. At each point the temperature should be held constant until the LVDT reading comes to equilibrium.

13. When comparing the expansion of similar materials, satisfactory results can be obtained with constant heating and cooling rates of up to $5\ K\ min^{-1}$.

NOTATION

Symbol	Definition
a, b, c	Lattice spacing constants
a_0, a_1, a_2, a_3	Constants for a third-power polynomial
A	Calibration constant for dilatometer
B	Calibration constant for dilatometer
C	Electrical capacitance
C_c	Capacitance of an electrical circuit
d	Lattice separation
f	Resonance frequency
h, k, l	Miller indices for crystal-lattice planes
L	Length of specimen at any temperature
L_0	Length of specimen at reference temperature
$\mathscr{L}$	Electrical inductance of a coil
P	Pressure
T	Temperature
α	Coefficient of linear thermal expansion (expansivity) or angle between lattice spacings b and c
α_m	Mean coefficient of linear thermal expansion
β	Angle between lattice spacings a and c
γ	Angle between lattice spacings a and b
δ	Uncertainty of a measured quantity
Δ	Difference of two measured quantities
λ	Wavelength
ω_i	Angle between the direction of measurement and the ith crystallographic axis
ω_3	Angle between the direction of measurement and the c-axis
θ	Angle of incident ray
∂	Partial derivative operator

REFERENCES

1. Y.S. Touloukian, R.K. Kirby, R.E. Taylor, and P.D. Desai, *Thermophysical Properties of Matter*, The TPRC Data Series, Vol. 12, *Thermal Expansion of Metallic Elements and Alloys*, IFI/Plenum Press, New York (1975).
   Y.S. Touloukian, R.K. Kirby, R.E. Taylor, and T.Y.R. Lee, *Thermophysical Properties of Matter*, The TPRC Data Series, Vol. 13, *Thermal Expansion of Nonmetallic Solids*, IFI/Plenum Press, New York (1977).
2. J.F. Nye, *Physical Properties of Crystals*, Oxford University Press (1960).
3. T. Preston, *The Theory of Heat*, MacMillan and Co., London (1904).
4. H.L. Callendar, "On the Practical Measurement of Temperature," *Philos. Trans. R. Soc. London* **178**, 161 (1887).
5. L. Holborn and A. Day, "The Air Thermometer at High Temperature," *Ann. Phys.* **307**, 505 (1900).
6. W. Souder and P. Hidnert, "Measurements on the Thermal Expansion of Fused Silica," Sci. Pap. BS **21** (S524) (1926).
7. B.D. Rothrock and R.K. Kirby, "An Apparatus for Measuring Thermal Expansion at Elevated Temperatures," *J. Res. Natl. Bur. Stand.* **71C**, 85 (1967).
8. M.H. Fizeau, "Memoir on the Expansion of Solids by Heat," *Compt. Rend.* **62**, 1101 (1866).
9. D.B. Fraser and A.C. Hollis-Hallet, "The Coefficient of Thermal Expansion of Various Cubic Metals Below 100 K," *Can. J. Phys.* **43**, 193 (1965).
10. S.J. Bennett, "The NPL Interferometric Dilatometer," *J. Phys. E* **10**, 525 (1976).
11. K. Becker, "An X-Ray Method to Determine the Thermal Expansion Coefficient at High Temperature," *Z. Phys.* **40**, 37 (1926).
12. F.A. Mauer and L.H. Bolz, "Problems in the Temperature Calibration of an X-Ray Diffractometer Furnace," in: *Advances in X-Ray Analysis*, Vol. 5, p. 544, Plenum Press, New York (1961).
13. P. Hidnert and W.T. Sweeney, "Thermal Expansion of Magnesium and Some of Its Alloys," *NBS J. Res.* **1**, 771 (1928).
14. W.A. Plummer, "Differential Dilatometry, A Powerful Tool," AIP Conf. Proc., No. 17—Thermal Expansion, p. 147, American Institute of Physics, New York (1974).
15. J.M. Shapiro, D.R. Taylor, and G.M. Graham, "A Sensitive Dilatometer for Use at Low Temperatures," *Can. J. Phys.* **42**, 835 (1964).
16. K. Andres, "The Measurement of Thermal Expansion of Metals at Low Temperatures," *Cryogenics* **2**, 93 (1961).
17. P.W. Sparks and C.A. Swenson, "Thermal Expansion from 2 to 40 K of Ge, Si, and Four III-V Compounds," *Phys. Rev.* **163**, 779 (1967).
18. R.H. Carr, R.D. McCammon, and G.K. White, "The Thermal Expansion of Cu at Low Temperatures", *Proc. R. Soc. London* **A280**, 72 (1964).
19. A.P. Miiller and A. Cezairliyan, "Thermal Expansion of Iron During the Phase Transformation by a Transient Interferometric Technique," in: *Thermal Expansion 8*, p. 245, Plenum Press, New York (1984).
20. G. Ruffino, "Thermal Expansion Measurement by Interferometry," in: *Compendium of Thermophysical Property Measurement Methods*, Vol. 1, *Survey of Measurement Techniques*, p. 689, Plenum Press, New York (1984).
21. C. Kittel, *Introduction to Solid State Physics*, p. 648, 3rd edn., Wiley, New York (1966).
22. W.E. Schoknecht and R.O. Simmons, "Thermal Vacancies and Thermal Expansion," AIP Conf. Proc., No. 3—Thermal Expansion, p. 169, American Institute of Physics, New York (1972).
23. R.G. Merryman and C.P. Kempter, "Precise Temperature Measurement in Debye–Scherrer Specimens at Elevated Temperatures," *J. Am. Ceram. Soc.* **48**, 202 (1965).

24. F.A. Mauer and T.A. Hahn, "Thermal Expansion of Some Azides by a Single Crystal X-Ray Method," AIP Conf. Proc., No. 3—Thermal Expansion, p. 139, American Institute of Physics, New York (1972).

25. W.E. Prytherch, "A New Form of Dilatometer," *J. Sci. Instrum.* **9**, 128 (1932).

26. J.L. Haughton and F. Adcock, "Improvements in Prytherch's Capacity Dilatometer," *J. Sci. Instrum.* **10**, 178 (1933).

27. D. Bijl and H. Pullan, "A New Method for Measuring the Thermal Expansion of Solids at Low Temperatures," *Physica* **21**, 285 (1955).

28. J.F. Kos and J.L.G. Lamarche, "Thermal Expansion of the Noble Metals Below 15 K," *Can. J. Phys.* **47**, 2509 (1969).

29. G.K. White, "Measurement of Thermal Expansion at Low Temperatures," *Cryogenics* **1**, 151 (1961).

30. R.K. Kirby, "Thermal Expansion of Ceramics," NBS Special Publ. 303, *Mechanical and Thermal Properties of Ceramics*, p. 41 (1969).

31. G.R. Hyde, L.P. Dominques, and L.R. Furlong, "Improved Dilatometer," *Rev. Sci. Instrum.* **36**, 204 (1965).

32. P.S. Gaal, "Universal Graphite Dilatometer for High Temperature Studies," in: *Thermal Expansion 6*, p. 145, Plenum Press, New York (1978).

33. G. Ruffino, "A Recording Dilatometer and Measurement of Thermal Expansion Coefficient of Polyethyl Methacrylate (Lucite) Between 90 and 273 K," ASME Prog. Int. Res. Thermodynam. and Transp. Prop., p. 185, Academic Press, New York (1962).

34. R.W. Meyerhoff and J.F. Smith, "Anisotropic Thermal Expansion of Single Crystals of Thallium, Yttrium, Beryllium, and Zinc at Low Temperatures," *J. Appl. Phys.* **33**, 219 (1962).

35. G.K. White, "Thermal Expansion of Platinum at Low Temperatures," *J. Phys.* F2, L30 (1972).

36. F.C. Nix and D. MacNair, "The Thermal Expansion of Pure Metals," *Phys. Rev.* **61**, 74 (1942).

37. T.A. Hahn and R.K. Kirby, "Thermal Expansion of Platinum from 293 to 1900 K," AIP Conf. Proc., No. 3—Thermal Expansion, p. 87, American Institute of Physics, New York (1972).

38. G.K. White, "Reference Materials at Low Temperatures," AIP Conf. Proc., No. 17—Thermal Expansion, p. 1, American Institute of Physics, New York (1974).

39. T.A. Han, "Thermal Expansion of Copper from 20 to 800 K," *J. Appl. Phys.* **41**, 5096 (1970).

40. F.C. Nix and D. MacNair, "The Thermal Expansion of Pure Metals," *Phys. Rev.* **60**, 597 (1941).

41. R.O. Simmons and R. W. Balluffi, "Measurements of Equilibrium Vacancy Concentrations in Aluminum," *Phys. Rev.* **117**, 52 (1960).

42. R.B. Roberts, "Expansivity of Silicon from 293 to 775 K," in: *Thermal Expansion 6*, p. 187, Plenum Press, New York (1977).

43. H. Ibach, "Thermal Expansion of Silicon and Zinc Oxide," *Phys. Status Solidi* **31**, 625 (1969).

44. B. Yates, R.F. Cooper, and A.F. Pojur, "Thermal Expansion at Elevated Temperatures," *J. Phys. C* **5**, 1046 (1972).

45. J.B. Wachtman, T.G. Scuderi, and G.W. Cleek, "Linear Thermal Expansion of Aluminum Oxide and Thorium Oxide from 100 to 1100 K," *J. Am. Ceram. Soc.* **45**, 319 (1962).

46. A. Schauer, "Thermal Expansion, Gruneisen Parameter, and Temperature Dependence of Lattice Vibration Frequencies of Aluminum Oxide," *Can. J. Phys.* **43**, 523 (1965).

47. T.A. Hahn, "Thermal Expansion of Single Crystal Sapphire from 293 to 2000 K," in: *Thermal Expansion 6*, p. 191, Plenum Press, New York (1977).

48. P.S. Gaal, "Graphite Thermal Expansion Reference for High Temperature," AIP Conf. Proc., No. 17—Thermal Expansion, p. 102, American Institute of Physics, New York (1974).

20

Recent Thermal Expansion Interferometric Measuring Instruments

G. RUFFINO

1. INTRODUCTION

Interferometric methods, as well as general information on instruments currently used to measure thermal expansion, were dealt with in Chapter 18 of the first volume of this Compendium. The present chapter describes in more detail interferometric instruments, data acquisition and treatment, and finally analyzes the errors typical of this method.

The measurement of thermal expansion is carried out with a system which performs the following three functions:

- Generation of an interference fringe pattern, which requires a monochromatic light source, beam conditioning and steering, and an interferometer. One optical path of the interferometer is dependent on the specimen length.
- Expansion measurement by means of fringe detection and counting.
- Sample-temperature control and measurement.

The pertinent aspects are examined in the subsequent sections.

2. SAMPLE

2.1. Shape and Size

Interferometric thermal expansion measurements are performed between two planes containing the end faces of the sample. The sample can have the

G. RUFFINO • Mechanical Engineering Department, Second University of Rome Tor Vergata 00173 Rome, Italy.

shape of a rectangular prism, a cylinder, or a tube. The sample is part of the interferometer which, basically, can be of the Fizeau or Michelson type. Contact or contactless measurements are possible. In the Fizeau interferometer the end faces can be wrung to optical flats. One of these flats supports the sample while the other is the semireflecting optical flat placed on top of the sample. It will be seen in the following section that this configuration can be used for both the Fizeau and Michelson interferometers. Wringing of the end faces and the exact definition of the reference length requires precise machining and polishing.

With contactless measurements, the sample end faces act as the moving mirrors of a single or double Michelson interferometer. Obviously they must be optically polished, a feature which is not possible with every material. Their flatness must be kept within one interference fringe, which yields a sample reference length uncertainty of one wavelength. The sample end faces are not easily kept optically flat when subjected to a large temperature change, at least for some materials.

Many modern interferometric systems use laser light sources. Since they are highly coherent, they allow good fringe contrast (visibility) even with large optical path differences (typically 2 m, far in excess of CTE measurement needs). Therefore, practically, there is no limit to the specimen length. Sample distortions are reduced by keeping an even radial temperature distribution, which is achieved by a proper design of the heating or cooling system. Causes and effects of sample distortions will be examined in Section 9.

Michelson interferometers allow contactless measurements using both sample faces, or at least one face (the top face of a sample placed on an optical flat). On the other hand, Fizeau interferometers only allow contact measurements. With this interferometer three configurations are used (see Fig. 13 on p. 701 of Vol. 1 of this Compendium). First, the sample is placed between two reflecting flats. In this case the transverse dimensions must be sufficient to achieve good mechanical stability. The second sample shape is a circular right tube with parallel ends placed in the same manner as the previous cylinder. A third configuration (see Fig. 13 on p. 701 of Vol. 1) involves the use of three specimens of equal length placed on the baseplate at the vertexes of an equilateral triangle, with an optical flat located on their top. In this case the samples must have parallel faces and equal length in order to correctly define the reference length. Higher stability is achieved at the expense of fabrication complications.

2.2. Supports

Correct measurement of thermal expansion ideally requires that only sample motion along the sample axis be allowed. This entails that all rotational degrees of freedom and the two translational ones that are normal to the optical path direction, i.e., to the sample axis, are suppressed. These conditions are realized by correct sample supports. Two cases are possible.

a. Contactless Measurements. In this case, sample and interferometer supports are attached independently to the instrument baseplate. Both kinds of support must be extremely stable, a feature achieved by proper choice of support material and correct design.

With contactless measurements both sample ends are exposed to light, and therefore the sample must be kept in position by side supports and gravity if we wish to avoid constraints on the sample. Residual constraint is kept to a minimum by using point contacts between the sample and the support. This is achieved by using V-shaped supports with rounded edges. Even in this case the supports may cause small stepwise displacements during measurement, giving rise to some difficulty in fringe counting.

Good vibration insulation should be provided for the interferometer by inserting spring-damper supports between the interferometer and the table and between the table and the floor. A usual source of vibration is a rotary vacuum pump employed to avoid temperature dependence of wavelength and reduce sample optical surface contamination. Therefore, either rotary pumps must be insulated against vibration, or absorption pumps, cryopumps, should be used during interferometer operation.

Some support materials often used are invar, superinvar, and ULE or zerodur (glass-ceramic materials with very low thermal expansion coefficient), for moderate temperature excursions around 0 °C (±100 °C). Bearing supports must be kept near room temperature and should therefore be isolated from the heating or cooling system.

An example of a sample mounting setup, designed by Wolff and Eselun,[1] is shown in Fig. 1. The temperature around the specimen is controlled (as described in Section 7) by a cooler and heater, surrounded by radiation shields. Sample supports enter the temperature-controlled space through holes drilled in the shields and heater/cooler system.

b. Contact Measurements. In this case one or both end faces of the sample are in optical contact with the interferometer reflecting flats: the sample is placed on one of them which is the upper end of the baseplate. The baseplate and its support must be very stable. Their distortion is harmful as it causes sample rotation, and hence an optical path difference variation which is not due to the thermal expansion. Therefore the baseplate support must be fabricated with very stable material and any temperature gradient in it, if unavoidable, should at least be directed only along its axis. This condition is met if the heater/cooler are cylindrically symmetrical around the support.

A good example of sample mounting for contact measurements with a Michelson interferometer is given in Fig. 2, derived from Wolff and Eselun.[2] In this case the sample is a long circular tube (A) (914 mm length, 125 mm

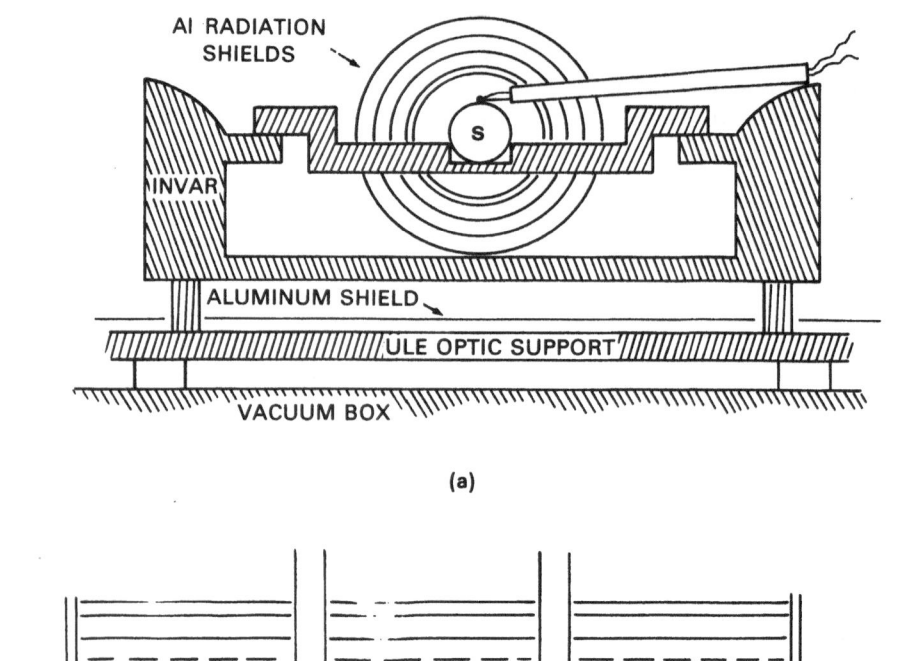

(a)

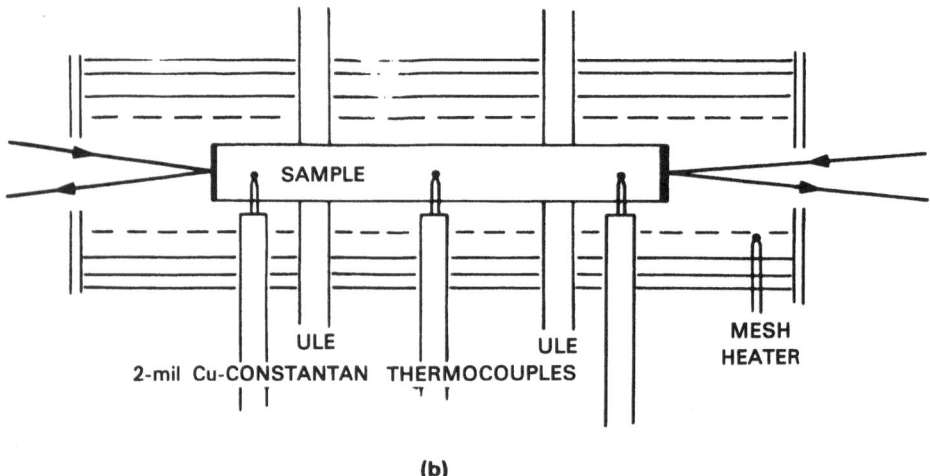

(b)

FIGURE 1. Sample mounting in a contactless interferometer.

id, and 2.5 mm thickness) of clear fused silica GE type 204. The support consists of a thicker quartz tube and a baseplate of the same material (B). In this case the optical flats are not directly attached to the test sample end faces but are the lower mirrored surfaces (M_1 and M_2) of two ULE plates. An invar rectangular plate (P_1) lying on the upper base of the sample and another one (P_2) standing on the baseplate support the mirrors. More details of the Michelson interferometer will be described in Section 3.1.

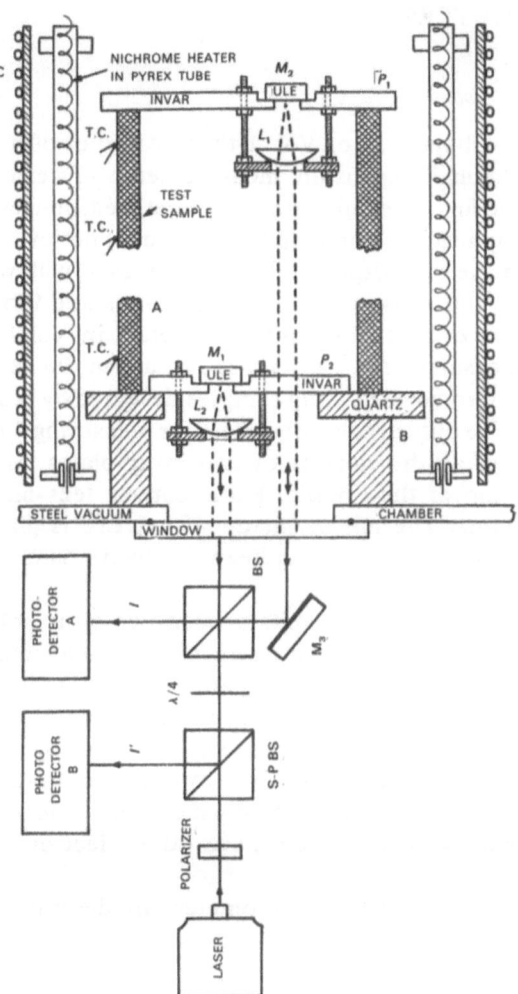

FIGURE 2. The contact Michelson interferometer of Wolff and Eselun.

Another type of support is presented in Figs. 13, 14, and 15 on pp. 701 and 703 in Vol. 1 of this Compendium. It consists of a sapphire interferometer baseplate and a vertical alumina stand made with a tube and a round plate on its top. The support is standing on the baseplate of the evacuated bell jar which contains the furnace and the interferometer. The upper end of the support is at the sample temperature and its lower end is at room temperature; it is placed in a temperature gradient which requires some care to avoid distortion.

3. INTERFEROMETERS

3.1. Interferometer Types

It was seen in Chapter 18 of Vol. 1 that two types of interferometer are mainly used in thermal expansion measurements—Fizeau and Michelson interferometers. The first type only allows contact measurements and generally requires more complicated sample shapes (either a hollow cylinder or three equal samples to make the "tripod"). It is mainly used with wide temperature ranges, both below and above 0 °C, within cryostats and furnaces.

The description of Michelson interferometers in Vol. 1, for contactless and one-contact measurements, can be used to aid in the design of a system. A two-contact interferometer for long samples designed by Wolff and Eselun[2] is shown in Fig. 2. The optical support system for measuring the relative motion of the end planes of a tube consists of two invar plates resting on a quartz baseplate and on top of the sample. Plano-convex lens–mirror systems are used as retroreflectors. The mirror, made with ULE, is placed in the focal plane of the lens so that the returning beam is always parallel to the incident beam.

A frequency-stabilized laser emits a beam which is polarized and split into two beams by beam-splitter B, and returns to form an interference pattern after travelling to M_1 and back and to M_2 and back. Photodetector D_1 receives the entire difference field (I) which varies sinusoidally with optical path difference (BM_2–BM_1). The inverted field (I') created by a half-wavelength plate and beam-splitter S-P BS and received by detector D_2 is used in a feedback loop. Changes in light levels or fringe contrast caused by optical contamination of the window or lens–mirror system have little effect on the measurement capability.

The optical elements of the interferometers are described in the following sections.

3.2. Optical Parts

Optical flats are generally coated to control their reflectance.

Transparent plates are generally used as beam-splitters, as in the Fizeau interferometer, where it is the upper plate lying on the specimen(s). In this case the beam-splitting surface is the one contacting the specimen. It should be made half-reflecting if the base surface is a metallic mirror, in order to have sufficient contrast in the interference pattern. However, metallization can be damaged by bringing the experimental cell to high temperatures, so it is more advisable to have both plates holding the specimen made with the same material. In order to avoid disturbing interference caused by the upper face of the top plate, this should be wedge-shaped (with an angle between 30' and

1°), so that the upper face reflects light out of the field. If the baseplate is transparent, only its upper face should be reflecting, the lower one being ground in order to be diffusing.

Mirrors are made by metallic coating of glass or other high-stability material (ULE or zerodur) with vapor deposition in vacuo. Usually, aluminum protected by a thin layer of silicon monoxide is used. Beam-splitters can be made with a plane-parallel transparent plate possessing one semireflecting face, a pellicle, or a cube. In modern Michelson interferometers the use of a beam-splitting cube is preferred to the beam-splitting and compensating plates. The cube consists of two total-reflection prisms cemented on their hypotenuses, one of them being half-reflecting. Light beams cross the faces with normal incidence. Pellicles are thin, half-reflecting collodion or polyester films stretched on a rigid, flat support frame. They are insufficiently precise and stable to be used in interferometers.

Optical flats are made with glass, fused silica, or sapphire. Glass is appropriate for use around room temperature. Wider temperature ranges require silica and at high temperatures, typically above 500 °C, sapphire is needed.

Flatness is measured in terms of interference fringes of yellow sodium light (589.3 nm) produced between the flat and a contacting reference flat (test plate).

Interferometric flats for dilatometry should be polished to a high, but not extreme, degree of flatness; 1/20 wavelength is available at a moderate price. Interferometers frequently require beam deviations of 90° and 180°: these can be done with mirrors, but they require very fine and critical adjustments. A ray deviation of 90° is performed by a pentaprism (Fig. 3), in which a ray entering through face AB, after two reflections within the prism, exits at right angles from face EA. Since the deviation does not depend on the incidence angle on AB, the prism does not require any delicate adjustment.

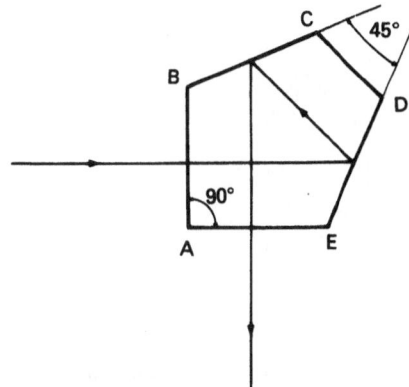

FIGURE 3. A pentaprism.

A light-team deviation of 180° is produced by retroreflectors. One of these is the corner cube, a system of three mirrors at right angles: a light ray entering at a small angle with the symmetry axis (Fig. 4), after reflection on three faces, exits in the opposite parallel direction, independent of the incident angle. The corner cube can be made with three glass plates mounted on a metallic frame, which assures their orthogonality; the plates have a reflecting coating on their faces inside the corner. A transparent right-triangular pyramid with mutually orthogonal lateral faces results when a cube is cut normal to its diagonal. This is the most common means of manufacturing a corner cube. As the 180° deviation does not depend on the angle between the ray and the symmetry axis, provided this is sufficiently small, the device has no need for fine adjustments.

Another sort of retroreflector is the so-called "cat's eye": it consists of a converging lens with a mirror on its focus (as in Fig. 2). A parallel light beam, incident on the lens, is focussed and reflected by the mirror and exits from the lens in the opposite parallel direction. This fact is not dependent on the inclination either of the beam or of the mirror on the optical axis. It is realized with a plano-convex lens and a thin beam, within the Gauss approximation, or with a spherically corrected lens and a wide beam. Also, in this case no precise angular adjustment is required, but a longitudinal one is needed in order to place the mirror exactly on the lens focus.

3.3. Polarizers and Compensators

Many interferometric dilatometers use polarized light. Among different kinds of polarizers, most convenient is the polarizing beam-splitter that functions both as polarizer and beam-splitter. This is made by cementing two right-angle prisms on their hypotenuse, which has been coated with a special multilayer dielectric film (Fig. 5). Monochromatic unpolarized light which is incident normally upon an external face of the resulting cube (internally incident at 45° upon the multilayer film) is separated into two polarized beams

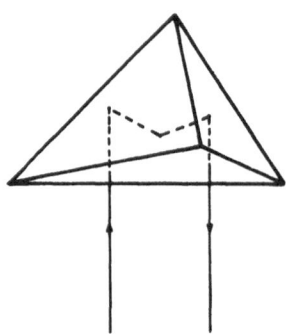

FIGURE 4. A corner cube.

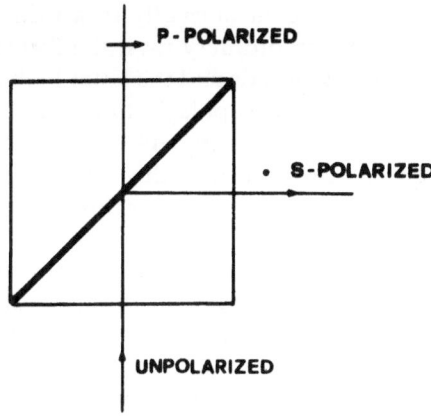

FIGURE 5. A polarizing beam-splitter.

emerging from the cube at right angles. The beam which passes straight through the cube emerges 98% linearly polarized, with the plane of the electric-field vector parallel to the plane of incidence on the multilayer film, namely, the plane passing on the incident ray and normal to the film (p-polarized). The other beam which emerges from the cube at right angles to the incident beam, by reflection on the multilayer film, is also 98% linearly polarized, but with the electric-field vector orthogonal to the incidence plane (s-polarized). This polarizer differs from the traditional ones (Nicol, Glan–Taylor, Glan–Thomson, Wollaston, all made with birefringent crystals) in that it polarizes light at only one wavelength, depending on the thickness and properties of the multilayer film.

Another important feature of the polarizing beam-splitter is that, when the incident beam is linearly p-polarized, it totally crosses the multilayer film, and when it is linearly s-polarized, it is totally reflected by it: in both cases the beams retain, after refraction or reflection, their polarization plane. When the entering beam is linearly polarized, with its electric vector inclined at less than 90° on the incidence plane, it gives rise to two mutually orthogonally polarized beams, whose intensities depend on the orientation of the original polarization plane. Thus, with proper choice of this orientation, we can obtain two beams of identical intensity polarized on orthogonal planes.

Birefringent crystalline materials, such as quartz and calcite (pure calcium carbonate), can also be used to obtain polarized light or to rotate the plane of polarization. A birefringent crystal splits an incident light ray into two linearly polarized rays traveling through the medium with different speeds; these are called ordinary and extraordinary rays, and generally propagate in different directions. Along the ternary symmetry axis of a quartz or calcite crystal there is no birefringence; the direction of this symmetry axis is called the optical axis. A plane parallel to the optical axis is named the principal plane. A ray incident normally to a principal plane gives rise to coincident ordinary and extraordinary rays; the former is polarized in the plane passing through the optical axis and the latter in the normal direction. As the refractive indexes along the two polarization planes are different, the plate will cause an optical path difference between the refracted rays. This corresponds to a phase difference of the emerging rays. Due to the ability of creating this phase difference, or retardation, birefringent plates are called retarders or compensators.

If h is the plate thickness, while n and n' are the refractive indexes of the ordinary and extraordinary rays, the optical path difference will be $(n - n')h$. The two emerging electromagnetic orthogonal vectors oscillating with different phases comprise an elliptically polarized vector; the light is said to be elliptically polarized. If the phase difference of the two vectors is $\pi/2$, then the light is circularly polarized; this happens when

$$(n - n')h = (m + \tfrac{1}{4})\lambda$$

where m, the order, is an integer and λ is the wavelength. In this case the compensator is called a quarter-wavelength plate.

The phase difference of multiorder retarders depends on the degree of collimation, angle of incidence, and temperature. A substantial reduction in this dependence is achieved by placing two compensators in series, with their direction of retardation opposed. The new retardation value is the net optical path difference of the two retarders. Such retarders are termed "zero order" and their positioning is much less critical than with multiorder ones.

When a light beam, linearly polarized at 45° to the optical axis, enters a quarter-wavelength plate, it generates circularly polarized light, with the direction of rotation depending on the retardation sign. If circularly polarized light enters a quarter-wavelength plate, it emerges as linearly polarized light, with the plane of polarization at a 45° angle to the optical axis.

The simultaneous use of a polarizing beam splitter and of a compensator highly enhances the interference fringe contrast, as it keeps the intensities of the two interfering beams nearly equal. This is used in Miiller and Cezairlyian's dilatometer.[3] Since this instrument is a good example of a polarized Michelson interferometer and involves the simultaneous use of almost all the optical elements described above, we reproduce here the detailed description provided by the authors (Fig. 6). The specimen is inserted in the measuring leg of the Michelson interferometer, so that two opposite faces, in connection with lenses L1 and L2, make two retroreflectors: in this way any (small) rotation of the specimen does not cause an appreciable variation in the optical path difference (for an error analysis of this rotation, we refer the reader to the appendix of the cited paper).

A polarizing beam-splitter PB1 separates the linearly polarized beam from the laser into two component beams. One component (s-polarized) is reflected around the specimen and into the detector by PB1, the pentaprism/lens combination PP1/L3/L4, plane mirror M1, and a second polarizing beam-splitter PB2, and serves as the reference beam. The other beam (p-polarized) is transmitted by PB1 and directed to the quarter-wave plate QP1, which has its optical axis oriented at 45° to the polarization plane. The emergent circularly polarized beam is retroreflected by lens L1 and the specimen face with a reverse sense of circular polarization. After a second pass through QP1 the beam is again linearly polarized, but now with s-polarization so that it can be reflected around the specimen by PB1, pentaprism PP2, and mirror M2. By a similar consideration of the optical elements PB2, QP2, and L2 one can show that, after reflection from the back surface of the specimen, the component beam ultimately emerges from the interferometer with its original, p-polarization. The light output from the interferometer consists of two superimposed beams which are polarized at right angles and cannot interfere, unless they are brought to the same polarization plane. The polarizing device which performs this operation also serves as the directional fringe counting device and will be described in Section 5.

Some unique features of this apparatus are now stated:

1. The optical path difference of the "specimen" leg is independent of rigid-body translation of the specimen owing to the successive front-surface/back-surface reflections. This fact is of vital importance in dynamic measurements, where translational motion may be generated by large current pulses through the specimen. But it is also very important in static, or quasi-static, measurements because such translations are unavoidable with temperature variation, due either to support deformation or to friction effects on the supports.

2. This interferometer uses parallel light and therefore it ideally produces "infinite breadth" fringes. In practice this is not the case, due to imperfect collimation and reflecting surface irregularities. The wavefronts of the two emerging beams can be brought to near equality by adjusting the distance between lenses L3/L4. Then, the fringes are broad enough to produce almost equal illumination on the sensor surface.

Finally, it should be noted that the two legs of the interferometer have different optical paths as it appears in Fig. 6. In addition, these optical paths are through air which may possess differences in refractive index that could potentially influence the length measurements. This fact is acceptable in the present arrangement, since the expansion measurements are completed in milliseconds during which time the index changes will be negligible. For experiments of longer duration, the design could be changed so that these paths would be in vacuo except where the beams are in coincidence.

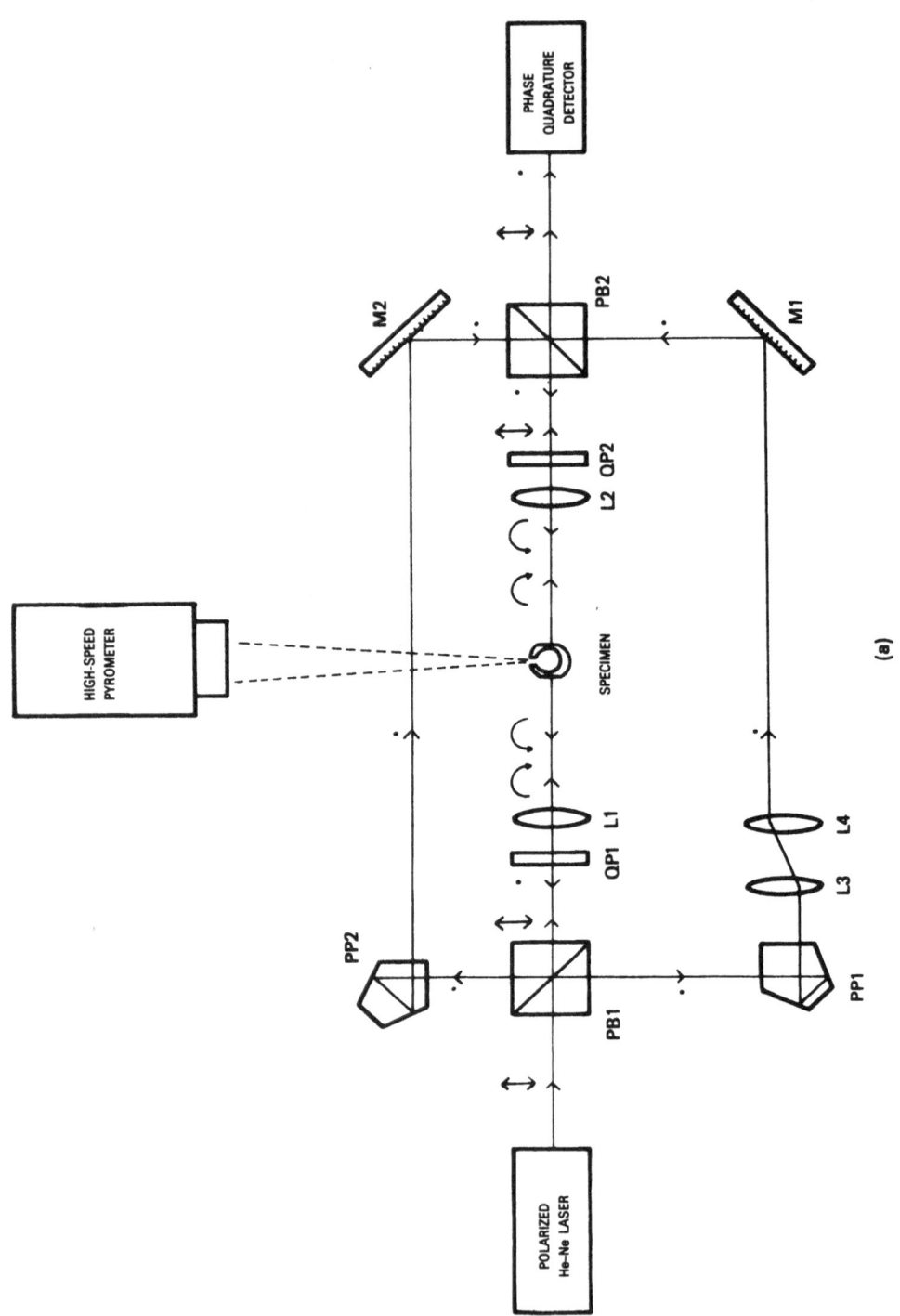

(a)

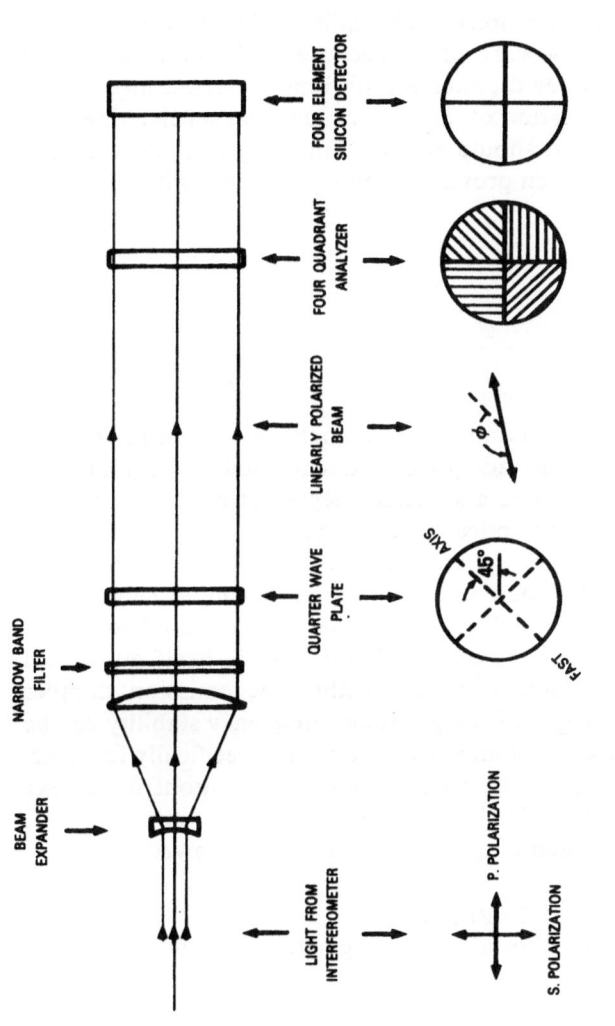

(b)

FIGURE 6. Schematic diagram of the Müller and Cezairliyan interferometric dilatometer: (a) interferometer, (b) phase quadrature detector.

3.4. Supports

Owing to the delicate nature of the interferometer, it is usually necessary to provide a vibration-free environment. This can be accomplished by proper design, and the use of commercial optical tables and supports for the furnace or cryostat that contains the interferometer. The individual optical components of the light system must also be very stable because of the potentially long measurement times necessary for thermal equilibrium. To avoid any motion of these components or deformation of the optical surfaces, special kinematic holders with three-point supports should be used. These are available commercially in various forms—some even providing adjustments that will aid optical alignment of the system.

4. LIGHT SOURCES

4.1. Lasers

Until the 1950s only spectral lamps with monochromators or filters were used in interferometry. They had adequate frequency stability for thermal expansion measurements but lacked a sufficient degree of coherence to give enough fringe visibility with long optical path differences.

With the development of lasers, possessing long coherence lengths, interferometer design has been extended to include greater sample lengths and differences in optical paths. The helium–neon laser is often used for interferometry, but it should be remembered that the laser is itself an optical interferometer and the frequency can vary within the emission doppler envelope depending on the length of the gas tube. Frequency stability can be achieved through automatic control and lasers designed specifically for interferometry are available commercially. Some typical specifications for a laser that may be used are listed in Table 1. Manufacturers' recommendations concerning warmup time and handling should be strictly followed.

TABLE 1
Typical Specifications for a Laser Designed Specifically
for Interferometry

Frequency	single stabilized TEMoo mode cw
Wavelength	632.8 nm
Polarization	linear
Beam diameter	1 mm
Beam divergence	<1 mrad full angle
Power	≈ 1 mW
Frequency stability	$< \pm 10$ MHz per 24 hours

4.2. Beam Expander and Pinhole

In many applications of a laser to optical interferometry the original beam diameter is too small. It can be enlarged by a beam expander, which is simply an inverted telescope. This is an afocal system, i.e., it consists of two lenses on the same optical axis and with their focal points in coincidence. The magnification is equal to the ratio of the focal lengths. Two configurations are possible.

(*a*) *Galilean Expander* (Fig. 7a). The first element is a divergent lens. If both elements are single thin lenses, this design provides an inexpensive solution, but does achieve moderate performance.

(*b*) *Positive Expander* (Fig. 7b). The first element is a convergent lens. Conveniently, this is a microscope objective lens (some manufacturers produce objectives corrected for one conjugate at infinity; these are preferable). The second element is an achromat lens corrected for spherical aberration. An advantage of this configuration is that a pinhole can be positioned on the common focus which acts as a spatial filter. If the pinhole has a certain diameter, it suppresses the amplitude "noise" in the intensity profile of the beam, giving a pure Gaussian intensity distribution across the cross section of the beam. Both the beam expanders and individual pinholes are commercially available.

In many applications the expanded beam must have a spherical wavefront; this is achieved by using either a divergent lens or a convergent lens with a pinhole. The pinhole is positioned at the focal point of the lens. Preston[4] employed this arrangement as a light source in a Fizeau interferometer.

5. DETECTOR AND FRINGE COUNTER

5.1. Detector

Early automatic interferometers used spectral lamps as light sources and, with the low radiant power, very sensitive detectors were necessary. Therefore

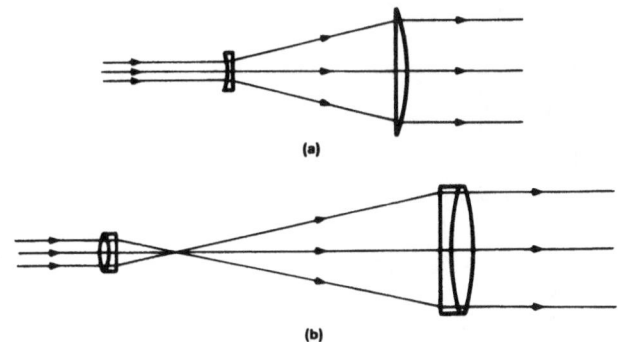

FIGURE 7. Beam expanders: (a) Galilean; (b) positive.

the only choice was the photomultiplier. However, the most popular detector now is the silicon photodiode. When used in association with a laser light source it has adequate sensitivity. It is compact and requires much simpler circuitry than the photomultiplier, without the inconvenience of a high-voltage power supply. A typical photodiode circuit is shown in Fig. 8. The negative voltage V is a bias, which can be omitted at the expense of response speed; A is an operational amplifier and the circuit gain is governed by the feedback resistor R, which can be selected over a considerable range in order to match the input of a counter used for fringe counting.

A single detector can be used, but it will not sense the fringe direction. This is acceptable only if the sample expands in just one direction and does not contract. It is much safer to use a bidirectional fringe counter in order to record both expansion and contraction. A bidirectional fringe counter needs a second signal in quadrature with the first signal.

Bidirectional counters have been made using several arrangements.[5,6] One system uses an optical quadrature signal and is particularly attractive for its simplicity and elegance. This was developed by R.J. Hocken of the National Bureau of Standards, USA (unpublished, but used by Miiller and Cezairliyan[3]).

With a minor change in the beam expander, the fringe sensor is presented in Fig. 6b. The superposed beams exiting from the interferometer are linearly polarized in mutually orthogonal planes and with a phase difference ϕ caused by the optical path difference of the two paths. The beams are expanded so as to fill the subsequent elements with a bundle of parallel rays. Thermal radiation from the specimen at high temperature is suppressed by an interference filter centered on the laser wavelength. The key optical components of the sensor are the quarter-wave plate with its optic, or "fast," axis at 45° to the polarization planes of the incoming light and the four-quadrant analyzer with optic axes rotated by increments of 45° in successive quadrants. The

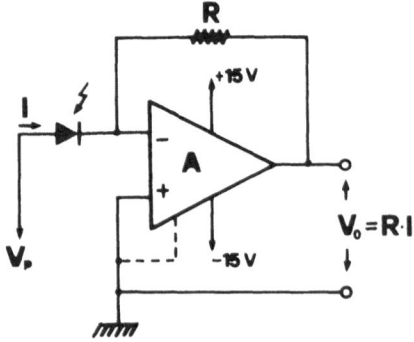

FIGURE 8. Electric circuit of a preamplifier for a silicon photodiode.

quarter-wave plate converts the linear polarizations of the two beams from the interferometer into left- and right-handed circular polarizations, respectively. If the intensities of the two beams from the interferometer are adjusted to be equal, the two circularly polarized beams combine to a single, linearly polarized beam whose polarization orientation is $\phi = \delta/2$, measured counterclockwise (ccw) from the direction of the fast axis of the plate.

Therefore, as δ changes by $+2\pi$ or -2π rad, the polarization plane of the light reaching the analyzer will rotate through $+\pi$ or $-\pi$ rad, and the light transmission through the four quadrants of the analyzer will undergo one cycle around the quadrants in a ccw or cw direction. This corresponds to a shift of one fringe into or out of the field of view.

The four beams from the analyzer pass to the corresponding four quadrants of a multiple silicon detector. The signals from opposite quadrants are amplified differentially, giving rise to two signals which are proportional to $\sin \delta$ and $\cos \delta$, respectively. They are therefore in quadrature.

The signals must be conditioned before being fed to the counter. They are amplified and clipped, or operate a Schmidt circuit to give two square waves in quadrature. These signals are sent to a commercial A-quad-B counter.

An electronic method for generating quadrature signals is based on fringe phase modulation and is described elsewhere by Eselun and Wolff.[7]

The interferometer detector produces a signal with alternating component

$$V = V_0 \cos \delta$$

in which δ is the phase given by

$$\delta = (4L\nu)/c$$

where L is the optical path difference, ν the frequency of light, and c the speed of light. If the phase is modulated with angular frequency ω_m, then the signal will be

$$V = V_0(\delta_0 + \phi \sin \omega_m t)$$

The right-hand side can be decomposed into a series of harmonics whose coefficients are Bessel functions of the first kind. The zero harmonic (or unmodulated term) and the first harmonic are selected because they contain the quadrature signal, the former as a DC signal, the latter by a phase locked amplifier. The derived signals are

$$V_x = V_0 J_0(\phi) \cos \delta_0 \quad \text{and} \quad V_y = -2V_0 J_1(\phi) \sin \delta_0$$

By proper gain adjustment the two amplitudes can be made equal and then the two signals are used for bidirectional counting.

Fringe modulation is usually achieved in one of two ways:

1. Geometrical modulation of the interferometer length with an oscillating mirror on a piezoelectric support driven by an AC voltage.
2. Modulation of the refractive index in one of the light beams with an electrooptic device.

In addition to fringe counting it is also necessary to optically monitor the fringes in order to align the interferometer and check its operation during measurements. A convenient and effortless way of observing the fringes is through the use of close-circuit television. The fringe pattern is duplicated (with a beam splitter); one image is used for counting, and the other is collected by a vidicon to be displayed on a TV monitor.

5.2. Fractional Fringe Counter

Very high sensitivity in the measurement of fringe motion is needed for measurements on low thermal expansion materials. In this case one half-wavelength detection may not give sufficient resolution.

A fringe can be measured to a small fractional order (typically $\lambda/100$) by optical and electronic means. The former method is based on multiple beam interferometry, which gives extremely sharp interference fringes as are produced in the Fabry–Perot interferometer. Such fringes can be easily interpolated with micrometric devices.

Electronic interpolation uses the two quadrature signals, which have equal amplitudes. Two techniques are available. One uses voltage multipliers to create $\sin^2 \delta$, $\cos^2 \delta$, $\sin \delta$, $\cos \delta$; these quantities yield $\sin 2\delta$ and $\cos 2\delta$. Now n successive stages continue doubling the phase so that the final signals are counted as before, but this time with the sensitivity increased to $\lambda/2^{n+3}$. This approach has the benefit of keeping the information digital.

The other technique operates a rectangular-to-polar conversion on the two signals, thus giving the phase angle on top of the integer counts. This method can be easily carried out with a minicomputer and is considered preferable.

6. THERMAL SYSTEM

In order to measure thermal expansion, the specimen must be submitted to a temperature variation over a well-defined range. Therefore, a thermal system is needed to heat the sample above or to cool it down below room temperature. Cooling is achieved by surrounding the specimen with a heat sink at low temperature; heating is produced either by a heat source or by self-heating of the sample using an electrical current. Except in this latter case, heat transfer to or from the specimen is provided by thermal radiation since the sample is usually kept in vacuo. Proper design of the thermal system is necessary to insure thermal equilibrium and to minimize thermal gradients so that the temperature can be measured with sufficient accuracy.

Measurements can be conducted using a ramp or a step temperature change. With the temperature ramp, simultaneous readings of fringe counts and temperatures are taken at certain time intervals. Simultaneity is achieved to a good approximation if the temperature is varied slowly and the two correlated readings are separated by no more than the acquisition time. The slope of the temperature ramp is controlled by the power input to the heater and must be adjusted to the thermal diffusivity of the sample material. Obviously, the lower the temperature rate the higher the temperature uniformity. A typical temperature rate is 1 K min^{-1}.

Temperature steps with long temperature holds potentially provide more accurate measurements and are generally used. However, when the sample exhibits phase changes that affect the expansion, it may be desirable to use the ramp technique.

Different techniques are necessary for temperature control, depending on the temperature range to be covered. Conventionally, the temperature ranges are classified as low, moderate, and high. Low, or cryogenic, temperatures have a conventional higher limit of 120 K. Moderate temperatures proceed from here up to 373 K. Interferometry has been applied in the moderate and high ranges. A few recent designs of interferometric apparatus will be reviewed in the following sections.

6.1. Moderate Temperature Ranges

This type of instrumentation is used for measurements on materials that are used near, or moderately below, room temperature, where the sensitivity of the interferometer is necessary because either highly accurate data are needed or the expansion of the material is very small.

A recent design for the measurement on low-expansion optical and composite materials by Wolff and Eselun[2] is shown in Fig. 2 for the ±100 °C temperature range. Cooling is provided by passing liquid nitrogen through copper coils that have been soldered onto a copper cylinder (C) that is coaxial with the specimen. The coils were positioned to achieve uniform cooling of the copper cylinder. Thermal transfer is enhanced by coating the cylinder with a high-emissivity material. In order to reach intermediate temperatures, eight Nichrome wire heaters were placed coaxially between the copper cylinder and the sample. Temperature control was maintained by varying the electrical power to these heaters, balanced with the flow of liquid nitrogen. For moderate temperatures above room temperature a second Nichrome heater was wound on a cylindrical support, which was placed coaxial with the sample. The whole assembly was thermally insulated from the outside using a number of radiation shields shown in Fig. 1.

6.2. High Temperatures

High sample temperatures are achieved either by heat transfer or by self-heating. In the first instance we need a furnace operated in vacuo or inert atmosphere up to $\simeq 1500\,°C$. As an example, we refer to the apparatus in Fig. 15 on p. 703 of Vol. 1 of this Compendium. The heater is a self-supported coiled tungsten, or molybdenum, strip held at its ends by two thick molybdenum rods entering through the bottom plate. Thermal insulation is maintained, first with coaxial metallic radiation shields G and then, at the outside, by a layer of graphoil sheets H. This material is a highly anisotropic pyrolitic graphite foil with very high thermal conductivity along the foil and very low thermal conductivity across the foil. Since it operates in vacuo, it acts mainly as a set of radiation shields. The whole furnace is contained within a metallic enclosure I, which is cooled by water circulated in copper tubes. In this way the solid outer shell J is maintained at room temperature and can act as a stable support for the specimen, which is part of the interferometer.

The power is supplied to the heater by a low-voltage, high-current source. As the thermal inertia of the heating coil is very low, the temperature ramp must be generated electronically with a long time constant integrator.

The self-heating technique is restricted to metallic conductors and uses Joule heating resulting from an electric current passing through the sample in metallic conductors. In recent innovative designs the specimen is heated by a current pulse of subsecond duration while the temperature and expansion readings are performed by fast automatic data acquisition systems.

Two systems have appeared in the literature. The first one, by Miiller and Cezairliyan,[3] was described in Chapter 18 of Vol. 1 of this Compendium with details added in this chapter (Section 5.1). A unique feature is that the expansion measurements are taken between two lateral faces, on a transverse section of the specimen where the temperature is supposed to be distributed uniformly.

The second design, by Righini et al.,[8,9] was tested successfully on a niobium sample. This is a variation of the pulse method in which the longitudinal expansion of a long specimen (cylinder, tube, or strip) is measured with a laser interferometer. At the same time, the temperature profile of the specimen is measured with a high-speed scanning pyrometer that was developed for these measurements.[10] The expansion and thermal profile are correlated to obtain the thermal expansion over the temperature range of the

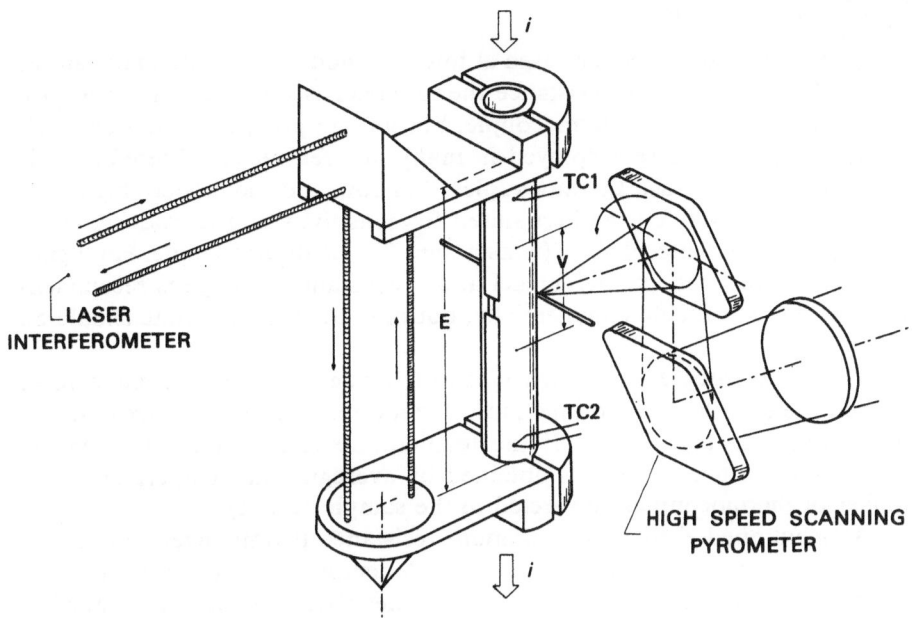

FIGURE 9. Interferometric pulse dilatometer of Righini *et al.*

experiment. A schematic diagram of the equipment is shown in Fig. 9. The electric current is fed to the sample through two clamps, one of which is fixed and the other free to move in the axial direction. A total reflection prism is attached to the fixed clamp and a corner cube retroreflector is attached to the mobile clamp. The prism and retroreflector are part of the measuring arm of a commercial Michelson interferometer. The distribution of the radiance temperature along the specimen is scanned by a rotating mirror attached to a high-speed pyrometer. The temperature of the sample ends is outside the pyrometer range and is measured by two thermocouples, TC1 and TC2.

The experimental data, temperature profile, and expansion are fitted using an analytical technique that yields a polynomial expression relating the thermal expansion to the temperature. The fitting technique may be found in the cited paper.

7. TEMPERATURE MEASUREMENT

The specimen temperature can be measured by either contact or contactless thermometers. The first group includes resistance thermometers and thermocouples, while the second group comprises spectral radiation thermometers. The choice is dictated by the temperature range.

7.1. Contact Thermometers

Contact thermometers are used at low and moderately high temperatures up to 700 °C. A platinum resistance thermometer is inherently more precise and reproducible than a thermocouple, but its size prevents its use on small specimens such as those employed in many interferometers. Therefore, it is common practice to use thermocouples. For low and moderately high temperatures (up to 400 °C) the best suited for sensitivity and reproducibility is the copper/copper–nickel alloy (constantan) T-type thermocouple. For higher temperatures the iron/copper–nickel alloy (constantan) J-type is recommended. These are available commercially, but only the most accurate should be used.

The thermocouples are attached so as to achieve the best thermal contact between the sample and the junctions. A good technique is to spot-weld the thermocouple to the sample. If the temperature is not uniform along the sample, a mean temperature is obtained by averaging the temperatures of a number of thermocouples connected to the sample (Fig. 1).

As mentioned earlier, the platinum resistance thermometer (prt) is the most precise sensor for measuring the sample temperature. It is proposed that the prt be stationed in a small copper block near the specimen and a number of differential thermocouples connected between this block and the specimen (Fig. 10). With this arrangement the thermocouple drift has much less influence on the overall accuracy, since the main measurement is carried out by the prt. Measurements are much simplified if the sensors are connected to the scanner

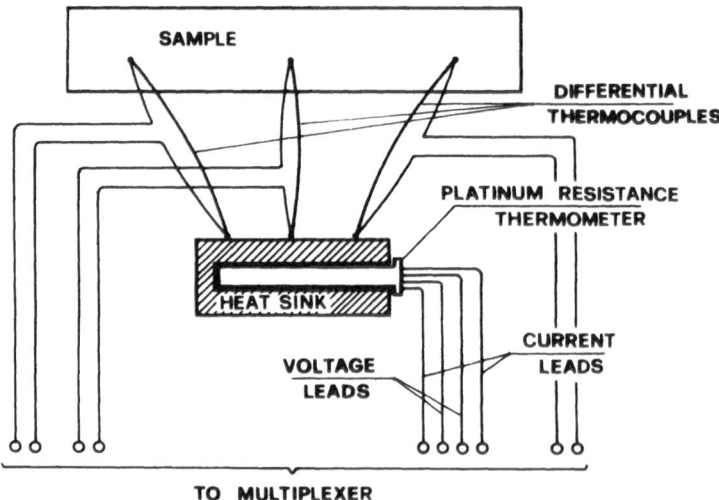

FIGURE 10. Temperature measurement with a platinum resistance thermometer and differential thermocouples.

of an automatic data acquisition system and data reduction is carried out by a computer.

7.2. Radiation Thermometer

The disappearing filament pyrometer, which is often used in both industrial and laboratory situations, requires a human operator to perform the readings and is therefore not well suited to automatic measurements, where photoelectric pyrometers are more desirable. Several instruments of this kind have been described in the recent literature, but not many are available commercially. Recent advances in radiation thermometry, in relation to the traceability of pyrometers to the national temperature scales, have been reviewed by this author.[11]

It is noteworthy that radiation thermometers measure true temperature only when they are focused on the aperture of a blackbody cavity. Blackbody conditions can be realized by drilling a hole in the sample with a small diameter-to-depth ratio ($\simeq 10$). A convenient arrangement, with the hole on the sample bottom, is shown in Fig. 13 on p. 701 of Vol. 1. In some other instances, as in Miiller and Cezairliyan's instrument, the hole is drilled through the sample wall. This arrangement is nearly a blackbody, since the sample is a hollow cylinder.

8. DATA ACQUISITION AND TREATMENT

Thermal expansion data are more useful if they are presented as a functional relationship between expansion and temperature. In the case of the ramp technique, both quantities must be read simultaneously. Traditionally, this has been achieved by simultaneously plotting temperature and expansion on an x, y recorder. More accurate results can be achieved using digital techniques. But simultaneity is achieved only within a certain approximation, which is greatly enhanced with the aid of an automatic data acquisition system. This is becoming more common practice in modern experimentation.

Advantages of the automatic data acquisition are the following:

- related measurements can be taken almost simultaneously;
- a great number of closely spaced data points can be taken, so that the functional relationship can be better defined and discontinuities detected;
- accidental errors are reduced by the great number of readings taken at each point;
- quantities affecting the readings can be recorded in addition to the expansion data, so that corrections can be introduced to restrict systematic errors;
- the data can be processed automatically.

These features cannot be achieved easily by manual operation which, moreover, is prone to errors in both transcription and calculation.

In an automatic data acquisition system a computer receives signals from sensors located in pertinent locations of the physical system being tested, converts them into physical quantities with suitable units, and determines their relationship. Between the sensors and the computer there are signal conditioners that transform the signals to make them compatible with the computer. Signals are multiplexed to the computer. Some sensors generate digital signals, as do optical detectors used in interferometry. In this case a counter sends a number of counts to the computer. More often sensors are analog, such as the emf of a thermocouple or the voltage drop in a resistance thermometer. These signals are converted into digital form by an analog-to-digital converter (ADC) or a digital voltmeter (DVM). The most convenient way of handling analog signals is to multiplex them to an ADC/DVM, which is connected to the computer by a bus. In this case one single ADC/DVM is sufficient to process many signals.

A schematic diagram of an automatic data acquisition and control system for thermal expansion measurement is shown in Fig. 11. It will be explained in the following sections.

8.1. Computer and Peripherals

The whole experimental system can be controlled by a computer. A desk computer is often sufficient for this task. A large memory is not needed—most modern computers have a central memory of 640 kbytes, which is more than sufficient. But the computer must be able to accept inputs from a number of peripherals and to communicate with the data acquisition and control system.

The computer must be equipped with the following peripherals:

- Keyboard, to interact with the operator.
- Hard disk and floppy disk drivers, to store the experiment program and data.
- Monitor, to display the commands and results of the operations.
- Printer, to output the results in tabular form and to write equations relating the physical quantities.
- Plotter, to show diagrams of strain vs temperature and coefficient of thermal expansion vs temperature.
- Control unit of the data acquisition and control system.
- Eventually, other instruments which are not under the control unit, such as the counter.

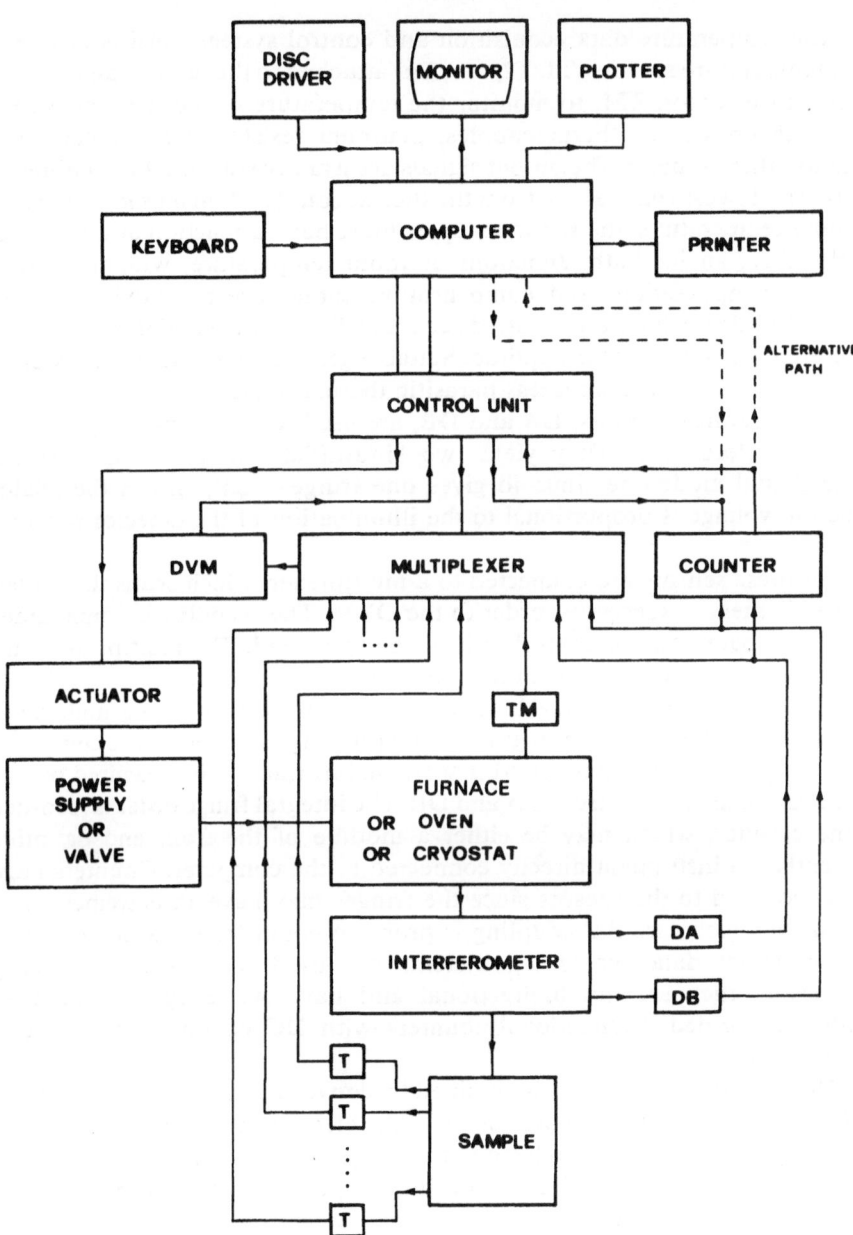

FIGURE 11. Automatic data acquisition and control unit for thermal expansion measurement.

8.2. Data Acquisition and Control System

The temperature data acquisition and control system consists of one or more temperature sensors (T1, T2, . . . , Tn) attached to the sample, and another temperature sensor, TM, to monitor the temperature of the thermal system. These sensors can be thermocouples, platinum resistor thermometers, or a radiation thermometer. The output signals are usually voltages, but of different levels. The lowest voltages are from the thermocouples. When used to measure absolute temperatures the thermocouples must have a junction at a reference temperature, an ice bath. Junctions at room temperature, with room temperature compensation, and compensation cables are not advisable since usually they do not have the required accuracy. With differential thermocouples no reference temperature is required. Suitable grounding of the thermocouples is necessary in order to suppress parasitic thermal emfs.

Electrooptical sensors, DA and DB, are used to detect the fringe motion in the interferometer. They yield two sinusoidal voltages in quadrature: in the digital mode one sinusoid gives one fringe count, and in the analog mode the voltage is proportional to the illumination of the detector sensitive area.

All these sensors are connected to a multiplexer, which scans the signals and sends them in successive order to the DVM. The scanning is commanded by the computer via the control unit of the front end. The multiplexer must be of the reed-relay type, which has low parasitic thermal emf.

The signals may have levels varying over a wide range, so a multirange DVM must be used with autoranging or with the range selected by the computer.

It was seen earlier (Section 5) that fringe information is measured by two quadrature signals via sensors DA and DB. The integral fringe order is recorded by the counter, which may be either a module of the front end or, more frequently, an instrument directly connected to the computer. Counters must be DC coupled to the sensors since the fringes may have an extremely slow motion, in which case AC coupling is prone to count losses. Many counters in commercial data acquisition systems are used to measure frequency and period, they are not bidirectional and have AC coupling, and they should be avoided. Bidirectional counters with DC coupling are available commercially.

The fractional fringe part is obtained by converting the quadrature signals (which give the x, y coordinates of the light electric vector) into polar form whose angular coordinate is related to the fringe fraction. These analog signals are sent via multiplexer and DVM to the computer, in order to perform the conversion automatically.

The thermal system—a furnace, an oven, or a cryostat—can also be controlled by the computer. In the ramp mode the computer triggers the temperature ramp by actuating the relay which switches on the heater, or the solenoid valve which controls the coolant. At the higher end of the temperature range the computer stops the experiment automatically, which also acts as a safety measure. In the step mode the computer drives the steps by outputting the set points to the temperature controller and taking measurements when the temperature has reached the steady state.

It is usually advantageous to purchase a commercial system owing to the complexity of interfacing the various parts.

8.3. Date Treatment

The data treatment consists of three parts: signal conversion into physical quantities, determination of the functional relationship, and displaying the results.

8.3.1. Signal Conversion

The sensor signals are converted into physical quantities by the computer via interpolating equations that have been fitted to the calibration data and are stored in the program. For thermocouples and platinum resistance thermometers, these equations have polynomial form with degree appropriate to the temperature range. The equations may be more complicated for radiation thermometers but are usually supplied by the manufacturer.

If the temperature measuring system consists of a set of differential thermocouples and a platinum resistance thermometer, the average of the differential temperatures must be added to that given by the resistance thermometer.

The thermal expansion strain is calculated by multiplying the fringe number by one-half of the wavelength (316.4 nm for a stabilized He–Ne laser) and dividing the expansion by the reference length of the specimen. The reference length is usually taken as the length at 20 °C. All data may be stored in a disk file during the test.

8.3.2. Functional Relationships

At the end of the experiment, strain and temperature data can be printed and a polynomial of suitable degree fitted to the experimental data. The derivative of the strain-vs-temperature function gives the coefficient of thermal expansion.

8.3.3. Data Plotting

It is also very helpful to have the results plotted. Both the fitted function and the data should be plotted in order to check the fit. Graphic programs to perform all these operations are usually available with most computers.

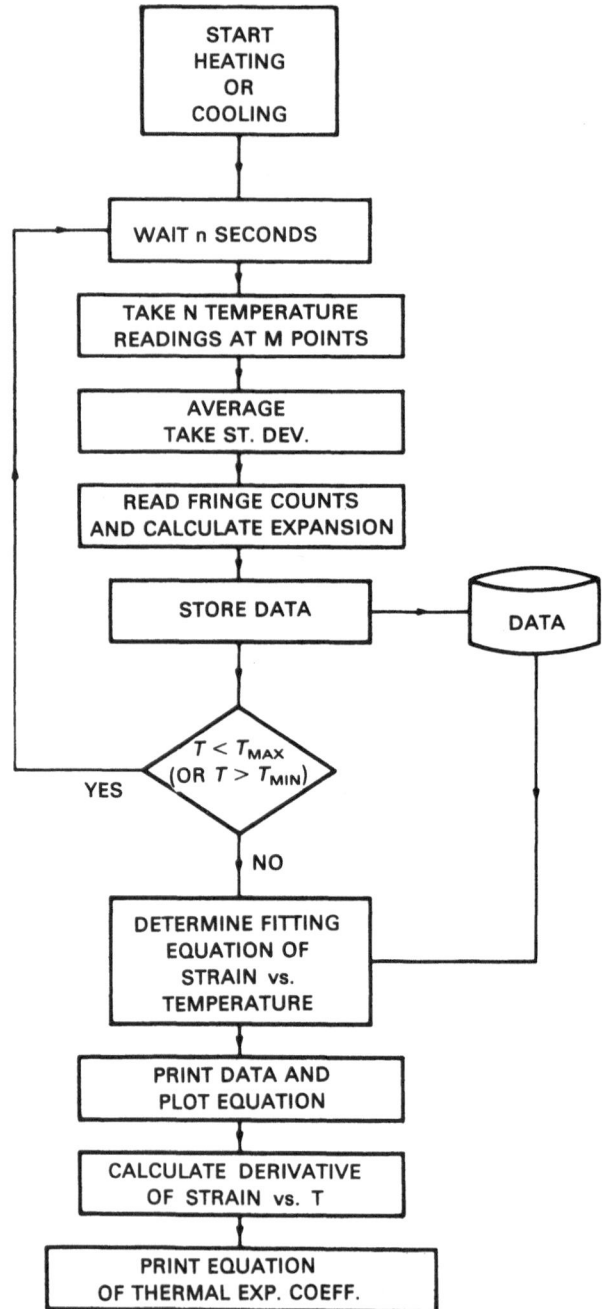

FIGURE 12. Flow chart of a program for automatic thermal-expansion measurement.

A flow chart of a program for automatic thermal-expansion data collection and computation is shown in Fig. 12.

9. ERROR ANALYSIS

The best way to avoid errors is to make sure all the subsystems of the apparatus are calibrated and working properly. A general procedure for detecting, analyzing, and evaluating errors is impractical, but major error sources will be listed, so the experimenter can check for them on his particular system.

Errors in thermal expansion determination originate from two sources: the temperature and the strain measurement.

9.1. Temperature Errors

For any temperature sensor there are two sources of error: a calibration error and sensor drift.

The traceability of any temperature calibration to the national standards is very important, since it assures the validity of the results on an international basis. It is advisable for any research laboratory to have secondary standards that are periodically calibrated with national standards and which in turn are used periodically to calibrate the working instruments. In this manner, the absolute accuracy of the measurements can be ascertained and any drift in the instrumentation can be followed over time in order to determine the required frequency of recalibration.

Another type of error depends on how well the sensor is actually measuring the sample temperature. This error occurs, for example, if there is a bad thermal contact between a thermocouple and the sample, or there are heat leaks along the thermocouple wires. In the case of a pyrometer, the deviation from blackbody conditions would also result in similar errors. Good experimental design and attention to detail are necessary to avoid these errors. Good design is also required to avoid thermal gradients in the sample, because these are also a possible source of error.

Miiller and Cezairliyan[3] estimated the temperature errors arising from all the aforementioned sources. With their radiation thermometer, in a pulse experiment, the error amounts to 5 K at 2000 K and to 10 K at 3000 K. This may give some idea of the errors that can be met in actual experimentation, although with a steady-state technique errors might be smaller.

9.2. Strain Errors

Although the interferometric measurement of expansion is intrinsically very sensitive and accurate, errors can arise from secondary effects. The following are some sources of error.

Reference length of the specimen. This value is taken at 20 °C and can be very precise. However, if measurements are carried out at high temperature, then annealing, recrystallization, or even evaporation may cause the reference length to vary. This effect can be detected and evaluated by remeasuring the sample length after the experiment.

Departure from flatness of the specimen end faces, besides causing fringe deformations, is responsible for lack of definition of the length.

Support deformation. Temperature gradients may be present within the sample supports so causing thermal strains that deform the specimen. Again, good design is of great importance.

Sample distortion. Chemical or physical inhomogeneity of the sample material can cause shape deformation, even with a uniform temperature distribution. In this case rotation of the end faces of the specimen may arise causing false fringe counts. Similar distortions are generated in homogeneous materials if the temperature distribution is not uniform.

Sample rotation. Support deformation causes sample translation and rigid rotation: the former has no effect on fringe counting since it does not affect the optical path difference, but the second rotates the end faces thus changing the fringe pattern and generating spurious counts.

The effects of deformation and rotation can be monitored by displaying the fringe pattern on a closed-loop television system.

Optical path changes. Variation in the optical path outside the sample chamber can also be caused by temperature changes. Optical components placed in separated branches of the interferometer may deform and change in refractive index. Both effects result in optical path change and therefore fringe motion, which simulates sample deformation. This effect is present in windows, prisms, lenses and retroreflectors, corner cubes, or cat's-eyes, and

TABLE 2

Errors on Cooling a Quartz Sample from 298 to 150 K

Source of errors	Value	
	ΔL (nm)	$\Delta L/L_0$ (10^{-8})
Air path	±53	±6
Mirror supports at different temperatures	−23.5	−2.6
Bending of mirror supports	−67.9	−7.4
Interference pattern changes		
(a) motion of lens screws	+4.9	+0.5
(b) refractive index changes of lens	−7.5	−0.8
Lens temperature changes	+194	+21
Nonparallel sample end faces	±82	±9
Root-mean-square		25

can in principle be calculated if the temperature change, thermal expansion coefficients, and temperature dependence of the refractive indexes are known. Mirrors can also be responsible for optical path changes due to deformation of their supports. Mirror rotation can cause deformation of the fringe pattern or even misalignment of the interferometer, resulting in a loss of interference.

Also related to this source of error is the variation in the refractive index of air within different branches of the interferometer, due to temperature, pressure, and humidity changes. To avoid problems, the components of the interferometer can be placed in an evacuated, temperature-controlled chamber.

An example of error evaluation given by Wolff and Eselun[2] for the equipment shown in Fig. 2 is summarized in Table 2.

REFERENCES

1. E.G. Wolff and S.A. Eselun, *Proc. SPIE* **192**, *Interferometry*, p. 204 (1979).
2. E.G. Wolff and S.A. Eselun, *Rev. Sci. Instrum.* **50**, 502 (1979).
3. A.P. Miiller and A. Cezairliyan, *Int. J. Thermophys.* **3**, 259 (1982).
4. S.D. Preston, *High Temp. High Pressures* **12**, 441 (1980).
5. W.E. Rowley, *IEEE Trans. Instrum. Meas.* **IM15**, 146 (1966).
6. J. Dyson, *Interferometry as a Measuring Tool*, Hunt Barnard Printing Ltd, Aylesbury (1970).
7. S.A. Eselun and E.G. Wolff, "Advances in Interferometric Signal Analysis," *24th Int. Instr. Symp.*, ISA, Albuquerque, New Mexico (1978).
8. F. Righini, R.B. Roberts, A. Rosso, and P.C. Cresto, *High Temp. High Pressures* **18**, (1986).
9. F. Righini, R.B. Roberts, and A. Rosso, *High Temp. High Pressures* **18**, (1986).
10. F. Righini, A. Rosso, and A. Cibrario, *High Temp. High Pressures* **17**, 153 (1985).
11. G. Ruffino, "Modern Radiation Thermometers: Calibration and Traceability to National Standards," paper presented to the *International Conference on High Temperatures and Energy Related Materials, IUPAC*, Rome (May 1987).

21

Apparatus for Continuous Measurement of Temperature Dependence of Density of Molten Metals by the Method of a Suspended Pycnometer at High Temperatures and Pressures

E.E. SHPIL'RAIN, K.A. YAKIMOVICH, and
A.G. MOZGOVOI

1. DESCRIPTION OF THE SUSPENDED PYCNOMETER METHOD

The hydrostatic weighing method and the pycnometric method are employed widely to investigate the density of liquids and, in particular, of molten metals.

The hydrostatic weighing method is based on the determination of the density of a liquid under investigation by the expulsive force acting on the immersed float of known mass and volume. This method is found to be suitable, since it enables the temperature dependence of the density to be determined over a wide range of temperatures within a single experiment. However, the hydrostatic weighing method reveals serious shortcomings when applied to the study of substances with elevated pressures of the saturated vapor, because the substance evaporating from the open surface in the zone of the working temperatures condenses on the suspension filament of the float and misrepresents the measurement results. Moreover, the surface tension at the boundary of the float suspension filament and the surface of the substance under investigation also adversely affects the readings of the instruments measuring the mass.

When using the pycnometric method, which usually gives more accurate data, one should know the mass and volume of the studied liquid contained

E.E. SHPIL'RAIN, K.A. YAKIMOVICH, and A.G. MOZGOVOI • Institute for High Temperatures, USSR Academy of Sciences, Moscow 127412, USSR.

in the pycnometer. However, the need for the volume measurement results in the fact that either the density can be derived in one experiment only at one temperature, or fairly complicated methods must be applied to record the initial volume of the liquid and its thermal expansion (dilatometric method).

The technique elaborated by the authors of the present chapter[1-3] is known as the suspended pycnometer method and combines the merits of both the aforementioned methods, namely, the high accuracy provided by the pycnometric method, and the possibility of deriving the temperature dependence of the density of the liquid under examination in one experiment, which is characteristic of the hydrostatic weighing method.

The essence of the suspended pycnometer method consists in the following (Fig. 1). The pycnometer containing the liquid under investigation is essentially a cylindrical ampoule whose top cover is closed, while its bottom terminates in a long capillary tube. The pycnometer is confined in a massive thermostatting block of a high-temperature resistance furnace with minor gap and is suspended by its top cover from an analytical balance via a long filament, so that the outlet of the capillary tube is below the pycnometer at a temperature slightly

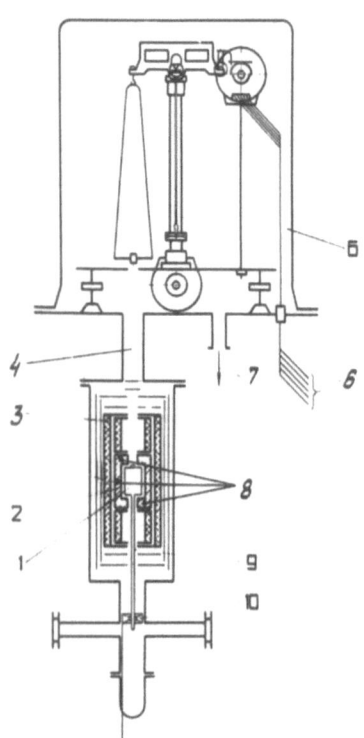

FIGURE 1. Key diagram of measurement by the suspended pycnometer method: 1—pycnometer; 2—thermostatting block; 3—high-temperature heating furnace; 4—suspension filament; 5—remote-control analytical balance; 6—remote-control system of balance; 7—vacuum; 8—thermocouples; 9—capillary; 10—gas cutter; 11—inert gas.

exceeding the melting point of the substance under investigation. Practically, this situation excludes any vaporization of the substance from the open surface of the pycnometer.

When heated to the temperature at which the experiment is carried out, the liquid under study starts dripping out the open end of the capillary in drops of a particular size. Prior to weighing the pycnometer by an analytical balance, drops which appear at the tip of the capillary tube are removed by a special gas-type cutter.

The analytical balance is positioned far from the heating furnace to avoid the effect of the latter. The pycnometer temperature is measured by thermo-couples situated at three points of the thermostatting block along the vial and the same along the capillary tube.

The density of the liquid being investigated is determined by the equation

$$\rho = \frac{m_0 - \Delta m_t - m_{cap}}{V_0(1 + \bar{\alpha}t)^3} \tag{1}$$

where t is the operating temperature, m_0 the initial mass of the liquid under investigation at the initial temperature of filling the pycnometer with the liquid, Δm_t the change in mass of the liquid due to its outflow from the pycnometer when heating to the operating temperature, m_{cap} the mass of the liquid in the capillary tube exposed to temperature varying along the length of the tube, V_0 the initial inner volume of the vial at 0 °C, and $\bar{\alpha}$ is the mean linear thermal expansion coefficient of the vial material.

The final result of the measurement will involve all the necessary corrections, the expulsive force of the gas, the ballast volume of the capillary, and so on. Therefore, when employing this method to find the density of the liquid under investigation at a preset temperature, one requires the mass of the substance contained in the pycnometer in the initial state, measured by weighing the substance at filling; the mass of the liquid dripping out of the pycnometer under the effect of heating, measured by weighing in the course of the experiment; and the inner volume of the pycnometer, determined preliminarily by calibration with due regard for thermal expansion.

The distinguishing feature of the technique lies in the fact that it is applicable for operation at temperatures exceeding the normal boiling temperature of the substance under investigation. This involves the high-temperature furnace, the pycnometer with the metal under investigation, and the analytical balance in a joined gas-tight enclosure. In order to prevent boiling of the liquid in the pycnometer, the pressure in the working volume created by an inert gas should exceed the pressure of the saturated vapor of the substance under investigation; at high temperatures it may reach dozens of atmospheres.

2. BASIC COMPONENTS OF THE EXPERIMENTAL SETUP

2.1. *The Pycnometer*

The pycnometer consists of an ampoule and a capillary tube through which the substance under investigation is inserted into the pycnometer before the experiment; the excess metal is discharged with rise of temperature in the process of the experiment (Fig. 2). As the pycnometer experiences no pressure differences during operation in the setup, it is rather simple in construction and its fabrication is very easy. Walls of 1-mm thickness are quite sufficient.

When selecting the dimensions for the pycnometer one should seek a compromise between the wish to have a pycnometer with a larger inner volume (for improving the accuracy of the results) and the necessity to keep the furnace volume within reasonable limits. Besides, it is essential that the inner volume of the pycnometer capillary tube is rather small compared to the inner volume

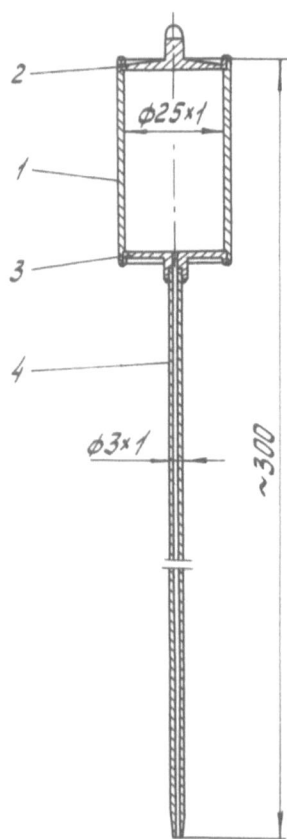

FIGURE 2. The pycnometer: 1—casing; 2—cover; 3—bottom; 4—capillary (dimensions are given in mm).

of the ampoule, because the capillary tube is placed in a nonisothermal zone and the respective correction should be introduced when processing the experimental results. The lower the mass of the substance in the capillary tube as compared to that in the ampoule, the more negligible will be the cited correction.

For experiments at temperatures up to 1300–1400 K the pycnometers are made of stainless steel; for experiments at higher temperatures they are made of molybdenum.

The ampoule body is made in the form of a thin-walled cylindrical shell. The closed top cover of the ampoule is provided with an eye to fasten the suspension filament. The bottom cover of the ampoule is provided for reliable welding of the capillary tube. The open end of the capillary tube is tapered so as to minimize adhesion of the molten metal drop (due to surface tension over the liquid–solid interface) at the end of the capillary, and to properly remove a hanging drop before weighing the pycnometer.

The pycnometer covers, the body, and the capillary tube are joined by welding under vacuum. After welding it is good practice to submit the pycnometer to annealing in order to relieve thermal stresses, which can distort the isobaric expansion coefficient of the pycnometer material. This coefficient is used to process the results of density measurements [see equation (1)], so correct information as to its actual value is of great importance for obtaining accurate results. The most reliable way to avoid an error in this case is to measure in the same laboratory the expansion coefficient of a sample of the material from which the pycnometer ampoule is fabricated.[4–8]

2.2. The High-Temperature Furnace

Provision is made for a high-temperature resistance furnace (illustrated in Fig. 3) in order to heat the pycnometer to the prescribed temperature. The pycnometer filled with the metal to be investigated is placed inside a massive block made of molybdenum or niobium, with a radial gap of some 2 mm. The block is closed at the bottom with a cover and serves to provide an isothermal space around the pycnometer.

The end covers of the thermostatting block are furnished with molybdenum tubes of diameter 10 mm and wall thickness 0.5 mm, the pycnometer capillary tube and the suspension filament passing inside the tubes.

Grooves of approximately 5-mm diameter are situated in the side wall of the block and arranged circumferentially at 120° along the block axis. One of the grooves extends to the center of the block in height; the remaining two are located at a distance of about 30 mm above and below this level. The grooves accommodate the thermocouples confined in straw-like alumina shells. The thermocouples are arranged such that the hot junctions fit snugly into the groove bottom thereby ensuring fine thermal contact. The thermocouples are

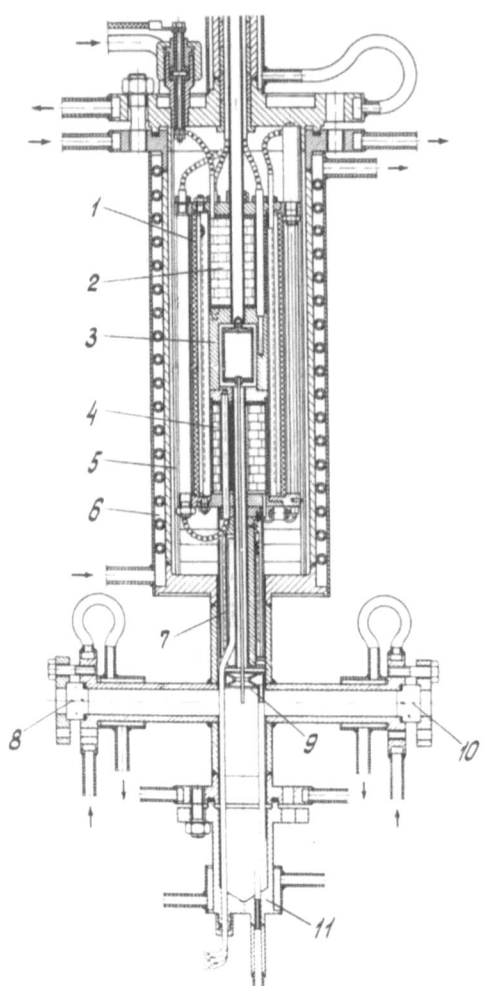

FIGURE 3. The high-temperature heating furnace: 1—main heater; 2—upper protective heater; 3—thermostatting block; 4—lower protective heater; 5—shields; 6—cooling jacket; 7—protective heater of capillary; 8, 10—viewing window; 9—gas cutter; 11—collector of investigated metal flowing out of the pycnometer.

intended for measuring the axial temperature distribution of the block. At temperatures of 1300–1400 K platinum–rhodium thermocouples were used. The thermostatting block was placed inside the heating furnace.

The main heater of the furnace is essentially a thick-walled tube made of aluminum oxide (Al_2O_3). The outer surface of the tube contains a special double-start screw-type groove in which a molybdenum electric-current-conducting wire of 1.0–1.2 mm diameter is laid bifilarly. The bifilar winding of the heater is used to avoid the influence of a scattered electromagnetic field upon the indication of the thermocouples. To prevent short circuiting between the heater loops it is enclosed in a protective alundum jacket, which at the same time serves as the first high-temperature heat shield. In addition to this shield the furnace is provided with seven radial shields made of 0.15–0.20 mm foil arranged in the following sequence (starting from the alundum one): four molybdenum shields, one zirconium shield, and then two stainless-steel shields.

In order to compensate for end losses of the main heater, i.e., to provide a uniform temperature field, the effective section of the furnace is equipped with two additional (upper and lower) end heaters of molybdenum wire 0.5 mm in diameter. Such a combination of main heater and end heaters allows the creation of a sufficient isothermal zone around the pycnometer ampoule. As a result the difference between the readings of the thermocouples is not in excess of 0.1–0.2 K (the temperature level of the sample being 1000–2000 K). Provision is also made for a special heater to maintain the capillary-tube temperature slightly above (by 10–20 K) the melting point of the metal under investigation.

The assembled high-temperature furnace is a cylinder-shaped structure of outer diameter 100 mm and some 300 mm in length. The power of the furnace amounts to 3 kW at a supply voltage of 127 V, and is terminated via supply wires and thermocouple leads for which a special airtight feed has been designed (see Section 2.5).

The high-temperature furnace is housed in an autoclave whose walls are capable of withstanding an internal pressure of the order of 15 MPa. The autoclave is provided with a cooling water jacket. The upper flange of the furnace casing is sealed with a fluoroplastic gasket.

The autoclave bottom incorporates a sleeve enclosing the pycnometer capillary tube and the capillary tube heater. Welded to the lower part of the sleeve near the capillary tube end and perpendicular to its axis are two flanged coaxial tubes with glass viewing windows. One of them is used to illuminate the end of the capillary, the other serves for visual inspection of the capillary end. To prevent undesirable heating of the glass windows by the neighboring heaters, the sections of the tubes adjoining the glasses are cooled with water.

A collector to catch the metal dripping out of the pycnometer capillary tube is attached to the lower flange of the furnace autoclave. The collector comprises a cylindrical container fitted with a cooling jacket fabricated as an integral part of the flange. The leads of the three thermocouples monitoring the temperature throughout the capillary tube are brought out via the lower flange. Besides, a metal pipe is welded to this flange for the delivery of inert gas to the gas cutter, which removes the metal remnants from the capillary tube end prior to taking the readings as soon as the predetermined temperature is reached in the setup (see Section 2.4).

2.3. The Remote-Controlled Analytical Balance

The suspended pycnometer method allows one to measure directly the mass of metal under investigation during the experiment as soon as the pycnometer attains the working temperature. One of the most widely used, and at the same time very precise, instruments is an analytical beam balance. This apparatus in particular was required in the setup for measuring the density of the substance by the above method, but to this end it was necessary to solve at least three serious problems:

(a) the balance together with the furnace and various other components of the experimental setup need to be sealed in a gas-tight enclosure, because it was intended for operation under high pressures;

(b) the zone in which the balance is located at room temperature must be separated from the furnace zone at high temperature;

(c) remote control of the balance is a necessity and includes the weighing process (loading with weights), checking the state of the balance, recording the readings, and so on.

All these problems were solved as follows. A type ADV-200, analytical beam balance with an ultimate load of 200 g and a measurement accuracy of $\pm(0.1-0.2)$ mg was selected as the basis of the mass-measuring apparatus. A special mechanism was designed for loading and unloading the weights fabricated in the form of rings with calibrated masses. The mechanism was composed of interrelated strips controlled by means of shaped cams. Each cam was linked to the shaft of an electric motor (synchrodrive). The synchros are provided with actuating mechanisms which are linked mechanically to the respective cam shaft that is directly responsible for turning it through the prescribed angle. In addition, the synchros possess a control mechanism through which the command for the required turn angle is conveyed over the electric circuit to the actuating mechanism.

The balance was mounted on a rigid platform and covered with a thick-walled housing. The environmental space of the balance was connected to the space of the furnace through a metal water-cooled pipe. The pycnometer suspension filament is connected to the analytical balance through the pipe.

The synchrodrive actuating mechanism was accommodated inside the housing together with the balance. The control mechanism was installed outside. The addition of a magnetic damper and a remote-controlled arrester had been developed for the analytical balance. The housing was furnished with two tubes possessing viewing windows, one of which was used to illuminate the microscale of the balance and the other for observing the scale and taking the readings at the instant of a balanced state.

The construction of the synchrodriven remote-controlled analytical balance is illustrated in Figs. 4 and 5.

The assembled balance with the suspended pycnometer was calibrated to avoid the effect of the Archimedes force at different pressures of the inert gas in the setup due to a continual difference between the balance beam arms and unequal volumes of the parts on the right and left sides of the knife-edged balance support.

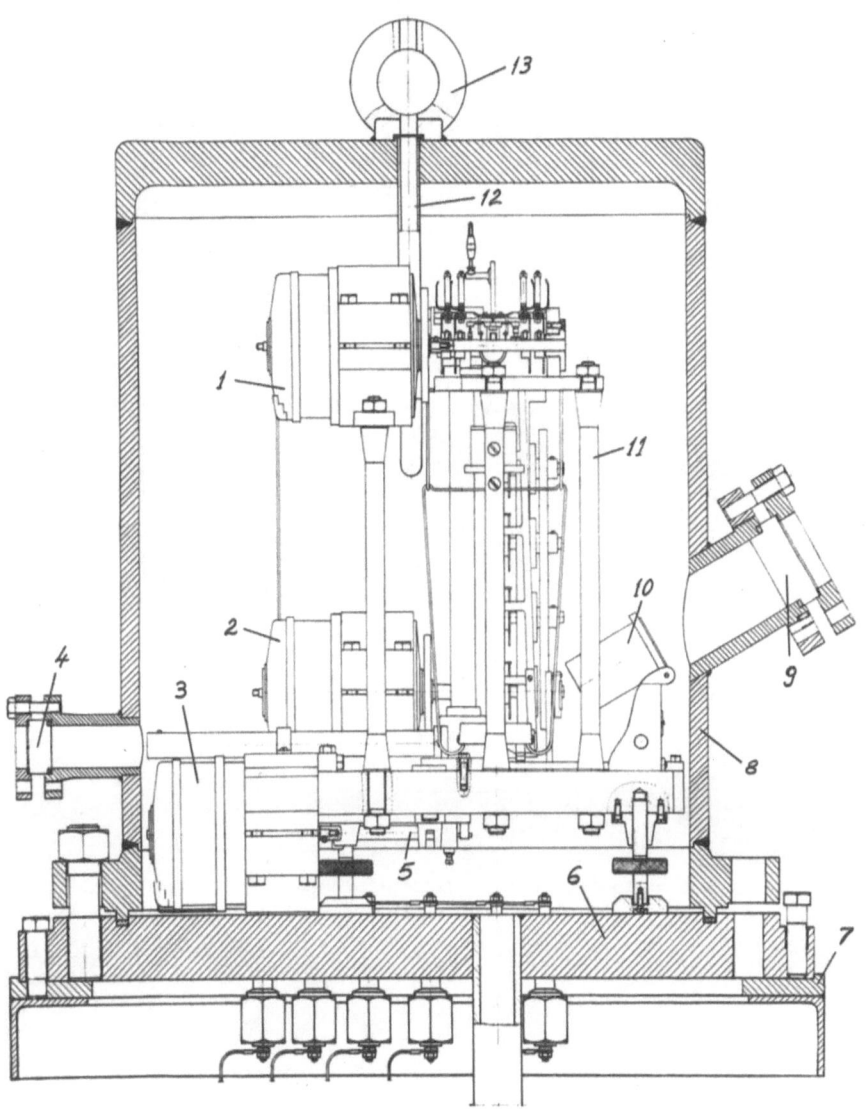

FIGURE 4. The remote-controlled analytical balance with synchrodrive (side view): 1, 2, 3—receiving synchros; 4—window for illumination of balance microscale; 5—balance arrester; 6—base; 7—bearing plate; 8—casing; 9—viewing window; 10—optical reading system of balance; 11—supports; 12—pocket for thermometer; 13—suspension eye.

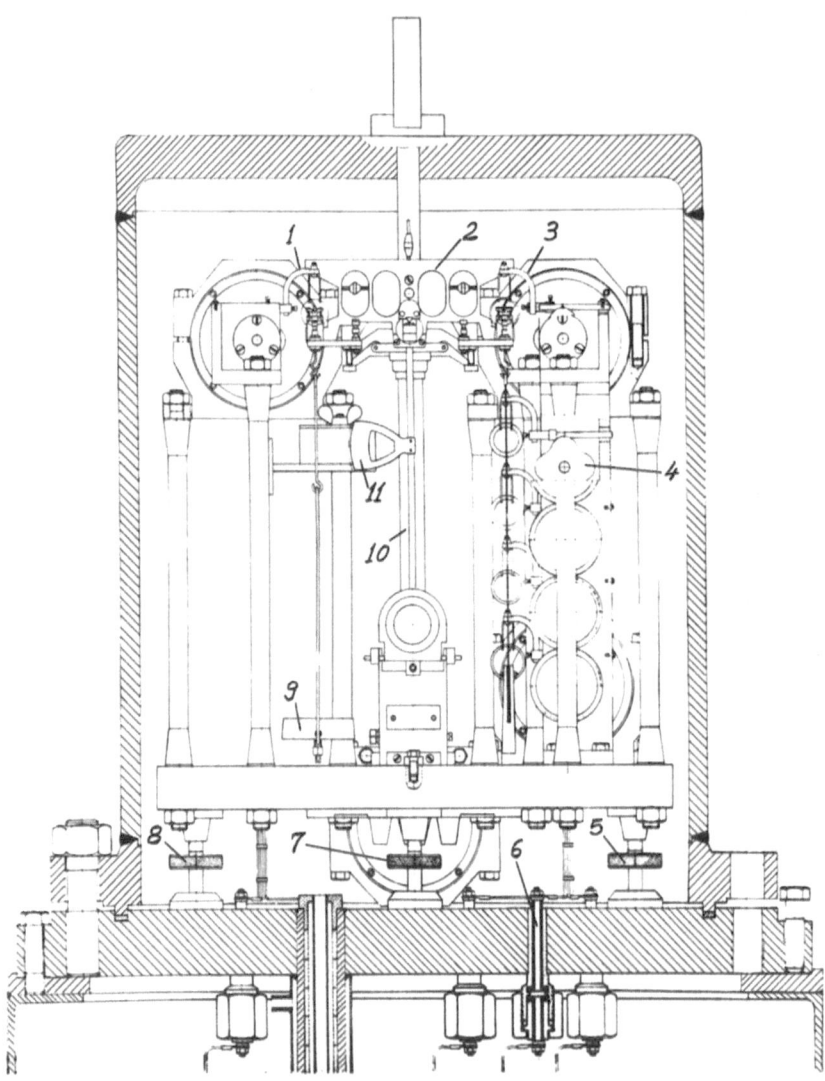

FIGURE 5. The remote-controlled analytical balance with synchrodrive (front view): 1—mechanism for delivering weights in multiples of 10 mg; 2—balance beam; 3—mechanism for delivering weights in multiples of 100 mg; 4—mechanism for delivering weights in multiples of 1 g; 5, 7, 8—adjusting screws; 6—power-supply lead-in; 9—load pan of balance; 10—balance column; 11—magnetic damper.

The remote-controlled balance under consideration proved to be reliable and handy and therefore made it possible to measure a mass to within ±(0.1–0.5) mg (depending on the density of the inert gas in the setup).

2.4. The Gas Cutter

A special gas cutter was developed for removing the molten metal drop from the outlet of the capillary tube before taking the readings of the balance. This was necessary, since the instant at which the working temperature of the pycnometer was attained and the instant at which the drop of metal dripping out of the capillary tube becomes detached from the capillary tube end were unlikely to always coincide. The cutter is essentially a hollow metal ring (collector) comprising a casing of intricate configuration (Fig. 6) and a thin-walled shell. The pycnometer capillary runs along its axis. Drilled in the lower slanting wall of the casing throughout the circumference are three rows of holes, 1 mm in diameter. The spacing between the rows of holes is 2 mm, the hole-to-hole spacing in each row being 3 mm.

A delivery pipe reaches the collector along the generatrix in order to supply the inert gas at a pressure somewhat exceeding (by 0.2–0.3 MPa) the working pressure in the setup. The gas cutter is mounted in the lower part of the furnace so that the cutting gas jet from the ring holes is directed to the edge of the capillary tube end at a height of about 5 mm above the end edge. The drops of molten metal and its traces were thereby reliably removed from the end of the pycnometer capillary tube.

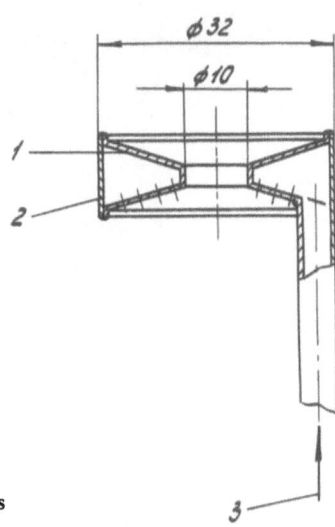

FIGURE 6. The gas cutter: 1—casing; 2—shell; 3—inert gas (dimensions are given in mm).

2.5. *The Feed-through*

The next problem to be solved in designing the setup was how to supply electrical power to the air-tight space and bring out the leads of the measuring thermocouples. With this in mind the current and thermocouple lead-ins were developed; their construction[9] is shown in Fig. 7. The current and thermo-couple feed-through are sealed in the casing by the use of fluoroplastic gaskets via a collar box and a union nut. Depending on the current density the power electrode is fabricated either of copper or of steel. The electrodes are insulated from the feed-through with the aid of textolite bushings. In order to prevent twisting of the thermoelectrodes when screwing up the union nut, provision is made for special flats milled on the end of the collar box that permits using a wrench. The feed-through is secured to the casing of the experimental setup by welding, however the joint may be detached, if necessary, and sealed with a gasket. The feed-through under discussion was employed for several years in a vacuum of the order of 1×10^{-4} Pa and a pressure of up to 15 MPa and was found to be reliable in operation. The temprature range of application was restricted with respect to the heat resistance of textolite and fluoroplastic.

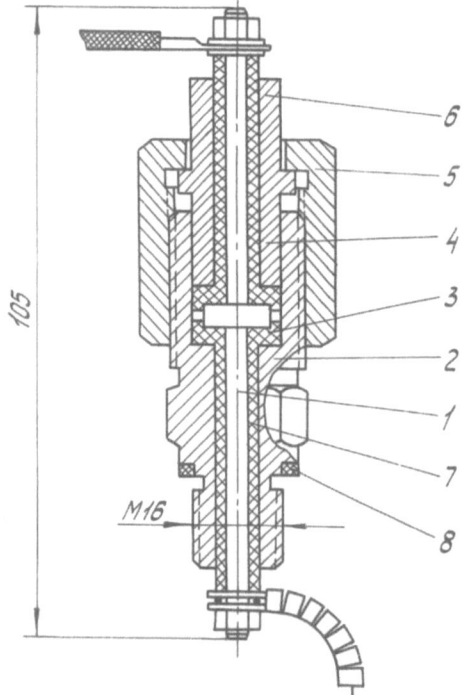

FIGURE 7. The feed-through: 1—electrode; 2—casing; 3, 8—fluoroplastic gaskets; 4—collar box; 5—union nut; 6, 7—insulating bushes (dimensions are given in mm).

3. THE EXPERIMENTAL SETUP

The circuit of the experimental setup (Fig. 8) comprises (a) an evacuation system designed for evacuating the component assemblies and systems of the setup to a vacuum of 5×10^{-4} Pa, (b) a system responsible for purifying the inert gas and filling some components of the unit with this gas, and (c) a power-supply system and an electrical measuring system. The evacuation and gas system includes a preevacuation pump and an oil-diffusion pump. The upper flange of the latter carries a liquid nitrogen-cooled trap for freezing the condensed vapors.

In line with the proposed method, it is essential that the pressure of the inert gas in the working space of the experimental setup is maintained at a level somewhat exceeding the saturation pressure of the metal under investigation. The latter pressure is about 8–10 MPa at a temperature of the order of 1800–2000 K.

The inert-gas purity in this case should be high enough (the total content of impurities should not exceed 0.001% in volume) to provide normal operation of the molybdenum heaters of the high-temperature furnace and preclude

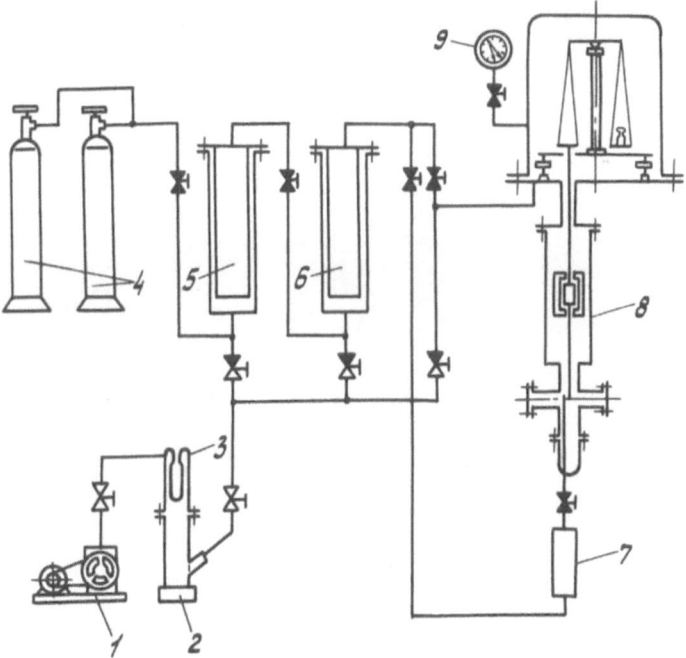

FIGURE 8. Schematic diagram of the experimental setup: 1—preevacuation pump; 2—oil-diffusion pump; 3—cooled trap; 4—inert gas bottles; 5, 6—absorption columns; 7—thermocompressor; 8—experimental setup; 9—pressure gauge; B1–B6—pressure valves; B1–B4—vacuum valves.

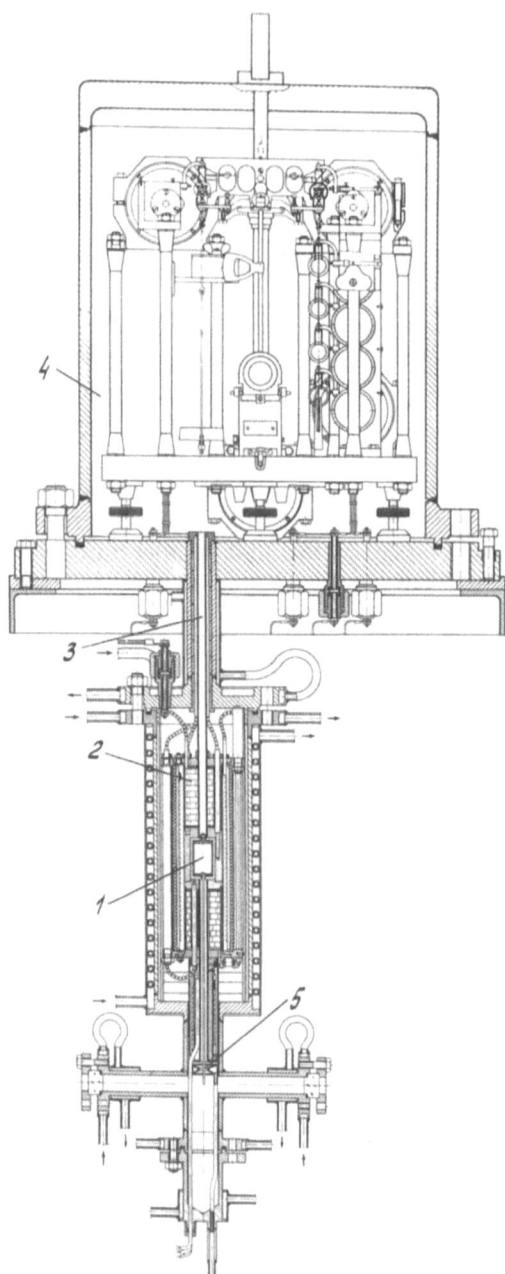

FIGURE 9. Construction of the experimental setup: 1—pycnometer; 2—high-temperature heating furnace; 3—suspension filament; 4—remote-control analytical balance; 5—gas cutter.

oxidation of the metal at the open end of the pycnometer capillary tube which may result in clogging of the capillary. Moreover, in order to remove the metal drop under investigation from the capillary tube tip by the gas cutter before weighing the pycnometer the inert gas must be delivered to the gas cutter at a pressure exceeding that of the experimental setup by 0.2–0.3 MPa. The gas is heated to 300–400 K to provide reliable removal of the metal drop without

FIGURE 10. General view of the experimental setup.

disturbing the uniform temperature field over the pycnometer height. The gas system is designed to meet these requirements.

The gas system is operated from two standard bottles of inert gas at a pressure of approximately 15 MPa. The inert gases used are argon at an experimental temperature of up to 1300–1400 K (i.e., in the temperature range suited for application of steel pycnometers) and helium at higher temperatures. This accounts for the necessity of altogether preventing gas convection, which tends to develop at high temperatures in the gap between the pycnometer and the thermostatting block wall and causes an increase of errors in measuring the mass of metals under study in the pycnometer by means of the remote-control analytical balance. If the inert gas is insufficiently pure, it is necessary to purify it additionally, for example, by the use of synthetic zeolite or calcium chips.

The construction and general view of the setup are illustrated in Figs. 9 and 10, respectively.

The power supply and electrical measuring systems include various standard electrical equipment to control the heating rate of the main and additional heaters, to supply power to other consumers, and also to incorporate a series of potentiometers for measuring the emf of the thermocouples.

4. EXPERIMENT: PREPARATIONS AND PROCEDURE

4.1. Calibration of Inner Volume of Pycnometer

Determination of the initial inner volume of the pycnometer is a very important operation and should be performed prior to the actual experiment as thoroughly as possible. It is found as the difference between the external volume and the volume of the pycnometer wall.[10] The external volume of the pycnometer can be found by weighing the pycnometer in water and in air with the capillary outlet end closed (Fig. 11). The volume of the pycnometer walls is derived proceeding from the known mass of the empty pycnometer and the density of the material from which the pycnometer is fabricated. In order to avoid excessive errors, the density of the pycnometer material should best be determined at the same laboratory using a sample of the pycnometer parent material.

Computation of the inner volume of the capillary tube can be derived from the total volume of the pycnometer by employing the values of the measured length and inner diameter of the capillary tube. The probable error in this case may be disregarded in estimating the accuracy of the results of measuring the density of the melt under investigation, because the inner volume of the capillary is approximately equal to 1% of the inner volume of the pycnometer.

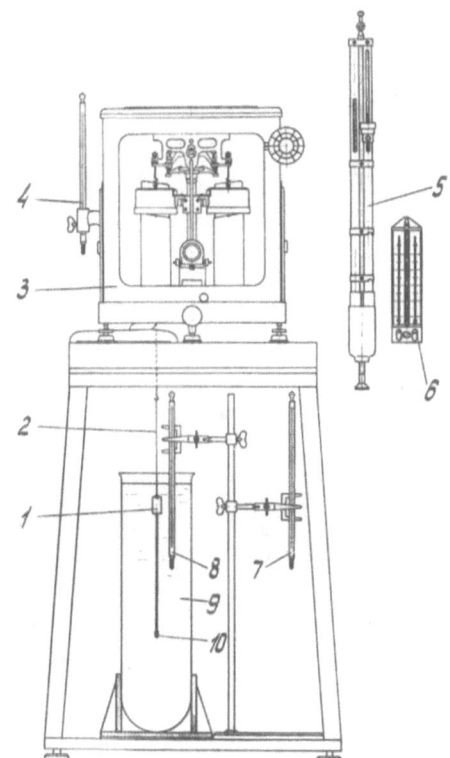

FIGURE 11. Apparatus for calibrating the internal volume of a pycnometer: 1—pycnometer; 2—suspension filament; 3—analytical balance; 4, 7, 8—thermometers; 5—barometer; 6—psychrometer; 9—vessel with twice-distilled water; 10—fluoroplastic plug.

4.2. Filling the Pycnometer with the Substance under Investigation

Filling the pycnometer with the substance to be investigated is a very important operation in preparing the experiment. It is evident that the present design version of the experimental setup allows the pycnometer to be filled only with molten metal and only through the capillary tube. The most important requirement in this case is that no gas enters the pycnometer ampoule from the outside. The most complicated operation is filling the pycnometer with a chemically active substance, such as an alkali metal.

Let us now examine the procedure of filling the pycnometer[11] with the metal under investigation, such as potassium, whose density has been studied by the suspended pycnometer method up to a temperature of 2000 K (Fig. 12). The starting potassium is supplied in sealed glass ampoules. The main unit intended for filling the pycnometer with metal is a stainless-steel vessel of inner diameter 50 mm, closed with a gasket-sealed cover. Built in the lower part of the vessel is a metallic cone facilitating the subsequent process of opening the ampoules.

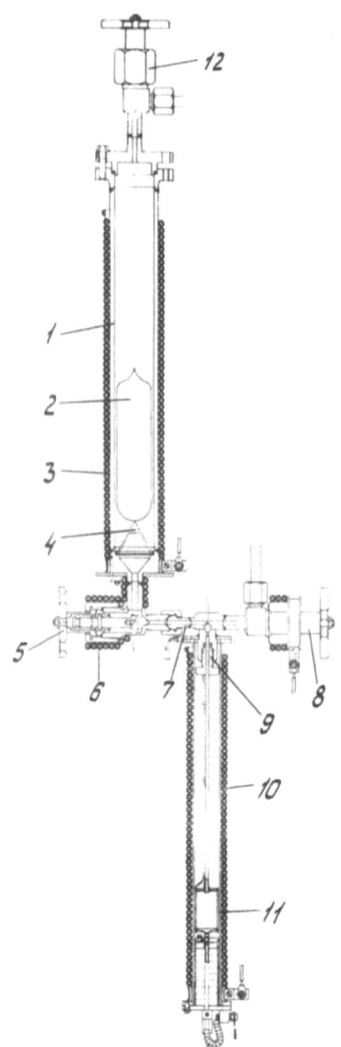

FIGURE 12. Apparatus for filling a pycnometer with metal: 1—melting pot; 2—ampoule with metal under investigation; 3, 6, 10—heaters; 4—striker; 5, 8, 12—silphon valves; 7—T-valve; 9—vacuum seal of capillary; 11—pycnometer.

The steel vessel communicates with the vacuum–gas system and with a special T-valve[12] to which the pycnometer is connected (through a vacuum rubber seal), with its capillary tube facing up.

The entire filling system, including the vessel with glass ampoule, bypass valves, and pycnometer, is equipped with electric heaters for heating the pycnometer metal under investigation and all the relevant channels to a temperature exceeding the melting point of the substance being studied. The system is also provided with monitoring thermocouples.

The pycnometer is filled with the substance to be examined in the following order. After assembling the apparatus, the filling system is evacuated to a vacuum of the order of 5×10^{-4} Pa and simultaneously heated to the required temperature (approximately 300 K). Then the glass ampoule is broken in the closed vessel and the metal flows into the pycnometer. To prevent glass fragments of the broken ampoule from getting into the channel, provision is made for a stainless gauze filter to be placed at the bottom of the vessel containing the glass ampoule.

Owing to the high vacuum of the system, the molten metal readily flows into the pycnometer and fills it completely. At the final stage of the filling operation, one may apply a small amount of the inert gas through the upper flange of the vessel containing the starting metal, thereby providing a sort of pusher for the flowing metal. The heaters are then switched off, the system cooled down, the pycnometer extracted from the rubber gasket, while the end of the capillary is sealed with picene.

The pycnometer is now regarded as filled with the metal to be investigated and is ready for use in the experimental setup. The mass of the substance filling the pycnometer is determined by weighing the empty and filled pycnometer.

4.3. Technique of the Experiment

When the appropriate preparatory operations have been completed, the components of the experimental setup are assembled in the following sequence:

- The pycnometer is installed in the thermostatting molybdenum block, and then this assembly is inserted into the heating furnace.
- The thermocouples are arranged in the grooves of the thermostatting block; the main and end heaters are assembled.
- The pycnometer is suspended from the analytical balance on a thin molybdenum filament; the gap between the pycnometer and the inner walls of the thermostatting block should be 2-4 mm.
- The power supply wires and thermocouple leads are connected to the respective terminals.
- The housing of the furnace is fitted, the flanges are closed, and the setup is hermetically sealed.

After checking the mechanical assemblies and electrical connections, the experimental apparatus is subjected to evacuation in order to obtain a vacuum of the order of 5×10^{-4} Pa. Then the setup is filled with the inert gas.

Measurement of the temperature dependence of the density of the metal under investigation by the suspended pycnometer method is practicable only in the direction from low to higher temperatures. As for the pressure, there are two approaches to the problem.

In the first approach the experiment is carried out along the line of liquid–vapor phase equilibrium of the substance under investigation. In this case it is necessary to increase the pressure in the setup in compliance with the rise in temperature of the pycnometer, exercising great care to provide some excess of the gas pressure (by at least 0.1–0.2 MPa) over the equilibrium vapor pressure of the molten metal under study (so as to reliably prevent boiling of the substance under investigation in the pycnometer).

In the second approach the density of the melt under investigation is measured along the isobar. In this case a constant preset pressure of the gas should be maintained in the setup.

As soon as the initial working pressure of the inert gas is attained in the experimental setup, the remote-control analytical balance should be checked again for proper adjustment and functioning, and the balance readings noted under the experimental conditions, i.e., the kind of gas medium (argon or helium) in which the pycnometer is weighed, its temperature and pressure. The balanced state of the analytical balance is assumed to be the zero, i.e., the reference point for determining the mass of metal flowing out of the pycnometer due to thermal expansion developed in the course of heating to the experimental temperature.

The experiment actually starts with heating the capillary of the pycnometer to a temperature somewhat exceeding the melting point of the substance under investigation. This is effected by means of a system of internal and external heaters.

The temperature of the capillary zone is monitored by three chromel-alumel thermocouples accommodated in a special protective enclosure. It is found practicable to maintain a constant temperature over the entire length of the capillary tube throughout the experiment, with the exception of the section adjoining the ampoule.

The lower protective heater of the furnace is then switched on. The main heater is switched on after the metal being studied becomes molten in the region where the capillary tube joins the ampoule. It is indicated by the respective monitoring thermocouple. The end and main heaters of the furnace raise the temperature of the pycnometer ampoule so that the substance under investigation attains the melting point within an extremely short period of time.

By operating the furnace heaters in the above manner along with a high rate of heating the metal under investigation (about $10 \deg \min^{-1}$) in the premelting temperature section (from room temperature to melting temperature), it is possible to eliminate inelastic strain of the pycnometer ampoule resulting from thermal expansion of the solid substance being studied. Otherwise, an uncontrollable deformation of the vial may arise while the furnace temperature is increasing.

Possible errors in determining the temperature of the metal under investigation as indicated by the monitoring thermocouples can be avoided by maintaining the rate of temperature rise in the final stage of attaining the prescribed temperature of the pycnometer ampoule at a level of the order of 0.01 deg min^{-1}.

While creating the rated temperature conditions of the pycnometer (ampoule and capillary tube), thermal expansion causes the molten metal to flow out of the pycnometer through the open outlet of the capillary tube in the form of spheroidal drops. The initial drops feature substantial dimensions (about 5 mm cross section) and mass, come off the capillary tube, and form a specific neck. The surface of the drops may be oxidized at the expense of residual oxygen impurities in the inert gas.

The instant at which the drop leaves the capillary tube on its own does not always coincide with the moment when the required temperature of the ampoule is attained. Therefore, prior to weighing the mass of metal dripping out of the pycnometer at the measured temperature of the experiment, the last drop should be removed from the pycnometer capillary end artificially with the aid of the gas cutter. So as not to disturb the uniform temperature field over the height of the pycnometer ampoule, wherever possible the inert gas delivered to the cutter should be heated preliminarily in the thermocompressor up to the working temperature of the pycnometer.

As the experimental temperature increases, the amount of desorbed gases tends to diminish materially while the surface of the outflowing drops remains unoxidized and features a metallic luster. The drops themselves resemble pellets of diameter 1–2 mm and leave the capillary tube end more uniformly and frequently. From an ampoule temperature of 1100–1200 K, the instant at which the drop comes off independently usually coincides with the moment that the required ampoule temperature is attained, so the gas cutter may be dispensed with.

When a uniform temperature field has been attained over the height of the pycnometer, the readings of the monitoring thermocouples (including the chromel–alumel thermocouples arranged over the capillary tube) are taken and the pycnometer then weighed by means of the remote-control analytical balance. At the given pycnometer temperature, which can be kept unchanged for 5–8 min, the aforesaid operations are repeated at least five times each.

The pressure of the inert gas in the setup, the temperature of the inert gas in the environmental space of the analytical balance, as well as the barometric pressure and temperature of the ambient air are measured simultaneously. These data are required for calculating the correction to the weighing results arising from the gas expulsive force exerted on the pycnometer with the suspension filament, the standard weights, and the movable parts of the balance. It should be emphasized that while weighing the pycnometer with the analytical balance, it is necessary to keep a permanent watch on the capillary end to ensure that a drop or even traces of the melt flowing out of the capillary do not remain.

If, during transition to subsequent temperature conditions, the pressure of the inert gas in the setup increases, it is necessary to prevent an accidental drop in pycnometer temperature when a new portion of the cold gas is applied, otherwise the melt is drawn from the capillary tube into the vial, followed by entry of the gas. If this happens there is no use in continuing the experiment and it should be stopped. The experiment should be conducted only when the temperature is rising (even very slowly). Therefore, it is better that the inert gas be supplied to the setup through two lines: cold gas (at room temperature) from the collector through the sleeve at the balance bottom, and gas warmed to the required temperature (1000–1100 K maximum) from the thermocompressor via the delivery pipeline of the gas cutter. The flow rates of the gases or, more important, their ratio should be selected experimentally.

Upon completing the measurements at the maximum temperature, the experiment reaches its conclusion. The heaters should be switched off in a sequence enabling the capillary to be the first to cool down quickly, so that the melt when solidified separates the inner zone of the pycnometer from the gas in the setup. As the melt in the pycnometer ampoule cools down, free space appears in it, and therefore it is necessary to regulate the gas pressure in the setup to prevent crumpling of the pycnometer wall.

5. ACCURACY OF THE MEASURED RESULTS

With a view to minimizing the measurement errors, each operation of the experiment (calibration of the pycnometer volume, weighing the starting mass of the sample, determination of the coefficient of thermal expansion of the pycnometer material, measuring the mass of the melt flowing out of the pycnometer during heating, measuring the pycnometer temperature and a number of other related measurements) was performed very carefully. Table 1 presents the absolute errors of the basic measured parameters.

TABLE 1
Absolute Errors of Basic Measured Parameters

Parameters measured	Absolute measurement errors
Internal volume of pycnometer at 0 °C, V_0	$\pm 1 \times 10^{-6} \, \text{m}^3$
Starting mass of substance under investigation, m_0	$\pm 1 \times 10^{-6} \, \text{kg}$
Change of mass of liquid flowing out of pycnometer, Δm_t	$\pm 1 \times 10^{-5} \, \text{kg}$
Mass of liquid in capillary, m_{cap}	$\pm 1 \times 10^{-5} \, \text{kg}$
Temperature of liquid, t:	
up to 1300 K (Pt–Pt + 10% Rh)	0.2–0.7 °C
up to 2100 K (Pt + 6% Rh–Pt + 30% Rh)	1–3 °C

TABLE 2

Confidence Error of Density Measurements

Temperature range of experiment	Confidence error (%)
T_{mp}-1000 K	0.15
1000–1500 K	0.20
1500–2000 K	0.25
2000–2200 K	0.30

The relative confidence error of the density measurements is indicated in Table 2. The data were obtained from experiments with liquid potassium.[3,13,14]

The additional, important characteristic of the experimental quality is the divergence of the derived experimental data (deviation from the approximative curve). The maximum deviation of the experimental data during operation of the setup under consideration amounts to ±0.20% (also obtained by experiments with liquid potassium). The mean-square deviation within the temperature range of T_{mp}-1000 K is ±0.10%, while in the range of 1000–2100 K it is 0.15%.

6. SUMMARY

The suspended pycnometer method applied to experiments monitored with suitable instruments and facilities enables measurements of molten metal densities to be conducted with a high degree of precision from the melting points up to the near-critical temperature. The pressure range is limited only by the strength of the casing of the experimental setup. A single series of measurements enables the temperature dependence of the density to be obtained over a wide range of temperatures.

The experimental setup examined in this chapter and tested practically is rather simple to fabricate and handle. The accuracy obtained, particularly the accuracy of the data on the density of potassium in the temperature range up to 2000 K and at pressures up to 10 MPa made it possible not only to find the temperature dependence of density, but also to obtain very reliable information on the compressibility of this particular metal.

The proposed method appears most suitable for investigating the density of substances that feature high chemical activity, toxicity, and high pressure of saturated vapors, since the only place where the ambient atmosphere (the outlet of the pycnometer capillary) is restricted to the contact area and always remains within a zone of fairly low temperatures.

REFERENCES

1. E.E. Shpil'rain and K.A. Yakimovich, "Experimental Setup for Investigation of Density of Molten Metals" (in Russian), *Teplofiz. Vys. Temp.* **1**(2), 173–176 (1963).
2. K.A. Yakimovich and A.G. Mozgovoi, "Method of Experimental Investigation of Density of Molten Alkali Metals at Temperatures above 1000 °C" (in Russian), in: *Properties of Molten Metals,* Theses of Reports at Symposium of Academic Council, Academy of Sciences of the USSR; *Physicochemical Principles of Metallurgical Processes,* pp. 37–38, A.A. Baikov Metallurgy Institute, Academy of Sciences of the USSR, Moscow (1976).
3. E.E. Shpil'rain, K.A. Yakimovich, and A.G. Mozgovoi, "Experimental Investigation of the Density of Liquid Potassium at Temperatures up to 1800 K and Pressures up to 10 MPa" (in Russian), *Teplofiz. Vys. Temp.* **14**(3), 511–521 (1976).
4. V.A. Petukhov, V.Ya. Chekhovskoi, and A.G. Mozgovoi, "Experimental Investigation of Thermal Expansion of Steel 1X18H9T" (in Russian), *Teploenergetika,* No. 3, pp. 64–65 (1976).
5. V.A. Petukhov, B.Ya. Chekhovskoi, and A.G. Mozgovoi, "Experimental Investigation of Thermal Expansion of Some Construction Materials. Molybdenum and Molybdenum Alloy VN-2M" (in Russian), *Teplofis. Vys. Temp.* **15**(1), 204–207 (1977).
6. V.A. Petukhov, V.Ya. Chekhovskoi, and A.G. Mozgovoi, "Experimental Investigation of Thermal Expansion of Some Construction Materials. Tantalum and Tantalum–Tungsten Alloy TV-10" (in Russian), *Teplofiz. Vys. Temp.* **15**(3), 534–538 (1977).
7. V.A. Petukhov, V.Ya. Chekhovskoi, V.G. Andrianova, and A.G. Mozgovoi, "Experimental Investigation of Thermal Expansion of Some Construction Materials. Niobium and Niobium Alloy 5VMTcl" (in Russian), *Teplofiz. Vys. Temp.* **15**(3), 670–673 (1977).
8. V.Ya. Chekhovskoi, V.A. Petukhov, and A.G. Mozgovoi, "Experimental Investigation of Thermal Expansion of Construction Materials, and Alkali Molten Metals Corrosion-Resistant at High Temperatures" (in Russian), Trans. IVTAN, pp. 33–44, IVTAN, Moscow (1978).
9. E.E. Shpil'rain, K.A. Yakimovich, and A.G. Mozgovoi, "Unitized Thermocouple and Power Supply Lead-ins" (in Russian), *Teplofiz. Vys. Temp.* **15**(3), 681 (1977).
10. E.E. Shpil'rain, K.A. Yakimovich, A.G. Mozgovoi, and A.K. Chelebaev, "Apparatus for Calibration of Internal Volume of Pycnometers" (in Russian), *Teplofiz. Vys. Temp.* **23**(2), 378 (1985).
11. E.E. Shpil'rain, K.A. Yakimovich, and A.G. Mozgovoi, "Apparatus for Filling Pycnometers with Alkali Metal" (in Russian), *Teplofiz. Vys. Temp.* **23**(3), 550 (1985).
12. E.E. Shpil'rain, V.A. Fomin, V.A. Savchenko, and A.G. Mozgovoi, "Unsealed Sylphon Valve" (in Russian), *Zavodskaya Laboratoriya,* No. 6, p. 717 (1976).
13. E.E. Shpil'rain, K.A. Yakimovich, and A.G. Mozgovoi, "Experimental Investigation of the Density of Liquid Potassium at Temperatures of 1740–2030 K" (in Russian), *Teplofiz. Vys. Temp.* **15**, 1104–1106 (1977).
14. K.A. Yakimovich, A.G. Mozgovoi, and V.V. Dubinin, "The Density of Liquid Potassium at Near-Critical Temperatures," *High Temp. High Pressures* **11**(5), 543–550 (1979).

VI

REVIEW OF THERMOPHYSICAL PROPERTY REFERENCE MATERIALS

22

Reference Materials for Thermophysical Properties

R.K. KIRBY

Reference materials for use in calibrating either the temperature scale of equipment or a physical property measured by the equipment as a function of temperature are available worldwide. Reference materials are useful in developing new test methods as well as ensuring the compatibility of measurements both within and among laboratories. The reference materials that are described in this chapter include those for thermal conductivity, electrical resistivity, heat capacity, and thermal expansion.

A certified reference material (CRM) is a representative sample from a homogeneous lot of material that has been carefully measured and certified by an authorized agency to have a stated composition, structure, and/or property. The value certified is generally the best estimate of the "true" value for that lot of material and is as free of random and systematic errors as is practical for its intended use. Certification generally includes the estimated uncertainty of the certified value.

Other reference materials that are recommended by groups such as the International Union of Pure and Applied Chemistry (IUPAC) can be obtained from a variety of sources as long as they meet the purity requirements that are specified. The recommended values for these materials are based on reliable experiments that are fully described in the open literature.

R.K. KIRBY ● Retired from National Institute of Standards and Technology, Gaithersburg, Maryland 20899, USA.

CRM's for thermophysical properties are available from the following sources[1]:

Bundesanstalt fur Materialprüfung	BAM
Unter den Eichen 87	
D-1000 Berlin 45	
Germany	
Community Bureau of Reference	BCR
Directorate General XII, CEE	
200, rue de la Loi	
B-1049 Brussels	
Belgium	
Gosstandart of the USSR	GOST
9, Leninsky Prospekt	
117049, Moscow	
USSR	
National Institute of Standards and Technology	NIST
(Formerly the National Bureau of Standards)	
Office of Standard Reference Materials	
Gaithersburg, MD 20899	
USA	
National Physical Laboratory	NPL
Office of Reference Materials	
Teddington, Middlesex TW11 0LW	
UK	
National Research Laboratory of Metrology	NRLM
Tsukuba	
Ibaraki 305	
Japan	
Service des Matériaux de Référence	SMR
1, rue Gaston Boissier	
75015 Paris	
France	

CERTIFIED REFERENCE MATERIALS

Tables 1–5 provide short descriptions of CRM's for thermometric and thermophysical properties according to the best information made available to the author.

TABLE 1
Thermometric Fixed Points

Source	Material		Nominal temperature (K)
NIST	Superconducting device		
	cadmium	(sct)[a]	0.5
	zinc	(sct)	0.9
	aluminum	(sct)	1.2
	indium	(sct)	3.4
	lead	(sct)	7.2
	niobium	(sct)	9.3
SMR	Helium	(tp)[b]	2
NRLM	Neon	(tp)	25
SMR	Oxygen	(tp)	54
NRLM, SMR	Argon	(tp)	84
NIST	Mercury	(tp)	234
NPL	Water triple point cell	(tp)	273
NIST	Gallium	(mp)[c]	303
NPL	Ethylene carbonate	(mp)	309
NIST	Rubidium	(mp)	312
NIST, NPL	4-Nitrotoluene	(mp)	325
NIST	Biphenyl-succinonitrile	(mp)	331
NPL	Naphthalene	(mp)	353
NPL	Benzil	(mp)	368
NPL	Acetanilide	(mp)	388
NPL	Benzoic acid	(mp)	396
NPL	Diphenylacetic acid	(mp)	420
NIST, NPL, NRLM	Indium	(mp)	430
NPL	Anisic acid	(mp)	456
NPL	2-Chloroanthraquinone	(mp)	483
NIST	Tin	(mp)	505
BAM, NPL	Carbazole	(mp)	519
NPL	Anthraquinone	(mp)	558
BAM	Cadmium	(mp)	594
NIST	Lead	(mp)	600
BAM, NIST	Zinc	(mp)	693
BAM	Sulfur	(mp)	718
BAM, NIST	Aluminum	(mp)	933
BAM	Ag–Cu	(ep)[d]	1053
BAM, SMR	Silver	(mp)	1235
BAM, SMR	Gold	(mp)	1338
GOST, NIST	Copper	(mp)	1358
SMR	Palladium	(mp)	1827
GOST, NIST	Alumina	(mp)	2326
GOST	Molybdenum	(mp)	2900

[a] Superconducting transition.
[b] Triple point.
[c] Melting point.
[d] Eutectic point.

TABLE 2
Thermal Conductivity (Resistivity)

Source	Material	Nominal conductivity (Resistivity)	Temperature range (K)
NIST	Fumed silica board	$(1.2 \text{ m}^2 \text{ K W}^{-1})$	293
BCR, NIST, NPL, SMR	Fibrous glass board	$(0.8 \text{ m}^2 \text{ K W}^{-1})$	100–330
NIST	Fibrous glass blanket	$(0.6 \text{ m}^2 \text{ K W}^{-1})$	100–330
NPL	Perspex slab	$(0.3 \text{ m}^2 \text{ K W}^{-1})$	273–343
NPL	Nylon	$(0.1 \text{ m}^2 \text{ K W}^{-1})$	263–353
NPL	Polythene	$(0.1 \text{ m}^2 \text{ K W}^{-1})$	263–353
GOST	Cr–V–Mg alloy	$9 \text{ W m}^{-1} \text{ K}^{-1}$	293–960
NIST, NPL	Stainless steel	14	2–1200
NPL	Inconel 600	15	323–1023
GOST	Low carbon steel	65	293–900
BAM, SMR	Platinum	73	273–1273
NIST	Iron	76	2–1000
NIST	Graphite	91	5–2500
GOST	Molybdenum	138	400–2150
NIST	Tungsten	173	2–3000
SMR	Copper	375	230–850

TABLE 3
Electrical Resistivity

Source	Material	Nominal resistivity	Temperature range (K)
NIST	Tungsten	$0.05 \ \mu\Omega \text{ m}$	4–3000
NIST	Iron	$0.10 \ \mu\Omega \text{ m}$	6–1000
NIST	Stainless steel	$0.80 \ \mu\Omega \text{ m}$	5–1200

TABLE 4
Heat Capacity and Enthalpy

Source	Material	Temperature range (K)
NIST, SMR	Copper	1–1300
GOST, NIST	Sapphire	5–2700
GOST, NIST	Molybdenum	273–2800
NIST	Polystyrene	10–350
NIST	Polyethylene	5–360
GOST	Tungsten	1200–2800
GOST	Graphite	298–4500
GOST	Glass	173–673
GOST	Quartz	293–800
SMR	Platinum	298–2000

TABLE 5
Thermal Expansion

Source	Material	Nominal coefficient of thermal expansion	Temperature range (K)
SMR	Invar	$0.3\ 10^{-6}\ K^{-1}$	273–350
NIST, SMR	Fused silica	0.6	80–1300
NIST, GOST	Tungsten	4	80–2000
NIST	Glass	5	80–680
GOST	Molybdenum	5	293–2500
GOST	Sapphire	5	293–2000
NIST	Stainless steel	10	293–780
SMR	Magnesia	10	293–1800
GOST	Stainless steel	15	293–1650
NRLM, SMR	Copper	18	300–873
SMR	Pyros	(Not known)	293–1400
SMR	Zircon	(Not known)	293–1600

RECOMMENDED REFERENCE MATERIALS

Reference materials that are recommended (not certified) by CODATA,*[2] IUPAC,[3] and the Calorimetry Conference are indicated in Tables 6–9.

TABLE 6
Thermal Conductivity

Group	Material	Temperature range (K)
CODATA	Aluminum	1–900
CODATA	Copper	1–1300
CODATA	Iron	1–1000
CODATA	Tungsten	1–3000
IUPAC	Platinum	273–373

*The Committee On Data (CODATA) is an interdisciplinary Committee of the International Council of Scientific Unions, which deals with data of importance to science and technology, their compilation, critical evaluation, storage, and retrieval.

TABLE 7
Electrical Resistivity

Group	Material	Temperature range (K)
CODATA	Copper	20–1300
CODATA	Iron	20–1700
CODATA	Platinum	2–2000
CODATA	Tungsten	20–3400

TABLE 8
Heat Capacity and Enthalpy

Group	Material	Temperature range (K)
CODATA	Copper	10–1250
CODATA	Iron	5–1800
CODATA	Tungsten	10–3400
CODATA	Alumina	10–2300
IUPAC	Copper	1–300
IUPAC	Molybdenum	273–2900
IUPAC	Platinum	298–1500
IUPAC	Alumina	10–2250
IUPAC	Benzoic acid	10–350
IUPAC	2,2-Dimethylpropane	4–254
IUPAC	Diphenyl ether	10–300
IUPAC	Heptane	10–182
IUPAC	Hexafluorobenzene	10–278
IUPAC	Naphthalene	10–350
IUPAC	Polystyrene	10–360
IUPAC	Polyvinylchloride	10–350
Cal Conf[a]	Sapphire	0–1200
Cal Conf	Benzoic acid	14–350
Cal Conf	n-Heptane	0–182

[a]The materials recommended by the Calorimetry Conference can be obtained from the Office of Standard Reference Materials, National Institute of Standards and Technology, Gaithersburg, Maryland 20899, USA.

TABLE 9
Thermal Expansion

Group	Material	Temperature range (K)
CODATA	Copper	10–1200
CODATA	Silicon	10–1000
CODATA	Tungsten	10–3400
CODATA	Alumina	20–2000

REFERENCES

1. *Directory of Certified Reference Materials* (CRM): Sources of Supply, ISO/REMCO No. 51 (1982) and ISO/REMCO No. 113 (1986), The International Organization for Standardization, Case Postale 56, CH-1211, Geneva 20, Switzerland.
2. *Thermophysical Properties of Some Key Solids* (Guy K. White and Merrill L. Minges, eds.), CODATA Bulletin No. 59 (December 1985). CODATA Secretariat: 51 Boulevard de Montmorency, 75016 Paris, France.
3. *Recommended Reference Materials for the Realization of Physicochemical Properties* (IUPAC) (Ken Marsh, ed.), Blackwell Scientific Publishers, Oxford, UK (1987).

Index